Advances in
MARINE BIOLOGY

VOLUME 32

Advances in
MARINE BIOLOGY
The Biogeography of the Oceans

Series Editors

J. H. S. BLAXTER

Dunstaffnage Marine Research Laboratory, Oban, Scotland

and

A. J. SOUTHWARD

Marine Biological Association, The Laboratory, Citadel Hill,
Plymouth, England

Guest Editors

A. V. GEBRUK

P. P. Shirshov Institute of Oceanology, Moscow, Russia

E. C. SOUTHWARD

Marine Biological Association, The Laboratory, Citadel Hill, Plymouth, England

P. A. TYLER

Department of Oceanography, University of Southampton, Southampton, England

ACADEMIC PRESS

San Diego   London   Boston
New York   Sydney   Tokyo   Toronto

Academic Press Inc.
525 B Street, Suite 1900, San Diego, California 92101-4495, USA

Academic Press Limited
24–28 Oval Road, London NW1 7DX, UK

ISBN 0-12-026132-4

A catalogue record for this book is available from the British Library

Typeset by Keyset Composition, Colchester, Essex
Printed and bound in Great Britain by MPG Books Ltd, Bodmin, Cornwall
97 98 99 00 01 02 EB 9 8 7 6 5 4 3 2 1

CONTRIBUTORS TO VOLUME 32

S.V. GALKIN, *P.P. Shirshov Institute of Oceanology, Russian Academy of Sciences, Nakhimovsky Prospekt 36, Moscow 117851, RUSSIA*

A.V. GEBRUK, *P.P. Shirshov Institute of Oceanology, Russian Academy of Sciences, Nakhimovsky Prospekt 36, Moscow 117851, RUSSIA*

A.N. MIRONOV, *P.P. Shirshov Institute of Oceanology, Russian Academy of Sciences, Nakhimovsky Prospekt 36, Moscow 117851, RUSSIA*

L.I. MOSKALEV, *P.P. Shirshov Institute of Oceanology, Russian Academy of Sciences, Nakhimovsky Prospekt 36, Moscow 117851, RUSSIA*

K.N. NESIS, *P.P. Shirshov Institute of Oceanology, Russian Academy of Sciences, Nakhimovsky Prospekt 36, Moscow 117851, RUSSIA*

N.V. PARIN, *P.P. Shirshov Institute of Oceanology, Russian Academy of Sciences, Nakhimovsky Prospekt 36, Moscow 117851, RUSSIA*

H.J. SEMINA, *P.P. Shirshov Institute of Oceanology, Russian Academy of Sciences, Nakhimovsky Prospekt 36, Moscow 117851, RUSSIA*

M.N. SOKOLOVA, *P.P. Shirshov Institute of Oceanology, Russian Academy of Sciences, Nakhimovsky Prospekt 36, Moscow 117851, RUSSIA*

A.J. SOUTHWARD, *Marine Biological Association of the UK, Citadel Hill, Plymouth, PL1 2PB, UK*

A.L. VERESHCHAKA, *P.P. Shirshov Institute of Oceanology, Russian Academy of Sciences, Nakhimovsky Prospekt 36, Moscow 117851, RUSSIA*

M.E. VINOGRADOV, *P.P. Shirshov Institute of Oceanology, Russian Academy of Sciences, Nakhimovsky Prospekt 36, Moscow 117851, RUSSIA*

N.G. VINOGRADOVA†, *P.P. Shirshov Institute of Oceanology, Russian Academy of Sciences, Nakhimovsky Prospekt 36, Moscow 117851, RUSSIA*

O.N. ZEZINA, *P.P. Shirshov Institute of Oceanology, Russian Academy of Sciences, Nakhimovsky Prospekt 36, Moscow 117851, RUSSIA*

†Deceased

CONTENTS

Some Problems of Vertical Distribution of Meso- and Macroplankton in the Ocean

M. E. Vinogradov

Ecology and Biogeography of the Hydrothermal Vent Fauna of the Mid-Atlantic Ridge

A.V. Gebruk, S.V. Galkin, A.L. Vereshchaka, L.I. Moskalev and A.J. Southward

Biology of the Nazca and Sala y Gómez Submarine Ridges, an Outpost of the Indo-West Pacific Fauna in the Eastern Pacific Ocean: Composition and Distribution of the Fauna, its Communities and History

N.V. Parin, A.N. Mironov and K.N. Nesis

Gonatid Squids in the Subarctic North Pacific: Ecology, Biogeography, Niche Diversity and Role in the Ecosystem

K.N. Nesis

Zoogeography of the Abyssal and Hadal Zones

N.G. Vinogradova

Biogeography of the Bathyal Zone

O.N. Zezina

Trophic Structure of Abyssal Macrobenthos

M.N. Sokolova

An Outline of the Geographical Distribution of Oceanic Phytoplankton

H.J. Semina

PREFACE

This volume of *Advances in Marine Biology* is devoted to the geographical and vertical distribution of life in the open oceans, including the great depths, a subject of many important researches carried out by the former USSR. It was a feature of the Soviet system that extensive facilities for marine research could be provided on a worldwide scale, comparatively lavishly, using very large research vessels. Russian biologists, unlike those in other countries, were thus able to visit and sample practically all parts of the World Ocean, allowing them a global overview of life in the deep sea. Most of the large number of samples that form the basis of these reviews are now deposited in the Institute of Oceanology in Moscow, although some others, especially the Pogonophora, are archived in the Zoological Institute in St Petersburg. It is important for knowledge of global diversity that these samples continue to be curated and preserved.

The proposal to distil the knowledge of the Russian investigations, and make it known to a wider audience in Europe and North America, came from the principal guest editor of this volume, Andrey Gebruk, after stimulating discussions with a number of western European marine biologists. Andrey Gebruk's doctoral thesis concerned abyssal sea-cucumbers (holothurians), their distribution and possible pathways of global evolution. From this basis, and because of his multi-lingual capacities, he was able to initiate the preparation of this volume and help the other Russian authors prepare their contributions.

The individual contributions have, as their main purpose, the task of bringing Russian thinking about the geographical and depth distribution of the ocean biota to a wider audience. It may be that these reviews do not represent a truly international viewpoint, but this is partly a result of the problems of obtaining western journals in Russia, both under the previous regime and now as a result of financial problems. From time to time some English, German and French contributions were made available in translation in Russia, but these were only a small proportion of the work being done. At the same time, Russian contributions published in *Okeanologia* and several other journals were issued in translation by American groups, and the Israel translation programme also issued versions of some Russian works published in book form. Unfortunately, these translations were also only a small part of the available Russian literature. Thus, there has been an unavoidable language barrier between Russian and western marine biogeographers, and only limited cross-fertilization has been possible. This book is designed to help bridge the gap.

The contributions range very widely, from plankton and squid to the bottom fauna of the bathyal, abyssal and hadal zones. Michael Vinogradov has drawn on his encyclopaedic knowledge of the vertical distribution of large zooplankton to present new syntheses and outline new problems, including data based on recent observations with electronically operated closing nets and from submersibles. Andrey Gebruk and his co-authors describe the results of 10 years of studies of the hydrothermal vents along the Mid-Atlantic Ridge, including otherwise inaccessible Russian results, analysing the ecology and biogeography of the ridge fauna based on submersible observations. Nikolai Parin, Alexander Mironov and Kir Nesis give a detailed study of the peculiar biogeography of the seamounts and rises of the Nazca and Sala y Gómez ridges in the southeastern Pacific, using material obtained during extensive trawling expeditions by several Russian research vessels, and show the relationship of most of the fauna to that of the western Pacific, with only a small element related to the South American plate fauna. Kir Nesis describes the biology, taxonomy and geographical distribution of the large gonatid (commander) squids of the north eastern Pacific, based on many trawling expeditions, and demonstrates the important role of these animals in the ecosystem of the region. Nina Vinogradova draws on great experience of benthic sampling to deal with the vertical and horizontal distribution of the true deep-sea benthos, below 3000 m, including the specialized hadal fauna from deeper than 6000 m, where there are many endemic species limited to single trenches or to close groups of trenches. On the basis of a large database of Brachiopoda and other faunal elements, Olga Zezina describes the varied biogeographical patterns of the bathyal zone, which extends from the edge of the continental shelf down to the beginning of the abyssal zone at 3000 m. This region shows fewer faunistic divisions of the benthos than shallow water regions, but is much more diverse than the abyssal zone and contains many relict species; it may have acted as a reserve for recolonization of both the abyssal regions and the shelves after global cataclysms. Marina Sokolova reviews Russian thinking about the effect of nutritional conditions on the vertical and horizontal distribution of the deep-water fauna, based on examination of the food contents of large numbers of trawl-caught macrobenthos; she shows the relationship to ocean hydrology and to climate, in describing the global pattern of eutrophic and oligotrophic regions in the abyssal. The last contribution, by Halina Semina, shows the patterns of distribution of the larger phytoplankton, diatoms and dinoflagellates, living in the euphotic zone, and their relationship to nutrient levels and ocean gyres. Where possible the authors quote original literature, most of it published in Russian, and they also refer to their own reviews of the subject. In the references the titles of articles published in Russian are shown in English, with a note in parentheses that the article is in Russian. It is to be noted that many of the articles quoted from Russian journals are available in cover-to-cover English translations made in U.S.A. There was no space to quote all the translations, which are in any case not always available at many libraries in Western Europe and elsewhere owing to present day financial problems. Where possible, the authors quote translations of their reviews, which may be more widely available.

This volume pays tribute to previous Russian biogeographers of the ocean, especially George Belyaev, Zinaida Filatova and Lev Zenkevitch. Lev Zenkevitch expanded Russian deep-sea researches from small beginnings; George Belyaev was a scholar, well known for his incisive thinking; and Zinaida Filatova was the driving force behind much of the benthic sampling programme of the USSR in the 1950s and 1960s. Zinaida Filatova is well remembered in the west as being chief scientist of several expeditions at a time when British and North American organizations were reluctant to allow women scientists on board their vessels.

This volume has been completed by intensive combined efforts of the series editors and the guest editors, who strove to make it understandable to the non-specialist, and who also had to overcome the problems of communication that develop when telephone links with Moscow can be interrupted for weeks on end. Diskettes of the contributions and copies of the illustrations were supplied to Alan Southward during his visit to Moscow in September 1996 for an INTAS workshop, and updates and corrections were sent by e-mail and facsimile at frequent intervals thereafter, supplemented by personal transport of documents and disks. The editors had to overcome many problems of English style, in translation of technical and other terms, in checking of references and in conversion of text transmitted in various word-processor formats that were not immediately compatible with Microsoft Word 6, involving very tortuous transfer systems. They are indebted to several biologists in the Institute of Oceanology in Moscow for helping maintain the electronic data links, notably Alexander Mironov, Nick Nezlin and Kir Nesis.

Sadly, one of the contributors, Nina Vinogradova, died just as the MS were being completed, but by her valiant efforts while ill, assisted by her son George, she was able to complete the final revision. George Vinogradov has greatly helped the editors by critical checking of the MS during the final stages of going to press, and in reading the proofs.

Additional thanks are due to Hjalmar Thiel, of the Institute for Polar and Marine Research, Bremerhaven, as one of the originators of the idea for these reviews, for support; and also to Alexei Kuznetsov, the present head of the Benthos Laboratory at the Institute of Oceanology, and Lev Moskalev for their interest and advice. The International Association for the Promotion of Cooperation with Scientists from the Independent States of the Former Soviet Union (INTAS) provided partial financial assistance for several of the participants in the volume.

The authors wish to dedicate this volume to the memory of Nina Vinogradova, whose contribution to it is her last scientific publication. Nina loved the deep sea and her colleagues loved her.

Some Problems of Vertical Distribution of Meso- and Macroplankton in the Ocean

M. E. Vinogradov

P. P. Shirshov Institute of Oceanology, Russian Academy of Sciences, Nakhimovsky Prospekt 36, Moscow 117851, Russia

ABSTRACT

The plankton in the whole water column depends on photosynthetic production in the upper layer, the euphotic or epipelagial zone, where there are zooplankton that remain in this layer all the time. Below is the deep-sea zone, with animals that

ADVANCES IN MARINE BIOLOGY VOL. 32
ISBN 0-12-026132-4

show different vertical distributions in the different climatic regions of the ocean. As defined here, mesozooplankton has a size range of 200 μm to 30 mm length, but including also chaetognaths and polychactes up to 50 mm. Macroplankton ranges up to 150 mm, and includes large shrimps and the gelatinous animals such as medusae, siphonophores and ctenophores; fish and squid in this size range are conventionally classed as micronekton.

Methods of sampling depend on the size range. Most mesoplankton is taken with vertical closing nets and large water bottles, while macroplankton requires horizontal or oblique tows of large nets and trawls, but both types of gear require closing devices or depth telemeters to control the sampling depth. Manned submersibles have extended the range of quantitative observations on plankton, and show that both meso- and macroplankton samplers underestimate the true abundance of many species.

Patterns of mesoplankton distribution are listed for the different depth layers and the importance of small scale spatial variability is emphasized. Comparisons are then made of the distribution and abundance of mesoplankton over the whole water column in the open oceans.

Macroplankton is more difficult to assess quantitatively than mesoplankton. Its composition and depth distribution are summarized and the importance of gelatinous animals emphasized.

A scheme is presented for classifying the vertical distribution of plankton, based mainly on biological criteria. Regional and climatic variations affect the boundaries of the different zones.

The final section reviews the special conditions near the sea bottom where there are distinct communities of benthopelagic animals. Nutritional conditions are very variable near the bottom and there can be a high proportion of dead animals. Chemosynthetic input from mid-ocean ridge vent communities is relatively unimportant to the pelagic community.

1. INTRODUCTION

The key feature of pelagic communities is that the population of the whole water column subsists on the primary organic matter produced by phytoplankton within an upper layer that may be as thin as a few tens of metres and at most 100–150 m thick. Only within this thin layer do animals occur permanently with the plant producers. The population of the rest of the water column subsists solely on organic matter produced in the surface zone. Their food is the remains of animals and plants sinking from the upper layers together with faecal pellets and animals migrating downward. This separation between the sites of production and consumption is the most important feature of the oceanic pelagic community. The division of the water column into surface or epipelagic (producing) and deep-water (consuming) zones is beyond question (Bogorov, 1948; Ekman, 1953; Bruun, 1956; Vinogradov, 1968).

A water mass occupies a certain part of the ocean area, is characterized by definite hydrological conditions and origin, and contains a typical population. This water mass might be considered as the movable biotope of the plankton community. The plankton population may serve as a characteristic feature of a water mass in addition to physical and chemical peculiarities (Vinogradov, 1968; Beklemishev, 1969). The vertical distribution of zooplankton depends to a much lesser degree on the distribution of water masses. Vertical migrations through layers occupied by different water masses appear to have evolved to give the migrating species some biological advantage. The whole water column, or a large part of it, appears to serve as a biotope for such migratory species. A continuous exchange of populations between communities in the different layers unites these communities. The interzonal migrants (that is to say the animals migrating back and forth between the upper zone and a deeper one) form up to 40% of the total biomass in both the mesopelagic and bathypelagic zones (1000–3000 m) (Vinogradov, 1968; Vinogradov and Arashkevich, 1969). Naturally, the migrants include components of both surface and deep-water communities. From this standpoint, the autonomous communities of the upper layers and the deep-water ones dependent on them should be considered as parts of one large community that occupies all or almost all the water column as a single ecosystem. Nevertheless, the vertical distribution of plankton in the oceans shows a well-defined stratification, which cannot be fully explained by gradients in the physical environment (Banse, 1964; Paxton, 1967; Vinogradov, 1970, 1972). Stratification is preserved despite mixing by turbulence. It is probable that inherent characteristics of the community itself favour the maintenance of a sharp stratification in population distribution in defiance of external disturbing factors (Vinogradov and Tseitlin, 1983).

The biological mechanism that maintains the stratified plankton distribution is not always evident. It seems to be significantly related to competition among ecologically dominant, closely related species with similar feeding spectra. Layers of dominance of these species separate according to G. F. Gause's concept of competitive exclusion of species with similar ecological properties (Vinogradov, 1968).

Although all the animals inhabiting the ocean water column depend on organic matter produced in the surface euphotic layer, they differ in their trophic coupling. One group feeds directly on phyto- and bactcrioplankton or animals and never leaves this layer (surface plankton or epiplankton). Another group feeds mainly in the surface zone and spend part of its life in deep water (interzonal plankton), thus transporting organic matter from the upper to the deeper layers; these can be divided into upper- and lower-interzonal species, according to where their young feed and how intensively the species use the surface or the deep food supply (Vinogradov, 1968). A third group lives exclusively in the deep-water zone (deep-water plankton) and its relationship with surface species is through intermediate links of the trophic chain or is limited to grazing on descending organic particles.

2. METHODS OF EVALUATING THE DISTRIBUTION OF PLANKTON

This review considers only the vertical distribution of meso- and macroplankton, the first including animals in the size range 200 μm to 30 mm (chaetognaths and polychaetes up to 50 mm) and the second, larger animals of body length up to 10–15 cm, mainly gelatinous animals, but also shrimps and similar animals. Fish and squid of the same size range are conventionally classified as micronekton.

Naturally, organisms ranging over three orders of magnitude in size cannot be collected by a single type of sampling equipment. Many different types of sampling equipment have been designed to collect and estimate the quantity of meso- and macroplankton of various sizes, including the simplest samplers, complex multi-functional units like "BIONESS", "MOCNESS", and mid-water trawls. There are drawbacks inherent in all of them. Unfortunately, this problem has not been completely solved. All types of plankton nets scare away animals and, due to resistance to water flow in the filtering cone, they do not filter the whole volume of water entering the mouth and they always underestimate plankton biomass. This drawback is improved to some extent by metering the filtered water in various ways. Bottle samplers, even large ones, do not give reliable measures of the concentration of the large mesoplankton or of its distribution between the sampled depths. Pumps also seem to be ineffective in collecting mesoplankton. Direct counts of plankton animals from submersibles are inevitably subjective and depend on size and visibility.

Several reviews have considered methods and equipment used for quantitative estimation of zooplankton, their pros and cons (e.g. UNESCO, 1968; Vinogradov, 1983b; Omori and Ikeda, 1984). This review discusses in detail the data collected by Russian expeditions, above all those mounted by the P. P. Shirshov Institute of Oceanology. Thus, it seems appropriate to describe briefly the main methods of data collection and analysis, not all of which are well known to western readers.

2.1. Mesoplankton

Small mesoplankton (0.2–3 mm, chaetognaths up to 5 mm) is usually collected by large water bottles of 130–180 l (Figure 1) at standard depths and at extreme points of the continuous profiles of various parameters obtained by instruments. These parameters, including temperature, salinity, density, density gradient, oxygen,

Figure 1   Sampling with the 150-1 water bottle.

chlorophyll, turbidity and bioluminescence, can influence plankton distribution or are related to it. A series of large bottle samples within the 0–200-m layer usually consists of 15–20 samples. This allows a detailed description of plankton distribution. Deeper bottle samples are collected less frequently. The large bottle samples are filtered through a sieve with a mesh size about 70 μm. Usually about 100–150 l of water are used for measurement of plankton concentration. The rest of the water volume is used for chemical analysis, estimation of phytoplankton, bacterioplankton and microheterotroph biomasses, primary production and chlorophyll concentration.

Large mesoplankton (body size 3–30 mm) is usually collected with the help of vertical tows by closing nets BR 113/140 (Juday net type) with an upper non-filtering cone and a lower filtering sieve cone (mesh size 530 μm) (Figure 2). The mouth area of this net is 1 m². The net is closed either mechanically with the help of a messenger or with an electronic closing device. In some cases, especially when the total water column is investigated down to 2000–4000 m and deeper (to 8000–9000 m), all the mesoplankton samples are collected by this net. Usually the following layers are sampled: 0–50, 50 m – thermocline depth, thermocline depth – 100 m, 100–200, 200–300, 300–500, 500–750, 750–1000, 1000–1500, 1500–2000, 2000–2500, 2500–3000, 3000–3500, 3500–4000, 4000–5000, 5000– 6000, 6000–7000, 7000–8000 (9000) m. Some layers may be combined or, conversely, split into subdivisions. Near-bottom horizons are sampled in more detail. In this

Figure 2   Collecting samples by BR 113/140 closing plankton net with a 1-m² mouth area.

case the nets (as well as the large bottles) are equipped with pingers, which allow the determination of distance from the bottom. The towing speed is 0.8–1.0 m s^{-1}.

For the collection of plankton samples within the upper 500–1000 m layer standard vertical tows are taken with a closing JOM net with mouth area 0.5 m² and filtering cone mesh size 180 μm. These nets usually have electronic closing devices, which allow the collection of samples from layers of 10–20 m width and the lowering of the net to a defined depth. Thus, on some cruises such an electronic closing device combined with a depth indicator on the net enabled sampling of the upper 500 or even 750 m of the water column with an accuracy of 10 m.

The biomass values obtained by the BR and JOM nets are compared with the biomass values obtained by the 150-l bottle on the basis of a large number (> 100) of simultaneous series of samples. It can be demonstrated that in the net samples the biomass of the plankton size groups as a whole is underestimated by two to three times, especially in the upper layers (Shushkina *et al.*, 1980; Tutubalin *et al.*, 1987; Vinogradov *et al.*, 1987).

The total plankton volume in the net samples is determined with the help of a voluminometer (substituted volume). Then the samples are, if necessary, divided into subsamples (1/5–1/100 part of sample volume) and the animals identified, measured and counted in subsamples. In the case of big bottle samples, usually the total samples are analysed. The animals are assigned to species, age stages, and sometimes larger trophic and taxonomic groups. The weight of every animal is calculated from body length and shape with a help of nomograms (Chislenko, 1968) or specially fitted empirical equations. Then these values are summed up and the wet biomass of different groups and species and the total biomass of the sample calculated. After this analysis all the macroplankton animals are removed from the sample, measured and weighed.

Bulk determinations of the organic carbon/wet weight ratio are made for various animals in different regions of the ocean. With the help of these values the wet biomass values can be recalculated as organic carbon biomass.

2.2. Macroplankton

Animals in the size range 3–10 (15) cm are collected by BR 113/140 plankton nets, a non-closing mid-water Isaacs–Kidd trawl, or a similar trawl of Aseev–Samyshev modification. The latter modification has a bag 25 m long (not 17 m) of uniform 5 mm mesh size. Hence, the samples give more quantitative information (supposing that the whole water volume that passes through the trawl mouth is filtered through the net). Most of the collected animals are not seriously damaged, though the gelatinous forms (ctenophores, siphonophores, some jellyfish, etc.) are partly or totally destroyed and it is not possible to measure their biomass. On some expeditions (Rudjakov and Zaikin, 1990) special near-bottom tows were made with the Aseev–Samyshev trawl equipped with electric cable. Horizontal or multi-step tows were also made (usually down to 500–1000 m depth). Organisms caught during the trawl descent and ascent were discounted. Later, when trawl catches were compared with the results of direct counts from submersibles, it was evident that the biomass obtained by trawls was also an underestimate by several times (Vinogradov and Shushkina, 1992).

All these methods have the same drawback: the gaps between samples (the distances between the depths of horizontal tows of pelagic trawls, or the thicknesses of the sampled layers during vertical or oblique plankton net tows) are so wide that it is impossible to detect layers of sharply diffcrent plankton

Figure 3 The deep-water submersibles "Mir 1" and "Mir 2" on board the R/V "Akademik Mstislav Keldysh".

concentration. These layers are often only 2–10 m thick. As a result the impression is given that the pattern of vertical distribution of plankton is more uniform than it really is. According to modern data (Vinogradov and Shushkina, 1976; Longhurst, 1981) it is non-uniform plankton distribution that supports the existence of a community where food is scarce, either in the oligotrophic regions of the upper zone or in the food-deficient ocean depths.

The ability to determine the real pattern of plankton distribution was achieved in the 1960s–1980s with the employment of deep-water manned submersibles. These submersibles have enabled direct counts of plankton animals to be made with high depth precision.

During Russian expeditions the submersibles "Argus" (in the Black Sea) "Mir-1" and "Mir-2" have been used for plankton investigations (Figure 3). The "Mir" submersibles have made it possible to count plankton organisms in the whole water column down to 6000 m depth. Such investigations were made during about 20 descents in various regions of the Pacific and North Atlantic oceans and the Norwegian Sea.

The animals are counted from the submersibles either within the cone of light from specially arranged lamps (Bernard, 1958; Pérès, 1958a,b; Barham, 1966; Vinogradov and Sagalevich, 1994), or by taking into account the distance between the specimens in the aggregations (Hamner and Carleton, 1979; Alldredge *et al.*, 1984). Russian expeditions have developed a method using wire frames

(Vinogradov and Shushkina, 1982, 1994; Shushkina *et al.*, 1991; Vinogradov and Sagalevich, 1994). The method is as follows: a cube made from wire with a side area of 0.6 or 1.0 m^2 is held by a manipulator in front of the window. In one corner of this cube is another with a side area of 0.04 or 0.06 m^2. As the submersible descends, the viewer counts the animals that pass through the horizontal side of the cube, using a tape-recorder. Simultaneously, the navigator names the depth every 2, 5 or 10 m. Thus, the numbers of different plankton species that pass through the frame are recorded against the background of a depth scale and the plankton concentration within the layers of 2, 5, 10, 20, 50 or more metres can be calculated. The velocity of the submersible's descent is adjusted according to the concentration of animals. In some cases (for instance, in near-bottom layers) the animals are counted during steady horizontal progress of the submersible. In this case the animals that pass through the front side of the cube are counted. The numerous animals of 3–5 mm length (mainly large Calanidae such as *Neocalanus cristatus* or *Eucalanus inermis*) are counted in the small cube, the others being counted in the large one.

In order to evaluate the size of macroplankton animals correctly, especially the various gelatinous ones, the wires of the cube are marked every 5 cm. Taking into account the sizes of the animals and their shape their biomass can be estimated.

It should be emphasized that direct counts from submersibles have made it possible to obtain not only the patterns of the different scales of stratification of plankton distribution, but also to evaluate the role of delicate gelatinous animals, such as large appendicularians, ctenophores (Lobata, Cestida), siphonophores, physophorids and other animals that are destroyed when any other sampling equipment is used.

As mentioned above, each of the described methods has its own restrictions and is adapted to count some distinct group of animals. Evidently, the simultaneous use of all these methods provides the best understanding of the pattern of the distribution of the meso- and macroplankton.

3. PATTERNS OF MESOPLANKTON DISTRIBUTION

3.1. Distribution in the Surface Productive Zone and in the Subsurface Layers

The upper layers are inhabited by three major types of pelagic fauna: arctic–boreal, tropical and antarctic, corresponding to the main climatic zones. Each of these regions is characterized not only by general faunistic traits, but also by the common ecology of the communities, which live in a more or less uniform environment.

The most clear-cut distinction in the distribution of the plankton is the division into cold-water (polar and subpolar) and tropical areas. The major criterion of this classification is the presence or absence of marked seasonal changes during the year; that is, the phytoplankton either grows more or less uniformly or exhibits distinct cyclical growth. Differences in the vertical distribution and migrations of the zooplankton of high and low latitudes can be explained as a tendency toward more efficient utilization of the phytoplankton.

Temperate and high latitudes. In the polar and subpolar regions the winter minimum of phytoplankton growth is followed by the spring bloom. The formation of a seasonal thermocline results in the bulk of phytoplankton being concentrated within the near-surface layers. The quest for food (phytoplankton) results in the concentration of all the phytophagous species of surface-living and upper-interzonal zooplankton within the near-surface water layer, where they exploit the spring phytoplankton bloom.

In summer the seasonal thermocline descends deeper, the layer of phytoplankton abundance becomes wider and the total amount of phytoplankton in the open regions decreases. As this takes place, the bulk of the phytophagous population also disperses within the extended water column; the dominant species with similar nutritional spectra seem to alternate their layers of their maximal abundance so that competition is reduced (Vinogradov, 1968).

During winter the increasing homothermy of the upper zone results in a sharp decrease in phytoplankton abundance. The environmental conditions within the surface zone become unfavourable and this induces the descent of the animals into the deeper layers.

The variability of vertical stratification of the surface waters and the long periods of homothermy seem to hinder the formation of groups of species connected with the distinct layers of the upper zone.

Low latitudes. Over most of the tropical region stratification is more or less stable all the year round, the intensity of solar radiation is sufficient for photosynthesis, and the entry of nutrients into the upper layers is not dependent on seasonal cyclicity. Low temporal variability in stratification and in the hydrological regime results in the formation of species groups inhabiting distinct (sometimes very narrow) layers of the upper water zone. Tropical regions are also characterized by a low concentration of epiplanktonic species near the surface. Peculiar communities of neuston and pleuston organisms develop near and on the surface film.

The existence of species groups restricted to the middle and lower layers of the subsurface zone is possible only in the tropical region (Heinrich, 1993), because there phytoplankton can develop within a wider depth range than in temperate waters.

In view of these fundamental differences in the functioning of the ecosystems of the temperate and tropical regions, these areas will be discussed separately with particular reference to some characteristic features. Thus, the plankton of high and middle latitudes will be described in terms of seasonal changes in the vertical distribution of several dominant species that determine the nature of the whole community. On the other hand, the description of tropical plankton will include an outline of the vertical distribution of the whole plankton biomass as affected by the hydrographic structure; that is, in the oligotrophic zones and in the productive areas of divergences and upwellings.

3.1.1.   *Cold Regions of the Ocean*

The zooplankton in all cold-temperate regions is adapted to the annual cycle of phytoplankton development. In the spring the mesoplankton biomass sharply increases due to the ascent from deeper layers of a few interzonal phytophagous species, mainly of the family Calanidae, and to a lesser degree the Eucalanidae. Usually three or four species in each region exhibit such a pattern of seasonal movements. At the same time the abundance of the small opportunist species rapidly increases. In autumn the interzonal calanids descend into the deep layers and go into diapause (Conover, 1988), and the populations of small opportunistic species die off.

Let us consider in detail the pattern of these processes in the North Pacific, Central Arctic Basin, North Atlantic and in the Southern Ocean.

3.1.1.1.   *The North Pacific* During the growth period the bulk of mesoplankton biomass (up to 90%) of the upper 500 m layer consists of interzonal filter-feeding copepods: *Neocalanus plumchrus* in the west and the related species *N. marshallae* in the east; with *N. cristatus* and *Eucalanus bungii*. Species of the epiplankton are more abundant only in some areas, especially along coasts, where they may contribute over 50% of the total plankton biomass of the upper 25-m layer. Prominent members of this group are *Oithona similis*, *Pseudocalanus elongatus* and some neritic forms near the coasts. Only a few species are confined to the deeper part of the surface zone and the underlying strata without rising above the 50–100 m level.

In the north-west Pacific (the Kurile–Kamchatka region) during the winter season the water column becomes isothermal (2.5–1.0°C) to a depth of 120–180 m and this layer gradually deepens during the winter months. Below the isothermal layer is a warm intermediate stratum with a temperature of about 3°C extending from 200 to 800 m depth. During the spring a warming and slight freshening of the surface creates a seasonal pycnocline layer, which sinks in summer to a depth of 25–30 m. At the same time, the cold-water masses formed in winter are dispersed by the warmer and saltier waters of the underlying more warm layer. The cold intermediate layer contains low plankton biomass all summer. In autumn (September–October) gradual cooling down of the surface layer begins, which effaces the thermocline and creates the isothermal conditions of winter.

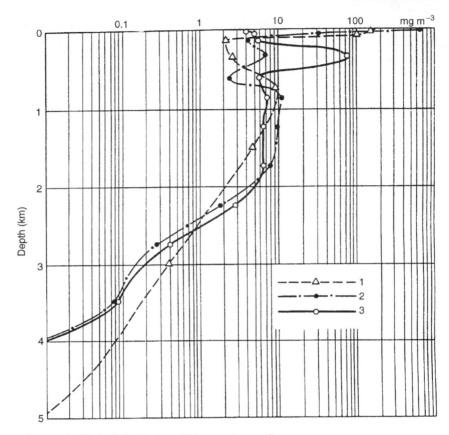

Figure 4 Vertical distribution of biomass (mg m^{-3}) of *Neocalanus cristatus* (measured by BR 113/140 plankton net) in the north-western Pacific. The decrease in the cool intermediate layer and the maximum in the warm intermediate layer during July–August are evident. (From Vinogradov and Arashkevich, 1969.) 1, May–June, 1953, five stations averaged; 2, July, 1966, three stations averaged; 3, August, 1966, three stations averaged.

In the north-western Pacific the mesoplankton biomass is dominated by *Neocalanus cristatus*. During May–June within the upper 100-m layer it averages 25–40% of the total mesoplankton biomass. In July the biomass in the upper layers sometimes reaches 70–80%, but early in August the main mass of the stage V copepodids descends deeper than 300 m into the warm intermediate layer; during this period its percentage of the total biomass of the plankton of the surface layer decreases to 2–3% (Figure 4).

Direct observations from the submersible "Mir" have shown, that at that time the population of *N. cristatus* forms a very narrow and highly contracted layer (300

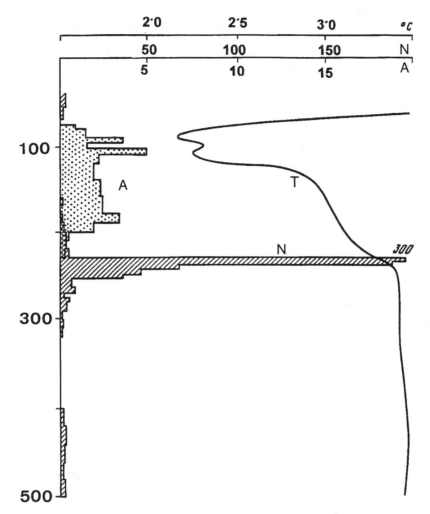

Figure 5 Vertical distribution in abundance (specimens m^{-3}) of the copepod *Neocalanus cristatus* (N) and the medusa *Aglantha digitale* (A) in middle August at the station near south-western Kamchatka, estimated on the basis of direct counts from "Mir" submersible. The vertical profile of temperature (T) is indicated to the right (from Vinogradov and Shushkina, 1994).

specimens m^{-3}) close to the boundary of the warm intermediate layer at 250 m depth (Figure 5).

During the winter *N. cristatus* is absent above the boundary of winter convection. During this time its population consists mainly of the stage V copepodids and mature females and is concentrated within the 300–2000 m layer, some specimens descending to 4000–5000 m. Hence, a large proportion of the

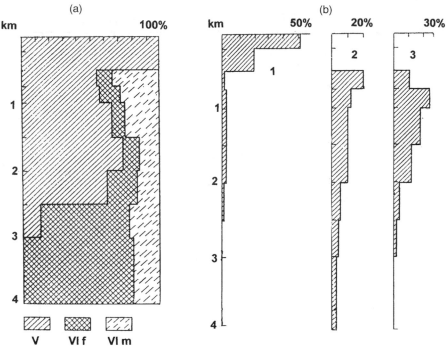

Figure 6 Variation with depth of the age structure of the population (a) and relative abundance of the different copepodid stages (b) of N. cristatus in July–August in the north-western Pacific (six stations averaged) (from Vinogradov and Årashkevich, 1969). 1, V copepodid stage; 2, VI (females); 3 VI (males).

mature specimens remains within the 200–500 m layer all the year round. The reason is that at least some of the population has a 2-year cycle of development. In general, it may be assumed that 90% of the population of this species inhabits the waters between the surface and 3000 m depth.

Naturally, not only the biomass and abundance, but also the age structure of the population changes with depth. This problem was specially studied in *N. cristatus* by Vinogradov and Årashkevich (1969, Figure 6a,b). Another species, *N. plumchrus* is resident in the upper zone for a longer time. As early as May–June it constitutes about 14–22% of the total mesoplankton biomass within the upper 100-m layer and, in some regions, surface aggregations of the stage V copepodids reach up to 6 g m^{-3}. It starts to descend later in the autumn, in September–October. The mature stages are concentrated at 300–750 m depth, but only individual specimens descend deeper than 1000 m and 90% of the population biomass inhabits the 0–750 m layer.

Eucalanus bungii inhabits colder waters than both species described above. Its maximum abundance (up to 35 specimens m^{-3}) occurs within the cold intermediate layer. Mature specimens occur from the surface layer, but the most

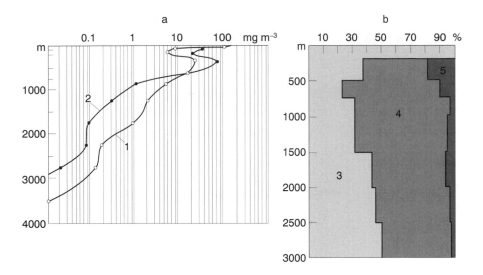

Figure 7 (a) Variation with depth of biomass (mg m^{-3}) of *N. plumchrus* (1) *E. bungii* (2) and the percentage of *N. plumchrus* copepodite stages (b) in the north-western Pacific in July–August (6 stations averaged) (from Vinogradov and Årashkevich, 1969). 3-V copepodite stages; 4-VI females; 5-VI males.

important population is within the 200–500 m layer; less than 5% of the population occurring deeper than 750 m (Figure 7).

Thus, the bulk of the populations of common upper-interzonal copepods (above 90%) occurs within the layers above 3000 m (*N. cristatus*), 1000 m (*N. plumchrus*) or 750 m (*E. bungii*). Nevertheless, the minor numbers dispersed deeper than the main body of the population are an essential component of the quantitatively poor deep-water plankton. As indicated in Table 1, *N. cristatus* is extremely important in the deep layers. In general, even in summer, when the bulk of the populations of upper-interzonal species is concentrated in the upper zone, they contribute more than one-third of the total plankton biomass in the bathypelagic zone (1000–3000 m) and more than 10% of the biomass in the upper layers of the abyssopelagial (3000–4000 m layer). In winter, when the populations of this species descend from the surface layers, their role in the deep-water plankton seems to increase even more. Thus, the seasonal variations of the vertical distribution of mesoplankton within the upper 200 m layer in the North Pacific might be described as follows: during early spring, in March–April, the animals that spent the winter in the warm intermediate layer ascend to the upper water layers for feeding and reproduction. Some species (*N. cristatus*, *N. plumchrus*) breed below the upper zone and the young stages ascend to the surface. At the end of June or a little earlier the typical summer pattern of vertical distribution is

Table 1 The role of different upper-interzonal copepods in the total biomass of net mesoplankton in the Kurile–Kamchatka Region, as %. Average of six stations in July–August.

Species	Depth, m										
	0–50	50–100	100–200	200–500	500–750	750–1000	1000–1500	1500–2000	2000–2500	2500–3000	3000–4000
Neocalanus cristatus	53.6	15.6	4.3	15.5	4.4	22.3	30.3	36.8	14.3	9.9	7.3
Neocalanus plumchrus	16.9	10.7	5.8	9.9	17.6	12.6	8.4	4.6	1.1	2.6	0.6
Eucalanus bungii	4.4	20.7	23.9	3.1	17.4	2.4	1.4	0.4	0.4	0.3	0.1
Total	74.9	47.0	34.0	28.5	39.4	37.3	40.1	41.8	15.8	12.8	8.0

established. The bulk of the plankton is concentrated near the surface. The maximum biomass is usually within the 10–25- or 25–50 m layer. Deeper, in the cool intermediate layer, the plankton biomass sharply decreases.

In August the bulk of the population of *N. cristatus*, and then that of *N. plumchrus*, starts to descend gradually into the deeper layers. At this time their populations divide into two groups, one of them located above the cold intermediate layer, and the other one below it. The interchange between these two groups seems to be very slight. This is evident for *N. plumchrus*, because in some regions the surface group consists of *N. plumchrus f. plumchrus*, which occurs in the surface layers only. *E. bungii* as usual has two maxima of abundance – above the cool intermediate layer and below it. In some regions, where the cool intermediate layer is less evident, the upper maximum is absent and a large amount of *E. bungii* occur in the cool intermediate layer.

In autumn, during September and October, along with the cooling of the surface layers the intensive descent of *Neocalanus* and *Eucalanus* populations continues. With the cooling of the surface layers more and more plankton descends deeper than the boundary of the surface zone and, finally, by the time the winter homothermy is established its distribution takes the winter form.

It should be noted that some interzonal species that perform intensive diel migrations in summer (*Metridia pacifica*, *Themisto japonica*), also ascend into the upper layers during winter, but compared with other seasons, their abundance there is low.

On the basis of average data from many stations in the dichothermal waters of the Kurile–Kamchatka region and the southern part of the Bering Sea, Figure 8

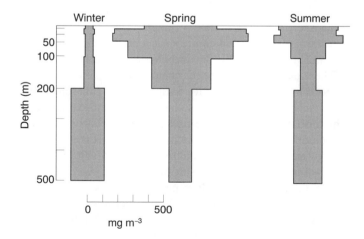

Figure 8 Seasonal variations of vertical distribution of mesoplankton biomass (mg m^{-3}) in the north-western Pacific and southern part of the Bering Sea measured by Juday net (67 stations averaged) (from Vinogradov, 1968).

shows seasonal changes in the vertical distribution of the total plankton biomass.

The more sharply defined is the cold intermediate layer, the poorer is its plankton. This is evident from a comparison of the vertical distribution of the plankton in the north-western Pacific, where the cold intermediate layer is relatively warm, with the distribution in the Bering Sea and especially the Sea of Okhotsk, where the cold intermediate layer is more sharply outlined and almost invariably has below zero temperatures.

The north-western part of the ocean consists of dichothermal waters. Eastward, the cold intermediate layer tapers off and finally disappears. At the same time there is a corresponding change in the vertical distribution of the plankton, the density minimum at the depth of the cold layer becomes less pronounced and gradually vanishes (Bogorov and Vinogradov, 1955).

A question arises about the mechanism of ontogenetic migrations. Rudyakov (1986) concluded that the active mechanism of regulation of the inhabited depth works only at some stages of ontogenesis, other developmental stages dispersing vertically due to differences in the rate of sedimentation, the level of the individual's activity and the mobility of the environment. The increase of the average depth of distribution with age could be the result of dispersion of the individuals beyond the subsurface layer where reproduction took place. Thus, the greater the distance of the individuals from that layer, the greater their average age. Dispersal is a stochastic process, the result of chaotic movement of the animals, turbulent mixing and the range of the sinking rates of the animals, the last value being the result of the relationship between swimming activity and buoyancy. The sinking rate seems to increase with age due both to the specific weight increase (Kils, 1979) and decreasing the capacity of ageing animals to compensate for the sedimentation rate.

Thus, the near-surface population consists of individuals most suited for existence in the pelagial by the balance between their rates of sinking and ascending. These animals have a better chance to complete development and to support the population due to better food supply and higher temperature at upper levels. The lower part of their vertical distribution seems to contain less viable animals which do not take an active part in reproduction. As Vinogradov and Arashkevich (1969) noted, "their occurrence there seems not to be determined by the necessity of the existence of population" (pp. 495–496). Consequently, the vertical distribution is not only the result of individual reactions to the situation during the sampling, but is the result of previous events.

Similarly, the descent of the animals during diel migrations is, according to Rudyakov (1986), due to both decrease of activity after the night ascent and gravitational sinking. However, it is well known that the mechanism of the autumn descent of populations of many copepods is connected with the fact that, while the diel vertical migrations at this time of the year become more intensive, a larger and larger part of the population remains at their daytime depth and does not

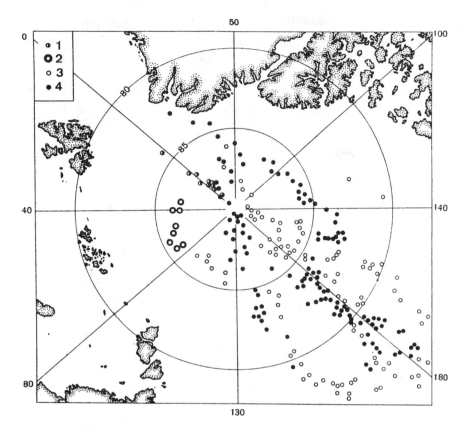

Figure 9 Locations of the plankton stations in the Central Arctic Basin; samples were collected from drifting ice stations during Russian expeditions in 1937–1974. 1, 1937; 2, 1937–1939; 3, 1950–1957; 4, 1968–1974 (from Pavshtics, 1980).

ascend to the surface at night. Thus, seasonal migration is the result of modification of the diel migrations of a population. There are experimental and field observations of the active movements of animals, including copepods, during their morning descent from the upper layers (Bainbridge, 1952; Vinogradov *et al.*, 1996a; etc.).

3.1.1.2.   *The Central Arctic Basin*   The vertical distribution of plankton over the Central Arctic Basin was studied during the drift of numerous ice stations, "North Pole" (Figure 9). As in the cyclic communities of the northern Pacific, the zooplankton develops during spring and summer months. The zooplankton consists mainly of interzonal species such as the phytophagous *Calanus hyperboreus, C. glacialis*, the carnivores *Pareuchaeta glacialis, P. polaris, Themisto libellula* and the fine species such as *Oithona similis* and *Oithona*

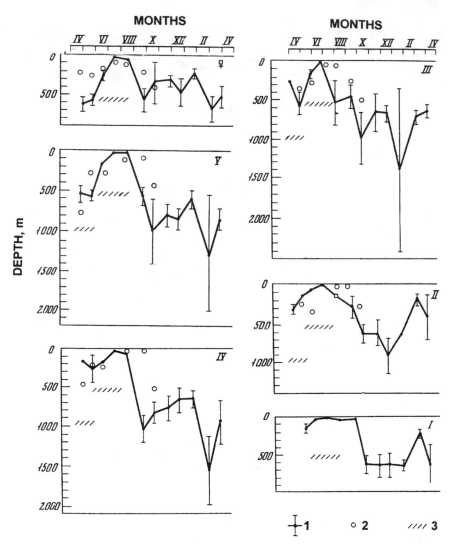

Figure 10 Seasonal variation of the depth occupied by different copepodid stage of *Calanus hyperboreus* in the Central Arctic Basin. 1, Median of the vertical distribution of abundance and its standard error according to observations during 1950–1951; 2, the same according to observations during 1975–1977; 3, depth of the bottom, if the depth of the station is less than 1000 m (from Heinrich *et al.*, 1980). The horizontal scale is for months of the year (January to December). The symbols in the right hand top corner of each graph refer to mature females (female symbol) and copepodid stages (I to V).

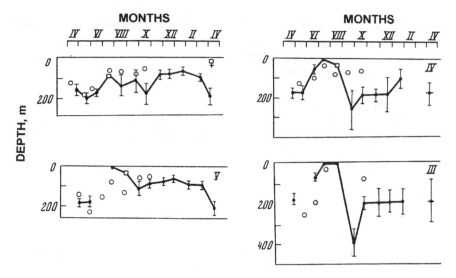

Figure 11   Seasonal variation of the depth occupied by *Calanus glacialis* in the Central Arctic Basin. The labels and symbols are the same as in Figure 10 (from Heinrich *et al.*, 1980).

borealis. According to Pavshtics (1980) between June and September from 50% to 90% of the zooplankton is concentrated near the surface and under the ice cover, within the 0–25 m layer. The biomass in this layer reaches 200–500, sometimes 1000 mg m⁻³.

Heinrich *et al.* (1980) studied the seasonal variations of the common species of the Arctic Basin. Figure 10 illustrates the vertical distribution of the *C. hyperboreus* population. It is evident that mature females and stage V copepodids ascend to the surface in July–August, and then most of them descend to the 200–500 m layer (females) and 500–1000 m layer (stage V). The distribution of younger copepodids is about the same, but stages I–II are concentrated near the surface for a longer period in May to October. During this period they have two maxima: at 20–50 m and at 300 m. The peculiarities of the vertical distribution of this species arise from its long life cycle (2–3 years).

The population of the other common species, *C. glacialis* (Figure 11), inhabits shallower depths, the main aggregations of females never occurring in the upper layers.

Unlike these species *Metridia longa* performs extensive diel migrations, and the dominant age groups of the population (stages V and VI) concentrate within the upper layers essentially during winter (polar night).

Hence, the magnitude of seasonal migrations of these species (calculated from the median of population abundance) is less then 600 m (*C. hyperboreus*), 170 m (*C. glacialis*), and 250 m (*M. longa*).

Similar seasonal variations in vertical distribution are typical for the small species *Microcalanus pigmaeus*. Aggregations of stages VI, V and IV copepodids of that species change in depth during the year from 500–1000 m to near-surface horizons, the juveniles inhabiting upper layers all the year round (Kosobokova, 1980).

Unlike the species already discussed, the depth inhabited by common fine species varies little from one season to another. Thus, the bulk of the *Oithona similis* population is concentrated within the upper 100 m, or the 50 m layer all the year round. The seasonal migrations of *Onceaea borealis* are limited to the upper 200 m layer of arctic waters. *Oncaea notopus* inhabits the 100–200 m layer almost permanently (Kosobokova, 1980). Hence the role of these species in the total plankton biomass is small.

3.1.1.3. *The North Atlantic* In the North Atlantic in winter the saline surface waters sink to a considerable depth as they cool down. Isothermal conditions are not confined to the 0–150 (200) m layer as in the Pacific, but may extend to a depth of 400–700 m and in some regions down to the near-bottom layer.

Year-round observations in the Norwegian Sea on the south-eastern slope of the Lofoten Depression (weather station "M") have shown that winter convection begins there in October and is confined at first to the zone above the thermocline (0–50 m). By February all the waters to a depth of 2000 m (demersal horizons) are affected. In April thermal stratification begins, which by midsummer causes a gradual rise of temperature in the layer from 600–800 m to the surface. These times and depths vary considerably according to latitude and other factors, but the general patterns of the seasonal hydrological cycle remain essentially the same over most of the area.

The zooplankton of this part of the North Atlantic consists mainly of copepods – *Calanus finmarchicus* s. l. and *C. hyperboreus* (the latter species is essentially abundant in the coldest waters). Of lesser importance are *Metridia longa*, *M. lucens*, *Pseudocalanus elongatus* s. l., *Oithona similis*, *O. atlantica*, *Pareuchaeta norvegica*, the chaetognath *Sagitta elegans* and the tunicates *Oikopleura labradoriensis*, *O. vanhoeffeni* and *Fritillaria borealis*. However, only *C. finmarchicus* s. l. is dominant everywhere (except in parts of the North Sea); the proportion of the remaining species varying considerably from one region to another.

In winter (November–January) most of the *C. finmarchicus* population of the deep-sea regions of the ocean collects at depths greater than 400–700 m, while in shallower regions they concentrate in the demersal waters of the slope and trenches.

In the early spring (February–March) the overwintering population begins to ascend. Soon (March–early April), the entire *C. finmarchicus* population move upward, so that by April–May it is concentrated above 200 m, particularly between 0 and 50 m. *C. finmarchicus* reaches its peak biomass after the spring bloom of phytoplankton. In the second half of June, the stage IV and V copepodids, having accumulated lipids (reserve of energy) begin to descend; in deeper regions they sink as deep as 500 and even 1000 m. The whole population descends by August, and in October–November the distribution assumes the characteristic winter pattern. The wintering stock consists mainly of copepodid stages IV and V, which attain sexual maturity at the beginning of the upward migration in the spring (Sömme, 1934). Further south in the North Atlantic, *C. finmarchicus* produces two generations – a short-lived one in summer, and a prolonged one in winter. The winter generation performs more pronounced seasonal migrations of the type described above.

Another dominant species, *C. hyperboreus*, inhabits cold waters, but can be advected with these waters far to the south. For instance, high numbers of this species have been found in the frontal zone of the deep flow of the Labrador Current and Gulf Stream (41°40'N, 50°W), where its biomass was as high as 10 mg m^{-3} at 1000–1800 m depth and more than 30 mg m^{-3} at 1800–2300 m. According to the range of ontogenetic migrations and the depth of the stage V and VI copepodid concentrations, this species is similar to *Neocalanus cristatus* in the Pacific Ocean, whereas *C. finmarchicus* is similar to *N. plumchrus*. Seasonal changes in the vertical distribution of this species are shown by data collected at weather station "M" in the Norwegian Sea (Østvedt, 1955); other sources are the works of Wiborg (1954) and Sömme (1934).

The bulk of the *C. hyperboreus* population, mostly stage V and VI (females), winters at a depth of 1000–2000 m; that is, deeper than the wintering stock of *C. finmarchicus*. Only isolated specimens of *C. hyperboreus* rise to 600–1000 m. In late February–mid-March, a partial ascent of the population begins, and *C. hyperboreus* is found not at 600–1000 m, but in the 100–600 m layer, where it spawns. After a month and a half, the stage I and II copepodids appear at a depth of 0–100 m. Other copepodids rise at the same time to the upper layers after hibernation in deep waters. However, only part of the *C. hyperboreus* population reaches the surface, in contrast to *C. finmarchicus*. At least one-third of the *C. hyperboreus* population remains at depths of more than 1000 m. In other words, the wintering stock of *C. hyperboreus* consists of two physiologically different groups (Sömme, 1934; Wiborg, 1954; Conover, 1988), which suggests a 2-year developmental cycle.

Thus, *C. hyperboreus* appears in the upper 100-m layer of the Norwegian Sea as late as May–June, but leaves the surface zone in July and becomes concentrated at a depth of more than 600–1000 m by August (Figure 12). In this case both the adult copepods and the young are concentrated in deep water (Figure 13).

The populations of surface-dwelling (epipelagic) species (*Oithona similis, O. atlantica, Oikopleura*) are small in winter, but rather evenly spread down to a depth of 600 m or more, although limited quantities of the animals penetrate deeper. In the regions where winter homothermy spreads to great depths, these species penetrate more deeply than in the Pacific. In spring, during the spring phytoplankton "bloom", the bulk of the population of epipelagic species is concentrated in a relatively thin surface layer.

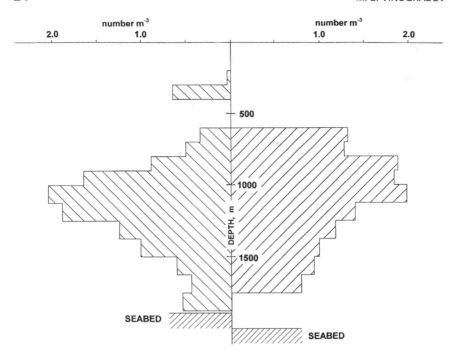

Figure 12   Vertical distribution of abundance (specimens m^{-3}) of the older copepodids
(V and VI) of *C. hyperboreus* measured by direct counts from submersible "Mir" in the
Norwegian Sea at 73°42'N, 13°15'E. To the left is 22.07.1994; to the right is 28.07.1994
(from Vinogradov *et al.*, 1995).

Thus, the pattern of seasonal migrations of common plankton species in the
North Atlantic is similar to the migrations of their counterparts in the North
Pacific. It is natural that the general pattern of seasonal variation of the vertical
distribution of plankton biomass does not differ greatly in the two regions. It
should be mentioned, however, that in the Atlantic Ocean low biomasses are never
observed in the 100–200 m layer, this phenomenon being characteristic of summer
in the northwestern Pacific.

3.1.1.4.   *Polar and Subpolar Regions of the Southern Ocean*   The waters
south of the Subtropical Convergence may be classed with the polar and subpolar
regions of the southern hemisphere. The location of this convergence varies in
different regions of the Southern Ocean and during the various seasons, ranging
from 37–38° to 52°S; the average boundary is at 40°S.

The vertical distribution of plankton is far from uniform over the vast area from
40°S to the coast of Antarctica. Heterogeneity arises from differences in
hydrological conditions and in the composition of the plankton. In waters of the
westward drift, south of the Antarctic Divergence, the whole water mass from the

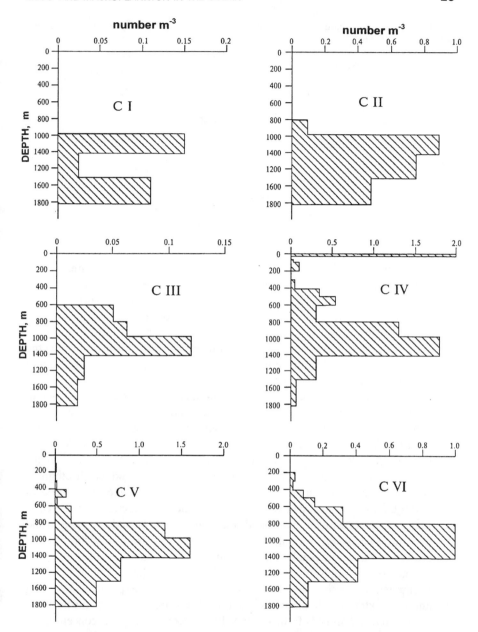

Figure 13 Vertical distribution of abundance (specimens m^{-3}) of the different copepodid stages of *C. hyperboreus* measured by plankton net BR (73°42'N, 13°15'E, 21.07.1994) (from Vinogradov *et al.*, 1995).

Table 2 Seasonal changes of the mean plankton biomass (g m^{-2}) at different depths (after Foxton, 1956).

Depth, m	Summer	Autumn	Winter	Spring
	XII, I, II, III	IV, V	VI, VII, VIII, IX	X, XI
		Subantarctic waters		
0–250	11.25	5.0	4.25	10.6
250–1000	10.9	9.8	13.1	10.9
0–1000	22.15	14.8	17.3	21.5
		Antarctic waters		
0–250	10.5	10.25	5.4	8.6
250–1000	11.7	15.0	20.5	11.7
0–1000	22.2	25.2	25.9	20.3

surface to a depth of 200–300 m has a negative or near-zero temperature, even in summer. The cold waters move northward from the divergence. Sometimes, a comparatively thin layer of warmed summer water appears at the surface above a cold intermediate layer at a depth of 50(100)-200(300) m.

The split thermal structure of the North Pacific is such that waters below the cold intermediate layer are comparatively warm and there is a steep gradient at the lower border of the cold layer. In the Antarctic, on the other hand, temperature increases slowly and gradually with depth.

As in the subarctic region, only a limited number of species are represented in the main vertical distribution of plankton in subantarctic and antarctic waters. In the high polar waters south of the Antarctic Convergence these species are *Calanus propinquus*, *Calanoides acutus* and *Metridia gerlachei*. North of the Antarctic Divergence, the dominant species are *C. propinquus*, *C. acutus*, *Calanus similimus*, *Rhincalanus gigas*, *Calanus tonsus* in the Pacific, and the chaetognath *Eukronia hamata*. In some areas (e.g. Drake Passage), *Rhincalanus gigas* accounts for 75% of the whole copepod fauna, but in the Pacific subantarctic *Calanus tonsus* dominates. Together with the euphausiids *Euphausia superba* and *E. vallentini* and the hyperiid *Parathemisto gaudichaudi*, these species exemplify the vertical distribution of the whole planktonic population in the Antarctic.

In winter (June–August) and early spring (September), the bulk of the plankton is concentrated at a depth of more than 200–250 m. During the spring an increasing amount of plankton rises to the surface layers. This process begins in the northern part of the area and spreads further south. In October, the highest concentration of plankton is found at a depth of 100–250 m, while in summer the 50–100 or 0–50 m layers are richest. During the summer, surface currents carry the plankton northward and concentrate it near the Antarctic Convergence. In April–May the plankton leaves the surface zone and gathers at a depth of more than 200 m. These changes are summarized in Table 2.

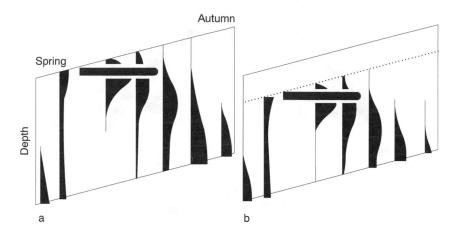

Figure 14 Seasonal changes in vertical distribution of populations of interzonal species in the Antarctic waters. (a) regions with no sharp hydrological boundaries; (b) regions with sharp thermocline (from Voronina, 1972a).

This scheme is a rough presentation of seasonal changes in the vertical distribution of the planktonic biomass. The seasonal fluctuations do not appear simultaneously in the entire antarctic water area. The spring begins earlier in areas where the ice melts sooner, whether in northern waters or coastal zones. Conversely, the specific hydrometeorological conditions of other areas can preserve the ice for a longer period and thus delay the onset of spring (Vinogradov and Naumov, 1961). These features determine the timing of the seasonal redistribution of the plankton biomass.

The pattern of distribution of the total biomass is the result of seasonal migrations of the common interzonal species. This pattern was studied in detail by Voronina (1972a,b; 1974, 1976, 1984, etc.). On the basis of extensive field observations she traced the seasonal reorganizations of the vertical structure of populations of common species and proposed a generalized scheme of this process: ascent during the spring biological season, maximum abundance (young copepodids) in the upper layers and the descent of the population during summer–autumn. The spring ascent of populations can take place in regions with a different level of water stratification either as far as the surface or up to the level of seasonal pycnocline (Figure 14).

Moreover, the ascent of the most common interzonal species takes place in a seasonal succession. First of all the population of *Calanoides acutus* ascends, then *Calanus propinquus*, and, finally, *Rhincalanus gigas*. The maximum density of the latter occurs in the upper layers when the population of *C. acutus* starts to descend. As a result, depending on the exact situation in each particular region, either one

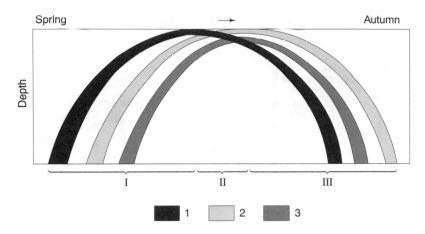

Figure 15 The pattern of seasonal variations of vertical structure of the community in the Southern Ocean. The location and relative thickness of the "nuclei" of the populations (the part of a population within the 25% percentiles) are depicted. 1, *Calanoides acutus*; 2, *Calanus propinquus*; 3, *Rhicalanus gigas* (Voronina, 1972a).

or the other species can reach the most favourable feeding conditions and produce the most young (Figure 15). As with all the seasonal events in the Southern Ocean, the ascent of populations starts in the north and then gradually shifts southward, though local hydrographic conditions may disturb the smooth progression. Hence, seasonal variations do not occur simultaneously over all the area of the antarctic waters. Spring starts earlier in the regions where the ice cover disappears earlier. Conversely, where the cover is prolonged due to hydrometeorological conditions, spring comes later. The time of seasonal transfer of plankton biomass changes according to the timing of these events (Vinogradov and Naumov, 1961). As a result, at any instant and in every location rather complex patterns of vertical distribution can occur, depending on the season, the location of the region and what common species of interzonal copepods dominate the plankton.

The most common euphausiid species of antarctic waters, *Euphausia superba*, is characterized by a similar type of seasonal migrations. According to Fraser (1936), near South Georgia the metanauplii of *E. superba* appear in large quantities in February–March and are concentrated below 500 m depth. At the calyptopis stage (I–III) these crustaceans start to undertake intensive diel migrations ranging through about 750 m, their maximum concentration occurring in the 50–100 and 0–50-m layers. The young furcilia (I–IV) also continue to perform intensive diurnal vertical migrations. However, the older furcilia (V–VI) and adult animals do not migrate but remain concentrated in the upper layers.

On the whole, the vertical distribution of the antarctic and subantarctic plankton closely resembles the pattern observed in subarctic waters. In both cases, the plankton is composed largely of upper interzonal species showing considerable seasonal migrations (rise in summer, descend in winter), but rather modest diel movements (*Calanus* and *Eucalanus* in the north, *Calanus*, *Calanoides* and *Rhincalanus* in the south). Other plankters make very rapid diurnal migrations of an amplitude of 250–500 m (species of *Pleuromamma*, *Metridia* except *M. gerlachei*, *Pareuchaeta*, etc.), but practically no seasonal movements. This category accounts for a small fraction of the plankton.

Thus, in subarctic areas and the Antarctic, the bulk of the plankton concentrates in summer (day and night) in the upper 50–100 m of water. Regions of phytoplankton blooms show the greatest concentration of plankton in the surface layers. In winter, on the other hand, the plankton gathers deeper than 200 m, leaving the surface zone sparsely populated.

Another important difference is that in the Antarctic, there is hardly ever an impoverishment of the plankton of the cold layer. On the contrary, the concentration of plankton is often greatest in this layer, at a depth of 100–200 m. Indeed, antarctic species spend most of their life cycle in a water mass that includes this layer, Moreover, the gradients at the lower boundary of the cold layer are smaller in the Antarctic than in the northwestern part of the Pacific.

These differences, however, do not upset the basic similarity in the distribution of the plankton in the whole non-tropical part of the World Ocean.

3.1.1.5. *General regularities* The vertical distribution of the plankton in all the cold regions can be summarized as follows:

- A small number of dominant species determine the vertical distribution of the whole planktonic biomass.
- The plankton is dominated by upper interzonal species that spend part of their life span in the surface zone and part at great depths. Hence, the vertical migrations of the plankton cross depths of hundreds or thousands of metres.
- The epiplankton is only slightly stratified. It contains only a few species that live entirely in the lower part of the surface zone.
- The whole pattern of vertical distribution changes from one season to another. In the spring and summer, when phytoplankton reaches its peak, the concentration of zooplankton in the surface waters increases many times over values observed in winter, when the drastic depletion of phytoplankton results in many dominant species descending to deeper layers.
- There is also a seasonal change in the intensity of the diel migration of the bulk of the zooplankton. Migration intensity reaches its peak in the late summer and autumn, and drops to a minimum in winter. This is due, on the one hand, to a seasonal increase in the migrating intensity of species characterized as relatively weak migrators and, on the other hand, to the increased role of intensive migrators during the summer–autumn period.

3.1.2. *Tropical Zone of the Ocean*

The oceanic regions of the tropical zone are not uniform in water structure and, consequently, not a uniform environment for plankton. On the one hand, there are the regions of anticyclonic gyres, characterized by descending surface waters and a deeply located main pycnocline layer. Such a situation hinders the penetration of nutrients into the euphotic layers and results in very low productivity. On the other hand, there are regions of divergent currents, equatorial divergence and coastal upwellings. These regions are characterized by the ascent of deep waters, sometimes almost to the surface.

    3.1.2.1. *Oligotrophic tropical regions of the ocean*   According to Shushkina *et al.* (1995) the area occupied by oligotrophic waters with a chlorophyll concentration of less than 0.1 mg m^{-3} is as large as 50% of the total area of the World Ocean and 70% of the total area of the tropical and subtropical regions (between 40°N and 40°S). These regions can be termed "ocean desert" and correspond to the arid zones of the land. The plankton communities of these regions are characterized by a high degree of maturity and a low biomass. In the layer above the thermocline a quasi-stable density microstructure is formed. This structure causes stratification of plankton distribution (Heinrich, 1961; Vinogradov, 1968), which is stabilized by trophic relationships between consumers and their prey (Vinogradov and Shushkina, 1976). Thus, the vertical distribution of plankton in the upper mixed layer seems to be essentially stratified.

    Stable hydrodynamic conditions (i.e. permanent stratification of the water layers) enable an extended succession of plankton communities in the oligotrophic layers, compared with the duration of succession in any other region of the ocean. The communities of oligotrophic waters of the global anticyclonic gyres reach the most complete stage of succession maturity and are near the climax stage (Vinogradov and Shushkina, 1987).

    Small-scale wind-driven circulation and mesoscale eddies within the euphotic zone and internal waves cause the development of ephemeral fronts and increase the irregularity of plankton distribution. Local nutrient enrichment can result in local "rejuvenation" of a community, secondary succession and increase of biomass. Hence the distribution of plankton in gyres is non-uniform. Patches occur with biomass several times higher than that of the background.

    Mesoplankton distribution patterns vary in different regions. The lowest biomass was observed in the southern part of the southern anticyclonic gyre of the Pacific (Figure 16a); where the general mesoplankton biomass (B_z) in the 0–200 m layer was only 2.1 g m^{-2}. The biomass throughout this water column fluctuated from 5 to 10 mg m^{-3} and increases only near the top of the thermocline, to 27 mg m^{-3}. The mesoplankton was a little richer ($\sim$3 g m^{-2}) south of Hawaii (Figure 16b) and in the southern oligotrophic waters of the Indian Ocean (Figure 16c). The mesoplankton was rather evenly distributed vertically. In the upper 10-m

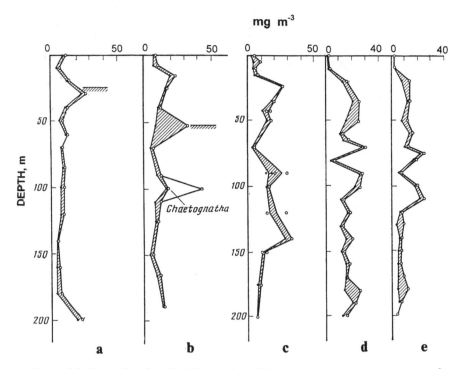

Figure 16 Examples of vertical distribution of the biomass of mesoplankton (mg m^{-3}) at stations in tropical oligotrophic regions (150-1 water bottle). Daytime. Shadowed area indicates the biomass of predators. Horizontal shaded line indicates the top of seasonal thermocline. (a) south central gyre of the Pacific; (b) central gyre southward from the Hawaii Is; (c) southern oligotrophic region of the Indian Ocean; (d, e) Mascarene Basin, Indian Ocean (from Vinogradov and Musaeva, 1989).

layer, in daytime, the biomass declined to 4–5 mg m^{-3}. Against an average background of 10–15 mg m^{-3}, a maximum of up to 25–50 mg m^{-3} appears rather regularly near the upper part of the pycnocline. In this maximum, predators are very important. The Mascarene Basin region, which is near richer areas, is also poor (Figure 16d,e).

In the oligotrophic tropical regions diel changes of plankton biomass in the upper 200 m layer are extremely sharp. The biomass at night can increase as much as 1.5–2.5 times due to the ascent of plankton from deeper layers (Figure 17). In contrast, in more productive regions the intensity of migrations decreases. This phenomenon has been mentioned repeatedly by many authors, and has been analysed in detail (Vinogradov, 1968). New data obtained recently confirm the previous conclusions.

The spatial irregularities typical of oceanic plankton are most pronounced in

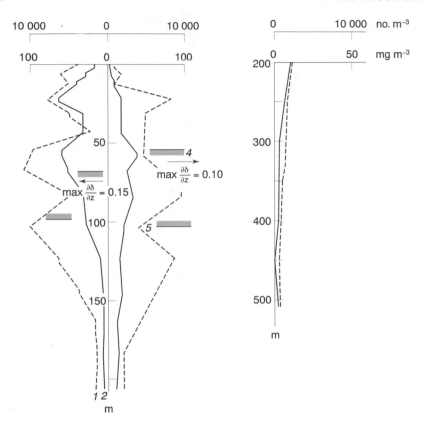

Figure 17 Vertical distribution of abundance and biomass of mesoplankton in the oligotrophic waters of the northern subtropical gyre of the Atlantic (23°22'N, 30°54'W, October, 1991). The samples were collected by 150-l water bottles. Night sample is depicted to the left, day sample is depicted to the right. 1, Mesoplankton abundance (specimens m^{-3}); 2, mesoplankton biomass (mg m^{-3}); 4, 5, upper and lower boundaries of the temporal thermocline; arrows indicate the depth of the main maximum of the specific density gradient in the pycnocline (after Vinogradov *et al.*, 1993).

oligotrophic regions. Figure 18 illustrates temporal variation in the biolumines-cence field intensity caused by plankton organisms. It is obvious that the layers of higher plankton concentration consist of patches, denser in that layer, than above or below. Such layers are evident in mesotrophic waters. In oligotrophic regions the distribution of such patches is more random. That seems to be a reason

Figure 18 Microvariations of distribution of the bioluminescent field observed from a drifting vessel. (a) Mesotrophic region; (b) oligotrophic region. The degree of shading intensity indicates the intensity of bioluminescence (from Gitelzon and Levin, 1983).

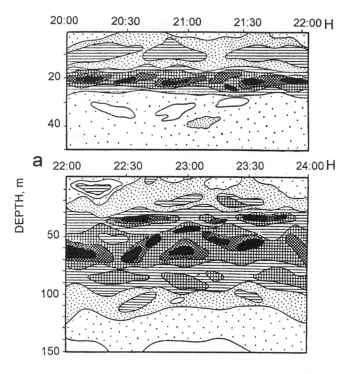

a

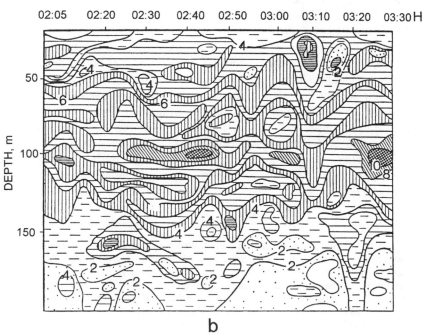

b

for the irregularity of the peaks of biomass distribution in oligotrophic regions, mentioned above.

The mesoplankton of oligotrophic regions can be characterized as follows:

- The total biomass in the upper 200-m layer is about 5 g m^{-2} or less.
- The decrease of biomass with depth in the upper 200-m layer is scarcely evident
- The maximum biomass, two to five times above the background values, occurs near the top of the thermocline.
- Additional irregular maxima, two to three times above the background values, are present in the upper mixed layer and in the lower thermocline. Aggregations in the upper mixed layer consist of nanophages, while carnivores in the daytime are predominant in the upper thermocline and below it.

3.1.2.2. *Divergences, upwellings and adjacent regions*   Seasonal (temporal) variations are slight over most parts of the tropical zone, though remote sensing data on chlorophyll concentration reveal regular variations. However, in the regions of coastal upwellings (Peruvian, Californian, Benguela, Moroccan, etc.) temporal community changes are very marked, periods of activity alternating with periods of relaxation. Real cyclic communities develop, but typically the periods of their active and relaxed periods are not usually fixed. As in temperate cold-water communities, interzonal phytophagous animals are dominant, particularly copepods of the genus *Calanus* and some phytophagous euphausiids. During the relaxation of upwelling (e.g. off Peru) interzonal copepods, such as the phytophagous *Calanus australis* and *Eucalanus inermis*, and the euphausiids *Euphausia mucronata* and *E. tenera* form quasi-permanent dense aggregations (> 500–600 mg m^{-3}) over the continental slope in the 300–600 m depth range. These aggregations are located at the top and in the middle of the oxygen minimum layer, so the animals have to decrease their respiration rate. They do not migrate to the upper layers at this time (Flint and Kolosova, 1990). These aggregations are in a state of diapause and are similar to the "wintering stock" of interzonal phytophages in cold-water regions (Longhurst, 1967; Heinrich, 1993). During periods of intense upwelling these species concentrate during the day above 100–150 m depth (*C. australis*, *E. inermis*) or make intensive diel migrations to the surface (*E. mucronata*) (Timonin and Flint, 1984, 1985).

Figure 19 Examples of vertical distribution of mesoplankton in waters of different production level in the region of Peru (sampled by 150-l water bottle). (a,b) Young community of the eutrophic waters at the shelf off Peru; (c,d) the zone of frontal convergence near the continental slope; (e) mature community in the mesotrophic waters offshore from the shelf. (B) total biomass of mesoplankton; (Bf) biomass of nanophagous species; (Bv) biomass of gross filtering species; (Bs) biomass of carnivores; T° – vertical profile of temperature (from Vinogradov *et al.*, 1984).

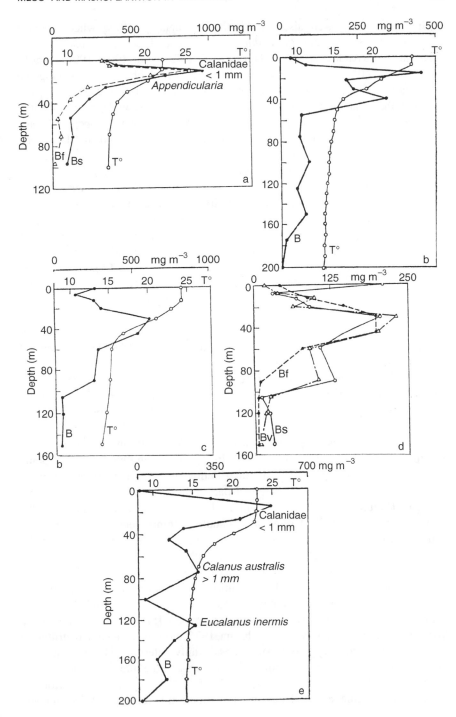

Similar processes were discovered in the Benguela upwelling, where *Calanus carinatus* dominates and *Rhincalanus nasutus* and some euphausiids are subdominants (Andronov, 1976; Thiriot, 1978; Timonin, 1991, 1992). Arashkevich *et al.* (1996) have shown that during a period of active upwelling the population of *C. carinatus* consisted of two subpopulations: at the surface, over the shelf, the population included all developmental stages, while in deeper water, offshore, the population was dominated (90–95%) by diapausal copepodid stage V. Over the shelf, phytoplankton concentrations were high and the daily food intake of *C. carinatus* (40–58% of body carbon) was much greater than carbon loss due to respiration (6.2–7.5% of body carbon) (Timonin *et al.*, 1992). Thus, surplus energy could be allocated to "structural" population growth either by rapid development of the gonads or by accumulation of reserve lipids. Copepodid stage V was characterized by a high content of lipids and undeveloped small gonads. It transformed surplus energy into reserve lipids and might be considered as a diapausal part of the *C. carinatus* population. These "fat" copepods with undeveloped gonads descended to depths of more than 300 m. There they formed the diapausal stock; their respiration rate decreased dramatically. Moulting of diapausal stage V copepodids of *C. carinatus* into adults took place in deep water.

During periods of intensive upwelling in coastal waters over the shelf, vertical water advection and high primary production result in the development of plankton communities, which are then carried offshore. As this takes place the community passes through certain stages of succession and the trophic level of the corresponding waters changes from eutrophic to oligotrophic (Vinogradov, 1977b; Frontier, 1978; Vinogradov and Shushkina, 1984, 1987; Vinogradov *et al.*, 1990a).

During this process the general pattern of vertical distribution of plankton in the upper productive layer also changes. In "young" communities in eutrophic waters of "primary" succession, the distribution of the total biomass is usually unimodal with a sharp peak in the upper part of the thermocline or near its upper boundary (Figure 19a,b). This maximum is almost entirely formed of nanophages: small calanids, appendicularians and nauplii. Omnivorous and carnivorous animals are of secondary importance, and their maximum is located deeper, in the lower part of the thermocline.

Young communities in "secondary" succession are typical of divergence zones in the open ocean. The vertical distribution of plankton in these regions, for example at the equatorial divergence, has been analysed in detail by several authors (Vinogradov and Voronina, 1964; Gruzov, 1971; Flint, 1975; Timonin and Voronina, 1975; Vinogradov and Shushkina, 1976; etc.). The plankton distribution in regions of "secondary" succession is essentially determined both by the type of community that occupies the upwelling waters and the intensity of upwelling. High polymixity of communities also results in a variety of possible distribution patterns, in particular a bimodal biomass distribution in the upper 0–100 m layer.

The pattern of vertical distribution of plankton in zones of frontal convergence is extremely complex and diverse, it is often polymodal (Figure 19c,d).

The mesoplankton biomass distribution of mature communities in mesotrophic regions is diverse, differing according to water structure, depth and structure of thermocline, intensity of turbulent mixing, species and age structure of the community and many other factors. Nevertheless, some general regularities may be noted, which are typical of mesotrophic regions with a fairly high mesoplankton biomass (over 10–20 g m^{-2} in the 0–100 m layer). The pycnocline is usually sharp in these regions, with its upper boundary at 10–50 m depth. The bulk of the mesoplankton is concentrated during the day in the upper part of the thermocline or immediately above it (Figure 19e). The biomass decreases unevenly with depth. A second biomass maximum often occurs in the lower part of the thermocline. Usually this peak consists of euryphages: large calanids and cyclopoids or planktivorous copepods (*Euchaeta*, etc.) and chaetognaths. Layers of high plankton concentration also occur below the thermocline, but the biomass in these aggregations is less than in the layer above the thermocline.

The characteristics of the vertical distribution of mesoplankton in tropical regions, in waters of various productivity levels, have been described in detail in recent papers, using data obtained by plankton nets and 150-l water bottles (Vinogradov, 1968; Timonin, 1975; Timonin and Voronina, 1975; Vinogradov and Musaeva, 1989; Vinogradov et al., 1984, 1990b, 1993).

The patterns of vertical distribution of total mesoplankton biomass and the biomasses of its main trophic groups were compared for communities of various maturity levels and differing in productivity. Some general regularities were noted. For example, regions influenced by intensive upwelling are occupied by "young" communities. The bulk of the mesoplankton biomass is concentrated in the upper layers, above the thermocline or within its upper part, where the main aggregations of phytoplankton and protozoa are located. Hence, the food requirements of zooplankton are best satisfied there. Such distribution is typical in upwelling regions, both in shelf or coastal regions and in areas of divergence in the open ocean. As a community matures, the pattern of vertical distribution of the mesoplankton changes. As the total biomass of the community increases, the filter-feeders and omnivores (copepods and euphausiids) with long life cycles, become more and more dominant. These animals aggregate in the lower part of the thermocline or below. During certain periods when phytoplankton develops in the upper layers the bulk of the population of these animals can ascend to feed.

3.1.3. *The Basic Differences Between the Mesoplankton Distribution in Cold-water and Tropical Regions*

The differences between the distribution of plankton in the upper zone of the subpolar and tropical regions of the ocean have been analysed in detail in several papers (Vinogradov, 1968, 1977a, 1983a). Some peculiarities were noted that determine the essentially different kinds of connections between surface and deep-water communities in the temperate cold-water and the tropical regions of the ocean.

In the open ocean in tropical regions the stratification of water layers varies little with time. Stratification results in the formation of plankton groups located in certain (sometimes very thin) layers of the upper zone. Phytoplankton development is rather uniform, the ecosystems are more stable than in cold-water regions, and the physical conditions are less variable. This results in a more or less permanent vertical distribution of plankton. There are no reasons for plankton animals to leave the upper layers; seasonal migrations are absent or slight. Hence, the role of upper-interzonal species is minor.

The common small interzonal species (the main food source for macroplankton of the middle depth range in cold-water regions) are absent there. The bulk of middle-depth macroplankton species in tropical regions perform regular (diel) migrations into the near-surface zone. Hence, the role of the low-interzonal, mainly macroplankton, species is greater in the tropics than in the productive regions of high latitudes.

In cold-water regions the vertical range of interzonal species is very wide. The bulk of their populations inhabits the deep layers, down to 2000–3000 m, for long periods. They are an important component of the deep-sea community and play the role of distinctive "pseudo-primary producers". The pelagic communities of the ocean depths depend for energy on the surface communities. They need for their development the organic resources in these pseudo-producers.

In the tropical zone the upper-interzonal species never descend to such depths. Only occasional individuals occur deeper than 1000 m. Such a difference is a result of the predominance of seasonal migrations in cold-water regions, and of diel migrations in tropical regions. Temperature stratification should also have a significant influence on the process.

All this results in alternative patterns of organic matter flow, with animals migrating from the surface productive zone into the deep-water layers of the ocean (Figure 20).

3.2. The Small-scale Vertical Spatial Variability of Plankton Distribution

During recent years much attention has been given to studies of unevenness of plankton distribution, particularly in its vertical stratification. Plankton sampling with closing or multi-depth nets, or by big water bottles, and direct observations from submersibles, have shown the regular occurrence of relatively thin layers of extremely high concentrations of one or few species against a background of low or very low abundance. The existence of such layers seems to be an essential

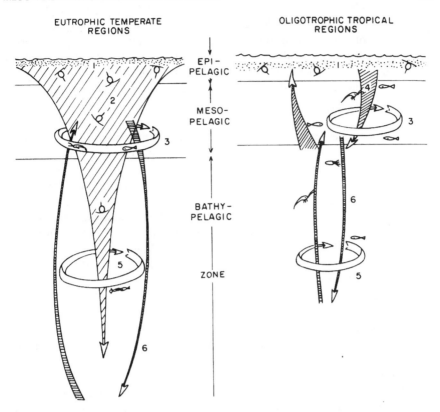

Figure 20 Main pathways of active transfer of organic matter to meso- and bathypelagic layers in the cold-water (left) and tropical regions (right): (1) Zone of development of phytoplankton and feeding herbivores; (2) ontogenetic migrations of phytophagous upper-interzonal copepods down to meso- and bathypelagic depths; (3) mesopelagic carnivorous species living on interzonal species migrating to mesopelagic waters; (4) migrations (mainly diurnal) of carnivorous mesopelagic macroplankton; (5) bathypelagic carnivorous species living on interzonal species migrating to bathypelagic waters; (6) migrations (mainly ontogenetical) of bathypelagic species to overlying layers richer in food (from Vinogradov and Tseitlin, 1983).

element of the spatial structure of the plankton community (Vinogradov, 1968; Longhurst, 1981).

From the ecological point of view, several main types of stratification can be distinguished:

- Aggregation in the main feeding zone of the animals as a result of their search for food or escape from predators.
- Aggregation at the depth of day time occurrence by species performing active diel migrations.

- Aggregation of animals in deeper layers, after grazing in the upper layers, at the onset of unfavourable conditions (i.e. the "wintering stock" at temperate and high latitudes). Similar layers may be formed by phytophages in tropical upwelling regions during the cessation of upwelling.
- Aggregations at various depths for reproduction, or other features of the life cycle of the animals.

In all these cases the aggregations often (but not necessarily) coincide with layers of increased gradient of density, oxygen concentration or other physical or chemical characteristics.

The most permanent aggregations occur at the depth of the day time or seasonal location of migrating animals. Their thickness may be 10–20 m or more, but in other cases (particularly aggregations of euphausiids and ctenophores) it can be as low as 1–3 m. The highest density in the aggregations occurs when migrations are limited by the bottom (Omori and Ohta, 1981; Alldredge et al., 1984; etc.) or by oxygen-depleted waters or the presence of hydrogen sulphide (Vinogradov and Shushkina, 1982; Vinogradov et al., 1992a; etc.).

A detailed study of such layers was undertaken in the Black Sea. The layers are located in the upper part of the main pycnocline, directly under the very sharp oxycline and at the same time at a distance of 20–40 m or less from the upper boundary of water that contains hydrogen sulphide (H_2S concentration < 0.1 mg l^{-1}) (Figures 21 and 22).

It was assumed that the layer of high mesoplankton concentration in the Black Sea forms during the day due to migration of the animals to deeper water, blocked by the layer of oxygen-depleted water. Thus, the animals concentrate at a depth where the oxygen is sufficient to allow them to survive, though it is near the lethal minimum. However, measurements of the oxygen concentration in the 150-l water bottles in which the plankton samples were collected have shown a wide range of values in the layer of maximum aggregation of plankton (from 0.2 to 0.06 ml O_2 l^{-1}). The intensive feeding of some animals in this aggregation layer also refutes the suggestion that they are in a zone of sublethal oxygen concentrations (Vinogradov et al., 1992a).

Later studies gave an insight into the connection between the lower layers of mesoplankton concentration in the Black Sea and certain values of water density, and showed that the locations of the concentration maxima reflect the pycnocline microstructure. It turned out that the layer of plankton aggregations coincides with the maximum gradients of the main pycnocline. Some species (the ctenophore *Pleurobrachia pileus* and the protozoan *Noctiluca miliaris*) are connected with the maximum density gradient in the upper boundary of the pycnocline, others (the copepods *Calanus euxinus*, *Pseudocalanus elongatus* and the chaetognath *Sagitta setosa*) are located in the secondary density gradient layer (Vinogradov et al., 1990a, 1991a, 1992b,c). The aggregations of *C. euxinus* consist of two ecological groups. The higher one ($\sigma_t = 15.5$–15.7), is located during the day at an oxygen

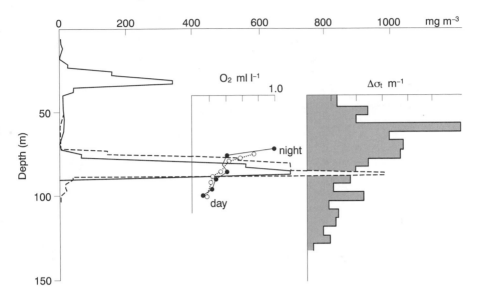

Figure 21 Vertical distribution of biomass (mg m^{-3}) of *Calanus euxinus* (CV and CVI) in the central part of the Black Sea in August. Solid line indicates night, dashed line indicates day (samples of 150-l water bottle). The vertical profiles of oxygen (Broenkov method, samples taken from the same water bottles) and density gradient, smoothed by moving average with 5-m intervals (m^{-1}), are depicted to the right.

concentration of 0.8–0.2 ml l^{-1}, and at night it migrates to the surface. The second group is a "reserve stock"; it is concentrated deeper (σ_t = 15.8–16.1) and does not perform diel migrations. It exists permanently at oxygen concentrations of 0.3–0.15 ml l^{-1}. The animals of this group are in a peculiar kind of anabiosis and do not die during their long stay in water with oxygen concentrations 0.06 ml l$^{-1(1)}$ (Vinogradov *et al.*, 1992a).

This layer ascends with the pycnocline in cyclonic gyres and descends in anticyclonic gyres and near the continental slope. The upper boundary of the pycnocline in the Black Sea is located at a specific density (σ_t = 14.9–15.0). Hence the plankton concentrations coincide with a distinct isopycnic layer. The layers of high concentration usually have very sharp upper and, especially, lower boundaries (Vinogradov *et al.*, 1992b,c).

In the eastern tropical Pacific, narrow layers of high plankton concentration are connected primarily with the oxygen minimum layer (Longhurst, 1967; Alldredge

---

[1]Some cyclopoids in freshwater lakes may be abundant at the same oxygen concentration (0.06 ml O$_2$ l^{-1}) (Brandt, 1951).

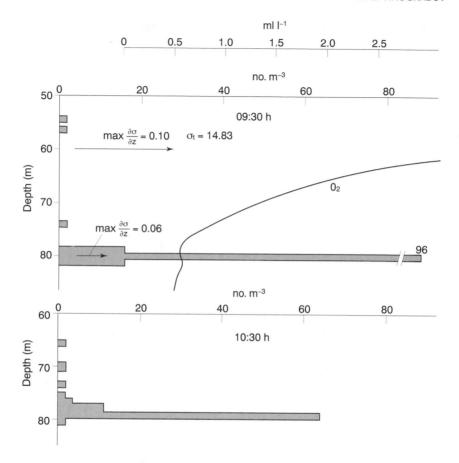

Figure 22 Vertical distribution of abundance of the ctenophore *Pleurobrachia pileus* in the eastern cyclonic gyre of the Black Sea (measured by direct count from submersible "Argus"). Winter. The depths and values of the 1st and the 2nd maxima of density gradient in the layer of main pycnocline and the profile of oxygen concentration are indicated. The first observation was at 09.30, the second one was at 10.30 (from Vinogradov *et al.*, 1992c).

et al., 1984; Flint and Kolosova, 1990; Vinogradov *et al.*, 1991b; Shushkina and Vinogradov, 1992; Vinogradov and Shushkina, 1994). The species capable of living in anaerobic conditions usually form thin aggregations in the upper and (or) lower parts of the oxygen minimum zone, where the oxygen concentration can be no higher than in the centre of the layer (Sameoto, 1986; Vinogradov *et al.*, 1991b). Such layers are formed by euphausiids (as many as 10 000 specimens m^{-3} – Alldredge *et al.*, 1984; Sameoto *et al.*, 1987; Flint and Kolosova, 1990), by lantern fishes and gonostomiatid fishes (Pearcy and Laurs, 1966; Clarke and Wagner, 1976; Vinogradov et al., 1991b; Shushkina and Vinogradov, 1992; etc.), some ctenophores, small medusae and chaetognaths.

This phenomenon is most pronounced for some interzonal copepods. Thus, Alldredge et al. (1984) observed at 450 m depth, near the coast of California (34°N), dense aggregations of stage V copepodids of Calanus pacificus, the animals being in diapause. The abundance was enormous, up to 24×10^6 specimens m^{-3} (on average 14×10^6 specimens m^{-3}) in a layer with oxygen concentration of 0.2 ml l^{-1}.

Similar aggregations connected with the layer of oxygen depletion are formed in the eastern Pacific by Eucalanus inermis (Sameoto, 1986). The distribution of this copepod was studied in detail during Russian expeditions in the region of the Costa Rica Dome (Vinogradov et al., 1991b; Vinogradov and Shushkina, 1994). There are two oxygen minimum layers in this region: one, less evident, at 80–160 m (0.5 ml O_2 l^{-1}) and a more pronounced lower one at 350–800 m depth, with a minimum oxygen concentration of 0.22 ml l^{-1} at about 500 m depth.

The zooplankton aggregations occurred in the upper and lower parts of the main oxygen minimum, at 350 m (biomass 290 mg m^{-3}) and 600 m depth (130 mg m^{-3}). Between these aggregations, in the core of the oxygen minimum layer the zooplankton biomass decreased to 4–5 mg m^{-3}. The aggregations of mesoplankton in the upper and lower parts of oxygen minimum layer consist almost entirely of stage V and VI copepodids of Eucalanus inermis (90–95% of total mesoplankton biomass). According to our data, outside these maxima stages V and VI E. inermis were absent.

Counts using the submersible's standard cage were done in 10-, 5-, and even (in the greatest aggregations) 2- and 1-m layers. Figure 23 presents the pattern average for layers 10 m thick. Using such averaging, the abundance of Eucalanus within the layer of concentration fluctuated from 5 to 64 specimens m$^{-3(2)}$. In fact, the internal zonation was finer, and when counts were made for 1–2-m layers, the abundance varied from 5 to 128 specimens m^{-3} (Figure 24). The layer of eucalanid aggregations appears to be divided into two parts: an upper layer extending from 350 to 400 m depth, and a narrow lower layer (410–420 m). This suggests the possibility that there are ecologically non-uniform hemipopulations of E. inermis.

The layer of concentration of E. inermis in the lower part of the oxygen-minimum zone has approximately the same thickness as the upper part, covering the depths 550–640 m; the abundance of copepods varies (counted at 10-m intervals) from 3 to 71 specimens m^{-3}. It also has an internal fine-structure and is divided into two layers: an upper layer (560–610 m) with abundances up to 71 specimens m^{-3} and a weakly developed lower layer (630–650 m) with abundances

---

[2]It is interesting to note that in one study (Sameoto, 1986), catches by BIONESS nets in 50-m layer for these depths yielded a concentration of E. inermis of 68 specimens m^{-3}.

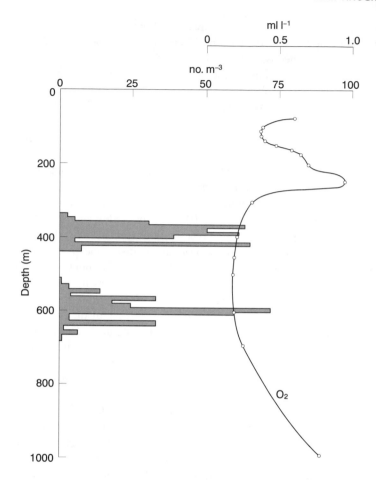

Figure 23   Distribution of large (6–10 mm) *Eucalanus inermis* in the region of Costa Rica Dome (February, 1987) based on direct counts from "Mir" submersible, averaged across 10–20 m layers (shaded). The distribution of oxygen concentration (ml $O_2 l^{-1}$) measured in the same water-bottle samples is indicated to the right (from Vinogradov *et al.*, 1991b).

up to 32 specimens m^{-3}. It is interesting to note that at the peak concentrations of *Eucalanus* (340–400, 410–420, 580–610 m) virtually all other animals (copepods, chaetognaths, polychaetes, etc.) were absent from the mesoplankton, and the number of macroplankton forms was minimal. In general, fishes were absent at these depths.

   E. inermis can live in layers where the oxygen concentration is below 0.1 ml l^{-1} and it can survive for up to 12 h in water devoid of oxygen (Boyd and Smith, 1980).

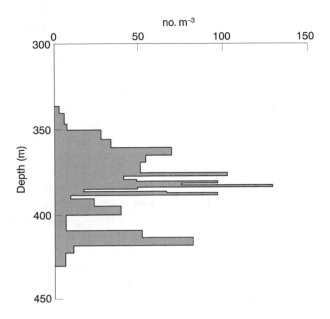

Figure 24   The fine structure of vertical distribution of large *Eucalanus inermis* in the upper layer of concentration at the same station as at Figure 23. Raw data, not averaged (from Vinogradov and Shushkina, 1994).

According to both Sameoto's (1986) data, based on BIONESS net tows, and our direct observations from a submersible, aggregations of *E. inermis* occurred in the upper and lower parts of the oxygen minimum layer, but in the centre of this layer within a depth range of at least 200–300 m the copepods were absent. The reasons for this absence, where oxygen concentrations varied (at the investigated stations) from 0.1 to 0.18 ml l^{-1}, are still unknown. These concentrations are just suitable for the survival of *E. inermis* and only a little lower than the oxygen concentration measured at the depths where aggregations of *E. inermis* occurred. It is interesting that in the Gulf of California *E. inermis* lives in the centre of the oxygen-minimum layer, where the oxygen concentration is about 0.1 ml l^{-1} (Shushkina and Vinogradov, 1992).

In some regions, for example in the Gulf of California, the decrease of plankton biomass in the core of the oxygen minimum layer (O_2 concentration < 0.1 ml l^{-1}) is less evident. According to data collected using 150-l water bottles and BR-nets, the plankton biomass is about 4.5 mg m^{-3} (Shushkina and Vinogradov, 1992), and according to Wiebe *et al.* (1988) the plankton biomass in the oxygen minimum layer at 600–700 m depth is 11–17 mg m^{-3}. Immediately above the upper aggregation of *E. inermis* at 280–340 m, in the uppermost part of the deep oxygen minimum (0.40–0.25 ml l^{-1}), aggregations of euphausiids were observed with a biomass of 50–70 mg m^{-3}. Among them *Euphausia distinguenda* and *E. eximia* were dominant.

Above the centre of the oxygen minimum, under the layer of euphausiids, at 390–450 m depth, layers of micronektonic fish (mainly Myctophidae) occurred in daytime. In the centre of the oxygen minimum these fishes were practically absent, but under it, a little deeper than the aggregations of *E. inermis*, at an oxygen concentration of 0.3–0.7 ml l^{-1}, new aggregations of myctophids occurred. The upper aggregations consisted mainly of juveniles of 1–2 cm length, the lower aggregations consisted of larger fishes of 2–7 cm length. Deeper than 900–950 m Myctophidae were practically absent.

Such occurrences of aggregations of micronekton fish (mainly Myctophidae) in the daytime in the upper and lower horizons of the oxygen minimum layer and their absence in its centre are typical for other regions of the eastern Pacific. Shushkina and Vinogradov (1992) observed this in the Gulf of California, the oxygen deficit in the intermediate waters being very sharply defined (Figure 25). However, the most evident layer was observed in Monterey Bay, where the maximum concentrations of micronektonic fishes occurred at 390–400 m depth (1 specimen m^{-3}) in the upper part of the oxygen minimum. They were almost absent below 400 m and they reappeared below the centre of the oxygen minimum, at 510–600 m depth. A similar layer has been observed near the coast of Oregon (Pearcy and Laurs, 1966). Two distinct layers of myctophid concentration, above the oxygen-deficient layer and below it, are typical for the regions of the north Pacific, where the oxygen-minimum layer is less evident (Vinogradov and Shushkina, 1992).

Thus, lantern-fishes form aggregations both above and below the oxygen minimum layer, within a rather wide range of oxygen concentrations (from 1.5 to 0.2 ml l^{-1}). In some regions myctophids disappear when oxygen concentration decreases to 0.5 ml l^{-1}, whereas in other regions (with a more pronounced oxygen minimum in the core of the layer) they can form their maximum concentrations at an oxygen concentration as low as 0.2 ml l^{-1}. The main factor limiting the distribution of both myctophids and *E. inermis* seems to be the difference between the oxygen concentration in the layer where they live and in the core of the oxygen-minimum layer, rather than the absolute values of oxygen concentration. It is evident that the presence of an oxygen-minimum layer increases and aggravates an innate feature of populations of interzonal animals, both phytophages (copepods, euphausiids) and carnivores (fishes, chaetognaths, ctenophores, etc.). This feature is the drive to form thin layers of dense aggregations during diel or seasonal (ontogenetic) migrations. Different age groups and populations in different physiological states form aggregations at different depths.

The possible ecological and biocoenological reasons for the formation and the peculiarities of layers of increased concentration of certain species and groups have been discussed in detail (Vinogradov, 1968, 1970). At this point it is worth noting one more peculiarity in the formation of layers of increased concentration.

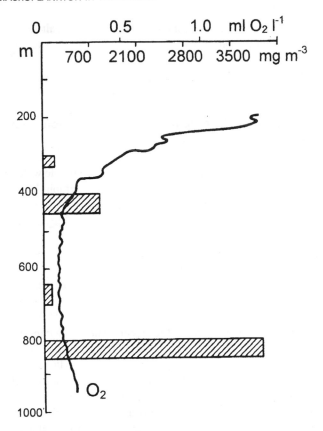

Figure 25 The distribution of micronekton fishes biomass in the Gulf of California measured by visual estimations from the "Mir" submersible. The vertical profile of oxygen concentration measured by CTD probe directly from the "Mir" submersible is indicated to the right (after Vinogradov and Shushkina, 1994).

Layers of carnivores (fishes, ctenophores, medusas) are located just below the layers of high concentration of their prey (i.e. copepods or euphausiids). The predators seem to consume descending damaged and disabled animals.

It has long been known that, in surface oligotrophic waters with mesoplankton biomass of units and tens of mg m^{-3}, carnivores can only obtain sufficient energy if they are concentrated in microaggregations of their prey, not if the carnivores and their prey are distributed evenly. This seems to be the cause of the typical stratification of various trophic groups of animals in the epipelagic zone (Vinogradov and Shushkina, 1976) and is more evident in oligotrophic waters.

The formation of layers of increased concentration of certain species and groups, alternated with extensive layers where they are almost absent, is responsible for the clear vertical zonation of the fauna against a background of smoother change with depth of the environmental factors, such as pressure and temperature.

3.3. Distribution of Mesoplankton Over the Whole Water Column of Open Ocean Regions

The peculiarities of mesoplankton distribution over the whole water column down to 8000–9000 m depth were analysed in detail in a monograph (Vinogradov, 1968). Here we take a quick look at more general features.

3.3.1. *Quantitative Distribution of Mesoplankton*

The biomass of net plankton in the deep-ocean layers is primarily determined by its biomass (or production) in the upper zone. The ratio between the quantity of net mesoplankton in the surface (0–500 m) and deep (500–4000 m) layers is rather stable over the whole area of the ocean. The mesoplankton in the upper 0–500 m layer is about 2/3 (65%) of the total biomass in 0–4000 m layer. This ratio is different only in those places where there is advection of deep water from adjacent regions that differ greatly in their production of surface plankton. If this flow comes from less productive regions, the percentage of deep-water plankton decreases; if the flow comes from more productive waters, the biomass of deep-water plankton increases. For instance, in the frontal zone between the Gulf Stream and the Labrador Current rich arctic deep-water plankton (mainly *Calanus hyperboreus*) is advected from north to south at 600–3000 m depth. Its biomass in the 1800–2300-m layer is as high as 50 mg m^{-3}. As a result the 0–500 (600) m layer contains only 40–50% of the total biomass of net mesoplankton (Figure 26).

The total zooplankton biomass changes with depth in a similar way over almost all the ocean. Deeper than 500–1000 m in both tropical and temperate cold-water regions this regularity can be described by the exponential equation $y = ae^{-kx}$, where y is the plankton biomass (mg m^{-3}), x is depth (m), a is the coefficient reflecting the abundance of plankton in the upper layers, and k is the coefficient of plankton biomass decrease with depth.

The variation in the coefficient a is correlated with differences in the abundance of the surface plankton, and indicates the considerable poverty of deep-sea plankton of the open waters of the tropical ocean compared to that of the marginal tropical seas (Bougainville Trench), not to mention eutrophic subpolar areas (Kurile–Kamchatka Trench).

The rate at which the biomass decreases with depth in tropical areas (Marianas

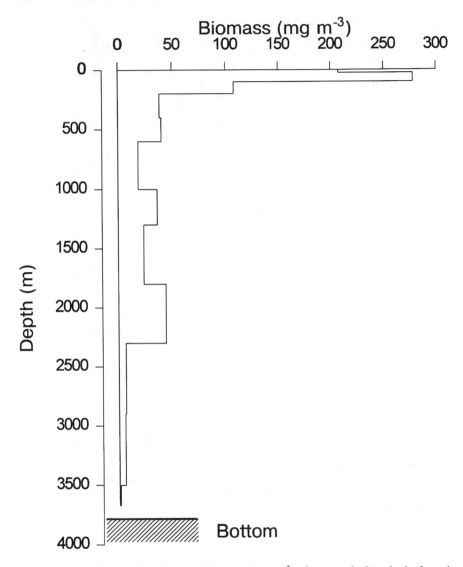

Figure 26   Vertical distribution of biomass (mg m^{-3}) of net zooplankton in the frontal zone between the Gulf Stream and the Labrador Current (41°39'N, 49°58'W) (two stations averaged, plankton net BR used) (from Vinogradov *et al.*, unpublished data).

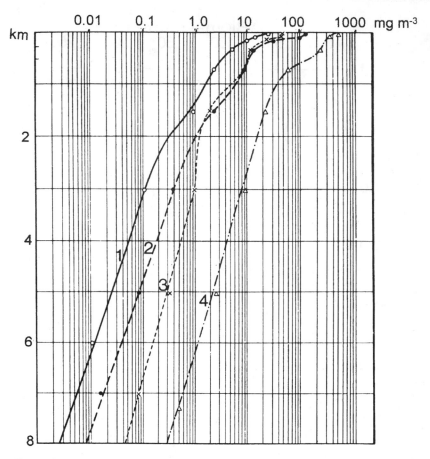

Figure 27 Vertical distribution of the biomass (wet weight) of net mesoplankton (mg m^{-3}) in different regions of the Pacific Ocean. 1, Mariana Trench, two stations averaged; 2, Bougainville Trench; 3, Kermadec Trench; 4, Kurile–Kamchatka Trench, six stations averaged, May–June (from Vinogradov, 1968).

and Bougainville trenches) is uniform ($k = 8.5 \times 10^{-4}$) and higher than in the subpolar area (Kurile–Kamchatka Trench, $k = 6.5 \times 10^{-4}$). The plankton biomass of the 4000–8000 m layer in the area of the Kurile–Kamchatka Trench is 160 times smaller than that of the 0–500-m layer; the corresponding ratio for the Bougainville and Marianas trenches is 760–780 times. In the Kermadec Trench the biomass decreases at the same rate as in subpolar waters, and not according to the tropical pattern; probably this is due to the inflow of antarctic waters (Figure 27). The reasons for these differences seem to be connected with the different patterns of transport of organic matter from the upper productive zone, already mentioned in the previous chapter. The values are tabled below:

$$\text{for Kurile–Kamchatka Trench } y = 56.2 \cdot e^{-6.5 \cdot 10^{-4} \cdot x},$$

$$\text{for Kermadec Trench } y = 7.75 \cdot e^{-6.5 \cdot 10^{-4} \cdot x},$$

$$\text{for Marianas Trench } y = 1.82 \cdot e^{-8.5 \cdot 10^{-4} \cdot x},$$

$$\text{and for Bougainville Trench } y = 5.74 \cdot e^{-8.5 \cdot 10^{-4} \cdot x},$$

The result of more sophisticated treatment is that the biomass decrease with depth is less uniform. There are layers characterized by both higher and lower gradients. They could be illustrated by the example of the Kurile–Kamchatka Trench (Figure 28). The curve of vertical distribution of biomass can be divided into three divisions. The first one corresponds to the layer from the surface to 500–750 m. The distribution of plankton biomass in this layer is extremely non-uniform and changes seasonally. However, the whole of this layer can be characterized by biomass values of the same order (hundreds of mg m^{-3}). Deeper down, from 1000 to 2500 m, the biomass decreases slowly, about two times. The order of magnitude remains the same (tens of mg m^{-3}). The third and longest division covers the water column deeper than 3000 m. Plankton biomass values there are from a few to less than 1 mg m^{-3}. The biomass change over the huge depth range is less than 10 times. Between the divisions the biomass changes more rapidly: within the relatively narrow layer at 500–1000 m it decreases about 10 times, in the 2500–3000 m layer it decreases nine times. These divisions or types of vertical distribution are well shown in Figure 29.

The surface pattern of distribution is characterized by changes of plankton biomass in waters inhabited by mainly upper-interzonal species, feeding in the surface zone. The middle-depth pattern is typical for intermediate layers, where the quantity of interzonal animals decreases but is still comparatively high, while the dominating group consists of autochthonous carnivores. The abyssal distribution type occurs in the waters poorest in plankton, deeper than 3000 m. The part played by carnivores is negligible, and decreases to zero at the greatest depths where carnivores practically disappear from the plankton and give way to small detritivores and species of mixed feeding type.

Similar divisions can be distinguished in the curve of vertical distribution of biomass in tropical regions: surface, middle-depth and abyssal types of plankton distribution. However the boundaries between them are located at depths other than those typical for boreal regions. The surface type of distribution meets the middle-depth type in the 100–200 m layer, and the middle-depth type gives way to the abyssal one in 1500–2500 m layer. This pattern corresponds to the change with depth of the feeding types of the main mass of plankton (Arashkevich, 1972).

In subpolar regions a significant proportion of the interzonal species performs seasonal migrations over 2000–2500 m range. In tropical regions interzonal

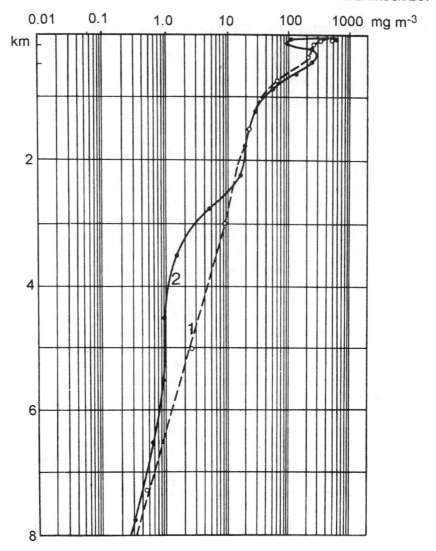

Figure 28 Vertical distribution of wet weight of mesoplankton biomass (mg m⁻³) in the Kurile–Kamchatka region of the Pacific Ocean. 1, Six stations with long vertical tows averaged (May–June); 2, nine stations with more discrete vertical tows averaged (July–August) (from Vinogradov, 1968).

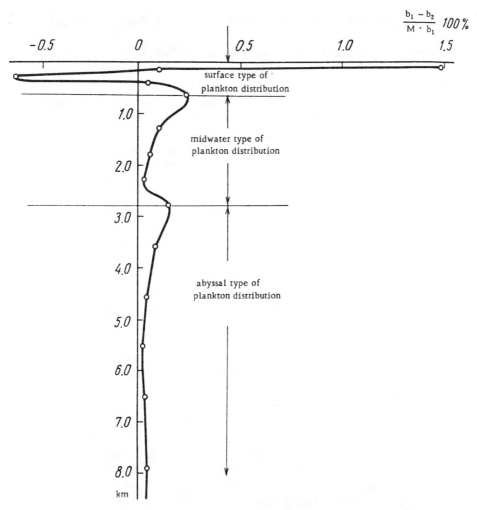

Figure 29 Intensity of changes with depth of net mesoplankton biomass in Kurile-Kamchatka region of the Pacific Ocean (nine stations averaged, July–August). b_1, mean biomass of the overlying layer; b_2, mean biomass of the underlying layer; M is the distance (m) between the centres of the compared layers (from Vinogradov, 1968).

species undertake mainly diel migrations, of much smaller range. The middle-depth type of distribution is connected with the presence of a significant amount of animals migrating from the upper zone. Thus, it is natural that in the subpolar regions this distribution type occupies a greater depth range in the water column and penetrates deeper than in the tropical regions.

3.3.2. Depth-dependent Changes in the Composition of Plankton Groups

As depth increases and plankton concentration decreases, the feeding type of the planktonic animals also changes. This is shown by the different ratios of various systematic groups characterized by different food types.

Within pelagic communities occupying relatively similar biotopes, ecological differences are displayed mainly in the feeding types of the species that compose the communities. Feeding competition is the basic factor determining divergence of ecological niches at the same site. Competitive exclusion should lead to sharper changes with depth in the qualitative composition of the fauna than might be expected if judged from abiotic conditions only.

When food is very abundant, it is possible for ecologically and trophically similar species to co-exist (Fryer, 1957; Vinogradov, 1968). As food gets scarcer, food competition intensifies. First the trophic ranges of species with close ecological niches partially overlap. In this case, competition between trophic groups may be low, while niches of certain species may overlap completely, leading to elimination due to competition. As the food supply gets even less, competition occurs between species with more dissimilar food types and, eventually, between species groups belonging to the same trophic type.

In the pelagial, as the depth increases and the concentration of food decreases, the composition and distribution of taxonomic groups becomes adapted to various types of food and feeding. This process was specially discussed by Heptner (1996): in the rich surface layers organisms spend less time finding and consuming food and can perform diel vertical migrations into deeper waters, where the 'pressure' of carnivores is lower (Figure 20). In deeper layers below the productive zone, carnivores and scavengers dominate and consume the descending herbivores. Variation in the food supply also leads to differences in the proportion of carnivores within eutrophic and oligotrophic areas of the ocean. In eutrophic boreal areas, carnivores are abundant at depths down to 3000–4000 m, while in tropical areas their biomass is significant only at depths of less than 2000 m. But even there, the slowly moving Euaugaptilidae, Heterorhabdidae, or medusae dominate (Arashkevich, 1972; Vinogradov et al., 1996b).

In deeper waters, in the zone of the "abyssal" type of biomass distribution (Figure 29), the proportion of animal corpses and detritus increases, and food becomes scarce. Under these conditions, the energetic cost of searching for and pursuing the prey becomes too high. Active carnivores are replaced by ambush carnivores, as reflected in their morphology and physiology (Marshall, 1960; Vinogradov, 1968). Passive carnivores cannot provide themselves with food, and omnivores dominate the communities, while the proportion of scavengers increases in the near-bottom layers.

The feeding type of some higher taxonomic groups (several calanoid families) has been shown to be reflected in their vertical distribution (Heptner, 1996). Families with species of similar feeding type (Euchaetidae) are characterized by species distribution curves close to normal, while groups that contain heterogeneous feeding types (Aetedidae, Heterorhabdidae) have asymmetrical species distribution curves.

The highly productive Kurile–Kamchatka region of the Pacific Ocean (Table 3) exemplifies the vertical distribution of various taxa. In the upper part of the productive zone, the upper-interzonal herbivorous filter-feeders *Neocalanus cristatus*, *N. plumchrus* and *Eucalanus bungii* dominate. The carnivorous chaetognaths, mainly *Sagitta elegans*, are subdominant. In the lower part of the productive zone, in the "cool intermediate layer", the concentration of *N. cristatus* and *N. plumchrus* sharply decreases. Owing to this fact, the total average biomass of the copepods does not exceed $40 \, \text{mg m}^{-3}$. The concentration of the chaetognaths (carnivores) does not change ($40 \, \text{mg m}^{-3}$ in average); therefore, their proportion of the total plankton biomass increases to about 45% and sometimes reaches 60%. Thus, the cool intermediate layer is inhabited mainly by carnivores.

In the upper layer of the warm intermediate waters, at 200–500 m, the main plankton bulk is composed of interzonal herbivores, their biomass exceeding $100 \, \text{mg m}^{-3}$. The chaetognath concentration remains equal to that above 200 m depth, but due to the increase of the total plankton biomass, the proportion of chaetognaths decreases (Table 3). The species composition of chaetognaths also changes: in the 200–300 m layer, *S. elegans* and relatively abundant small *Eukrohnia hamata* are present, the latter usually being the dominant species in the 300–500-m layer. At depths of 300–500 m, rare specimens of deep-water species appear.

In the lower part of the warm intermediate waters and below, in the 500–1000 m layer, the plankton composition differs significantly from that in the 200–500 m layer. The herbivorous copepods (mainly *Neocalanus cristatus*, as abundances of *N. plumchrus* and *Eucalanus bungii* decrease) and chaetognaths, represented almost exclusively by small *Eukrohnia hamata*, are dominant. In this zone, the shrimps *Hymenodora frontalis* and *Gennadas borealis* (mainly juveniles), mysids *Eucopia grimaldii* and juvenile *Gnathophausia gigas* (Table 3), are regularly present. The 500–1000 m layer is the dwelling zone of the main part of the population of the medusa *Crossota brunnea*.

Deeper than 1000 m, the community structure changes, but is relatively constant over the depth range 1000–3000 m. The main bulk of the plankton is composed of carnivores, among which chaetognaths, shrimps and mysids dominate. Copepods are subdominant. The biomass of chaetognaths, low in the 1000–1500 m layer (about $1.5 \, \text{mg m}^{-3}$), increases to a maximum (about $8 \, \text{mg m}^{-3}$) in the 2000–2500 m layer. Deep-water chaetognath aggregations are formed by a single species, *Eukrohnia fowleri*, that does not ascend above a depth of 1000 m. The shrimp biomass reaches its maximum in the 1000–1500 m layer; at this depth, *Hymenodora frontalis* is dominant. Deeper, the decapod concentration and their proportion in the plankton slightly decrease and then increase only

Table 3 Biomass of various systematic groups (mg m^{-3}) in the plankton of the Kuril–Kamchatka region of the Pacific Ocean (average nine stations of BR-net samples).

Depth	Chaetognatha	Polychaeta	Ostracoda	Copepoda	Mysidacea	Amphipoda	Euphausiacea	Decapoda
0–50	36.8	0.17	0.34	461.0	0.05	11.4	14.8	0.02
50–100	33.6	0.12	0.25	78.3	+	3.6	4.6	0.07
100–200	38.7	0.55	0.58	38.9	+	4.5	4.6	0.76
200–300	33.1	0.46	1.1	203	0.15	7.2	3.5	0.05
300–500	28.4	2.0	3.8	152	0.75	5.2	2.1	0.26
500–750	15.4	1.1	1.0	57.4	3.0	0.92	0.2	1.0
750–1000	6.4	0.38	0.54	33.5	2.7	0.30	+	0.6
1000–1500	1.4	0.23	0.34	17.6	1.8	0.24	0.02	2.9
1500–2000	5.6	0.13	0.14	8.6	0.4	0.23	0.06	0.83
2000–2500	7.8	0.04	0.12	4.6	1.2	0.21	0.017	1.4
2500–3000	2.0	0.008	0.08	1.8	0.02	0.03	0.70	0.56
3000–4000	0.06	0.01	0.05	0.78	0.05	0.02	>0	0.25
4000–5000	0.01	0.006	0.02	0.41	0.25	0.02	0.03	0
5000–6000	0.002	0.002	0.005	0.18	0.13	0.1	0	0
6000–7000	>0	0.014	0.007	0.15	0.05	0.08	0	0
7000–8000	+	0.01	0.006	0.06	0	0.03	0	0

in the 2000–2500 m layer, where *H. glacialis* dominates. Among the mysids, *Eucopia grimaldii* and *E. australis*, the latter replacing the former species in the deeper layers, are the most abundant. Down to depths of 2500–3000 m, small fishes, mainly *Cyclothone atraria*, Melamphaeidae and juvenile Macruridae, are present in the samples.

In the 1000–3000 m layer, a relatively low plankton biomass, not exceeding 20–30 mg m^{-3}, was observed. Carnivores are more numerous than herbivores, upper-interzonal suspension feeders dominating in the latter group. The species diversity is high and reaches the maxima for most of the groups.

At depths of 2500–3000 m, the plankton biomass decreases sharply. It is noticeable that the decrease in the plankton biomass in the 500–750–1000-m layer is accounted for by a drop in the copepod concentration (an abrupt decrease in the higher-interzonal filter-feeder biomass); in the 2500–3000 m layer the rate of decrease in the copepod biomass remains constant, but the concentration of carnivores sharply decreases, while chaetognaths and fishes completely disappear. Only Decapoda, Amphipoda and Coelenterata (both medusae and siphonophores) are still present but scarce in the 3000–4000 m layer; they too practically disappear at greater depths.

From 3000–3500 m to the greatest trench depths, plankton biomass and abundances are very low and decrease smoothly with depth. The diversity is poor, the number of species decreasing with depth.

The main bulk of the plankton population is copepods, the proportion of which ranges from 30 to 50% of the total biomass of net plankton in the 4000–5000 m layer and from 25 to 30% at greater depths, 5000–6000, 6000–7000 and 7000–8000 m layers. It is paradoxical, but the proportion of herbivorous filter-feeders is relatively high. They are represented mainly by the mysids of the genus *Boreomysis* (most abundant *B. incisa*), their gut contents consisting of large diatoms and tintinnids, caught probably in the near-surface layers. *Boreomysis incisa* composes 20–30% (sometimes 55%) of the total plankton biomass in the 4000–6000 m layer.

Mysids are not found deeper than 6000–7000 m. Gammarids and polychaetes, partly benthopelagic groups, are subdominant.

Depth-related changes in the dominant species are especially characteristic for carnivores, which form a high proportion of the deep-sea plankton. Each of the groups dominates in certain defined layers, the proportion being significantly low in between. Among the decapods, Chaetognatha and Coelenterata, layers of dominance are formed at various depths by different species, each maximum usually being composed of a single species, the proportion of other species of the same group being low. In some groups, such as Amphipods, which are partly carnivorous and partly semi-parasitic, distribution patterns seem to be absent; no clear pattern was found for copepods either (Vinogradov, 1968).

Among carnivores that have partly overlapping spectra of prey, a tendency towards vertical divergence of their maximum layers was observed. For instance, in the layers where chaetognaths dominate, the proportion of decapods decreases. This pattern is not observed for groups that do not have overlapping prey spectra.

The vertical distribution of planktonic animals is determined by elimination due to competition. Mutual elimination is determined mainly by the similarity of food type. The scarcer the food store is, the more prominent becomes the elimination of ecologically similar species, and the clearer the layers of dominance. Owing to vertical water movements and the ability of planktonic animals to move, the complete elimination of species and complete vertical divergence are not observed. The principle of elimination due to competition is displayed mainly by the vertical divergence of dominance layers of ecologically similar species with similar prey spectra. Thus, even gradual depth-related changes in the food supply, on reaching a certain level, lead to sharp changes in the plankton community structure.

4. VERTICAL DISTRIBUTION OF MACROPLANKTON

4.1. Quantitative Distribution of Macroplankton

The vertical distribution of the macroplankton differs significantly from the distribution of the net mesoplankton. Macroplankton animals are able to move faster and form more prominent aggregations in the layers with richer food supplies.

Studies on the sound-scattering layers using echo-sounders, horizontally towed nets (with large mesh size), deep-water pelagic trawls and, finally, visual observations from manned submersibles revealed that macroplanktonic animals form aggregations that are relatively narrow in vertical extent. Therefore, estimations of macroplankton biomass based on horizontal tows are very uncertain. Nevertheless, the first attempts to estimate the vertical distribution of macroplankton quantitatively, by means of sampling with horizontally towed nets (Jespersen, 1935), revealed the highest concentration of larger planktonic animals at intermediate depths, between 500 and 1500–2000 m. Later, Vinogradov (1968) analysed catches of macroplanktonic animals in samples taken in vertical hauls of large plankton nets (mouth diameter 113 cm) and averaged the data for numerous locations. This author concluded that, in oligotrophic tropical areas, the maximum macroplankton biomass is situated in the 500–1000 m layer (about 25% of the total plankton biomass). In more productive equatorial waters macroplankton is abundant in the 1000–2000 m layer, the maximum being found in this layer (up to 74% of the total biomass of the net plankton); in the subpolar regions, the main bulk of the macroplankton is recorded from the 500–1000 m layer, but its proportion of the total zooplankton biomass is greatest at 2000–3000 m depth (Table 4).

Table 4 Mean values of the biomass (mg m^{-3}) of the mesoplankton (B$_z$) and macroplankton (B$_g$) in selected boreal, tropical, and frontal areas. Samples taken with BR 113/140 plankton nets.

Depth, m	Kurile–Kamchatka region (July–Aug 1966, 9 locations averaged)			Frontal zone between Labrador and Gulf Stream currents (Aug 1995, 2 locations averaged)			Equatorial region of the Pacific and Indian oceans (22 locations averaged)			Oligotrophic tropic regions (40–12°N; 12–40°S, 20 locations averaged)		
	Bz	Bg	Bg/ Bg + Bz, %	Bz	Bg	Bg/ Bg + Bz, %	Bz	Bg	Bg/ Bg + Bz, %	Bz	Bg	Bg/ Bg + Bz, %
0–50	626	0	0	161	0	0	63.5	0	0	27.6	0	0
50–100	109	0	0	–	–	–	52.3	15.2	22.5	25.9	0	0
100–200	108	0	0	93.2	0	0	18.8	0.3	1.6	14.7	0	0
200–500 (600)	247	0.3	0.1	43.8	15.6	26.3	7.8	2.2	22.3	6.8	0.8	10.5
500–1000	77	3.4	4.2	29.6	68.2	69.7	5.2	2.7	34.2	4.8	1.6	25.0
1000–1500	27	1.3	5.6	23.2	145.6	86.2	1.2	5.9	74.0	1.8	0.3	14.2
1500–2000	19	–	–	21.3	44.3	67.5	–	–	–	–	–	–
2000–2500	17	1.3	11.8	43.9	40.8	48.2	0.23	0.02	8.0	0.4	0	0
2500–3000	5	–	–	6.3	10.8	63.2	–	–	–	–	–	–
3000–4000	1.6	0.1	6.9	2.5	4.8	65.7	–	–	–	–	–	–

Macroplankton is most abundant and plays a most significant role in the communities of large-scale frontal regions, for example in that between the Labrador Current and Gulf Stream. The rich plankton, mainly *Calanus hyperboreus* and *C. finmarchicus*, is advected by the Labrador Current from the Arcto-boreal region in both the surface and deep-water layers. In the zone proximal to the Gulf Stream "wall", the plankton dies and abundant planktonic scavengers and carnivores, mainly shrimps of the genus *Acanthephyra*, live upon this resource. They create a "living reticulum" across the path of the plankton advected southward. Consequently, their biomass in certain layers (500–2000 m) is higher than that of their prey, the mesoplankton (Table 4).

In the same manner, a layer of macroplankton animals develops at depths of 500–2000 m, under the surface productive layers. These aggregations create a "living filter" (Barham, 1963), or "living reticulum" under the productive zone of the ocean and consume a significant part of the sinking organic matter from the layers above. Owing to this filter, very little of the organic matter enters the deeper water layers (Figure 30).

In oligotrophic tropical areas, the layer of carnivorous macroplankton is relatively thin, and the biomass is low, while in more productive equatorial areas the layer occupies a significant part of the water column, and the biomass is greater.

Horizontal water samples, taken with Isaacs–Kidd trawls and their modifications, have revealed regional patterns of macroplankton distribution in various oceanic areas, mainly in the Pacific Ocean and mainly in the upper 1000 m (Legand *et al.*, 1972; Aizawa, 1974; Vinogradov and Parin, 1973; Parin, 1975).

The vertical stratification of the macroplankton is much more obvious than that of mesoplankton. All the data available show quantitative heterogeneity in the vertical distribution of the main taxonomic macroplankton groups, characterized by two or three concentration maxima. The distribution patterns are substantially influenced by local hydrographic and trophic factors. Analysis of the vertical distribution of cephalopods demonstrates that two abundance maxima exist in tropical waters: (i) in the lower epipelagial and the upper mesopelagial, at 50–200 m (formed by the juveniles of numerous species living entirely in the upper layers, and interzonal migrants at night); and (ii) in the lower mesopelagial, near the upper border of the bathypelagial, from 500–700 m to 800–900 m (800–1500 m in the subpolar areas) (formed by the non-migrating deep-sea species and the interzonal migrants in the daytime). The latter maximum is less prominent in oligotrophic central waters.

Two maxima were also recorded for pelagic shrimps, but only at night (Foxton, 1970a,b; Legand *et al.*, 1972; Vinogradov and Parin, 1973; Aizawa, 1974). Unlike the cephalopods, the shrimps rarely ascend to the near-surface water layers at night. The upper maximum of shrimp abundance and biomass at night is situated in the lower epi- and upper mesopelagial, usually at depths of 100–300 (500) m, and is formed by rapidly migrating species, mainly sergestids and oplophorids.

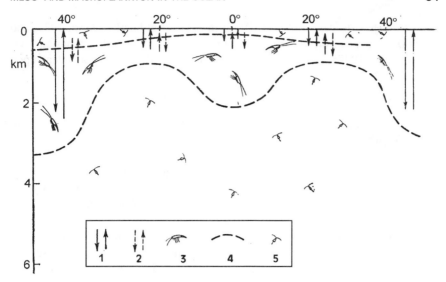

Figure 30 The distribution pattern of meso- and macroplankton at the meridional section across the ocean. 1, 2, The range of diurnal or seasonal migrations of mesoplankton; 3, the zone of macroplankton concentration; 4, the boundary of the zone of macroplankton concentration; 5, the zone of dominance of mesoplankton (from Vinogradov, 1968).

The lower maximum appears to be due to slowly migrating species. In the daytime, interzonal species descend to the depths, where non-migrating animals dwell, and only one maximum of shrimp abundance exists.

Finally, the night distribution of macroplanktonic fishes is characterized by three concentration maxima. The first, close to the surface, is created by the nycto-epipelagic myctophids. The second maximum, prominent in productive areas, but much less obvious in oligotrophic areas, is usually positioned near the upper border of the main thermocline or immediately underneath, at depths of 50–120 m. This maximum appears to be due to the interzonal myctophids: for instance, in the western tropical Pacific, the biomass of *Ceratoscopelus warmingi* and smaller specimens of the genera *Diaphus, Lampanyctus*, etc. reaches 10 mg m^{-3}; that is 25–50 or even 75% of the total macroplankton biomass (Vinogradov and Parin, 1973). Only in certain areas, characterized by specific conditions, do other groups become important in the maximum in question, for example *Bathylagus nigrigenys*, and other bathylagids, in the eastern equatorial Pacific (Parin, 1975). Below the second maximum the fish concentration generally decreases and then increases again in the lower part of the mesopelagic zone (the third maximum). In the eastern equatorial Pacific area, this maximum is easily recognized at depths of 300–400 m and accounted for by aggregations of comparatively large fishes (*Anoplogaster, Gonostoma*, etc.) and cyclothones. This maximum shifts to depths of 400–650 m in the central Pacific (Parin, 1975) and to 1000–1500 m, in the Philippine Basin and the semi-closed Indonesian Seas (Parin et al., 1977a). Changes in the fish biomass depend a good deal upon hydrological and trophic conditions.

4.2. Composition of the Macroplankton and its Dependence upon Depth

A general review of depth-dependent changes in macroplankton composition has been given by Parin et al. (1977b). It was noted that three basic groups of macroplanktonic animals exist in the open ocean: epipelagic, mesopelagic and bathypelagic.

Macroplanktonic animals living entirely in the epipelagic region (defined as the upper isothermal layer) are not abundant; significant numbers of near-surface species are found also in the upper horizons of the mesopelagial, down to 400–450 m. These include some penaeid shrimps, argonautoid octopuses, squid and fish juveniles, both those that are epipelagic when adult (Exocoetidae, Scomberesocidae, Carangidae, Scombridae, etc.) and those performing ontogenetic migrations (Gonastomatidae, Myctophidae, Ceratoidei, etc.). In addition, the interzonal nycto-epipelagic invertebrates and fishes regularly ascend to this layer. The species composition of the epipelagic macroplankton is comparatively scanty, especially outside the tropics.

The macroplankton of intermediate depths, from 100–200 to 1000–1500 m, is richer and more diverse. It is in the mesopelagial of tropical areas that the highest number of genera and species is recorded for many groups, in particular for the pelagic crustaceans, cephalopods and fishes. The shrimps (Sergestidae, Peneidae and Oplophoridae), squids (Enoploteuthidae, Octopoteuthidae, Histioteuthidae, Chiroteuthidae, Cranchiidae, etc.), some octopuses (Amphitretidae), fishes (Gonostomatidae, Sternoptychidae, Chauliodontidae, Stomiatidae, Myctophidae, Melamphaeidae, etc.) are especially characteristic of this zone. The species diversity of the fauna is extremely high (Parin, 1975). In subpolar areas, the mesopelagic plankton is generally represented by the same groups as in tropical waters (although the species are different), but is much less diverse. The macroplankton in the Arctic and Antarctic (south of the Antarctic convergence) is especially low in biodiversity.

The mesopelagic complex of the macroplankton is composed of species, both migrating and non-migrating, living entirely in the mesopelagial region, as well as eurybathic species distributed from the lower epipelagial to the upper bathypelagial and most abundant in the mesopelagial.

The mesopelagial of tropical areas is faunistically heterogeneous throughout the total depth range. Considering the vertical distribution of fishes, cephalopods, shrimps and euphausiids, Legand et al. (1972) point out that there is a boundary at a depth of about 450 m; non-migrating mesopelagic fish ("light" species of *Cyclothone, Melamphaes*) and euphausiid species remain and migrants aggregate in the daytime below this boundary. The lower boundary of the night ascent of interzonal macroplankton species is also at about the same depth. Thus, the upper layers of the mesopelagial (to depths of 400–500 m) have small populations in the daytime, mainly of eurybathic near-surface species and juveniles of some deep-sea animals. This fact permits distinction of epi- and hypomesopelagic faunas among the macroplankton of intermediate waters.

The species diversity of the bathypelagic macroplankton is much lower than that of the mesopelagic plankton. However, it includes almost all taxonomic groups of plankton: euphausiids (Bentheuphausiidae), mysids, shrimps, cephalopods (Vampyroteuthidae, Bathyteuthidae, Mastigiteuthidae, Grimaldi-teuthidae and Biolitaenidae), fishes ("dark" [black] species of *Cyclothone* of the Gonostomatidae, *Taaningichthys* of the Myctophidae, Cetomimidae, Cyemidae, Eurypharingidae, 13 families of angler fishes of the suborder Ceratoidei). This species complex is the richest in the waters of the tropical region.

Unlike the mesoplankton, the macroplankton is almost without specific abyssopelagic species, and the macroplankton fauna gets generally poorer below 3000 m. However, a special benthopelagic (near-bottom) complex of macro-planktonic animals is conspicuous immediately above the sea floor.

4.2.1. *Ontogenetic Migrations*

The vertical distributions of most of the macroplanktonic animals vary conspicuously during their life cycles and during the day. It is rare for large animals to inhabit the same depth throughout their lives. According to Parin *et al.* (1977b), three basic types of ontogenetic migration can be distinguished among macroplanktonic animals:

- Adults, living in the mesopelagial region, ascend and spawn in the upper layers, where embryonic and early postembryonic development takes place; as an animal grows, it gradually descends to greater depths. Such a type of migration is characteristic for some squids of the family Enoploteuthidae (*Watasenia scintillans, Abralia veranyi.* etc.) and some fishes (lantern-fishes of the near-surface genera, such as *Myctophum* and *Symbolophorus*). Larvae of these species live in the epipelagial region and perform slight daily migrations. As an animal grows, the range of the daily migrations increases and finally covers the whole mesopelagial, the animals ascending to the epipelagial and often reaching the surface at night. In the daytime, the animals may be found in a wide depth range down to 1000–1500 m.
- Adults live and spawn in the deep sea (in meso-, bathy-, or abyssopelagial regions), while eggs and larvae ascend and develop in the epipelagial or upper

mesopelagial; growing juveniles descend to the depth inhabited by the adult. The deep-sea angler fishes (Ceratoidei) exemplify this type of migration (Bertelsen, 1951). Many bathypelagic fishes spend their larval period in the epipelagial, among them the most abundant are cyclothones (*Cyclothone atraria, C. braueri, C. acclinidens*); eggs and larvae of these species have been caught near the surface (Mukhacheva, 1964; Gorbunova, 1975). On the other hand, the larvae of the near-bottom macrourids ascend only up to a depth of 200 m (Marshall, 1965b). Larvae of bathypelagic *Anoplogaster cornuta* develop also in the mesopelagic region (Grey, 1955). This type of ontogenetic migration is also characteristic of most of the oceanic cephalopods, for example for the meso- and bathypelagic squids of the family Cranchiidae (Nesis, 1973; Young, 1975).

• Adults live at lesser depths than their larvae and juveniles; this type of migration is much rarer. Some pelagic octopuses (*Argonauta* spp.) spawn near the surface, while their larvae and juveniles are caught in depths of 500–1000 m.

4.2.2. *Diel Migrations*

Many planktonic animals perform extensive diel vertical migrations, especially the mesopelagic species in tropical regions. The range of these migrations varies in different species. A detailed classification based upon the character and range of migration was proposed by Roper and Young (1975). Diel migrations of the various planktonic groups exemplify one of the most characteristic ecological patterns of the macroplankton of the upper 1000 m layer. At these depths, the migrants dominate, and their vertical movements at dawn and sunset lead to dramatic changes in the vertical distribution of the fauna (Figure 20).

4.3. Role of Gelatinous Animals in Plankton Communities

When considering macroplankton distribution the role of the gelatinous organisms needs to be emphasized, as represented by the carnivore: ctenophores, siphonophores, medusae; and the nanophagous salps and appendicularians.

Large scattered gelatinous animals are undersampled by traditional gear and are often destroyed, leaving little chance to judge their size, structure and abundance. Therefore, visual observations from manned submersibles are the only method for reliable studies, including estimation of their abundance, distribution and taxonomic composition.

The first data obtained using manned submersibles gave evidence of the abundance of gelatinous animals in the oceanic plankton of the meso- and bathypelagic regions and of their great role in the functioning of planktonic communities. For instance, they have been shown to create a "living filter" of carnivores under the productive zone, consuming a significant amount of the organic matter transported from the surface layers by migrating or dying animals (Barham, 1963, 1966). Other researchers, viewing from manned submersibles (Hamner *et al.*, 1975; Youngbluth, 1984; Mackie, 1985; Laval *et al.*, 1989), also reported high abundances of gelatinous animals.

Vinogradov and Shushkina (1994) specially estimated the role of gelatinous organisms both in the plankton counted from manned submersibles and in the total biomass of meso- and macroplankton (water bottles, plankton nets, trawls, submersibles) in various Pacific areas. Gelatinous animals appeared to contribute 92% (wet weight) or 37% (carbon weight) of the biomass of plankton counted from manned submersibles in the 0–1000 m layer in the temperate waters of the north-western Pacific (Kurile–Kamchatka region). Values are similar for the 0–500 m layer (Table 5).

Since small mesoplankton (shorter than 3 mm) is not counted from submersibles, the authors combined the data obtained with various samplers. For this purpose, non-gelatinous animals taken by 150-l water bottles, (length 0.2–3.0 m), plankton nets (3–30 mm) and organisms counted from manned submersibles (longer than 30 mm) were combined. Gelatinous animals from water bottles (length less than 10 mm) and those counted from manned submersibles were combined. According to this method, in the 0–500 m layer, where most of the dominant interzonal copepods live, the proportion of gelatinous animals in the total biomass of meso- and macroplankton was on average as high as 88% (wet weight) or 32% (carbon weight), similar to the values obtained directly from visual observations.

The upper 0–200 m layer of the oligotrophic tropical waters, contained a higher proportion of gelatinous animals than did the same layer in the temperate

Table 5 Average share (%), by wet weight (ww) and carbon weight (cw), of gelatinous animals in the zooplankton community according to visual observations from DSRV "Mir" (Vinogradov and Shushkina, 1994).

Depth, m	Northwest region, 5 locations		Central oligotrophic region, 3 locations		Eastern Pacific region, 4 locations		Average for 12 locations	
	ww	cw	ww	cw	ww	cw	ww	cw
0–200	97	43	98	88	81	53	92	58
200–500	81	30	96	27	95	49	89	34
500–1000	95	39	85	31	66	23	84	32
>1000	85	24	98	68	90	28	90	36

north-western part of the Pacific Ocean, the carbon weight reaching 88%. In the 200–500 m layer, the proportion was similar in all the locations studied, especially by carbon weight. However, in the 500–1000 m layer, the proportion of gelatinous animals was significantly lower in the eastern part of the ocean, where in the thick intermediate layer, characterized by low oxygen concentration, gelatinous animals were rather scarce.

It has been important to discover which groups of gelatinous animals are dominant in various regions and depths and what are the proportions of carnivores (coelenterates and ctenophores) and of fine suspension-feeders (salps). Results of this analysis for the northern Pacific (Table 6) demonstrate that in all the areas studied and at all depths the proportion of suspension-feeders was low and did not usually exceed 1–5% of the total biomass of gelatinous animals.

Among the gelatinous carnivores, medusae, consuming small prey, dominated in the productive temperate region (usually more than 50% of the total biomass of gelatinous animals). Large carnivorous siphonophores were significantly (half) less abundant than medusae.

In the oligotrophic areas, gelatinous animals in the upper 1000 m layer were half as abundant as in the productive temperate area. Among them, siphonophores, which consume larger and more active prey than medusae do, were dominant (40–70%). They formed an especially high proportion in the 200–500 m layer (72%). They make up a "living network" of carnivores, feeding on the migrants from the upper productive zone, as discussed above. In the 500–1000 m layer, and especially below 1000 m, the proportion of siphonophores decreased, and the ctenophores, consuming less actively swimming prey, became dominant.

In the productive East Pacific areas (Costa Rica Dome, Californian Upwelling, and Gulf of California), the total biomass of gelatinous animals exceeded that recorded from the northwest part of the Pacific Ocean. Around the Costa-Rica dome and in the Gulf of California, ctenophores dominated (58 and 42%, respectively); at other locations, as in the productive north-west region, medusae dominated (36–49%), although the proportion of ctenophores was significantly high (23–40%). The latter averaged 30% in the 0–1000 m layer. The proportion of siphonophores decreased to the values even lower than in the Kurile–Kamchatka region.

Thus, estimates of the composition and biomass of the gelatinous fauna, using direct observations from submersibles, has conspicuously enriched our knowledge of the role of these animals in the pelagic communities of the total water column, at least in the epi-, meso-, and bathypelagial regions. At depths of 3000 m and less, the gelatinous animals usually compose 80–98% of the total wet biomass of the plankton, counted from manned submersibles. If the plankton biomass is expressed in carbon units, the role of the gelatinous animals appears smaller, but even in this case the proportion ranges from 25 to 40% or more of the total plankton biomass.

The abyssopelagic communities, below 3000 m, have scarcely been studied

Table 6 Average proportion (%) of various groups of gelatinous animals of their total wet biomass for the northern Pacific. Visual observations from DSRV "Mir" (Vinogradov & Shushkina, 1994).

Region	Northwest region, 4 locations			Central oligotrophic region, 3 locations			East Pacific region, 4 locations		
Depth, m	0–500	500–1000	>1000	0–500	500–1000	>1000	0–500	500–1000	>1000
Total biomass of gelatinous animals (gm^{-2})	940	1650	160	670	420	500	3270	310	710
jellyfishes	41	67	76	24	19	14	29	41	26
Siphonophora	23	25	18	72	43	28	24	13	27
Ctenophora	32	1	1	4	36	46	33	29	42
salps	4	7	5	0.3	2	12	14	17	5

but, taking into account the general patterns of trophic relations in communities from different depths (Vinogradov, 1968), the proportion of carnivores in the abyssopelagial region is likely to be insignificant.

5. VERTICAL ZONATION OF THE PELAGIC FAUNA

The irregular depth-dependent changes in plankton composition, the presence of water layers with especially sharp faunistic boundaries caused both by physical and biological factors, the vertical changes in dominance of various trophic groups and other ecological features of the population, all make it possible to divide the water column into several biological zones. Vertical zonation, like any other biological division, can be based upon analyses of the biological data alone. A biological scheme should be based on the vertical succession of life forms of dominant groups of ecologically similar species, because similarity of life forms is determined by the ecological, morphological and other biological features of the organism. A concept of biological zonation, based on the changes in various dominant life forms, seems to be the most natural and satisfactory. However, our present knowledge of marine animals (especially those of the deep sea), is not sufficient to carry out an analysis of their distribution according to their life forms. Those species most closely related systematically have similar life forms. Therefore, analysis of the ecological composition of the population may roughly be approximated by an analysis of the taxonomic composition. Faunistic data should be used first in preparing a scheme of biological zonation. Of course, this does not exclude the need to confirm and explain the suggested schemes of vertical zonation using data on the physico-chemical patterns of the environment. Indeed, a lack of information on vertical changes in the faunistic composition of plankton has led some researchers to believe that it is easier to base a scheme on the analysis of physico-chemical parameters.

Bruun (1956) proposed a scheme of vertical zonation of the oceanic fauna that was based on the vertical temperature changes. He defined eutrophic and mesopelagic zones as having temperatures exceeding 10°C, a bathypelagic zone with temperatures of 4–10°C, and an abyssopelagic zone with temperatures less than 4°C. Bruun emphasized that the temperature may be considerably higher in the bathypelagic and abyssopelagic zones of certain semi-enclosed seas, for example inside the Malay Archipelago.

Earlier, Iselin (1936) found it possible to divide the water column into: (i) the surface layer (from the surface to the thermocline); (ii) the intermediate layer (from the thermocline to the upper border of the cool intermediate layer), called the thermocline layer, because the temperature drops there from 20°C to about 5°C at a depth of 1000 m (in tropical areas); and (iii) the deep waters. As indicated by Marshall (1960), Iselin's intermediate layer is the twilight zone for the penetration of light. The scheme proposed by Iselin was subsequently repeated by several authors. Among others, Beklemishev (1969) developed this scheme on the basis of the migration patterns of the sound-scattering layers. All these schemes are based on the classification of water masses and on the temperature gradient, and are not supported by biological data. Detailed biological studies are necessary to confirm them; that is, the establishment of independent schemes based on the analysis of the biological data.

Since the end of the last century, various authors have proposed dozens of faunistic schemes for different oceanic areas and animal groups. These schemes are generally similar to one another, which confirms once more the existence of objective patterns in the vertical distribution of the pelagic fauna. Differences between the schemes are mainly associated with different depths of the boundaries between vertical zones and with the degree of detail in the differentiation of some of these zones. A scheme of faunistic zonation of the total oceanic water column, from the surface to ultra-abyssal depths, was proposed by Birstein *et al.* (1954) on the basis of numerous catches with plankton nets and ring trawls in Kurile–Kamchatka region. Evolution of our knowledge of vertical zonation as well as appropriate schemes, recently adopted, have been discussed in detail in several publications (Hedgpeth, 1957; Belyaev *et al.*, 1959; Vinogradova *et al.*, 1959; Grice and Hulsemann, 1965; Vinogradov, 1968).

Extensive data on the vertical distribution of the main meso- and macro-planktonic groups have been presented here, and satisfactory and reliable schemes for the biological zonation of the ocean can now be proposed. Such a scheme for the pelagic fauna, based on previous publications (Vinogradov, 1968, 1977a), is shown in Table 7.

The types of vertical distribution of the plankton biomass described above agree well with this scheme and confirm the validity of its main divisions. The surface type of distribution corresponds to the surface zone and transitional layer; the midwater type of distribution nearly corresponds to the upper (bathypelagic) subzone of the deep-sea zone; the abyssal type of distribution corresponds to the lower (abyssopelagic) subzone of the deep-sea zone.

Of course, the details of this scheme differ somewhat in various regions of the ocean. For instance, the population of the surface zone of tropical areas is quite clearly stratified, making it possible to distinguish layers characterized by definite planktonic groups. In the subpolar regions, the distribution of the populations of the surface zone changes more gradually and depends upon the season. The isolation of the populations of the surface warmer waters from the populations of deeper and colder layers is much more pronounced in tropical than in the temperate latitudes, where there is less temperature difference between the surface and deep-sea layers.

In oligotrophic tropical areas, where the boundaries between the various types of vertical distribution of the plankton biomass and the layers of the dominance of various trophic groups are displaced upwards, a certain change is also observed

Table 7   Scheme of vertical zonation of plankton.

Surface (epipelagic) zone, 0–150 (200) m		
Transitional (mesopelagic) layer, from 150 (200) to 750 (1000) m		
Deep-sea zone, below 750 (1000) m	Upper (bathypelagic) subzone of the deep-sea zone, from 750 (1000) m to 2500 (3500)	
	Lower (abyssopelagic) subzone of the deep-sea zone, below 2500 (3500) m	Ocean depths, 2500 (3500) to 6000 m
		Ultra-abyssal (hadal) depths, below 6000 m

in the position of the faunistic boundaries described in Table 7. The boundaries also ascend somewhat higher than in the eutrophic subpolar regions. However, the materials available at present are not satisfactory for the evaluation of such changes.

The positions of the boundaries between different zones and subzones are not clear, because a large part of the data on the zooplankton distribution has been obtained with various plankton nets towed without closing or through rather wide vertical ranges. The exact actual depth of capture of an animal was not determined. Therefore, such terms as the range of catch, and the inhabited zone, have been used (Heptner, 1981).

In this respect, the patterns of animal distribution in the epi- and mesopelagic zones are much better known than those in the deep-sea zone, where sampling with plankton nets and various horizontally towed trawls has been scanty. The deep-sea layers are homogeneous physico-chemically and there are wide transitional layers between faunas of the different subzones, bathy- and abyssopelagial.

Direct visual observations from manned submersibles have allowed us to gather information about the vertical distribution of various species at the depths where they live. Recently, these observations have become serious planned studies. Studies on the vertical distribution of large meso- and macroplanktonic animals throughout the total water column, from 0 to 6000 (5948) m by means of direct visual observations from the DSRV "Mir" were undertaken by the P. P. Shirshov Institute of Oceanology in 1990 in the north part of the Pacific Ocean (Vinogradov and Sagalevich, 1994; Vinogradov and Shushkina, 1994; Vinogradov and

Tchindonova, 1994). Analysis was restricted to the most numerous species, because it was impossible to estimate the depth range of rare animals, seen once or twice per dive. Therefore, it is hard to reach conclusions about animal distribution at great depths, where the populations are very scattered. Observations are more reliable in the productive areas of the Kurile–Kamchatka region, where plankton abundances are an order of magnitude higher than in the oligotrophic tropical waters of the ocean. In addition, the deep-sea plankton of this area has been well studied faunistically, making it easy to identify the animals.

Figures 31 and 32 exemplify the distribution of selected gelatinous animals and that of certain identified crustaceans. Analysis of these figures and of some other observations suggests that there were three prominent layers in the distribution of the gelatinous animals: 400–500, 1000–1600, and 2600–3300 m. The animals displayed the following types of vertical distribution: (a) in the mesopelagic zone only (sometimes its upper part); (b) in the bathypelagic zone; the meso- and bathypelagic zones synchronously; and (c) in the abyssopelagic zone. It is remarkable that no medusae were found dwelling in the meso-, bathy- and abyssopelagic zones synchronously, or even in the bathy- and abyssopelagial synchronously, despite a prediction of such a distribution based on plankton net samples.

Analysis of the distribution of other animal groups (crustaceans, polychaetes) shows that, as in the case of medusae, several species groups may be distinguished: mesopelagic (sometimes penetrating the upper bathypelagic zone); bathypelagial eurybathic, inhabiting the whole deep-sea zone (meso, bathy- and abyssopelagial), and pure abyssopelagic species. Ranges of various species living in the different zones usually overlap, and the boundaries between zones are uncertain. Direct observations from manned submersibles gave no evidence of a more definite stratification than was discovered with the use of standard gear. At the same time, these direct observations confirm the existence of a prominent faunistic boundary at a depth of about 3000 m, in the 2500–3000 m layer. The fauna change at this boundary is sharper than that between the meso- and bathypelagial, at 750–1500 m. This fact illustrates once more the transitional nature of the mesopelagic zone, lying between the productive surface layers and the deep-sea communities that depend upon organic matter produced in the surface layers. Vinogradov and Tchindonova (1994) remarked that deviations in the distribution of macroplankton populations were rather high in the meso-, bathy- and, especially, the abyssopelagic zone. However, even rare deep-sea animals were represented by two or three examples in the count area, rather than singly. Similar observations were made by Tchindonova in the abyssopelagial for the amphipod *Cyphocaris richardi*, the polychaete *Flota* sp., the isopod *Eurycope murrei*, and other animals. This shows that deep-sea macroplanktonic animals are able to form micro-aggregations of single species, making sexual contact and other intra-specific relations easier.

In concluding this account, it is necessary to discuss the distinction of the ultra-

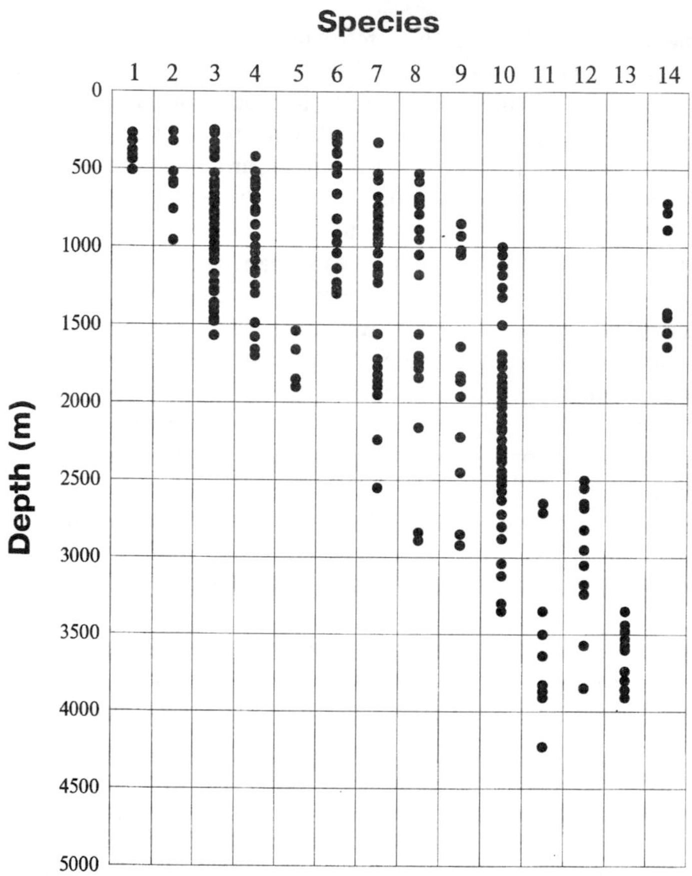

Species

Figure 31 Vertical distribution of the most abundant species of medusae and ctenophores in the Kurile–Kamchatka region of the Pacific Ocean, measured by the direct counts from the "Mir" submersible during July–August, 1990. The four dives in the Kurile–Kamchatka region of the Pacific Ocean averaged (44°21'N, 149°51'E, depth of the dive 4679 m; 54°59'N, 165°43'E, depth of the dive 5250 m; 55°34'N, 167°19'E, depth of the dive 4386 m; 53°06'N, 161°00'E, depth of the dive 5948 m). 1, *Aegina* sp.; 2, *Periphylla hyacinthina*; 3, *Solmissus* sp.; 4, *Crossota brunnea*; 5, *Catoblema vericarium* (?); 6, *Colobonema* sp.; 7, *Pantachogon haeckeli*; 8, *Turritopsis nutriculus* (?); 9, *Pelagia* sp.; 10, *Botrynema brucei*; 11, *Crossota* sp.; 12, ctenophora *Lampactena* sp.; 13, *Colobonema* sp.; 14, Rhizostomidae. The dots indicate the occurrence of the species within 50-m intervals of submersible movement (from Vinogradov and Tchindonova, 1994).

abyssal (hadal) fauna. An ultra-abyssal zone has been recognized for both its benthic and its pelagic fauna; comprising a species grouping not recorded from lesser depths (Birstein *et al.*, 1954; Belyaev *et al.*, 1959). However, as the pelagic fauna of the ocean depths became better explored, this conclusion had to be

Species

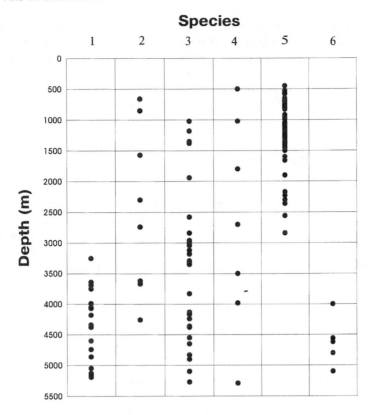

Figure 32   Vertical distribution in the Kurile–Kamchatka region of the Pacific Ocean of the crustaceans identified with confidence from the submersible. The data of four dives listed at Figure 31 were summarized: 1, adolescent and adult Cyphocaris richardi; 2, Nebaliopsis typica; 3, Eurycope murrei; 4, Gnathophausia gigas; 5, Eucopia grimaldi; 6, E. australis (from Vinogradov and Tchindonova, 1994).

modified (Vinogradov, 1968), so that the only endemic element of the ultra-abyssal pelagic zone was the hadal gammarids.

Gammarids are a mainly benthic group of Amphipoda, that explore the pelagial region intensively. A significant number of the gammarids found in the water column are actually benthopelagic, not truly pelagic (Vinogradov, 1992, 1995). Even gammaridean species considered to be hadal in the past are now being recorded from lesser depths. For example, the typical ultra-abyssal gammarid *Vitjaziana gurjanovae* from the Kurile–Kamchatka Trench, was found at a depth of 3500 m in the Atlantic (Vinogradov and Vinogradov, 1996), while *Scopelocheiros shellenbergi* from the same trench was recorded from a depth of 4000 m in the Antarctic (Vinogradov and Vinogradov, 1993). Although several species, for example the abundant benthopelagic *Hirondellea gigas* and the rare, probably also benthopelagic *Halice quarta*, *H. subquarta* and *Andaniexis subabyssi* have not yet been recorded from less than 6000 m, the reality of a separate pelagic fauna in the hadal region should be reconsidered. The distinction of a separate ultra-abyssal subzone for pelagic animals now seems doubtful.

6. PATTERNS OF PLANKTON DISTRIBUTION IN THE NEAR-BOTTOM LAYER

At the end of the last century, Fowler (1898) suggested that there might be specific populations in the 200 m layer above the bottom, termed hypoplankton and believed to be more closely related to the bottom, than to the upper water layers. Later, Bogorov (1948) also noted a special near-bottom zone inhabited by hypoplankton and occupying a layer 0–50 m above the bottom. However, these ideas were not based upon biological data. Only in the last 25–30 years have technical improvements allowed the study of the near-bottom layer, using towed nets and trawls of various types with special gear, big water bottles with pingers, bottom traps and, of course, manned submersibles.

6.1. Near-bottom Communities in the Ocean

The reality of specific near-bottom communities was first revealed in shelf areas (Bossanyi, 1957; Beyer, 1958; Oug, 1977), then deeper on continental slopes (Ellis, 1985; Vereshchaka, 1990, 1994, 1995; Vinogradov, 1990a) and seamounts (Andriyashev, 1979; Boehlert, 1988; Vereshchaka, 1994, 1995, Vinogradov, 1990a,b; Vereshchaka and Vinogradov, 1996). Similar near-bottom communities were described from abyssal depths (Grice and Hulsemann, 1970; Grice, 1972; Boxshall and Roe, 1980; Wishner, 1980a; Domanski, 1986; Childress *et al.*, 1989). A detailed review of these studies was published by Heinrich *et al.* (1993).

In order to understand the biological structure of the near-bottom layer, it is necessary to consider the physical and chemical conditions in this layer. Interaction between ocean waters and the sea floor produces a mixed benthic boundary layer which has proved to be ubiquitous throughout the world. This layer is several dozens of metres thick and is characterized by steep vertical gradients of current velocity and turbulence, equalizing temperature, salinity and density. At about 2 m above bottom (mab), turbulence has been shown to increase abruptly (Lozovatsky *et al.*, 1977). The benthic boundary layer is restricted by the bottom surface below and the zone of near-bottom thermo-, halo- and pycnoclines above.

In addition to the benthic boundary layer, a benthic nepheloid layer overlies the sea floor. This layer is several hundred metres thick, and is characterized by high concentrations of suspended particulate matter.

From a biological viewpoint, the near-bottom layer lies between the two principal oceanic biotopes: pelagial and benthal. Therefore, its population would be expected to be very diverse and to include a variety of ecological groups. There is no widely adopted terminology for the structure of near-bottom communities.

The main feature of the near-bottom population is the mixture of purely pelagic species with animals related to the bottom. The revelation of a fauna of benthopelagic scavengers, swimming above the bottom and aggregating near the sunken corpses of animals is one of the most interesting discoveries of modern oceanography. The scavenger communities include several species of fishes, mainly of the family Macrouridae. The most important invertebrates are large amphipods of the family Lysianassidae. Their mouthparts are adapted for biting, tearing and cutting. Their gut content shows that the remains of large animals are their usual food. Kamenskaya (1984) considered them to be a separate ecological group. These benthopelagic animals are characteristic of the ultra-abyssal depths of the oceanic trenches (Lemche *et al.*, 1976; Hessler *et al.*, 1978), the oligotrophic tropical areas of the oceans (Shulenberger and Hessler, 1974; Rowe *et al.*, 1986), and the open oceans (Thurston, 1979; Stockton, 1982; Vinogradov and Vinogradov, 1991). These amphipods seem to form a significant proportion of the near-bottom plankton, although they are poorly sampled by quantitative towed gear, nets and trawls. Various scavengers are able to swim up to different levels, tens, hundreds or thousands of metres above the bottom. This is one of the possible paths of the amphipod expansion into the pelagic realm (Vinogradov, 1992, 1995).

An unusually high proportion of dead animals among the near-bottom plankton is another feature of the near-bottom layers (Wishner, 1980b; Roe, 1988; Angel, 1990; Heptner *et al.*, 1990; Heinrich *et al.*, 1993). Nevertheless, the mesoplankton biomass in the near-bottom layer usually corresponds to the general patterns of vertical biomass changes in the water column (Wishner, 1980a; Berg and Van Dover, 1987; Beckmann, 1988; Roe, 1988; Childress *et al.*, 1989; Rudyakov *et al.*, 1990; etc.). However, there are numerous exceptions, either when the near-bottom plankton appears very enriched (e.g. Omori and Ohta, 1981; Alldredge *et al.*, 1984), or very poor (Pérès, 1958a,b; Wishner, 1980a; Rudyakov *et al.*, 1990). The latter phenomenon may be explained: Wishner studied plankton in the Red Sea where there are unusual conditions in the deep sea; Rudyakov *et al.* (1990) worked near Walters Shoals, where a high concentration of benthopelagic carnivorous fishes was recorded. After analysis of the data available, Heinrich *et al.* (1993) believe it is impossible to make definite conclusions about the general tendency of plankton biomass changes near the bottom, because of the small amount of data and its heterogeneity.

In fact, the actual situation can differ under different conditions. For example, the crew of a manned submersible recorded enriched plankton at 1–5 m above bottom at depths of 1020–1050 m (Angel, 1983), while Pérès (1958b) found a layer of extremely transparent water at 1–5 m above bottom at a similar depth of 1390 m. The absence of enriched near-bottom plankton was noted from a manned submersible at 1–30 m above bottom, in the Norwegian Sea (Vereshchaka and Vinogradov, 1996) and above a hydrothermal field (TAG) in the tropical Atlantic (Vinogradov *et al.*, 1996b).

However, the existence of numerous benthopelagic species that are most abundant near the bottom is well established. The depth ranges differ among the species, from several metres to two to three thousand metres. Table 8 shows the depth ranges of selected species (according to the medial values of the sampled horizons), reported by various authors to be most abundant near the bottom in various regions above abyssal depths (3000–6000 m).

There are different viewpoints on the subdivision of the near-bottom benthopelagic fauna. For example, Heinrich *et al.* (1993) came to the conclusion that the available data on the vertical ranges of animals abundant near the bottom show that the ranges differ significantly. Thus, the upper boundary ranges from centimetres to kilometres above bottom, the lower boundary may be inside the sediment. At the same time, there are no definite differences between benthic and benthopelagic animals, as all are able to swim and even those that spend most of the time on the bottom may be referred to either of these groups. Only non-swimming benthic animals lacking planktonic larvae, on the one hand, and planktonic animals spending all their lives in the water column, on the other hand, may be distinguished with certainty.

However, Vereshchaka (1995) thinks that two principal planktonic groups may live in the benthopelagic zone: pelagic and benthopelagic. The pelagic animals have been called pure planktonic (Bossanyi, 1957; Wishner, 1980a), "tychobenthos" (Beyer, 1958), or "hypoplankton" (Hesthagen, 1973). Their presence in the benthopelagial is accidental and episodic, they get close to the bottom as a result of water mass advection near sea-floor rises, passive sinking, or avoidance of warm upper waters.

Benthopelagic organisms are those obliged to spend at least part of their life in the benthopelagial. The term "benthopelagic" was first proposed by Marshall (1965a,b) for those fishes and other organisms that "freely and habitually" swim near the sea floor. Later, the term has been used both in the initial (Marshall and Merrett, 1977) and a broader sense to define the fishes living in the water column and related to tops of seamounts (Parin *et al.*, 1985).

Since benthopelagic animals migrate, they may be related to either of the neighbouring biotopes: pelagic and/or benthic. Three cases are possible, so three ecological subgroups can be expected among benthopelagic organisms (Vereshchaka, 1995): (i) hypobenthopelagic animals which spend part of their life in the benthopelagial and part deeper, in the benthal; (ii) epibenthopelagic animals spending part of their life in the benthopelagial and part higher, in the pelagial; and (iii) amphibenthopelagic animals which periodically live in all three biotopes; that is, pelagic, benthopelagic and benthic.

Table 8 Vertical distribution of selected species in the near-bottom layer at abyssal depths (Heinrich et al., 1993).

Species	Local depth, m	Layer sampled, m above bottom	Layer of maximal abundances, m above bottom	Range of distribution, m above bottom
COPEPODA				
Benthomisophria palliata	4000	5–1600	5–20	12–1600
DECAPODA				
Hymenodora acanthitelsonis	4040	0, 10–90, 540	40–55	18–47
Hymenodora glacialis	4040	0, 10–90, 540	10–27	18–540
Hymenodora glacialis	5440	0, 10–4530	1530–2110	257–3328
Acanthephyra microphthalma	5440	0, 10–4530	0	0–2540
Benthesicymus iridescens	5440	0, 10–4530	0	0–70
AMPHIPODA				
Eurythenes gryllus	5870	2–1000	2	2–1000
Eurythenes gryllus	4876	2–1800	2	2–1400
Eurythenes gryllus	4386	2–1800	10–20	2–800
Eurythenes gryllus	3715	2–1800	10	5–20
Abyssorchomene chevreuxi	4000	0–1500	0–20	10–750
Paralicella caperesca	4000	0–1500	0–200	10–375
Paralicella caperesca	5440	10–2330	10–25	18–930
Paralicella tenuipes	5440	10–2330	25–48	18–930
CHAETOGNATHA				
Heterokrohnia furnestinae	4000	10–550	10–25	18–550
HOLOTHURIOIDEA				
Scotothuria herringi	5440	0, 10–4530	11–55	20–3980
Enypniastes diaphana	5440	0, 10–4530	11–31	18–2935

Pelagic animals demonstrate no adaptation to a near-bottom life in their behaviour and feeding. Evolution of these species has led to loss of contact with the bottom and a situation where the animals are not able to distinguish the bottom from water. On the contrary, benthopelagic animals have this ability and, in the presence of the bottom, react quite deliberately and try to use it both as shelter and as supply source.

During migrations, benthopelagic animals rise from the bottom to distances of several dozens (hypobenthopelagic), hundreds (epibenthopelagic), and thousands of metres (amphibenthopelagic). Therefore, they may influence the biological processes in the whole pelagial. Their role in pelagic communities depends upon the distance to the sea floor; at several thousand metres above bottom, the proportion of benthopelagic animals does not exceed 1% of the total macro-plankton biomass, whilst at 200–400 mab and less this proportion goes well above 80% (Figure 33). The benthopelagic contact zone (BCZ) was proposed (Vereshchaka, 1995) to be the zone where benthopelagic animals are dominant (50% and more of the total biomass) at night and/or in the daytime. The BCZ usually occupies the benthopelagial and the lower part of the pelagic region.

The benthopelagial may well correspond to the benthic boundary layer, and the benthopelagic contact zone corresponds to the benthic nepheloid layer. This correspondence should be confirmed in the future. The boundaries within them (dozens and hundreds of mab) can be expected to lie close to each other.

Pelagic species are not obliged to inhabit the near-bottom layer. The depth at which they are abundant does not depend upon the local depth, provided the latter is great enough and does not restrict the animal's lower limit (Rudyakov, 1986). When their population is near the sea floor, local temporary aggregations can be formed, at any depth, from the shelf to the abyssal.

For example, near the Portuguese coast, abundant euphausiids were observed buried in ooze. At shallow depths in the Sea of Japan, the copepods *Calanus glacialis* and *Neocalanus cristatus* (V and VI copepodites) actively buried themselves in the ooze, at densities reaching 1300 individuals m^{-3} (Kos, 1969). A similar phenomenon was recorded for the deep-sea pelagic fish *Vinciguerria* sp. in the Red Sea (Heinrich *et al.*, 1993).

Pelagic animals do not need to make contact with the sea floor, since all their ontogenetic stages occur in the water column, and juveniles (at least of well-studied species) have not been recorded close to the bottom. Moreover, if these species reach the sea floor they may suffer a dramatic increase in mortality, due to the feeding activity of near-bottom carnivores (Longhurst, 1976).

6.2. Influence of Hydrothermal Ecosystems Upon the Surrounding Environment

The question of the influence of the enriched ecosystems of hydrothermal fields upon the surrounding environment is intriguing. Does the chemoautotrophic

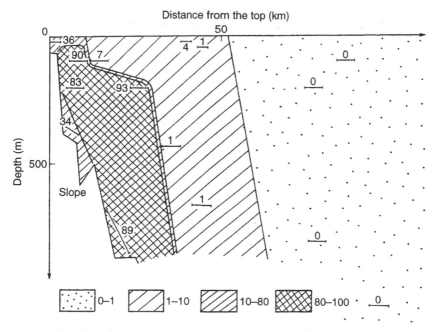

Figure 33   The share (%) of the benthopelagic animals in the total biomass of the mysids, euphausiids, shrimps and bottom-dwelling decapod larvae in the vicinity of the Walters Seamount (south-western Indian Ocean). Night samples (after Vereshchaka, 1995).

production by hydrothermal bacteria, which may reach 0.1–0.6% of total photosynthetic production (Jannasch, 1985; Vinogradov and Shushkina, 1987), lead to the enrichment of abyssal depths of the ocean (only about 1% of photosynthetic production penetrates there), or is the biological structure of the hydrothermal ecosystem quasi-closed, giving off almost nothing outside?

The hydrochemical influence of the hydrothermal fluids can be traced in the form of characteristic "plumes" over significant distances from the eruption sites (e.g. Lisitzin, 1981; Johnson *et al.*, 1986; Roe, 1986; Roth and Dymond, 1989; Burd and Thomson, 1994).

Plumes of sulphide-enriched waters support bacteria using biological processes of chemoautotrophic and methanotrophic synthesis of organic matter, at the sites of methane and hydrogen sulphide emission. In the hydrothermal plumes the concentration of bacteria appears to be significantly higher than in the background waters. This was reported from the hydrothermal fields of Juan-de-Fuca, in the Guaymas and Manus Basins, at the TAG site, etc. Outside the plumes, in the background waters, autotrophic CO_2 assimilation was not found (Winn *et al.*, 1986; Thomson *et al.*, 1991; Lein *et al.*, 1993).

Detecting the influence of the hydrothermal community upon the concentration of the near-bottom plankton is difficult, as near-bottom aggregations unrelated to hydrothermalism have been shown above to be a common feature of the distribution of near-bottom plankton. Therefore, the true influence of the hydrothermal communities is difficult to estimate. For example, in the Gulf of California, in the area of the Guaymas hydrothermal fields, a near- bottom increase in plankton concentration was recorded from the submersible "Mir". This increase was conspicuous in the bottom 300 m layer (and especially in the thin 30 m near-bottom layer). It was formed mainly of copepods, large ctenophores and, sometimes, shrimps and amphipods (Shushkina and Vinogradov, 1992). Whether this was the consequence of the hydrothermal field influence or the usual increase in plankton concentration in the near-bottom layer of the productive region was uncertain.

Studies by American researchers on the hydrothermal fields of the eastern Pacific (Smith, 1985; Smith *et al.*, 1986; Berg and Van Dover, 1987; Wiebe *et al.*, 1988) do not solve this question either. According to Wiebe *et al.* (1988) who used MOCNESS nets in the 100 m near-bottom layer, very little or no difference was found between near-bottom plankton concentrations above the hydrothermal sites and outside them. Wishner (1980a) sampled the plankton at a distance of 10–100 m above bottom and did not find high near-bottom concentrations. Smith *et al.* (1986) used small nets (0.16 m^2) towed by DSRV "Alvin" and recorded the absence of plankton enrichment above the Guaymas hydrothermal fields at a distance of 2–14 m above bottom. However, closer to the sea floor (1–5 m), the plankton biomass above the field was about six times as great as outside the fields. Similar differences (10 times increase) were found above other hydrothermal sites of the eastern Pacific by sampling with nets at 1–3 m above bottom.

Thus, a sharp increase in plankton concentration seems to occur very close to the bottom due to concentrations of local benthopelagic and nectobenthic species. A few dozen metres to a hundred metres away, near-bottom enrichment of the plankton cannot be detected.

In 1990–1992, Canadian researchers investigated the influence of hydrothermal fields upon the deep-sea oceanic plankton at the Endeavour segment of the Juan-de-Fuca Ridge (48°N, 129°W). The sound-scattering layer adjacent to the upper border of a hydrothermal plume at a depth of about 1900 m was found using a 153 kHz Doppler current profiler. Repeated sampling of this layer with nets showed that the plankton biomass immediately above the plume is several times as great as in the upper layers or inside the plume, the enrichment being caused by pelagic animals, not by the local benthopelagic fauna (Thomson *et al.*, 1991, 1992; Burd and Thomson, 1994). Later, they discovered that both deep-sea and mesopelagic species were present in the aggregations, the latter at a greater depth than in samples taken 15–50 km away from the plume area. Moreover, the total plankton biomass in the water column above the hydrothermal area was higher than that 15–50 km away from the fields (Burd and Thomson, 1994, 1995). This increase was attributed to the consumption of bacteria and organic particles in the plume by the planktonic suspension-feeders. Earlier, Roth and Dymond (1989) analysed material from seston traps and suggested the possibility of organic carbon transport from the hydrothermal fields. However, detailed analysis of the Canadian data reveals a remote possibility that the changes seen in the plankton distribution may be due to circulation patterns caused by the presence of the Endeavour Ridge, not to the hydrothermal plumes.

From this viewpoint, studies of the influence of hydrothermal ecosystems on background ecosystems, carried out by Vinogradov *et al.* (1996b) in the ultra-oligotrophic areas of the central Atlantic above the TAG and Broken Spur locations, are of interest. In this area, at depths greater than 1500–2000 m, the biomasses of the "water bottle" (small) and the "net" (medium-size) plankton were as high as 0.03 mg m^{-3} and 0.04–0.05 mg m^{-3}, respectively. However, near the plume (at depths around 2800 m), increases in detritus and "water bottle" mesoplankton were recorded (Figure 34). It is uncertain whether these were particles and animals ascending from the bottom with the plume waters, as proposed by the Canadian researchers for Endeavour Ridge (Roth and Dymond, 1989; Burd *et al.*, 1992; Thomson *et al.*, 1992; etc.), or descending particles, concentrating in the density gradient layer at the plume boundary. Since high concentrations of detritus and plankton feeding on the organic particles are not

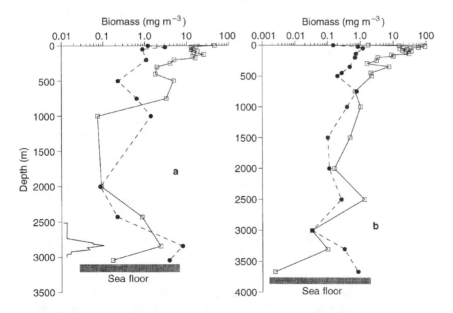

Figure 34   Biomass (mg C$_{org}$ m^{-3}) of mesoplankton (–□–) and detritus biomass (--•--) measured by water-bottle samples during the daytime at 29°N (hydrothermal field Broken-Spur (a)) and 26°N (hydrothermal field TAG (b)). The turbidity distribution at the depth of the hydrothermal plume is illustrated on the left. The symbols indicate the depth of samples collected by 150-l water bottle (Vinogradov *et al.*, 1996b).

observed inside the plume, only above it, the second suggestion is more likely. The larger "net" mesozooplankton was not influenced by the plume above the TAG and Broken Spur sites.

Thus, the very important question about transport of organic particles by hydrothermal plumes in the amounts necessary for enrichment of the oceanic "above-plume" plankton, as supposed by the Canadian researchers, remains open. The influence of stratification, leading to the local concentration of detritus in the otherwise homogeneous and deficient waters of the ocean depths, seems to be a more likely cause of the plankton concentrations above the plumes.

Studies on the ecosystems of hydrothermal fields show that either the communities do not give off fluxes of suitable organic matter, or the influence of the hydrothermal field is so local, due to its restricted area, that it cannot provide a stable food source for the multi-level trophic system in the extremely poor, almost lifeless depths of the oligotrophic areas of the ocean.

REFERENCES

Aizawa, Y. (1974). Ecological studies of micronectonic shrimps (Crustacea, Decapoda) in the Western North Pacific. *Report of the Ocean Research Institute, University of Tokyo* **6**, 1–84.

Alldredge, A.L., Robinson, B.H., Fleminger, A., Torres, J.J., King, J.M. and Hamner, W. M. (1984). Direct sampling and *in situ* observation at a persistent copepod aggregation in the mesopelagic zone of the Santa Barbara Basin. *Marine Biology* **80**, 75–81.

Andriyashev, A.P. (1979). On the zonation of the sea-floor fauna. *In* "Biological Resources of the World Ocean" (S.A. Studenetskiy, ed.), pp. 117–138. Nauka, Moscow. (In Russian).

Andronov, V.N. (1976). Some data about the development of *Calanus carinatus* (Copepoda, Calanoida) in the South-East Atlantic. *Trudy AtlantNIRO* **60**, 117–130. (In Russian).

Angel, M.V. (1983). Are there any potentially important routes whereby radionuclides can be transferred by biological processes from the seabed towards the surface? *In* "Ecological Aspects of Radionuclides Release" (P.J. Coughtrey, ed.), Vol. **3**, 161–176. Blackwell, Oxford (*Special Publication Series British Ecolological Society*).

Angel, M.V. (1990). Life in the benthic boundary layer: connections to the mid-water and sea floor. *Philosophical Transactions of the Royal Society of London* **331**, 15–28.

Arashkevich, E.G. (1972). Vertical distribution of different trophic groups of copepods in the boreal and tropical regions of the Pacific. *Okeanologia* **12**, 315–325. (In Russian).

Arashkevich, E.G., Drits, A.V. and Timonin, A.G. (1996). Diapause in the life cycle of *Calanoides carinatus* (Copepoda, Calanoida). *Hydrobiologia* **320**, 197–208.

Bainbridge, R. (1952). Underwater observations on the swimming of marine zooplankton. *Journal of the Marine Biological Association of the United Kingdom* **31**, 107–112.

Banse, K. (1964). On the vertical distribution of zooplankton in the sea. *Progress in Oceanography* **2**, 53–125.

Barham, E.G. (1963). Siphonophores and the deep-scattering layer. *Science* **140**, 826–828.

Barham E.G. (1966). Deep-scattering layer. Migration and composition observations from diving saucer. *Science* **151**, 1399–1403.

Beckmann, W. (1988). The zooplankton community in the deep bathyal and abyssal zones of the eastern North Atlantic. Preliminary results and list from MOCNESS hauls during cruise 08 of the R.V. "Polarstern". *Berichte zur Polarforschung* **4**, 57 pp.

Beklemishev, C.W. (1969). "Ecology and Biogeography of the Open Ocean". Nauka, Moscow. (In Russian).

Belyaev, G.M., Birstein, Ja.A., Bogorov, B.G., Vinogradova, N.G., Vinogradov, M.E. and Zenkevich, L.A. (1959). A diagram of the vertical biological zonality of the ocean. *Doklady Akademii Nauk SSSR* **129**, 658–661. (In Russian).

Berg, G.J. and Van Dover, C.L. (1987). Benthopelagic macrozooplankton communities at and near deep-sea hydrothermal vents in the eastern Pacific Ocean and the Gulf of California. *Deep-Sea Research* **34** (3A), pp. 379–401.

Bernard, F. (1958). Plancton et benthos, observés durant trois plongées en bathyscaphe au large de Toulon. *Annales de l'Institut Océanographique* **35**, 287–326.

Bernard, F. (1962). Contribution du bathyscaphe a l'étude du plancton: avantages et inconvenients. *Rapports et Procès-Verbaux des Réunions du Conseil Permanent International pour l'Exploration de la Mer* **153**, 25–28.

Bertelsen, E. (1951). The ceratioid fishes. *Dana Report* **7**(39), 1–276.

Beyer, F. (1958). A new, bottom-living trachymedusa from the Oslofjord. Description of the species, and a general discussion of the life conditions and fauna of the fjord deeps. *Nytt Magasin for Zoologi* **6**, 121–143.

Birstein, Ja.A., Vinogradov, M.E. and Tchindonova, Yu.G. (1954). The vertical zonation of the Kurile-Kamchatka Trench region. *Doklady Akademii Nauk SSSR* **95**, 389–392. (In Russian).

Boehlert, G.W. (1988). Current–topography interactions at midocean seamounts and the impact on pelagic ecosystems. *GeoJournal* **16**, 45–52.

Bogorov, B.G. (1948). Vertical distribution of zooplankton and vertical zonation of the Ocean. *Trudy Instituta Okeanologii* **2**, 43–59. (In Russian).

Bogorov, B.G. and Vinogradov, M.E. (1955). Main peculiarities of zooplankton distribution in the north-western part of the Pacific. *Trudy Instituta Okeanologii* **18**, 113–123. (In Russian).

Bossanyi, J. (1957). A preliminary survey of the small natant fauna in the vicinity of the sea floor off Blyth, Northumberland. *Journal of Animal Ecology* **26**, 353–368.

Boxshall, G.A. and Roe, H.S. (1980). The life history and ecology of the aberrant bathypelagic genus *Benthomysophria* Sars 1909 (Copepoda, Mysophrioida). *Bulletin of the British Museum (Natural History), Zoology* **38**, 9–41.

Boyd, C.M. and Smith, S.L. (1980). Grazing patterns of copepods in the upwelling system off Peru. *Limnology and Oceanography* **25**, 583–596.

Brandt, T. (1951). "Anaerobiosis in the Invertebrates". Isdatelstvo inostranii literaturi, Moscow. (In Russian).

Bruun, A.F. (1956). The abyssal fauna: its ecology, distribution and origin. *Nature* **177**, 1105–1108.

Burd, B.J. and Thomson, R.E. (1994). Hydrothermal venting at Endeavour Ridge: effect on zooplankton biomass through the water column. *Deep-Sea Research* **41**, 1407–1423.

Burd, B.J. and Thomson, R.E. (1995). Distribution of zooplankton associated with the Endeavour Ridge hydrothermal plume. *Journal of Plankton Research* **17**, 965–997.

Burd, B.J., Thomson, R.E. and Jamieson, G.S. (1992). Composition of a deep scattering layer overlying a mid-ocean ridge hydrothermal plume. *Marine Biology* **113**, 517–526.

Childress, J.J., Gluck, D.L., Carney, R.S. and Gowing, M.M. (1989). Benthopelagic biomass distribution and oxygen consumption in a deep-sea benthic boundary layer dominated by gelatinous organisms. *Limnology and Oceanography* **34**, 913–930.

Chislenko, L.L. (1968). "Nomograms for Estimation of the Weight of Aquatic Organisms from Body Size and Shape". Nauka, Leningrad. (In Russian).

Clarke, T.A. and Wagner, P.J. (1976). Vertical distribution and other aspects of the ecology of certain mesopelagic fishes taken near Hawaii. *Fishery Bulletin* **74**, 635–645.

Conover, R.J. (1988). Comparative life histories in the genera *Calanus* and *Neocalanus* in high latitudes of the northern hemisphere. *Hydrobiologia* **167/168**, 127–142.

Domanski, P. (1986). The near-bottom shrimp faunas (Decapoda, Natantia) at two abyssal sites in the Northeast Atlantic Ocean. *Marine Biology* **93**, 171–181.

Ekman, S. (1953). "Zoogeography of the Sea". Sidgwick & Jackson, London.

Ellis, C.J. (1985). The effect of proximity of the continental slope seabed on pelagic halocyprid ostracods at 49°N. *Journal of the Marine Biological Association of the United Kingdom* **93**, 171–181.

Flint, M.V. (1975). Trophic structure and vertical distribution of trophic groups of mesoplankton at the equator (97°W). *Trudy Instituta Okeanologii* **102**, 238–243. (In Russian).

Flint, M.V. and Kolosova, E.G. (1990). Mesoplankton of the Peruvian coastal waters. *In* "Ecosystems of the East Boundary Currents and the Central Regions of the Pacific" (M.E. Vinogradov and E.I. Musaeva, eds), pp. 213–230. Nauka, Moscow. (In Russian).

Fowler, J.K. (1898). Appendix to the foregoing report. *Proceedings of the General Meeting for Scientific Business of the Zoological Society of London for Year 1898* **3**, 544–549.

Foxton, P. (1956). The distribution of the standing crop of zooplankton in the Southern Ocean. *Discovery Reports* **28**, 191–236.

Foxton, P. (1970a). The vertical distribution of pelagic decapods (Crustacea: Natantia) collected on the SOND cruise 1965, I. The Caridea. *Journal of the Marine Biological Association of the United Kingdom* **50**, 939–960 and 961–1000.

Foxton, P. (1970b). The vertical distribution of pelagic decapods (Crustacea: Natantia) collected on the SOND cruise 1965, II. The Penaeidae. *Journal of the Marine Biological Association of the United Kingdom* **50**, 961–1000.

Fraser, F.C. (1936). On the development and distribution of the young stages of krill (*Euphausia superba*). *Discovery Reports* **14**, 1–192.

Fryer, G. (1957). The food of some freshwater cyclopoid copepods and its ecological significance. *Journal of Animal Ecology* **26**, 263–286.

Frontier, S. (1978). Interface entre deux écosystèmes: example dans le domaine pelagique. *Annales de l'Institut Ocèanographique Monaco* **54**, 95–106.

Gitelzon, I.I., Levin, L.A. (1983). Bioluminescence field. *In* "Modern methods assessing of the quantitative distribution of the marine plankton" (M.E. Vinogradov, ed.), pp. 10–28. Nauka, Moscow. (In Russian).

Gorbunova, N.N. (1971). Vertical distribution of fish eggs and larvae in the western tropical Pacific. *In* "Functioning of Pelagic Communities in the Tropical Regions of the Ocean" (V.E. Vinogradov, ed.), pp. 228–240, Nauka, Moscow. (In Russian. English translation (1973). Israel Program for Scientific Translations).

Gorbunova, N.N. (1975). Vertical distribution of fish larvae in the Eastern Equatorial Pacific Ocean. *Trudy Instituta Okeanologii* **102**, 295–312. (In Russian).

Grey, M. (1955). Notes on a collection of Bermuda deep-sea fishes. *Fieldiana, Zoology* **37**, 265–302.

Grice, G.D. (1972). The existence of a bottom-living calanoid copepod fauna in deep water with descriptions of five new species. *Crustaceana* **23**, 219–242.

Grice, G.D. and Hulsemann, K. (1965). Abundance, vertical distribution and taxonomy of calanoid Copepoda at selected stations in the Northeast Atlantic. *Journal of Zoology* **146**, 213–262.

Grice, G.D. and Hulsemann, K. (1970). New species of bottom-calanoid copepods collected in deep water by the DSRV Alvin. *Bulletin of Museum of Comparative Zoology, Harvard* **139**, 185–227.

Gruzov, L.N. (1971). Aggregation of zooplankton in the pelagic part of the Gulf of Guinea. *Trudy AtlantNIRO* **37**, 406–427. (In Russian).

Hamner, W.M. and Carleton, J.H. (1979). Copepod swarms: attributes and role in coral reef ecosystem. *Limnology and Oceanography* **24**, 1–14.

Hamner, W.M., Madin, L.P., Alldredge, A.L., Gilmer, R.W. and Hamner, P.P. (1975). Underwater observation of gelatinous zooplankton: sampling problems, feeding biology and behaviour. *Limnology and Oceanography* **20**, 907–917.

Hedgpeth, J.W. (ed.), (1957). "Treatise on Marine Ecology and Paleoecology". 1, Ecology, Memoir **67**. Geological Society of America.

Heinrich, A.K. (1961). Vertical distribution and diurnal migration of copepods south-east of Japan. *Trudy Instituta Okeanologii* **51**, 82–102. (In Russian).

Heinrich, A.K. (1993). "Comparative Ecology of the Plankton Oceanic Communities." Nauka, Moscow. (In Russian).

Heinrich, A.K., Kosobokova, K.N. and Rudjakov, Yu.A. (1980). Seasonal change in the vertical distribution of the some mass species of the Arctic Basin copepods. *In* "Biology of the Central Arctic Basin" (M.E. Vinogradov and I.A. Melnikov, eds), pp. 155–166. Nauka, Moscow. (In Russian).

Heinrich, A.K., Parin, N.V., Rudyakov, Yu.A. and Sazhin, A.F. (1993). Dwellers of the ocean near-bottom layer. *Trudy Instituta Okeanologii* **128**, 6–25. (In Russian).

Heptner, M.V. (1981). An attempt to determine vertical boundaries of the distributional areas of animals (exemplified by Copepoda, Calanoida). *Okeanologia* **21**, 1079–1083. (In Russian).

Heptner, M.V. (1996). The typology of vertical distribution of oceanic zooplankton. *Zhurnal Obsheij biologii* **57**, 44–66. (In Russian).

Heptner, M.V., Zaikin, A.N. and Rudyakov, Yu.A. (1990). Dead copepods in plankton: facts and hypotheses. *Okeanologia* **30**, 132–137. (In Russian).

Hessler, R.R., Ingram, C.L., Yayanos, A.A. and Burnet, B.R. (1978). Scavenging amphipods from the floor of the Philippine trench, *Deep-Sea Research* **25**, 1029–1047.

Hesthagen, J.H. (1973). Diurnal and seasonal variations in the near-bottom fauna – the hyperbenthos – in one of the deeper channels of the Kieler Bucht (Western Baltic). *Kieler Meeresforschungen* **29**, 116–140.

Iselin, C. (1936). A study of the circulation of the Western North Atlantic. *Papers in Physical Oceanography and Meteorology* **4**(4), 1–101.

Jannasch, H.W. (1985). The chemosynthetic support of life and the microbial diversity of deep-sea hydrothermal vents. *Proceedings of the Royal Society of London* **225**, 277–297.

Jespersen, P. (1935). Quantitative investigations on the distribution of macroplankton in the different oceanic regions. *Dana Report* **7**, 1–44.

Johnson, K.S., Beehler, C.L., Sakamoto-Arnold, C.M. and Childress, J.J. (1986). *In situ* measurements of chemical distribution at a deep-sea hydrothermal vent field. *Science* **231**, 1139–1141.

Kamenskaya, O.E. (1984). Ecological classification of deep-sea amphipods. *Trudy Instituta Okeanologii* **119**, 154–160. (In Russian).

Kils, U. (1979). Preliminary data on volume, density and cross section area of antarctic krill, *Euphausia superba*. *Bericht der Deutschen Wissenschaftlichen Kommission für Meeresforschung* **27**, 207–209.

Kos, M.S. (1969). About the discovery of some *Calanus* species in the benthos. *Zoologicheskii Zhurnal* **48**, 605–607. (In Russian).

Kosobokova, K.N. (1980). Seasonal change in the vertical distribution and age of the populations *Microcalanus pigmaeus, Oithona similis, Oncea borealis* and *O. notopus* in the Central Polar Basin. *In* "Biology of the Central Polar Basin" (M.E. Vinogradov and I.A. Melnikov, eds), pp. 167–182. Nauka, Moscow. (In Russian).

Laval, P. *et al.* (1989). Small-scale distribution of macroplankton and micronecton in the Ligurian Sea (Mediterranean Sea) as observed from the manned submersible "Cyana". *Journal of Plankton Research* **11**, 665–685.

Legand, M., Bourret, P., Fourmanior, P., Grandperrin, R., Gueredrat, I.A., Michel, A., Rangurel, P., Repelin, R. and Roger, G. (1972). Relations trophiques et distributions verticales en milieu plagique dans l'Océan Pacifique intertropical. *Cahiers ORSTOM. Serie Oceanography* **10**, 303–393.

Lein, A.Yu., Gal'chenko, V.F., Pimeniv, N.V. and Ivanov, M.V. (1993). The role of bacterial chemosynthesis and methanotrophy in ocean biogeochemistry. *Geokhimya* **2**, 252–268. (In Russian).

Lemche, H., Hansen, B., Madsen, E.I, Tendal, O.S. and Wolff, T. (1976). Hadal life as analyzed from photographs. *Videnskabelige Meddelelser fra Dansk Naturhisorisk Forening* **139**, 263–336.

Lisitzin, A.P. (1981). Impact of endogeneous matter in oceanic sedimentation. *In* "Litology: on the new stage of development of geological knowledge" (A.P. Lisitzin, ed.), pp. 20–45. Nauka, Moscow. (In Russian).

Longhurst, A.R. (1967). Vertical distribution of zooplankton in relation to the eastern Pacific oxygen minimum. *Deep-Sea Research* **14**, 51–63.

Longhurst, A.R. (1976). Vertical migration. *In* "The Ecology of the Seas" (D.H. Cushing and J.J. Walsh, eds), pp. 116–137. Blackwell, Oxford.

Longhurst, A.R. (1981). Significance of spatial variability. *In* "Analysis of Marine Ecosystems" (A.R. Longhurst, ed.), pp. 415–441. Academic Press. London.

Lozovatsky, J.D., Ozmidov, R.V. and Nihoul, C.J. (1977). Bottom turbulence in stratified enclosed seas. *In* "Bottom Turbulence" (C.J. Nihoul, ed.), pp. 49–58. Elsevier, Amsterdam.

Mackie, G.O. (1985). Midwater macroplankton of British Columbia studied by submersible "Pisces IV". *Journal of Plankton Research* **7**, 753–777.

Marshall, N.B. (1960). Swimbladder structure of deep-sea fishes in relation to their systematics and biology. *Discovery Reports* **31**, 1–121.

Marshall, N.B. (1965a). "The Life of Fishes". Weidenfeld & Nicholson, London.

Marshall, N.B. (1965b). Systematic and biological studies of the macrourid fishes (Acantini, Teleostei). *Deep-Sea Research* **12**, 299–322.

Marshall, N.B. and Merrett, N.R. (1977). The existence of a benthopelagic fauna in the deep sea. *In* "A voyage of Discovery". George Deacon 70th Anniversary Volume (M.V. Angel, ed.), pp. 483–497. Pergamon Press, Oxford.

Mukhacheva, V.A. (1964). On the genus *Cyclothone* (Gonostomidae, Pisces) of the Pacific Ocean. *Trudy Instituta Okeanologii* **73**, 93–138.

Nesis, K.N. (1973). Cephalopods of Eastern-Equatorial and South-Eastern Pacific. *Trudy Instituta Okeanologii* **94**, 187–242. (In Russian).

Omori, M. and Ikeda, T. (1984). "Methods in Marine Zooplankton Ecology". John Wiley, Chichester.

Omori, M. and Ohta, S. (1981). The use of underwater cameras in studies on vertical distribution and swimming behaviour of a sergestid shrimp *Sergia lucens*. *Journal of*

Plankton Research **3**, 107–121.

Østvedt, O.J. (1955). Zooplankton investigation from weather ship "M" in the Norwegian Sea. 1948–1949. *Hvalradets Skrifter* **40**, 1–93.

Oug, E. (1977). Faunal distribution close to the sediment of a shallow marine environment. *Sarsia* **63**, 115–121.

Parin, N.V. (1975). Change of pelagic ichthyocoenoses along the equator in the Pacific Ocean between 97 and 155°W. *Trudy Instituta Okeanologii* **102**, 313–333. (In Russian).

Parin, N.V., Becker, V.E., Borodulina, O.D., Karmovskaya, E.S., Fedoryako, B.I., Shcherbachev Yu.N., Pokhilskaya, G.N. and Tchuvasov, V.M. (1977a). Midwater fishes in the Western Tropical Pacific Ocean and the seas of the Indo-Australian Archipelago. *Trudy Instituta Okeanologii* **107**, 68–188. (In Russian).

Parin, N.V., Nesis, K.N. and Kashkin, N.I. (1977b). Vertical distribution of pelagic life. Macroplankton and necton. *In* "Okeanologia. Biology of the Ocean", Vol. 2 (M.E. Vinogradov, ed.), pp. 159–173. Nauka, Moscow. (In Russian).

Parin, N.V., Neyman, V.G. and Rudyakov, Yu.A. (1985). On the biological productivity in the vicinity of open ocean underwater rises. *In* "Biological Basis of Fisheries in the Open Ocean" (M.E. Vinogradov, ed.), pp. 192–203. Nauka, Moscow. (In Russian).

Pavshtics, E.A. (1980). Some Regularities of the Plankton Living in the Central Arctic Basin. *In* "Biology of the Central Polar Basin" (M.E. Vinogradov and I.A. Melnikov, eds), pp. 142–154. Nauka, Moscow. (In Russian).

Paxton, I.R. (1967). A distributional analysis of the lantern fishes (family Myctophidae) of the San Pedro Basin, California. *Copeia* **2**, 422–443.

Pearcy, W.G. and Laurs, R.M. (1966). Vertical distribution of mesopelagic fishes off Oregon. *Deep-Sea Research* **13**, 153–165.

Pérès, J.M. (1958a). Trois plongées dans le canyon du Cap Sicié effectués avec le bathyscaphe F.N.R.S. III de la Marine Nationale. *Bulletin de l'Institut Océanographique, Monaco* **1115**, 1–21.

Pérès, J.M. (1958b). Remarques générales sur un ensemble de quinze plongées effectuées avec le bathyscaphe F.N.R.S. III. *Annales de l'Institut Océanographique* N.S. **35**, 259–285.

Roe, H.S.J. (1986). Bathypelagic calanoid copepods from midwater trawls in the NE Atlantic. *Syllogeus* **58**, 634–635.

Roe, H.S.J. (1988). Midwater biomass profiles over the Madeira abyssal plain and the contribution of copepods. *Hydrobiologia* **167/168**, 169–181.

Roper, C.F.E. and Young, R.E. (1975). Vertical distribution of pelagic cephalopods. *Smithsonian Contributions to Zoology* **209**, 1–51.

Roth, S.E. and Dymond, J. (1989). Transport and settling of organic material in a deep-sea hydrothermal plume: evidence from particle flux measurements. *Deep-Sea Research* **36**, 1237–1254.

Rowe, G., Sibuet, M., Vangrieshem, A. (1986). Domains of occupation of abyssal scavengers inferred from baited cameras and traps on the Demerara Abyssal Plain. *Deep-Sea Research* **33**, 501–522.

Rudyakov, Yu.A. (1986). "Dynamics of Pelagic Animal Vertical Distribution". Nauka, Moscow. (In Russian).

Rudyakov, Yu.A. and Zaikin, A.N. (1990). 18th cruise of R/V "Professor Stockman" the next step in hydrobiological investigations of the ocean near-bottom layer. *Trudy Instituta Okeanologii* **124**, 5–14. (In Russian).

Rudyakov, Yu.A., Vereshchaka, A.L., Vinogradov, G.M. and Heptner, M.V. (1990). Biomass of the seston in the near-bottom layer in the south-western Indian Ocean. *Okeanologia* **30**, 114–120. (In Russian).

Sameoto, D.D. (1986). Influence of the biological and physical environment on the vertical distribution of mesoplankton and micronecton in the eastern tropical Pacific. *Marine Biology* **93**, 263–279.

Sameoto, D.D., Guglielmo, L. and Lewis, M.K. (1987). Day/night vertical distribution of euphausiids in the eastern tropical Pacific. *Marine Biology* **96**, 235–245.

Shulenberger, E. and Hessler, R.R. (1974). Scavenging abyssal benthic amphipods trapped under oligotrophic central North Pacific gyre waters. *Marine Biology* **23**, 185–187.

Shushkina, E.A. and Vinogradov, M.E. (1992). Vertical distribution of zooplankton in the Guaymas basin (Gulf of California). *Okeanologia* **32**, 881–887. (In Russian).

Shushkina, E.A., Vinogradov, M.E., Glebov, B.S. and Lebedeva, L.P. (1980). The use of 100-l water bottles for collecting micro- and mesoplankton. *Okeanologia* **20**, 552–557. (In Russian).

Shushkina, E.A., Tchindonova, Yu.G., Vinogradov, M.E. and Sagalevich, A.M. (1991). Investigations on oceanic zooplankton in the Kurile-Kamchatka region studied by the deep manned submersible "Mir" *Okeanologia* **31**, 609–615. (In Russian).

Shushkina, E.A., Vinogradov, M.E., Sheberstov, S.V., Nezlin, N.P. and Gagarin V.I. (1995). The characteristics of epipelagic ecosystems of the Pacific Ocean based on both satellite and field observation. The stock of plankton in the epipelagial. *Okeanologia* **35**, 705–712. (In Russian).

Smith, K.L. Jr. (1985). Macrozooplankton of a deep-sea hydrothermal vent: *in situ* rates of oxygen consumption. *Limnology and Oceanography* **30**, 102–110.

Smith, K.L., Carlucci, A.F., Williams, S.M., Hendrichs, S.M., Baldwin, R.J. and Graven, D.B. (1986). Zooplankton and bacterioplankton of an abyssal benthic boundary layer: *in situ* rates of metabolism. *Oceanologica Acta* **9**, 47–55.

Stockton, W.L. (1982). Scavenging amphipods from under the Ross Ice Shelf, Antarctica. *Deep-Sea Research* **29A**, 819–835.

Sömme, J.D. (1934). Animal plankton of the Norwegian coast water and open sea. I. Production of *Calanus finmarchicus* and *C. hyperboreus* in the Lofoten area. *Fiskeridirektoratets Skrifter, Serie Havundersøkelser Bergen* **4**(9), 163 pp.

Thiriot, A. (1978). Zooplankton communities in the West African upwelling area. *In* "Upwelling Ecosystems" (R. Boje and M. Tomczak, eds), pp. 32–60. Springer, Berlin.

Thomson, R.E., Gordon, R.L. and Dolling, A.G. (1991). An intense acoustic scattering layer at the top of a mid-ocean ridge hydrothermal plume. *Journal of Geophysical Research* **96**, C3, 4839–4844.

Thomson, R.E., Burd, B.J., Dolling, A.G., Gordon, L.R. and Jamieson, G.S. (1992). The deep scattering layer associated with the Endeavour Ridge hydrothermal plume. *Deep Sea Research* **39**, 55–73.

Thurston, M.II. (1979). Scavenging abyssal amphipods from the north-east Atlantic Ocean. *Marine Biology* **51**, 55–68.

Timonin, A.G. (1975). Vertical microdistribution of zooplankton in the tropical western Pacific. *Trudy Instituta Okeanologii* **102**, 245–259. (In Russian).

Timonin, A.G. (1991). Zooplankton distribution in relation to the hydrological conditions in the Benguela upwelling zone. *Okeanologia* **31**, 265–271. (In Russian).

Timonin, A.G. (1992). Zooplankton and environmental variability in the northern Benguela upwelling area. *Russian Journal of Aquatic Ecology* **1**, 103–113.

Timonin, A.G. and Flint, M.V. (1984). The spatial distribution of the net mesoplankton near the north-Peruvian shore at the 8°S section. *In* "Frontal Zone of the South-east Part of the Pacific (Biology, Physics, Chemistry)" (M.E. Vinogradov and K.N. Fedorov, eds), pp. 219–231, Nauka, Moscow. (In Russian).

Timonin, A.G. and Flint, M.V. (1985). The peculiarities of the mesoplankton structure in the Peruvian region. *In* "Biological Basis of the Commercial Effort in the Open Ocean" (M. E. Vinogradov and M.V. Flint, eds), pp. 155–165. Nauka, Moscow. (In Russian).

Timonin, A.G. and Voronina, N.M. (1975). Distribution of net zooplankton along the equator. *Trudy Instituta Okeanologii* **102**, 213–231. (In Russian).

Timonin, A.G., Arashkevich, E.G., Drits, A.V. and Semenova, T.N. (1992). Zooplankton dynamics in the northern Benguela ecosystem, with special reference to the copepod *Calanoides carinatus. South African Journal of Marine Science* **12**, 545–560.

Tutubalin, V.N., Uger, E.G., Vinogradov, M.E., Flint, M.V. and Shushkina, E.F. (1987). Statistical model for comparison of mesoplankton numbers estimated from collections with water bottles and with plankton nets. *Okeanologia* **27**, 507–512. (In Russian).

UNESCO (1968) "Zooplankton sampling". *Monographs on Oceanographic Methodology* **2**, UNESCO, Geneva, 174 pp.

Vereshchaka, A.L. (1990). Vertical distribution of euphausiids, pelagic decapods, and mysids in the near-bottom layer of the western Indian Ocean. *Okeanologia* **30**, 126–131. (In Russian).

Vereshchaka, A.L. (1994). Distribution of pelagic macroplankton (mysids, euphausiids, decapods) over continental slopes and seamounts of the western Indian Ocean. *Okeanologia* **34**, 88–94. (In Russian).

Vereshchaka, A.L. (1995). Macroplankton in the near-bottom layer of continental slopes and seamounts. *Deep-Sea Research* **42**, 1639–1668.

Vereshchaka, A.L. and Vinogradov, G.M. (1996). The plankton and its distribution in the near-bottom layer. *In* "Oceanographic research and Underwater Technical Operations on the site of the Nuclear Submarine *Komsomolets* wreck" (M.E. Vinogradov, A.M. Sagalevich and S.V. Chaetagurov, eds) pp. 179–184, Nauka, Moscow. (In Russian).

Vinogradov, G.M. (1990a). Amphipods in the near-bottom layer in the south-western part of the Indian Ocean. *Okeanologia* **30**, 121–125. (In Russian).

Vinogradov, G.M. (1990b). Amphipods (Amphipoda, Crustacea) in the pelagic zone of the south-eastern part of the Pacific Ocean. *Trudy Instituta Okeanologii* **124**, 27–104. (In Russian).

Vinogradov, G.M. (1992). The probable ways of the gammarids (Amphipoda, Crustacea) invasion in the pelagic zone: analysis of the life forms. *Journal Obschey Biologii* **53**, 328–339. (In Russian).

Vinogradov, G.M. (1995). Colonization of pelagic and hydrothermal vent habitats by gammaridean amphipods: an attempt at reconstruction. *Polskie Archivum Hydrobiologii* **42**, 417–430.

Vinogradov, M.E. (1968). "Vertical Distribution of the Oceanic Zooplankton". Nauka, Moscow. (In Russian). Translation (1970), Israel Program of Scientific Translation, Jerusalem. 339 pp. Publication of US Department of the Interior and National Science Foundation, Washington DC.

Vinogradov, M.E. (1970). Some peculiarities of the change in ocean pelagic communities with change of depth. *In* "Program and Methods of Investigation of Biogeooceanological Aquatic Environments" (L.A. Zenkevich, ed.), pp. 84–96, Nauka, Moscow. (In Russian).

Vinogradov, M.E. (1972). Vertical stratification of zooplankton in the Kurile-Kamchatka trench. *In* "Biological Oceanography of the Northern North Pacific Ocean" (ed. A.Y. Takenouti), pp. 333–340. Idemitus Shoten, Tokyo.

Vinogradov, M.E. (1977a). Vertical distribution of zooplankton. *In* "Oceanology. Biology of the Ocean" pt **1** (M.E. Vinogradov ed.), pp. 132–151. Nauka, Moscow. (In Russian).

Vinogradov, M.E. (1977b). A spatial-dynamic aspect of the existence of pelagic communities. *In* "Oceanology. Biology of the Ocean", Part **2** (M.E. Vinogradov, ed.), pp. 14–23. Nauka, Moscow. (In Russian).

Vinogradov, M.E. (1983a). Open-ocean ecosystems. *In* "Marine Ecology", Vol. 5, pt 2 (O. Kinne, ed.), pp. 657–737. Wiley, Chichester.

Vinogradov, M.E. (ed.), (1983b) "Modern Methods of Assessing the Quantitative Distribution of the Marine Plankton". (M.E. Vinogradov, ed.). Nauka, Moscow. (In Russian).

Vinogradov, M.E. and Årashkevich, E.G. (1969). The vertical distribution of interzonal copepod filter-feeders and their role in communities at different depths in the north-western Pacific. *Okeanologia* **9**, 488–499. (In Russian).

Vinogradov, M.E. and Musaeva, E.I. (1989). Peculiarities of plankton distribution in oligotrophic tropical ocean regions. *Okeanologia* **29**, 494–501. (In Russian).

Vinogradov, M.E. and Naumov A.G. (1961). Quantitative distribution of the zooplankton in the Indian and Pacific Oceans' Antarctic waters. *Okeanologicheskie issledovaniya* **3**, 172–176. (In Russian).

Vinogradov, M.E. and Parin, N.V. (1973). Some features of the vertical distribution of macroplankton in the tropical Pacific. *Okeanologia* **13**, 137–148. (In Russian).

Vinogradov, M.E. and Sagalevich, A.M. (1994). A study of the vertical distribution of pelagic and bottom fauna of the northern Pacific with the use of the deep manned submersibles "Mir-1" and "Mir-2". *Trudy Instituta Okeanologii* **131**, 6–15. (In Russian).

Vinogradov, M.E. and Shushkina, E.A. (1976). Some characteristics of the vertical structure of a planktonic community in the equatorial Pacific upwelling region. *Okeanologia* **16**, 677–684. (In Russian).

Vinogradov, M.E. and Shushkina, E.A. (1982). Estimation of the concentration of jellyfish, comb-jellies and *Calanus* in the Black Sea as observed from the manned submersible "Argus". *Okeanologia* **22**, 473–479. (In Russian).

Vinogradov, M.E. and Shushkina, E.A. (1984). Succession of marine epipelagic communities. *Marine Ecology Progress Series* **16**, 229–239.

Vinogradov, M.E. and Shushkina, E.A. (1987). "Functioning of the Plankton Communities in the Oceanic Epipelagial". Nauka, Moscow. (In Russian).

Vinogradov, M.E. and Shushkina, E.A. (1992). Characteristics of meso- and macro-plankton vertical distribution in the central tropical areas of the Northern Pacific ocean. *Okeanologia* **32**, 115–127 (In Russian).

Vinogradov, M.E. and Shushkina, E.A. (1994). A study of vertical distribution of the North-Pacific zooplankton based on quantitative estimations from the deep manned submersible (DMS) "Mir". *Trudy Instituta Okeanologii* **131**, 41–63. (In Russian).

Vinogradov, M.E. and Tchindonova, Yu.G. (1994). Notes on vertical zonation of the pelagic fauna based on direct observation from DSM "Mir". *Trudy Instituta Okeanologii* **131**, 64–75. (In Russian).

Vinogradov, M.E. and Tseitlin, V.B. (1983). Deep-sea pelagic domain (aspects of bioenergetics). *In* " The Sea", Vol. **8** (G.T. Rowe, ed.), pp. 123–165. Wiley, Chichester.

Vinogradov, M.E. and Vinogradov, G.M. (1991). Scavenging amphipods from a bottom-trap set on the Nazca underwater mountain ridge. *Zoologicheskii Zhurnal* **70**, 32–38. (In Russian).

Vinogradov, M.E. and Vinogradov, G.M. (1993). Notes on pelagic and benthopelagic gammarids in the Orkney Trench. *Trudy Instituta Okeanologii* **127**, 129–133. (In Russian).

Vinogradov, M.E. and Vinogradov, G.M. (1996). The discovery of a hadal pacific gammarid *Vitjaziana gurianovae* (Crustacea, Amphipoda) in the Atlantic Ocean and the problem of endemism in abyssopelagial animals. *Zoologicheskii Zhurnal* **75**, 45–51. (In Russian).

Vinogradov, M.E. and Voronina, N.M. (1964). Quantitative distribution of plankton in the upper layers of the Pacific equatorial currents II. Vertical distribution of some species. *Trudy Instituta Okeanologii* **65**, 58–76. (In Russian).

Vinogradov, M.E., Shushkina, E.A., Musaeva, E.I. and Akimova, A.F. (1984). Some characteristics of the vertical distribution of the mesoplankton and the biomass of its trophic groups in the epipelagic tropical regions of the Ocean. *In* "Frontal Zones in the South-east Part of the Pacific Ocean (Biology, Physics, Chemistry) (M.E. Vinogradov and K.N. Fedorov, eds), pp. 180–192. Nauka, Moscow. (In Russian).

Vinogradov, M.E., Flint, M.V., Shushkina, E.A., Tutubalin, V.N. and Uger, E.G. (1987). On the comparative catching ability of big volume water bottles and plankton nets for vertical hauls. *Okeanologia* **27**, 329–337. (In Russian).

Vinogradov, M.E., Musaeva, E.I. and Semenova, T.N. (1990a). Factors determining the position of the lower layer of mesoplankton concentration in the Black Sea. *Okeanologia* **30**, 295–305. (In Russian).

Vinogradov, M.E., Shushkina, E.A., Musaeva, E.I. and Nikolaeva G.G. (1990b). The distribution of the mesoplankton biomass in the eastern boundary current waters in the regions of the Peruvian and Californian upwellings. *In* "Ecosystems of the Eastern Boundary Currents and Central Regions of the Pacific" (M.E. Vinogradov and E.I. Musaeva, eds), pp. 191–213. Nauka, Moscow. (In Russian).

Vinogradov, M.E., Nikolaeva, G.G. and Musaeva, E.I. (1991a). Vertical distribution of the plankton in the open regions of the Black Sea (March–April, 1988). *In* "Variability of the Black Sea Ecosystem: Natural and Anthropogenic Influence" (M.E. Vinogradov, ed.), pp. 211–224. Nauka, Moscow. (In Russian).

Vinogradov, M.E., Shushkina, E.A., Gorbunov, A.E. and Shashkov N.L. (1991b). Vertical distribution of meso- and macroplankton in the Costa-Rica Dome region. *Okeanologia* **31**, 759–769. (In Russian).

Vinogradov, M.E., Arashkevich, E.G. and Ilchenko, S.V. (1992a). The ecology of the *Calanus ponticus* population in the deeper layer of its concentration in the Black Sea. *Journal of Plankton Research* **14**, 447–458.

Vinogradov, M.E., Sapozhnikov, V.V. and Shushkina, E.A. (1992b). "The Black Sea Ecosystem". Nauka, Moscow. (In Russian).

Vinogradov, M.E., Shushkina, E.A., Musaeva, E.I. and Nikolaeva, G.G. (1992c). Vertical distribution of the Black Sea mesozooplankton in winter 1991. *In* "Ecosystem of the Open Black Sea in Winter" (M.E. Vinogradov, ed.), pp. 103–119. P.P. Shirshov Institut Okeanologii, Moscow. (In Russian).

Vinogradov, M.E., Musaeva, E.I., Nikolaeva, G.G. and Choroshilov, D.S. (1993). Characteristics of the vertical distribution of the North Atlantic mesoplankton depending on the productivity of the waters. *Okeanologia* **33**, 711–716. (In Russian).

Vinogradov, M.E., Shushkina, E.A., Vereshchaka, A.L. and Nezlin, N.P. (1995). The plankton community in the Norwegian Sea at the position of the sunken nuclear submarine "Komsomolets". *Isvestiya RAS, ser. biologicheskaya* **5**, 612–623. (In Russian).

Vinogradov, M.E., Shushkina, E.A., Vereshchaka, A.L. and Nezlin, N.P. (1996a). On migrations of *Calanus finmarchicus* s. l. during the polar day in the Norwegian Sea. *Okeanologia* **36**, 66–70. (In Russian).

Vinogradov, M.E., Vereshchaka, A.L. and Shushkina, E.A. (1996b). Vertical structure of the zooplankton communities in the oligotrophic areas of the North Atlantic and influence of the hydrothermal vents. *Okeanologia* **36**, 71–79. (In Russian).

Vinogradova, N.G., Birstein, Ja.A. and Vinogradov, M.E. (1959). Vertical zonation in the distribution of the deep-sea fauna. *In* "Itogi Nauki. Dostizchenia okeanologii", Vol. **1** (L.A. Zenkevich, ed.), pp. 166–187, Isdatelstvo AN SSSR, Moscow. (In Russian).

Voronina, N.M. (1972a). Vertical structure of a pelagic community in Antarctica. *Okeanologia* **12**, 492–498. (In Russian).

Voronina, N.M. (1972b). The spatial structure of interzonal copepod populations in the Southern Ocean. *Marine Biology* **15**, 336–343.

Voronina, N.M. (1974). Vertical distribution of Antarctic copepods *Calanus propinquus* and *Calanus acutus*. *Trudy Vsesojusnogo gidrobiologicheskogo obstchestva* **20**, 246–249. (In Russian).

Voronina, N.M. (1976). Variability of ecosystems. *In* "Advances in Oceanography" (H. Charnock and G. Deacon, eds), pp. 221–244. Plenum Press, London.

Voronina, N.M. (1984). "Pelagic Ecosystems of the Southern Ocean". Nauka, Moscow. (In Russian).

Wiborg, K.F. (1954). Investigations on zooplankton in coastal and offshore waters of western and northwestern Norway. *Fiskeridirektoratets Skrifter, Serie Havundersøkelser* **11**, 246 pp.

Wiebe, P.H., Copley, N., Van Dover, C., Tamse, A. and Manrique, F. (1988). Deep-water zooplankton of the Guaymas Basin hydrothermal vent field. *Deep-Sea Research* **35**, 985–1013.

Winn, C.D., Karl, D.M. and Massoth, G.J. (1986) Microorganisms in deep-sea hydrothermal plumes. *Nature* **320**, 704–706.

Wishner, K.F. (1980a). The biomass of the deep-sea benthopelagic plankton. *Deep-Sea Research,* **27**, 203–216.

Wishner, K.F. (1980b). Aspects of the community ecology of deep-sea benthopelagic plankton with special attention to gymnopleid copepods. *Marine Biology* **60**, 179–187.

Young, R.E. (1975). *Leachia pacifica* (Cephalopoda, Teuthioidea); spawning habitat and function of the brachial photophores. *Pacific Science* **29**, 19–25.

Youngbluth, M.I. (1984). Water column ecology: *in situ* observation of marine zooplankton from a manned submersible. *Memorial University of Newfoundland Occasional Papers, Biology* **9**, 45–57.

Ecology and Biogeography of the Hydrothermal Vent Fauna of the Mid-Atlantic Ridge

A.V. Gebruk[1], S.V. Galkin,[1] A.L. Vereshchaka,[1] L.I. Moskalev[1] and A.J. Southward[2]

[1]*P.P. Shirshov Institute of Oceanology, Russian Academy of Sciences, Nakhimovsky Prospekt 36, Moscow 117851, Russia*
[2]*Marine Biological Association of the UK, Citadel Hill, Plymouth PL1 2PB, UK*

ADVANCES IN MARINE BIOLOGY VOL. 32
ISBN 0-12-026132-4

ABSTRACT

Six sites of hydrothermal activity with associated specialized fauna have been studied, lying between 14°45' and 37°51'N along the Mid-Atlantic Ridge (MAR). To date about 100 species have been recorded, but the number increases with each expedition. There are common patterns in the distribution of the animals. The community organization usually changes with distance from the centre of venting, and two to three zones can be distinguished, related to environmental factors and trophic relations. The Atlantic vent communities differ considerably from those in the Pacific, notably in the absence of vestimentiferan tube worms and in the abundance of shrimp, but there is a common three-level trophic structure.

The dominant animals are bresilioid shrimps; six species belonging to five genera from two families have been described so far. *Rimicaris exoculata* is one of the most abundant shrimps and the adult stage appears to depend on exosymbiosis with sulphur bacteria. It is not clear at what stage exosymbiosis is established and how the specific nature of the relationship is maintained. The other species show a graded series of morphological and feeding adaptations, indicating a less specialized diet. More studies are needed on the life history, dispersal and microscale distribution of the shrimps, and of other vent-adapted animals at MAR.

There are some similarities at the genus level between the hydrothermal vent and the cold seep faunas, indicating historic links, but this topic requires further study. There may also be historic links between vent faunas of the West and East Pacific and the Atlantic. Further studies are needed to assess the importance of prokaryotic/eukaryotic symbiosis, including the role of exo- versus endosymbiotic bacteria, in structuring the vent communities in the different oceans.

1. INTRODUCTION

During 10 years of exploration of the Mid-Atlantic Ridge (MAR), six deep-sea sites of hydrothermal activity with a characteristic fauna have been discovered, lying between 14° and 38°N (Table 1). These sites have been studied by American, French, Russian, British and Japanese expeditions, with the greatest activity shown in the period 1992–1995. Some sites have been studied more than others, and the TAG site seems to be relatively the best known. Snake Pit, Broken Spur and Logatchev (better known as the "14–45" site) have been visited several times, although the number of submersible dives at each of them is limited. The two shallower sites, Lucky Strike (1620–1730 m) and Menez Gwen (855 m) are at present the least known.

While a large amount of data has been obtained from the expeditions, our knowledge of the biology of hydrothermal systems is still at an early stage and there are more unsolved problems than answers. Some data exist on community composition, distribution, nutrition, physiological adaptation and microbiological processes. Many of the published data from US, French and British sources were analysed by Van Dover (1995). The major unsolved problems are: the role of invertebrate/bacteria symbioses in the trophic structure of the communities; the influence of depth on the community structure; reproduction and life history; dispersal of larvae and colonization; and gene flow between remote sites along the ridge. In addition, the rates of many of the biological processes remain unknown, especially bacterial primary production, while we can only speculate about the history of the MAR vent fauna in general.

This contribution utilizes results from Russian researches that have not been mentioned in previous reviews of the MAR fauna. Expeditions conducted by the P.P. Shirshov Institute of Oceanology (Moscow) and "Sevmorgeologiya" (St Petersburg) from 1988 to 1995 visited four out of six of the MAR deep-sea sites, the exceptions being the two northernmost, Lucky Strike and Menez Gwen. At Broken Spur, TAG and 14–45 samples of the hydrothermal fauna were collected during dives by Mir-1 and Mir-2 (RV "Akademik Mstislav Keldysh" cruises of 1988, 1991, 1994, 1995 and 1996). At 14–45 the fauna was collected with dredge and a monitored hydraulic grab (RV "Professor Logachev" cruises of 1994 – one sample, and 1995 – 11 samples). Table 1 gives details of the six sites.

We first give a general description of the Atlantic vent fauna known from the explored fields. Then the biology and taxonomy of the dominant faunal element, the bresilioid shrimps, are discussed. The final section reviews the peculiarities of the MAR hydrothermal fauna, compares it with that of other oceans, and notes the influence of trophic, tectonic and historical factors in structuring the hydrothermal communities.

2. ECOLOGY AND STRUCTURE OF THE COMMUNITIES

The MAR hydrothermal ecosystems are described here by means of a conceptual model termed the "landscape approach" in which the visual image of the system is given a major role. This method is based on recognition of the characteristic (i.e. "landscape-determining") animal groups, evaluation of their abundance, biomass, microscale distribution, level of aggregation, correlation of distribution with geomorphological structures, currents, temperature gradient, type of substrata, etc. As a result different faunistic associations can be distinguished related to certain biotopes. Schematic landscape reconstruction presents a visual model of the ecosystem that is more informative when the relief is complicated. The landscape approach was used in particular by T.A. Stephenson, who pointed out that on the broken facies typical of most rocky shores the human eye is better at discerning patterns of distribution and zonation than careful instrumental surveys and counts

Table 1 The known and studied hydrothermal systems on the mid-Atlantic Ridge. Primarily based on Rona and Scott (1993) and Van Dover (1995), with details added from the references quoted, and incorporating new data.

	Logatchev ("14-45")	Snake Pit (MARK)	TAG	Broken Spur	Lucky Strike	Menez Gwen
Coordinates N W	14°45' 44°58'	23°22' 44°57'	26°08' 44°49'	29°10' 43°10'	37°18' 32°16'	37°50' 31°31'
Depth, m	2930–3010	3420–3480	3625–3670	3050–3110	1620–1730	855
Approximate area, m^2	20 000	21 000	100 000	6000	150 000	200
Geological setting and hydrothermal activity	Hydrothermal mound, most of it inactive; 3 active hydrothermal fields with 3–4 black smokers each	4 active hydrothermal mounds with black smokers on top	Active mound with black smokers on top and different types of vents on slopes	Hydrothermal field with 5 black smokers and 7 inactive structures	7 hydrothermal fields with vents of different type, including black smokers	Hydrothermal field with smokers discharging transparent fluid
Dominant mega-fauna ("landscape-determining forms")	Mussel aggregations, also on inactive smokers; shrimps not abundant	Shrimp swarms; mussels present on two mounds	Shrimp swarms; no mussels	Shrimps abundant only on the BX16 structure; mussels are present	Aggregations of mussels; shrimps not abundant	Aggregations of mussels; shrimps not abundant
Recent expeditions reporting on benthic communities	"Professor Logachev", cruise 7 (1993–1994) – 1 monitored grab; cruise 10 (1995) – 10 grab and 1 dredge; "Akademik Mstislav Keldysh" (AMK), cruise 35 (1995) – 2 "Mir" dives; "Nadir" (1995) – 4 "Nautile" dives	"Professor Logachev", cruise 3/2 (1991), (photo transect, c. 3000 photos)	AMK, cruise 23 (1991) – 6 "Mir" dives; cruise 34 (1994, BRAVEX) – 4 dives	AMK, cruise 34 (1994, BRAVEX) – 8 "Mir" dives; cruise 39 (1996) – 6 "Mir" dives	"Nadir", 1994 (DIVA 1) – 12 "Nautile" dives; 1994 (DIVA 2) – 19 "Nautile" dives	"Nadir", 1994 (DIVA 1) – 2 Nautile dives; 1994 (DIVA 2) – 7 "Nautile" dives
Publications	Batuyev et al., 1994; Sudarikov & Galkin, 1994; Sudarikov & Galkin, 1995; Bogdanov et al., 1995a,b; Sagalevitch & Bogdanov, 1995; Krasnov et al., 1995, 1996	Sudarikov & Galkin, 1994; Sudarikov & Galkin, 1995	Lisitsyn et al., 1989, 1990; Zonenshain et al., 1989; Humphris et al., 1995	Nesbitt, 1995; Copley & Tyler, 1995; Gebruk et al., 1997	Fouquet et al., 1994; Desbruyères et al., 1994; Saldanha et al., 1996; Child & Segonzac, 1996; Van Dover et al., 1996	Fouquet et al., 1994; Desbruyères et al., 1994; Saldanha et al., 1996; Saldanha, 1996; Van Dover et al., 1996

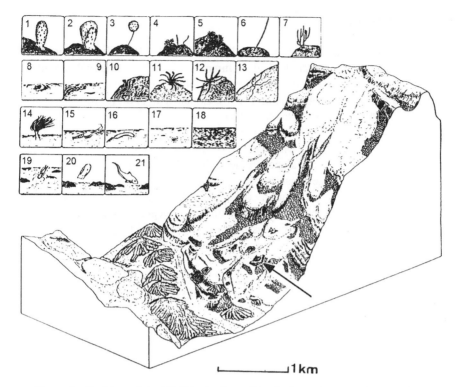

Figure 1 Position of the TAG hydrothermal mound on the eastern slope of the rift valley (arrow) and main groups of non-vent benthic megafauna (after Galkin and Moskalev, 1990a). 1 and 2, Pheronematidae (Hyalospongia); 3, *Hyalonema* sp. (Hyalonematidae, Hyalospongia); 4, Hydrozoa (several species); 5, Actiniaria, sp.1; 6 and 7, Gorgonaria; 8, Serpulidae (Polychaeta); 9, Crinoidea; 10, Brisingidae (Asteroidea); 11, Ophiuroidea; 12, Actiniaria, sp.2; 13, Decapoda, Macrura, Natantia; 14, Holothuroidea; 15–18, lebensspuren; 19, Galatheidae; 20, *Enypniastes eximia* (Pelagothuriidae); 21, *Coryphaenoides* sp. (Macrouridae). Groups 1–11 are inhabitants of basalt facies; 12–16 are inhabitants of sediment; and 19–21 are mobile and benthopelagic forms.

(Stephenson and Stephenson, 1972, p.16). In hydrothermal systems the relief is usually so complicated that landscape reconstructions often become the only means to describe a microscale distribution of organisms. In the Russian literature the landscape approach has been widely developed and was applied to aquatic ecosystems by Petrov (1971) and more recently by Preobrazhenskii (1980, 1984), Fedorov (1982), Arzamastsev and Preobrazhenskii (1990). It was used in studies of hydrothermal ecosystems by Fustec *et al.* (1987), Galkin and Moskalev (1990a,b) and Galkin (1993). For the present review we describe the appearance of the communities at each site separately, then discuss some of the general regularities of distribution of the vent fauna.

2.1. The TAG Site

The TAG site is one of the best studied among MAR hydrothermal faunal
communities. The geomorphological setting of the TAG hydrothermal mound and
characteristics of the non-vent (background) fauna were considered in part by
Rona et al. (1986), Thompson et al. (1988), Zonenshain et al. (1989), Galkin and
Moskalev (1990b) and Van Dover (1995).

The background benthic community at TAG was studied in detail during the
15th cruise of "Akademik Mstislav Keldysh" in 1988. In general terms, the
landscape of rift valley and slopes in the area of 26°N could be characterized as
a series of facies of broken basalts interchanged with pelagic sediments, occupied
by a clearly oligotrophic community (Figure 1). The dominant feeding strategy of
this background fauna seems to be passive filtration. Deposit-feeders and predators
depend on exploration of vast areas of the sea floor and they are rare. Bottom
currents are weak, food for benthic organisms is poor and evenly distributed
(Galkin and Moskalev, 1990a).

Analysis of the spatial distribution of the hydrothermal fauna on TAG is based
on data from "Akademik Mstislav Keldysh" cruises of 1988, 1991 and 1994. The
taxonomic groups are considered in the order in which they usually stand in
zoological lists.

Hexactinellid sponges, although common in the background community, were
not reported from the vent field, although Segonzac (1992) indicates *Asbestopluma*
sp. for the zone of diffuse emanations at the Snake Pit site. Sponges, gorgonians
and other suspension-feeders do not form any significant aggregations on the
periphery of the hydrothermal field, possibly owing to low productivity of
surrounding waters (see Galkin, 1992 for comparison with vent communities in
the Pacific). One of the dominant groups on the vent field is sea-anemones
(Actiniaria), two forms of which are present. The larger forms are white and up
to 10 cm in diameter (*Parasicyonis* sp.?). They form dense aggregations of up to
100 individuals m^{-2} and occupy considerable space in the zone of oxidized
structures and hydrothermal deposits. The distribution of the anemones is
obviously controlled by local hydrodynamics, and is especially clear in the
"Kremlin" and "shimmering-water" zones. Smaller forms, $c.1$ cm in diameter,
reach densities of 20–40 individuals m^{-2}. The taxonomy of this group is poorly
known and the small form, originally referred to Actiniaria, appears to be a
zoanthid.

Nematodes are extremely abundant in slurp-gun samples and may reach a
density of 400 individuals dm^{-3} Apparently, they play an important role in the
decomposition of organic matter in this community. Polychaetes of the family
Chaetopteridae often form dense populations on the surface of oxidized sulphides
away from black smokers. Live specimens are rare, although present in the
collections, and the species has never been identified, partly because of difficulties
of preserving this group. Polynoids (*Branchipolynoe* ?) were observed in the areas

of shrimp aggregations and clearly correlate with them. At least two species of limpets are present, usually in the zone of hydrothermal deposits, more rare in the zone of shrimp swarms. On the periphery of the community the gastropod *Phymorynchus moskalevi* is common, reaching a density of 20 individuals m^{-2} but also occurring singly. This species occurs in the areas of flatter bottom covered with loose deposits. A single specimen of a harpacticoid copepod was found in a slurp-gun sample from the active field. Shrimps (Alvinocarididae) are amongst the most abundant animals at black smokers and in areas of emission of heated water without entrained mineral particles (shimmering water). Their density has been estimated to reach 2500 individuals m^{-2} with a wet biomass of 1–1.5 kg m^{-2} (Segonzac, 1992; personal observations of L. Moskalev, S. Galkin and Y. Bogdanov). *Chorocaris chacei* is less abundant and often occurs away from active smokers, reaching a density of 50 individuals m^{-2} on the shimmering water zone. *Alvinocaris markensis* occurs in small groups or singly at the periphery of the field and may be caught in baited traps. The spatial distribution of different shrimp species correlates with their feeding strategies (see p. 111). Crabs, *Segonzacia* sp., often occur both in active and inactive zones of the hydrothermal mound reaching a density of 10 individuals m^{-2} in active zones. *Munidopsis* sp. (Anomura, Galatheidae) occur on the periphery of the field and at the base of the chimneys. These scavengers are also found in the background community and are easily identified on the black-and-white deep-tow photographs. Ophiuroids, *Ophioctenella acies*, are common on inactive parts of the mound and reach a density of 30 individuals m^{-2} (Tyler *et al.*, 1995, indicate up to 10 individuals dm^{-2}).

Despite the diverse topography of the TAG hydrothermal field (sulphide, anhydrite and beehive structures, "kremlins" and soft sediments), only two major faunistic associations could be distinguished, zoned around black smokers (Figure 2). One is marked by anemones and gastropods and the other by dense aggregations of shrimps *Rimicaris exoculata* and *Iorania concordia*. The two zones differ in composition of the macrofauna and infauna. Zones of shrimp swarms lack most sedentary organisms and are depleted in interstitial forms. The two associations correlate with the distribution of oxidized and unoxidized ore deposits. Oxidized sulphides, blocks and loose deposits are a typical substratum for the first type of association. Shrimp biotopes include shimmering water, black smokers, chimneys and "kremlins". The redox potential of the environment is presumably among the major factors limiting the distribution of animals. Mobile crustaceans are able to cross the border between the reduced and oxidized zones, which is a barrier for sedentary and low activity organisms. On the other hand, the distribution of shrimps close to black smokers is limited by temperature, 35–40°C being an upper limit for most metazoans. Outlets of smokers and beehive structures lack metazoan life though they are sometimes covered with bacterial mats. In addition to the central black smoker complex on the TAG hydrothermal mound, numerous hot vents occur on the mound slopes which complicate the zonation of the fauna and qualitative estimates. The active zones of the

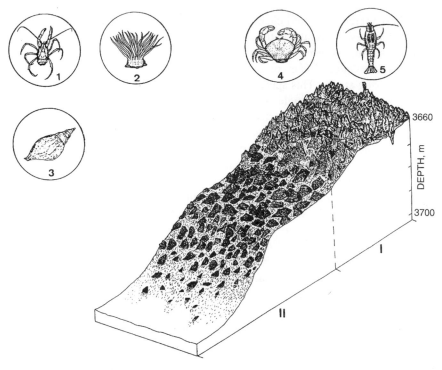

Figure 2   Prevailing substrata and typical representatives of fauna of the outer (1–3) and the inner (4–5) parts of the TAG hydrothermal field (transect length *c*. 100 m). I, zone of "kremlins", anhydrites, shimmering water and black smokers; II, zone of ochres, oxidized sulphides and inactive structures. 1, *Munidopsis* sp.; 2, *Parasycionis* sp.; 3, *Phymorhynchus moskalevi*; 4, *Segonzacia mesatlantica*; 5, Alvinocarididae.

hydrothermal mound, which is the potential shrimp biotope, comprise about 10% of the mound area. The maximum biomass of shrimps is estimated as 1–1.5 kg m^{-2} in swarms and up to 0.5 kg m^{-2} on the fields of sea-anemones.

2.2. Broken Spur

The Broken Spur site, discovered in 1993 (Murton *et al*., 1994) at 3080–3110 m depth, is a hydrothermal field of *c*. 6000 m^{2}, with five active smokers and seven non-active structures of various height from 5 to 40 m (Murton *et al*., 1995). The Broken Spur non-vent fauna has been discussed by Copley and Tyler (1995). This site was also studied in the 34th cruise of the "Akademik Mstislav Keldysh", during the joint Russian–British expedition BRAVEX. In total about 20 species

Figure 3 Distribution of fauna on the hydrothermal structure BX-16 (at Broken Spur) (based on "Mir-2" dive at St. 3425). Structural elements: I, mound about 15 m high; II, base of the structure; III, sulphide chimney. Hydrothermal vent types: A, diffuse venting and shimmering water; B, black smoker. Dominant fauna: 1, *Rimicaris exoculata* and *Mirocaris fortunata*; 2, *Chorocaris chacei*; 3, *Segonzacia mesatlantica*; 4, *Ophioctenella acies*; 5, Nematoda; 6, limpets (Lepetodrilacea); 7, *Parasicyonis ingolfi*. Scale bar – 1 m.

of benthic invertebrates have been found at this site so far. Each hydrothermal structure at Broken Spur seems to have a specific fauna with its own zonation. The distribution of the fauna was studied in most detail on the structure BX 16 (Mir-2, St. 3425, Gebruk *et al.*, 1997) (Figure 3). Community composition and structure have much in common with those of the TAG site although the total biomass is about one order of magnitude lower. The fauna is most abundant at the base of the structures, where shimmering water is discharged through crevices. The upper part of BX 16 is made up of cemented rocks and is devoid of life, and has outlets of black smokers and beehive structures. At the base of the structure chaetopterid polychaetes and white anemones are common. They co-occur with *Alvinocaris markensis* (?) and *Ophioctenella acies*, which are also found in the shimmering-water zone. The abundance of *Rimicaris exoculata* and *Iorania concordia* is lower at the base of the structure than on the slope, whereas *Chorocaris chacei* is more common at the base. Depressions of the substratum are filled with loose sulphide deposits inhabited by infauna similar to those at TAG. Harpacticoids are remarkably abundant – more than 50 individuals in one slurp-gun sample. Among the large forms, the gastropod *Phymorhynchus* sp. is common. In general, the population at the base of hydrothermal structures and the nearest surrounding area is similar to the faunistic association of oxidized sulphide deposits on the TAG mound. The decapods *Segonzacia* and *Munidopsis* are abundant in all biotopes at Broken Spur, as at TAG. The density of both forms reaches one to two individuals dm^{-2}, which is greater than at TAG. Compared to TAG, the Broken Spur community is distinguished by three features: (i) low biomass of benthos; (ii) large biomass of *Segonzacia* and *Munidopsis* compared to that of shrimps; and (iii) occurrence of *Bathymodiolus* in the zone of diffuse venting and at some distance from hydrothermal structures. In the latter case the abundance of live animals is low compared to that of dead shells. Murton *et al.* (1995) did not indicate any chemical anomalies which might have determined the low biomass in the Broken Spur community compared to TAG or Snake Pit. Quantitative dominance of carnivores compared to other forms suggests that the community is not in a climax condition. At the same time the Broken Spur system is not young, as indicated by large sizes of hydrothermal structures and abundance of dead mussels. It may be assumed that the peculiarities of the Broken Spur community are the result of recent short-term decrease in hydrothermal activity which has reduced the population of sessile symbiotrophic organisms, for example *Bathymodiolus*, and forced vagile forms to migrate. The ability of the mussels to switch to heterotrophy might have helped some of them to survive unfavourable conditions (Van Dover, 1995). There is indirect evidence of other hydrothermal systems in the Broken Spur area, possibly more active than the one studied (Murton *et al.*, 1994). If this is so an interchange of organisms between the nearby systems might be expected, when vagile forms, for example crustaceans, first occupy newly formed systems while declining populations of mussels at the periphery of the hydrothermal fields support abundant populations of carnivores, such as decapods and fishes.

2.3. Logatchev (the "14–45" Site)

The 14–45 site was discovered in 1994 on the seventh cruise of the RV "Professor Logachev" (Batuyev et al., 1994; Sudarikov and Galkin, 1994, 1995; Bogdanov et al., 1995a,b; Sagalevitch and Bogdanov, 1995) and recently has been given the name "Logatchev" (Krasnov et al., 1996). This site was revisited in 1995 during the 10th cruise of "Professor Logachev" when studies were made with a TV monitored grab and dredge. Two Mir dives were also conducted at this site on the 35th cruise of "Akademik Mstislav Keldysh" in 1995. The hydrothermal field lies at 2930–3010 m. A low hydrothermal mound on the east edge of the rift valley is 56 km south of the transform fault at 15°20′ on MAR. The mound is 200 × 100 m in diameter, 5–10 m high, with an area of c. 20 000 m^2 (Bogdanov et al., 1995a). Most of it is inactive. Hydrothermal deposits are covered with a thin layer (from one to tens of cm) of carbonate sediments. At the top of the mound at depths of 3005, 2960 and 2940 m three active hydrothermal fields up to 10 m in diameter have been identified. The periphery of each field is marked by three to four black smokers discharging dense black fluid. There are two types of smoker: (i) those spreading horizontally (creeping smoker); and (ii) those flowing vertically. The first type discharges from small funnels or small, bent chimneys less than 1 m long and the effluent is transported horizontally by the current. The second type is discharged from vertical pillar-like chimneys, about 3 m in height and 0.5–0.8 m in diameter (Figure 4). About 30 species of hydrothermal vent-specific animals have been identified so far from the hydrothermal fields at this site (however, only 12 species were included in Table 3). An inactive zone of the hydrothermal mound and the base of chimneys are marked by aggregations of dead mussels, with shells reaching 8 cm in length.

The sample with *Phymorhynchus* sp. (Sudarikov and Galkin, 1994) is presumed also to have been taken at the periphery of the field. The zone of shimmering water around the chimneys is marked by a 0.5 m wide belt of white anemones reaching a density of 20 individuals m^{-2}. In terms of biomass, the community is obviously dominated by mussels, which form dense aggregations slightly below the zone of shimmering water. This population of *Bathymodiolus* may be sustained by symbiotic bacteria that can oxidize methane as well as sulphur, since the concentration of methane gas in all samples from hydrothermal field is four orders of magnitude higher than in the ambient water (Bogdanov et al., 1995a,b). Occasional crabs (*Segonzacia* sp.) are present among mussels. The ophiuroid *Ophioctenella acies* is common on shells and on the walls of the chimneys where it reaches a density of over 10 individuals dm^{-2}. At least three species of shrimp live around the black smoker complexes. Their density is low and distribution very patchy. However, in the upper part of the pillar-like chimneys, in the zone of shimmering-water and bacterial mats, the density of shrimps in some places reaches 10 individuals dm^{-2}.

In general terms, the abundance of the fauna in the community at 14–45 is less

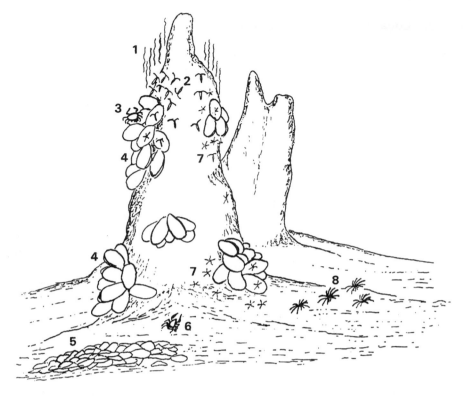

Figure 4 Distribution of fauna on the sulphide structure in the 14–45 area (depth 2967 m, structure *c*. 1 m high). 1, Warm fluid; 2, *Rimicaris* sp.; 3, *Segonzacia* sp.; 4, *Bathymodiolus* sp. (live); 5, valves of *Bathymodiolus*; 6, *Munidopsis* sp.; 7, *Ophioctenella acies*; 8, Actiniaria gen. sp.

than in other hydrothermal communities on the MAR. However, a zonation of the fauna (shrimps-mussels/anemones) can still be seen from the limited material available. Fishes were occasionally observed in this community: single individuals of *Benthosaurus* and brotulids were recorded on the video. The eel-like forms common in other hydrothermal sites on MAR were not found.

2.4. Snake Pit (MARK)

Discovered in 1985 (Kong *et al.*, 1985), this hydrothermal site lies in the middle part of the rift valley on the MAR at 23°N, depth of 3420–3480 m. It includes four active hydrothermal mounds 20–60 m in diameter and 20–26 m high, placed 50–100 m from each other. The mound tops are marked by black smokers (Mevel *et al.*, 1989; Fouquet *et al.*, 1993; Van Dover, 1995). This site was visited by the RV "Professor Logachev" using the deep-tow vehicle "Abyssal" and the 0.25-m^2 Okean grab (Sudarikov and Galkin, 1994, 1995). More than 2800 photographs from the series of photo-transects have been studied.

Analysis of mesoscale distribution of fauna in relation to the type of the sediment/substratum has revealed three facies in the zone around hydrothermal mounds (Figure 4):

- A central zone, located within a radius of 10–100 m from black smokers, consisting mainly of hydroxide–sulphide sediments with Fe sulphides predominant. This zone is also characterized by the presence of barite, which is not found outside this zone, and by the dominance of vent-specific fauna over non-specific.
- The intermediate zone lies 50–150 m from the black smokers and is marked by high-temperature springs. Sediments are rich in sulphide-hydroxide, Fe hydroxide and oxidized sulphide. The non-specific fauna shows maximum concentration here.
- The outer zone is located within 150–300 m of the smokers. Sulphides dominate in the sediment. Small suspension-feeders prevail in the fauna, a feature typical of pronounced oligotrophic communities. The abundance of the bacterial mats and hydrothermal sediments decreases towards the periphery of the hydrothermal field in the outer zone.

Microscale distribution of fauna immediately around the black smokers was studied by Segonzac (1992) who also presented a landscape reconstruction. The total number of invertebrates species reported in this zone is at least 48. The "indicator" taxon here is shrimps (Alvinocarididae). *Chorocaris chacei* and *Alvinocaris markensis* are more abundant at the base of chimneys and at some distance from hot springs, whereas *Rimicaris exoculata* is found in the close vicinity of the hot springs. Anemones *Parasicyonis ingolfi*, gastropods *Phymo-rhynchus* sp., galatheids *Munidopsis crassa* and ophiuroids *Ophioctenella acies* are among the dominant forms at the base of active chimneys. Chaetopterid polychaetes are also present here (Van Dover, 1995).

In general terms, the spatial structure of the Snake Pit hydrothermal community resembles that at TAG, although the latter lacks the mussel *Bathymodiolus puteoserpentis* that at Snake Pit forms aggregations on chimneys in the areas free from shrimps (Karson and Brown, 1988; Von Cosel *et al.*, 1994). According to Fornari *et al.* (1994) the temperature in the zone occupied by mussels is around 5°C, which corresponds to the shimmering-water biotope.

2.5. The Shallow Sites

The two northernmost sites on the MAR (Lucky Strike and Menez Gwen) with a characteristic hydrothermal fauna exist in shallower depths (1700–850 m) than

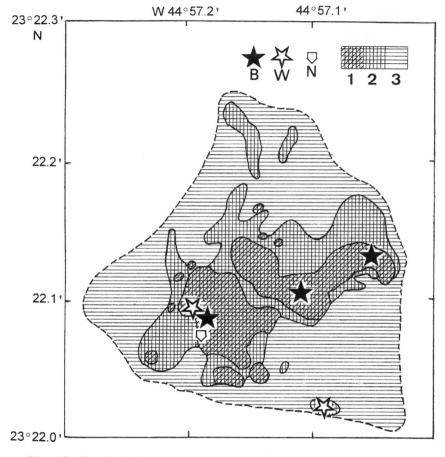

Figure 5 Geochemical zones on the Snake Pit hydrothermal field. 1, Central zone; 2, intermediate zone; 3, outer zone. The different vents are shown as: B, black smokers; W, white smokers; N, not active. Redrawn and simplified after Sudarikov and Galkin (1995).

the others. There are other hydrothermal fields in very shallow water in the Azores archipelago (R. Santos, personal communication). Some data on the fauna are given by Desbruyères *et al.* (1994), Saldanha and Biscoito (1996), Saldanha *et al.* (1996) and Van Dover *et al.* (1996). As we have no personal experience of these sites we are not attempting a summary of their ecology and biota. Van Dover *et al.* (1996) give an outline comparison of the fauna of these sites with that of the other MAR hydrothermal fields.

2.6. General Regularities of the Structure of the MAR Hydrothermal Ecosystems

Despite the varied geological setting, the different forms and scale of hydrothermal activity and differences in taxonomic composition, the hydrothermal sites on the MAR still show some common regularities in faunal distribution and community structure. As indicated earlier, the TAG site, despite the diverse topography and mosaic distribution of fauna, has only two major faunistic associations on the hydrothermal mound, zoned around hot springs. One is marked by aggregations of shrimps *Rimicaris exoculata* and *Iorania concordia*, the other by the anemone *Parasycionis ingolfi* and the gastropod *Phymorhynchus moskalevi*. The two associations correlate well with unoxidized and oxidized ore deposits correspondingly. This type of zonation is more or less distinguishable at all other known sites on the MAR, although the absolute and the relative sizes of zones, their clearness and location with regard to topography are different in each case and depend primarily on the form and power of hydrothermal activity, microrelief and local hydrodynamics.

Sometimes a third zone ("shimmering-water") can be distinguished between the first two. It is marked by *Rimicaris* and *Iorania* which do not, however, form swarms there, and also by *Alvinocaris*, *Chorocaris*, ophiuroids and occasional anemones. The "indicator" or marker taxon of this zone is the mussel *Bathymodiolus*, which is believed to have a temperature optimum at around 5°C. In the TAG community this mussel is absent and the zone is occupied by shrimps.

In this context "indicator" taxa are understood to be the most abundant and distinguished groups in an assembly, but they do not need to be entirely restricted to the zone they mark. For example, *Alvinocaris markensis*, *Chorocaris chacei* and *Segonzacia mesoatlantica* are quite common in the zone of *Rimicaris* and *Iorania* swarms, but they are much more obvious in the shimmering-water zone where there are no dense swarms of *Rimicaris* and *Iorania*. In turn, single anemones often occur in the shimmering-water zone although they reach their maximum density in inactive parts of the hydrothermal field, where they cover the surface of oxidized sulphides.

The relative abundance (ratio) of indicator taxa is different at different sites. Thus, 14–45, Menez Gwen and Lucky Strike are clearly dominated by mussels, although shrimps are also present there. Snake Pit is characterized by abundant shrimps; mussels were found on two of four hydrothermal mounds. At Broken Spur an abundant shrimp population occurs only on the BX16 structure, at a density an order of magnitude less than on TAG. Mussels are rare at Broken Spur and are not found on TAG which is completely dominated by shrimps, with anemones at the periphery.

At present, we lack precise measurements of environmental parameters that would allow correlation of the animal distribution with such factors as temperature, H₂S, CH₄ and O₂. The extreme thermo-chemical conditions of the hydrothermal habitat do not appear to form a barrier for penetration of the biota and colonization. The near hot spring neighbourhood is inhabited not only by vent-specific organisms, but also by elements of the background fauna, for example *Munidopsis* and Actiniaria. It seems likely that localization of most organisms within the hydrothermal field depends on their trophic specialization, or degree of adaptation for consumption of chemosynthetically derived food. Thus, *Rimicaris exoculata*, known to feed on bacterial matter, occurs in very close proximity to the black smokers. Even in this zone the concentration of H₂S is relatively low (6 mmol l⁻¹ at the TAG site) compared to hydrothermal systems in the Pacific. Presumably, these factors and the extremely narrow trophic specialization of *Rimicaris* force shrimps to live at the very edge of the hot discharges. As a result many animals in the population are damaged by scalding in the hot fluids.

The probable mixotrophic type of feeding of *Bathymodiolus* (suspension feeding and symbiotrophy) allows these mussels to survive at some distance from the hot water – in the shimmering-water zone and even on inactive parts of hydrothermal structures. A similar ability is shown by mussels at hydrothermal sites in other regions (Page *et al.*, 1990). There appear to be two types of symbionts in MAR *Bathymodiolus*: one is sulphur-oxidizing, the other methanotrophic (Saldanha *et al.*, 1996). A similar combination of symbionts is found at seeps in the Gulf of Mexico (Fisher *et al.*, 1987; Nelson and Fisher, 1995; Nix *et al.*, 1995). Such a dual symbiosis, added to the ability to feed on suspended particles, may help the mussels to survive periods of variation in flow of hydrothermal fluids. Some of alvinocaridid shrimps *(Chorocaris chacei)* common in the shimmering-water zone are also mixotrophic, able to feed on epibiotic bacteria and also act as scavengers/predators.

The dominant faunal elements of the inactive parts of MAR hydrothermal fields, such as the sea-anemones *Parasicyonis* and the chaetopterid polychaetes, are not specific (specialized) hydrothermal organisms, although they here reach high biomasses and large sizes. Slurp-gun and net samples of the sediment from the periphery zone contain nematodes, but arc also rich in shrimp exuvia and fragments. It seems likely that the anemones obtain a proportion of their nutrition requirements from chemosynthetically derived carbon through the food web. Ratios of ¹²C/¹³C in the anemones from TAG varied from −12 to −14 (unpublished observations) indicating that part of the food was derived from bacterial sources and not from photosynthetically derived carbon. The periphery of the vent fields is also rich in suspended organic matter, including some of bacterial origin, which is available to suspension-feeders. Much of the suspended organic matter seems to fall out at the periphery of the field (see also Karl, 1995) providing food for deposit-feeders (gastropods *Phymorhynchus*, polychaetes, etc.). Fish from the background fauna, such as macrourids (*Nematonurus armatus*) and chimaeras are also common at the periphery of the vents (S. Galkin, personal observations).

Active and mobile predators and scavengers, for example *Segonzacia*, *Munidopsis* and *Synaphobranchus*, occur in all zones but are most abundant on the bivalve "graveyards" and in shrimp aggregations. *Segonzacia* often occurs in the close vicinity of heated water, whereas the galatheids and fishes are more common on inactive parts of the field and outside it.

Thus, zones 1 and 2, situated close to hot springs, are dominated by organisms highly specialized for feeding on bacteria, though not all of them are symbiotrophs. At the periphery secondary consumers and suspension-feeders are more common. Organisms living at the periphery of hydrothermal communities usually lack pronounced morpho-functional adaptations for consuming bacterial organic matter and the background fauna is especially typical for this zone. The number of secondary consumers increases towards zone 3.

One of the general characteristics of the MAR hydrothermal communities is a weak development or complete lack of a zone of suspension-feeders (sponges, gorgonians, brisingid asteroids, etc.) at the periphery of the fields. Such aggregations are more common in Pacific hydrothermal systems, especially in the Manus Basin (Galkin, 1992) and on Piip's volcano (Sagalevitch *et al.*, 1992). Peripheral suspension-feeders, in addition to utilizing chemosynthetically derived organic matter, may also be able to subsist on organic matter entering the vent system from ambient waters through advective circulation driven by the plume (Enright *et al.*, 1981). If this is so, the low number of suspension-feeders in the MAR communities could be related to the extremely oligotrophic conditions in the surrounding ecosystem and the lesser number of active vents per ridge segment (Galkin and Moskalev, 1990b). Owing to the small (or extremely small) scale, the enhanced production of hydrothermal communities may not have much influence on the background ecosystem adjacent to the MAR, at least not the plankton (Vinogradov, this volume, p. 78; Vinogradov *et al.*, 1996). However, organic aggregates up to 1–2 mm in diameter have been observed at the TAG site in a layer at 2400–2600 m, that is, more than 1 km above the hydrothermal field (Vinogradov *et al.*, 1996). The concentration of aggregates increased towards the bottom. Some transportation of organic matter from vent communities to the deep sea in general may result from fallout from the hydrothermal plumes and also from the vagile background fauna such as fishes and crustaceans that enters the highly productive zones close to the vents.

3. BIOLOGY AND TAXONOMY OF HYDROTHERMAL VENT SHRIMPS

Shrimps constitute a large part of the biomass at MAR sites, in contrast to the well-known hydrothermal sites in the East Pacific where tube-worms with endosymbiotic bacteria are the major dominants of the communities (Tunnicliffe, 1988, 1991). They are thus of great importance in the MAR ecosystem.

3.1. Taxonomic Composition

Six species of carideans have been described so far from the MAR sites: *Rimicaris exoculata* (Williams and Rona, 1986), *Chorocaris* (=*Rimicaris*) *chacei* (Williams and Rona, 1986), *Alvinocaris markensis* (Williams, 1988), *Mirocaris* (=*Chorocaris*) *fortunata* (Martin and Christiansen, 1995), *Iorania concordia* (Vereshchaka, 1996) and *Mirocaris keldyshi* (Vereshchaka, 1997). In addition two undescribed species of *Chorocaris* are believed to occur at Menez Gwen (Desbruyères *et al.*, 1994) and Snake Pit (Segonzac *et al.*, 1993). Thus, the total number of shrimp species known between 14°N and 37°N in the depth range 800 m to 3800 m could be as high as seven or eight.

The taxonomy of this group is more difficult than at first thought. Vent shrimps were originally assigned to the family Bresiliidae Calman 1896 until a new family Alvinocarididae was established by Christoffersen (1986) for *Alvinoca lusca*. In 1990 Christoffersen also included the genus *Rimicaris* in this family. Later M. de Saint Laurent included in this family the genus *Chorocaris* as well (in Segonzac *et al.*, 1993), thus isolating the group of shrimps restricted to vent/seep habitats. This lead, however, was not followed by all (Holthius, 1993; Van Dover, 1995). Meanwhile, a new family Mirocarididae has been suggested for the genus *Mirocaris* (Vereschaka, 1997). Members of this group are also known from several sites in the Pacific: *Alvinocaris lusca* (Williams and Chace, 1982) from the Galapagos Spreading Centre, *A. longirostris* from the north-west Pacific (Kikuchi and Ohta, 1995), *Chorocaris vandoverae* (Martin and Hessler, 1990) from the Mariana Back Arc site and *Opaepele loihi* from the Loihi seamount (Williams and Dobbs, 1995). The group is not restricted to hydrothermal systems but appears also at hydrocarbon and cold seeps: in the Gulf of Mexico (*Alvinocaris stactophila*, Williams, 1988) and at the West Florida Escarpment (*A. muricola*, Williams, 1988).

Rimicaris exoculata is the most abundant and widely distributed of the MAR vent shrimps. This species usually forms dense swarms in the vicinity of black smokers or shimmering water. The smaller, "pink shrimp" *Iorania concordia* is another species known to form swarms. It was noticed in aggregations of *R. exoculata* on preliminary morphological analysis (Segonzac *et al.*, 1993, appendix) or from examination of the form of the dorsal organ (Van Dover, 1995; P. Tyler and M. de St Laurent, personal communication). Van Dover suggested that these small shrimps could be a new species of *Rimicaris*, and this view was followed by Nuckley *et al.* (1996). The description of this small shrimp (Vereshchaka, 1996) as a new genus and species was based on material obtained on the BRAVEX cruise ("Akademik Mstislav Keldysh", cruise No 34, 1994). Some workers (M. de St Laurent and others) still feel that these small pink shrimp correspond to one of the juvenile stages of *R. exoculata*. This opinion is based on: (i) the absence of females with eggs and mature males with developed *appendix interna* in samples of *I. concordia*; and (ii) *I. concordia* of carapace length 8–10 mm can occur in samples together with *R. exoculata* with a carapace length of 11 mm, suggesting the two are different stages of the same species. The identity of the two species is further suggested by allozyme analysis (Creasey *et al.*, 1996). However, in the samples from the 1994 cruise, among thousands of *R. exoculata* there are several tens with a carapace 7–11 mm long that are clearly different from several tens of *I. concordia* with carapace 8–11 mm long, and no transitional morphological forms exist between the two species (Vereshchaka, 1996). It seems possible there has been confusion of the two sets of 7–11 mm carapace specimens. To resolve this problem, a simultaneous analysis of allozymes and morphology is required.

Chorocaris chacei is a large shrimp, yellow in colour and clearly distinguished from other shrimps. *Mirocaris keldyshi* is another pink shrimp that lives at the periphery of hydrothermal fields, known at present from at least four hydrothermal systems. It was not at first recognized as a separate species, presumably because, as with *I. concordia*, there was morphological similarity to juvenile *R. exoculata*. *M. keldyshi* seems to occur in two subspecies (unpublished data), one living at Broken Spur and TAG and the other restricted to the 14–45 hydrothermal field. Detailed morphological analysis of a large amount of material has allowed another form, *Mirocaris fortunata*, originally assigned to the genus *Chorocaris* (Martin and Christiansen, 1995), to be transferred to the new genus *Mirocaris* (Family Mirocarididae) (Vereshchaka, 1997).

Alvinocaris markensis is one of the largest MAR vent shrimps; it occurs only at the periphery of hydrothermal fields and is known from Snake Pit, TAG and Broken Spur, although rarely seen. In the 35th cruise of the "Akademik Mstislav Keldysh" in 1995 two additional species of this genus were caught at the 14–45 site. Each one is represented only by a single specimen, one of them damaged, so that there are not enough data for reliable identification, However, it seems likely that one of the specimens is close to *A. muricola* known from the West Florida Escarpment, whereas the other one may be close to *A. longirostris* known from the north-west Pacific (Kikuchi and Ohta, 1995).

3.2. Feeding Strategies

The enormous densities of shrimps on the MAR (2500 individuals m^{-2} at the Snake Pit, Segonzac, 1992) and their dominance in some communities raise questions about how the populations are sustained. Van Dover *et al.* (1988) presented the first evidence for a bacterial diet of shrimps based on stable carbon isotopes, lipopolysaccharides and traces of bacteria in the gut. No evidence for endosymbiotic bacteria was found, although very much expected (Gal'chenko *et al.*, 1988). Instead, there was a "fouling" by epibiotic bacteria, most obvious on the mouthparts of *R. exoculata*. This species has been the subject of further studies, which we review here.

Three mechanisms have since been proposed for feeding of vent shrimps:

- Grazing on the surface of sulphide chimneys where free-living micro-organisms occur, with ingestion of sulphide minerals together with associated bacteria (Van Dover *et al.*, 1988; Van Dover, 1995). However, ingestion of epibionts was deemed likely as well. The evidence for grazing with ingestion of sulphide minerals included the presence of a large amount of sulphide crystals in the foreguts of all shrimp individuals examined.
- Suspension-feeding, with some contribution from epibiotic bacteria dislodged by shrimp's rapid movements (Jannasch *et al.*, 1991).
- Feeding on exosymbiotic bacteria growing on the mouthparts and the under surface of the carapace (Gebruk *et al.*, 1992, 1993; Pimenov *et al.*, 1992). First proposed by the Russian team, the idea of a symbiotic relationship and feeding on epibiotic bacteria was concurrently discussed by Segonzac *et al.* (1993), following earlier support for the grazing mechanism of feeding (Segonzac, 1992).

The arguments for feeding on the epibiotic bacteria can be summarized in two categories, morphological and behavioural (Gebruk, 1996). The morphological adaptations are:

- The shrimp mouthparts are highly specialized for carrying bacteria, par-ticularly the expanded exopods of maxilliped I and maxillae II, both densely covered with soft setae and quite atypical for normal bresilioid shrimps. A special term "bacteriophore" was introduced to emphasize this specialization (Segonzac *et al.*, 1993).
- There are unusually enlarged branchial chambers formed by the carapace, which wraps almost entirely around the body, isolating the bacteria and providing them with stable conditions.
- The small, scoop-shaped chelipeds are held within the branchial chambers, which makes them available for cleaning bacteria from the mouthparts and the under surface of carapace, but not for collection of particles from outside.

Behavioural adaptations include:

- swarming of shrimps in areas with conditions favourable for bacterial growth that is, where vent fluid rich in hydrogen sulphide is mixed with oxygenated sea water, within the temperature range of 10–30°C (Jannasch *et al.*, 1991; Gebruk *et al.*, 1992; Segonzac, 1992; Wirsen *et al.*, 1993; O'Neil *et al.*, 1995);

- submersible observations showing that shrimps stay motionless much of the time, grappling the substratum with their modified short and strong anterior appendages and legs, with no evidence for grazing movements (Gebruk *et al.*, 1993; Segonzac *et al.*, 1993).

Shrimps swarm close to the hot fluid, despite the possibility that this may be lethal – approximately 30% of the studied specimens were damaged (scalded) by the hot water, some of them severely. Most often affected are the antennae, legs, telson and even the *appendix interna*. One specimen had a hole in the carapace constituting 40% of the surface area. An additional morphological adaptation of vent shrimps is the hypertrophied antennular peduncles with articulation allowing primarily vertical movement which is probably a ventilation mechanism for the carapace chamber.

It should be noted that sulphide crystals in the gut are a common feature of many animals living in vent habitats with high H_2S levels or rich in suspended mineral particles; they do not necessarily imply a grazing mechanism. Sulphide particles are found in the gut of alvinellid polychaetes which are thought to have a close trophic relationship with epibiotic bacteria (Desbruyères *et al.*, 1983), but also occur in all other alvinellids that are supposed to be grazers (Juniper *et al.*, 1992). Abundant sulphide crystals were also found in the gut of amphipods *Ventiella sulfuris*, often associated with alvinellid colonies in the East Pacific hydrothermal systems (Vinogradov G., 1995).

Thus a strong case can be made for a "symbiotic" relationship between shrimps and bacteria, where the host is highly specialized for "farming" of bacteria which it ingests; it is supposed that the bacteria find stable conditions in the shrimp's branchial chambers advantageous and maintain a population in spite of being "cropped". The concept of stable conditions is usually regarded as an advantage gained by the endosymbionts that are common in Pacific vent communities. However, the alternative hypothesis of shrimps feeding on free-living bacteria is still a possibility and the problem of shrimp feeding biology thus remains challenging.

It seems possible that the mode of feeding of *R. exoculata* changes with age. Larval stages smaller than 8 mm are found only at a distance of several hundred metres from the hydrothermal vents (see below), where neither free-living bacteria occur nor suitable conditions for exosymbiotic micro-organisms exist.

Recent studies of the lipid composition of *R. exoculata* from TAG have revealed the occurrence of certain highly unsaturated fatty acids (HUFA), including eicosapentaenoic acid, that can only be synthesized from precursors produced primarily by phytoplankton (Allen *et al.*, 1996). These HUFAs are very evident in the juveniles from the plume. The adult fatty acid signatures were found to be consistent with their feeding primarily on bacteria, but some HUFAs were also present in the adults, possibly a relic of the juvenile feeding system or maybe a result of consumption of organic material sedimenting from the euphotic zone.

Other analyses of lipids of *Rimicaris* show that the individual fatty acids in the shrimps and the epibiotic bacteria are identical and are different from those in free-living bacteria (Rieley *et al.*, 1996). These two investigations may be taken as corroborating the hypothesis of symbiosis between the shrimps and bacteria and the symbiotrophic type of feeding in *Rimicaris exoculata*, but such studies need to be expanded to cover the other MAR shrimps.

It also seems likely that symbiotic bacteria attached to shrimps are used not only by the host but also by an associated fauna of copepods. Up to 10 copepods of different species were found on the mouthparts of shrimps (*Rimicaris* and *Chorocaris*) at Snake Pit and Lucky Strike (Humes, 1996).

It is anticipated that feeding in *Iorania concordia* will be similar to that suggested for *R. exoculata* since it also has morphological adaptations for "farming" of symbiotic bacteria (bacteriopohores, branchial chambers, etc.) and also occurs in swarms close to black smokers. Different types of feeding are suggested for other vent shrimp species (Casanova *et al.*, 1993; Segonzac *et al.*, 1993). *Alvinocaris markensis* usually behaves like a predator/scavenger. *Chorocaris chacei* seems to feed both on bacteria and as a scavenger/occasional predator, while *M. keldyshi* seems to be purely a scavenger (unpublished observations).

Besides the feeding specializations of the different vent shrimps, of special interest is a transition from "normal" type of feeding to bacteriotrophy, as shown by morphological adaptations and behavioural and distributional patterns. Interesting morphological trends occur in the series *Alvinocaris–Chorocaris– Rimicaris*: from normal setae to the bacteriophore type; from normal carapace to branchial chambers; from big claws to small scoop-shaped chelae; and from long and slender antennae to short and robust. Furthermore, the dorsal organ is lacking in adults of *Alvinocaris*, small in *Chorocaris* and well developed in *Rimicaris*. This dorsal organ is apparently derived from the normal compound eyes and seems to be adapted for detection of high-temperature vents either from low-level illumination, or black-body radiation (Van Dover *et al.*, 1989; Johnson *et al.*, 1995; O'Neil *et al.*, 1995; Nuckley *et al.*, 1996). *A. markensis* is a solitary predator keeping at the periphery of vent fields. *C. chacei* lives in small groups within the vent field often occurring together with *R. exoculata*, whereas the latter forms dense swarms around chimneys or in the areas of shimmering (heated) water and is known to detect chemical signals from sulphide (Renninger *et al.*, 1995). Epibiotic bacteria rarely occur on *A. markensis*, but are abundant on *R. exoculata*, with *C. chacei* occupying an intermediate position. Recently described new shrimp species and genera, based on morphological features and behavioural and distributional patterns, find their place remarkably well in the transitional series as follows: *Alvinocaris – Opaepele – Mirocaris – Chorocaris – Iorania – Rimicaris*. The species of *Alvinocaris* and *Opaepele* have the usual bresilioid appearance and behaviour, *Rimicaris* and *Iorania* appear highly specialized for the hydrothermal environment and *Chorocaris* and *Mirocaris* fall between these two pairs. This transitional series presents clear evidence (morphological and behavioural) of increasingly specialized adaptation to the hydrothermal environment.

Independently from studies of shrimp feeding biology, chemosynthetic primary production of the epibiotic bacteria on shrimps was demonstrated by Jannasch *et al.* (1991) and Wirsen *et al.* (1993). The epibiotic bacteria, most abundant on shrimp mouthparts and the inner surface of carapace but also appearing on legs, gills and antennae, form a complex community with various morphological forms dominated by two types of filamentous bacteria with different trichome width (0.2–0.5 μm and 1–3 μm); they have elemental sulphur inside the cells and attaching dics (Casanova *et al.*, 1993; Gebruk *et al.*, 1993). Chemosynthetic activity in bacterial mat on the shrimp carapaces was measured as 0.5–0.7 nmol CO_2 fixed mg protein min^{-1} (Jannasch *et al.*, 1991). Although these values are lower than the corresponding values for endosymbionts of *Riftia pachyptila* in the Pacific (45.8 nmol CO_2 fixed mg protein min^{-1}, as shown by Childress *et al.*, 1991), they nevertheless demonstrate high activity of CO_2 fixation in the epibiotic bacteria. High chemosynthetic activities have also been measured in the microbial film covering sulphide deposits and in bacterial suspensions in small plumes at the MAR. The activity in the MAR plumes was indeed two to four times greater than the maximum values obtained under similar conditions at various East Pacific Rise (EPR) sites (Jannasch *et al.*, 1991). At the same time, it was demonstrated by scanning electron microscopy (SEM) that filamentous bacteria at the TAG and Snake Pit sites seem to penetrate into the reduced sulphidic subsurface of iron oxide–hydroxide particles (Jannasch *et al.*, 1991). These results imply that free-living bacteria, although not forming extensive mats, may still be considered as a potential food source for shrimps.

Molecular biology also presents valuable evidence. Ribosomal 16s RNA sequences show that the bacterial community living on the shrimp surface at Lucky Strike is largely (up to 90%) a single bacterial type (phylotype), whereas in the free-living community on the sulphide minerals, the same type, although also a major component, is mixed with many other types (Polz and Cavanaugh, 1995a,b, 1996). This fact may reflect a symbiotic relationship of shrimps only with a specific type of bacteria.

3.3. Microscale Distribution of Shrimps

Hydrothermal fields on the MAR have a wide size range, from 200 m^2 to 1.5×10^5 m^2 (Table 1). Distribution of shrimps in the communities is irregular, although certain patterns exist. Microscale distribution was studied in detail at TAG during the 34th cruise of "Akademik Mstislav Keldysh" in 1994 (Vinogradov and Vereshchaka, 1995). Specimens were collected with the help of baited and pump traps and the slurp-gun.

It was found in particular that only two species, *R. exoculata* and *I. concordia*, occur in the closest vicinity to black smokers and shimmering water and form the dense swarms that are characteristic of the TAG site. At Snake Pit the density of the swarms has been estimated to reach 2500 individuals m^{-2} (Segonzac, 1992). At Lucky Strike and 14–45, however, the density does not exceed 20–30 individuals m^{-2} and is even lower for *C. vandoverae* in the West Pacific (Martin and Hessler, 1990). In terms of abundance, *R. exoculata* is always dominant: 70–80% in aggregations close to the black smokers and up to 90% in the shimmering-water zone (Table 2).

At the periphery of shrimp swarms, (i.e. about 5 m from the fluid discharge) the ratio of *R. exoculata* and *I. concordia* falls abruptly to *c*. 50% whereas *C. chacei* appears in large numbers. These large shrimps usually stay in small groups sitting on the substratum or swimming close to the bottom, sometimes attacking abundant *R. exoculata* in this zone. Enhanced numbers of *C. chacei* are also observed at the periphery of the shimmering-water zone where the density of this species reaches 50 individuals m^{-2}. Here, about 30 m from the mound top, the species composition changes abruptly once again. The dominant form here becomes *M. keldyshi*, absent in the swarm zone. These small shrimps, difficult to recognize on the video-tapes, do not form aggregations and always stay at the periphery of the field. Specimens of *C. chacei* in this zone are younger (smaller) than at the periphery of the swarm zone. Juveniles of other shrimp species, adults of which live close to the heated effluents in the centre of the field, are also common at the periphery of the field.

Shrimps may occur occasionally at a distance of several hundred metres from the mound top but have not been observed as far as 1 km away. In 1994 a baited trap was deployed 1 km to the north-west of the TAG field. After several days only typical background organisms had been caught: *Eurythenes gryllus* (Amphipoda) and *Thysanopoda* sp. (Euphausidae). Young, just-settled juveniles of hydro-thermal shrimps, with a carapace length of 5–6 mm occur occasionally in trawl samples taken several hundred metres away from the active mound.

3.4. Reproduction and Life History

Peculiarities of reproduction of MAR shrimps show some correlation with feeding behaviour. Forms with clearly pronounced scavenger/predator behaviour (species of *Alvinocaris*, *Mirocaris keldyshi* and *Chorocaris chacei*) that never form swarms do not seem to have synchronized reproduction. Studies of tens and hundreds of specimens of *Mirocaris fortunata* show females with eggs are always present. In addition to the Russian material collected on the MAR in September–October of 1994 and February 1995, this is also known from the French data obtained in July–September in different years (de St Laurent and Segonzac, personal

Table 2 Ratio of shrimp species (%) on TAG related to the distance from the mound top (m) in a south-west direction.

	0.5 m from the mound top	Centre of the swarm	Periphery of the swarm	15 m from the top, shimmering-water	30 m from the top	100 m from the top
R. exoculata	72	79	54	91	2	—
I. concordia	28	21	0	0	2	—
C. chacei	—	—	46	1	34	—
M. keldyshi	—	—	—	8	62	—

—, Not present in the sample; 0, present, but very scarce (occasional). A. markensis was not present in samples but observed on video-recordings.

communication). In contrast, among hundreds and thousands of specimens of *R. exoculata* and *I. concordia*, which appear largely dependent on epibiotic bacteria and living in swarms, only a few females carrying eggs are known from the extensive collections. Two were caught in September at the Broken Spur site: one in the BRAVEX cruise (P. Tyler, personal communication) and another one in the "Akademik Mstislav Keldysh" cruise 39 in 1996 (unpublished data). One more was taken by the French in July and two more in November. In the original description of *R. exoculata*, the authors, Williams and Rona (1986), noted a single female carrying eggs among about a hundred specimens caught in August. Thus, at least during February and July–October these two species do not reproduce. Alternatively, it might be suggested that fertilized females leave the swarms and stay outside the main populations.

It might be suggested that deep-sea shrimps, more dependent on chemo-synthetic bacterial production than on surface photosynthesis, lack a seasonal reproductive signal, reacting instead to an unknown trigger. However, the case for synchronous reproduction is supported in particular by the population size structure. In September–October 1994 the TAG and Broken Spur populations of *R. exoculata* showed two pronounced size peaks both for males and females: 9–10 and 14–15 mm, whilst *I. concordia* was represented by a single cohort with of carapace length 8–11 mm.

Larval stages of hydrothermal shrimps are not as well known as the adults. The earliest stages found in swarms are adults with a carapace length 8 mm and more. Younger stages have been found only at a distance of several hundred metres from the hydrothermal mound (TAG). Ten settling stages of *R. exoculata* and (?) *C. chacei* with carapace 5–7 mm long were caught by the Sigsbee trawl. One unidentified larva *c.* 5 mm long was observed in the water column about 50 m above the TAG hydrothermal field (observer M.E. Vinogradov). Juveniles have recently been caught in substantial numbers 200 m high in the vent plume, and identified by molecular methods into at least three species (Dixon and Dixon, 1996; Herring, 1996). During the BRAVEX cruise one larvae of *R. exoculata* 3-mm long was caught with a plankton net (BR113/140 – mesh size 0.3 mm) in the layer 30–500 m above the bottom (A. L. Vereshchaka, personal data).

These scattered data are not enough for the reconstruction of the whole life cycle, but some assumptions can be made. Hatched larvae, apparently lecithotrophic, rise in the water column where they stay for some time. Growing stages descend and at 5 mm carapace length they may already occur on the bottom, at the periphery of the hydrothermal field. Apparently, at this stage they feed as scavengers and avoid the hot water areas. Larger non-mature stages concentrate more in the typical biotopes of the species: *R. exoculata* and *I. concordia* in the vicinity of black smokers and the shimmering water; *C. chacei* at the periphery of the "swarm zone"; and the species of *Alvinocaris* and *Mirocaris* at the periphery of hydrothermal fields.

4. SPECIAL FEATURES OF THE MAR HYDROTHERMAL FAUNA

The deep-sea hydrothermal communities known so far on the MAR are characterized by swarms of shrimps or populations of mussels together with abundant sea-anemones. Long before the "hydrothermal era" several deep-sea expeditions reported finds in the area of the ridge in the Mid-Atlantic that appear now to be related to hydrothermal processes. Thus, the Swedish Deep-Sea Expedition in the "Albatross" in 1947–1948 reported a population of bivalves in the abyssal middle Atlantic, so dense that valves were sampled by the geological corer. The leader of the expedition, H. Pettersson wrote:

> In one of the sediment cores taken from a depth of 2925 m south-west of the Canary Islands we found large fragments of calcareous shells. We are not sure how these shells, which obviously belong to the shallower fauna, were transported that far away from the land and even islands. But there is nothing improbable in the hypothesis of significant submergence of the ocean bottom in this zone characterized by very unstable earth-crust (translated from the Russian edition of Pettersson, 1970).

An interesting comment on this point was given by the editor of the Russian translation of Pettersson's book, L.A. Zenkevich: "It seems more likely that the remains of shells found in the sediment sample were defecated by fish or whales". Now, 50 years after this record, it may be suggested that these were valves of mussels or clams, similar to those sampled by the "Sevmorgeologiya". However, dense populations of bivalves may possibly occur on the ridge away from hydrothermal vents. Large numbers of *Acesta excavata*, including a bed of dead shells, were observed in an area of the Reykjanes Ridge (59°N) without hydrothermal activity during the 27th cruise of the "Akademik Mstislav Keldysh", occurring together with patches on rocks completely covered by large barnacle populations (Crane *et al.*, 1992).

Hydrothermal discharges or the occurrence of plumes have been reported from other parts of the MAR in addition to the four sites described in Section 1 and the set of six sites listed in Table 1. For example, several sites of hydrothermal activity are known to the north and to the south of TAG (Nelsen, 1988). Seven sites have been reported between 36° and 38°N, including the Rainbow site which is believed to be the largest plume in the Atlantic and one of the largest in the world (Anon, 1995; Rudnicki *et al.*, 1995). Recent results suggest a minimum average frequency of one site with hydrothermal activity every 25–30 km between 36 and 38°N and one site per 100 km between 27 and 30°N on the MAR (German *et al.*, 1996). Hydrothermal activity has also been detected at Mohn's Ridge at 72°N and other sites near Iceland (Van Dover, 1995). The specialized hydrothermal fauna could thus be widespread along the MAR in addition to the sites indicated in Table 1. Intensive investigations of the Reykjanes Ridge section between 58° and 59°N conducted on the fourth "Akademik Mstislav Keldysh" cruise in 1982 (including 22 dives, 46 grab and 19 trawl samples) (Kuznetsov, 1985) and 27th cruise in 1992 (six dives), failed to find hydrothermal activity or specific vent communities but added to the studies of a ridge non-vent fauna. Additional information on the Reykjanes Ridge non-vent fauna has been presented by Copley *et al.* (1995, 1996).

The first recognized representatives of the hydrothermal fauna from the MAR were dredged in August 1985 and described shortly afterwards. These were shrimps (Williams and Rona, 1986; Williams, 1988) which, together with later described mussels and anemones, form the bulk of the biomass at vents along the MAR. After the first series of dives on the MAR a major difference emerged between hydrothermal communities in the Atlantic and Pacific. In the East Pacific, studied for more than 8 years before the discovery of the Atlantic vents (Jones, 1985; Grassle, 1986; Laubier, 1986), vestimentiferan tube worms and polychaetes of the family Alvinellidae ("pompeii worms" and "palm worms") form the bulk of the vent biomass, and total values exceed those now known from in the Atlantic. So far, tube-worms and alvinellid polychaetes have not been found along the MAR.

In 1987, hydrothermal fields of a different faunal composition were discovered in the West Pacific (Hessler *et al.*, 1988), characterized by populations of mussels and by gastropods of the family Provannidae. The bulk of the biomass at vents in the West Pacific is formed by two monotypical genera of the provannids, *Alviniconcha and Ifremeria* (= *Olgaconcha*). Provannids are not present on the MAR but together with vestimentiferans and alvinellids also occur in the East Pacific, though at lesser abundance than in the West Pacific (Warén and Bouchet, 1993).

Shrimps, assigned at present to the families Alvinocarididae (= Bresiliidae) and Mirocarididae occur also in the West Pacific, but only the family Alvinocarididae is found in the East Pacific. Thus the Atlantic and the West Pacific share the genus *Chorocaris*, while *Alvinocaris* is common to the Atlantic and both sides of the Pacific. Three monotypical genera of shrimps, *Rimicaris* and the recently described *Mirocaris* and *Iorania* (Vereshchaka, 1996, 1997), are endemic to the MAR.

4.1. Relationships and Origins of the Characteristic Dominants of the MAR Fauna

4.1.1. *Shrimps*

The group Decapoda, Macrura, Natantia, Carideae, to which all known hydrothermal shrimps belong, is a thriving modern assemblage, comprising 170 genera. This number is exceeded among Crustacea only by decapod crabs, Macrura, Reptantia and Brachyura, comprising 640 genera (Zarenkov, 1983).

Caridean shrimps are known from the late Jurassic (Glaessner, 1969, after Zarenkov, 1983), whereas other major hydrothermal faunal elements such as the Vestimentifera, Vesicomyidae and Alvinellidae, represented by fossils, are believed to have occurred at hydrothermal vents in the Phanerozoic, as early as the Silurian period (Avdonin, 1995; Little et al., 1995, 1997; Maslennikov et al., 1995; review in Russell, 1996).

In 1984, before they were recorded on the MAR, aggregations of shrimps were observed in the rift zone of the Gulf of Aden (Choukroune et al., 1988). Emission of heated fluid mixed with diffuse gel-like hydrated iron oxide was observed at 1400–1600 m depth. The fluid discharge was creating temperature anomalies of 0.5°C, marked by aggregations of shrimps, with sea-anemones and galatheids also abundant. These animals densely covered the surface of pillow-lavas and basalts. Shrimps were seen grappling the substratum by the posterior part of the abdomen, with possible filtration of water by the mouth parts (Juniper et al., 1990). Shortly before this, a similar community was observed on the Tadjura Rift (eastwards from the Gulf of Aden) at 1400–1450 m depth during the seventh cruise of the RV "Akademik Mstislav Keldysh" (dive Nr 116/12b, station 769, observer A.P. Lisitsyn). Some areas of basalt walls of the rift valley were covered by benthic organisms, one of which, c. 2 × 3 m, was completely covered by shrimps and other animals. Owing to problems with submersible manipulators neither biological nor geological samples were taken, and unfortunately there was no biologist with the expedition. Among 30 photos taken during this dive and analysed afterwards, a "carpet" of shrimps and associated anemones and crabs was seen. At these sites the abundance of shrimps, as estimated from submersibles, was an order of magnitude lower than in the Atlantic.

Isolated shrimps (or small groups), not attached to substratum, have also been observed at numerous sites: on the Tadjura Rift at the depth of 1270 m (Pisces-VII dive Nr 119/15, January 1984), on the lower part of the lava dome; on bacterial mats of Loihi volcano, south of Hawaii (Sagalevitch and Moskalev, 1992; Williams and Dobbs, 1995); at the gas hydrate site off Paramushir Is in the Sea of Okhotsk (Kuznetsov et al., 1987); at Piip's volcano in the Bering Sea (original data); and at some other hydrocarbon seeps and hydrothermal vent sites (Van Dover, 1995).

4.1.2. *Bivalves*

Populations or aggregations of *Bathymodiolus* are known from the sites in the North Atlantic except TAG (Van Dover, 1995; personal observations). At the 14–45 site mussels, including a 40-mm long live specimen, were sampled with a TV monitored grab. In one of the "Mir" dives, 15 specimens ranging from 8 to

90 mm were collected from sulphide pillars and a small field of empty valves, *c*. 5 m², was observed on the ancient mound. In aggregations mussels may cover 100% of the substratum (Bogdanov *et al.*, 1995a,b). Preliminary examination of specimens from 14–45 suggest (R. Von Cosel, personal communication) that this is not *Bathymodiolus puteoserpentis* known from Snake Pit (Von Cosel *et al.*, 1994). These specimens also differ from the Menez Gwen mussels, but are similar to those from Lucky Strike. Recent studies (R. Von Cosel, personal communication) suggest the species at Broken Spur may be the same as at Snake Pit.

Vesicomyid clams are much rarer than mussels in hydrothermal communities on the MAR. The few records include one specimen from Snake Pit (Segonzac, 1992), two valves south from Snake Pit (15°07,07'N, 44°50'W, depth 3424 m, Van Dover, 1995) and one valve of *Ectenagena* sp., 133 mm long, dredged by RV "Professor Logachev" in 1995 to the east of the 14–45 site (station D-51, start: 14°45,326'N, 44°58,507'W, depth 3010 m, end: 14°44,874'N, 44°57,699'W, depth 2780 m).

4.2. Trophic Structure of the MAR Fauna

The dominance of shrimps in the MAR hydrothermal communities is commonly recognized as a feature which strikingly differentiates vent communities in the Atlantic and the Pacific, and underlines the fact that hydrothermal communities in different parts of the ocean are often dominated by different taxa. Thus, vent communities in the West Pacific are often dominated by gastropods; in the East Pacific, usually by vestimentiferans, clams and polychaetes. In the Atlantic shrimps, mussels and anemones in different combinations make up the bulk of the fauna. In terms of the life style of dominant forms, there is also some difference between geographic regions: sessile forms in the East Pacific and some of the Atlantic vent communities, forms with limited mobility (gastropods) in the West Pacific, and vagile forms in some Atlantic systems. This is also true for cold-seep communities, dominated in different regions by vestimentiferans, mussels, clams and polychaetes, the latter in the Sea of Okhotsk (Kuznetsov *et al.*, 1987).

The trophic structure of the hydrothermal communities is thus of considerable interest. The term "hydrothermal (seep) community" has never been exactly defined, but is usually applied to communities in the areas of hydrothermal venting (methane/sulphide seeps). These communities are often characterized by three trophic levels which form a characteristic trophic pyramid (Figure 6A). The first level is bacterial production, which in the vent (seep) communities is mainly formed by chemosynthetic or methanotrophic bacteria. However, the primary role of chemosynthesis in hot vent communities has been recently questioned by Karl (1995), who has proposed several alternative mechanisms of primary production in these ecosystems. Bacterial organic matter available to vent communities usually exists in three forms: (i) bacteria attached to the hard substratum (rocks,

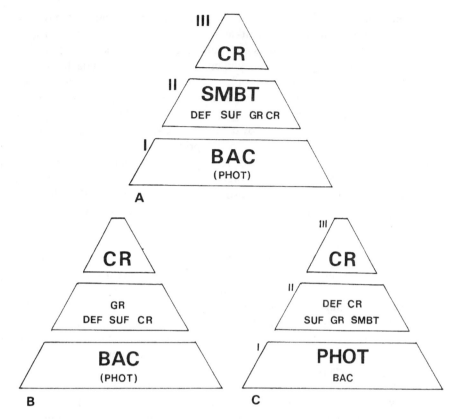

Figure 6 Trophic pyramids corresponding to different hydrothermal communities. A, "Classical" deep-sea hydrothermal community; B, MAR communities dominated by shrimps, assuming they are not symbiotrophs; C, shallow-water hydrothermal communities. I–III, trophic levels. BAC, bacterial primary production; PHOT, photosynthetic primary production; SMBT, symbiotrophs; DEF, deposit-feeders; SUF, suspension-feeders; GR, grazers, CR, carnivores. Size of the letters indicates the prevalence of a particular type of feeding or primary production.

tubes, etc.), including bacterial mats; (ii) symbiotic bacteria; and (iii) "bacterial soup" formed by suspended particles (see Jannasch, 1995; Karl, 1994). The role (ratio) of photosynthetically derived organic matter is unknown, but usually believed to be low, decreasing with increasing depth. On the second level are primary (first-order) consumers. In most deep-sea vent/seep communities the bulk of the biomass is formed at this level and usually by organisms that rely on bacterial production and often on symbiotic bacteria. This is usually a characteristic of the community – organisms that are specialized for symbiotic relations with bacteria and feed on them. These are also the organisms termed "specialized hydrothermal fauna". However, many other organisms may feed directly on free-living bacteria: deposit- and suspension-feeders, grazers and some carnivores, but their relative biomass is usually much lower (Galkin and Moskalev, 1990c). Species endemic to chemosynthetic environments include both symbiotrophic forms and non-symbiotrophic (e.g. *Alvinocaris*, many gastropods of the family Provannidae (Warén and Bouchet, 1993), copepods of the family Dirivultidae (Humes, 1996), etc.). The third trophic level (secondary, or second-order, consumers) is formed by carnivores (scavengers and predators), that feed on the remains (or whole bodies) of the primary consumers. This idealistic scheme is complicated by various trophic strategies and inter-relations but, as a rule, the indicated three levels are distinguishable (see also Galkin, 1992; Hessler and Kaharl, 1995). The complexity is also added to by the "background" deep-sea fauna, which may enter the hydrothermal community to feed on any of the above levels, and leave it.

On the face of it, the MAR community differs from the majority of vent communities by the apparent absence of first-order consumers with endosymbiotic bacteria. It is thus a major biological problem whether the vent shrimps (at least some of them) can be classed as symbiotrophs or not. As reported in Section 2, symbiotrophic feeding has been suggested for *Rimicaris exoculata* and is also expected for *Iorania concordia*. If they are not symbiotrophs, then the symbiosis between invertebrates and bacteria and the dominance of symbiont-containing forms would not be a necessary feature for existence of vent communities (Figure 6B). If the two shrimp species do feed on symbionts which they carry on the mouthparts and the under surface of the carapace, one remarkable difference between vent communities in the Pacific and the Atlantic would still remain: the bulk of the biomass in the shrimp-dominated communities would be formed by organisms with exosymbiotic bacteria, whereas in the Pacific the forms with endosymbionts dominate. Whatever is the case, different mechanisms appear to structure present-day vent communities in the two oceans, and much further analysis is required.

In addition to the deep-sea hydrothermal communities, a number of sites with shallow-water hydrothermal activity are known from the northern segments of the MAR (e.g. Kolbeinsey, Steinaholl; see Van Dover, 1995). Their general feature is that, although there is venting or bubbling of gas, and bacterial mats develop, no specialized hydrothermal fauna occurs there. The background fauna may concentrate around sites where the production of chemosynthetic or methanotrophic bacteria is added to the photosynthetic primary production, but specific prokaryotic/eukaryotic associations and hydrothermal trophic pyramids are often lacking (Figure 6C) and the level of endemism is usually low (see also Warén and Bouchet, 1993). This is also true for shallow-water gaso-hydrothermal communities in the West Pacific (New Zealand, New Guinea, the Kuril Is) (Tarasov *et al.*, 1990, 1993; Kamenev *et al.*, 1993) and shallow-water seep communities (Hovland and Judd, 1988; Hovland and Thomsen, 1989; Dando *et al.*, 1991). However, single symbiont-containing species may occur in shallow-water vent/seep communities, for example Thyasiridae in Kraternaya Bight (Zhukova *et al.*, 1991) and in the Skagerrak (Schmaljohann *et al.*, 1990; Dando *et al.*, 1994); and Pogonophora in New Zealand (Whale Is) (Malakhov *et al.*, 1992) and the Skagerrak (Flügel and Langhof, 1983; Schmaljohann and Flügel, 1987). Usually these examples belong to the non-vent/seep fauna that occurs also in other reducing habitats.

"Classical" hydrothermal communities fuelled by bacterial chemosynthesis or methane oxidation and dominated by specialized symbiotrophic organisms with a high level of endemism, are known only deeper than *c.* 800 m: Menez Gwen seems to be among the shallowest of them. Above this depth "transitional" hydrothermal communities occur with single specialized species which are not always dominant. Thus, abundant life has been found on the Piip's volcano (380–650 m), with spectacular bacterial mats and associations of *Calyptogena* sp. recorded at *c.* 400 m, but the whole community is dominated by the "background" suspension-feeders, primarily alcyonarians and actiniarians (Sagalevitch *et al.*, 1992). Seep communities dominated by specialized forms seem to occur at much shallower than hydrothermal communities. Thus, abundant vestimentiferan populations have been recorded at about 400 m and deeper in the Gulf of Mexico (Kennicutt *et al.*, 1988; Fisher, 1996; C.M. Young, personal communication) with *Bathymodiolus*-like mytilids appearing at 540 m (MacDonald *et al.*, 1989). Hydrocarbon seep communities were recorded at 470 m at Kaikata Seamount (Warén and Bouchet, 1993) and at 450–600 m on the slope off northern California (Kennicutt *et al.*, 1989). There is an outstanding record of a population of *Lamellibrachia* sp. at only about 80 m off Japan (Hashimoto *et al.*, 1993). However, the rest of the community present at this shallow site has not been described. The influence of depth on the structure of hydrothermal (seep) communities and interaction of depth with other physical and chemical factors and their combined effect on the bacterial and animal metabolism remains little known and is a challenging topic for further researches.

4.3. Geographic Patterns and Relations Between Vent and Seep Fauna

The total number of animal species known from the deep-sea MAR hydrothermal communities exceeds 100, many of them undescribed or identified on a preliminary basis to class or family level. The complete list of MAR hydrothermal fauna in Table 3 is based on published data (e.g. Galkin and Moskalev, 1990a,b; Segonzac, 1992; Segonzac and Vervoort, 1995; Sysoev and Kantor, 1995; Tyler *et al.*, 1995; Van Dover, 1995; Vinogradov, 1995; Bartsch, 1996; Desbruyères and Laubier, 1996; Bellan-Santini and Thurston, 1996; Humes, 1996; Saldanha and Biscoito, 1996; Saldanha *et al.*, 1996; Van Dover *et al.*, 1996; Vereshchaka, 1996, 1997),

Table 3 Fauna of the MAR hydrothermal communities, from published data (references in Table 1) and original observations.

	Logatchev ("14-45")	Snake Pit	TAG	Broken Spur	Lucky Strike	Menez Gwen
Depth (m)	2930–3010	3420–3480	3625–3670	3050–3110	1620–1730	855
DEMOSPONGIA		*Asbestopluma* sp.				
HYDROZOA	Gen. sp.	*Candelabrum serpentarii** *Halisiphonia arctica*	Gen. sp	Gen. sp	*Candelabrum phrygium*	
ACTINIARIA	Gen. sp.	*Parasicyonis ingolfi*	Gen. sp. 1; Gen. sp. 2; Gen. sp. 3 Gen. sp.?	*Parasicyonis ingolfi?* Gen. sp.?	Gen. sp.	
ZOANTHARIA		Gen. sp.			Gen. sp.	
NEMERTINA		*Lineus* sp.				
NEMATODA		*Chromadorita* sp. *Anticoma* sp. *Rhabdocoma* sp. *Araeolaimus* sp. *Acanthopharynx* sp. *Megadesmolaimus* sp.	Gen. sp. 1 Gen. sp. 2 Gen. sp. 3 Gen. sp. 4	Gen. sp.		
POLYCHAETA						
Spionidae		Gen. sp.				
Capitellidae		Gen. sp.				
Ampharetidae		Gen. sp.	Gen. sp.	Gen. sp.	**Amathys lutzi***	
Chaetopteridae		Gen. sp.	Gen. sp.	Gen. sp.		
Amphinomidae		Gen. sp.				
Polynoidae						
commensal		*Branchipolynoe seepensis*			*Branchipolynoe seepensis*	
free-living		Gen. sp.		Gen. sp.		
Maldanidae		Gen. sp.				
Serpulidae						Gen. sp.
Archinomidae			*Archinome* aff. *rosacea*			
Glyceridae					Gen. sp.	

Table 3 Continued

	Logatchev ("14-45")	Snake Pit	TAG	Broken Spur	Lucky Strike	Menez Gwen
GASTROPODA						
Scissurellidae		Sutilizona sp.				
Clypeosectidae		Pseudorimula midatlantica*	Gen. sp.			
Archaeogastropoda		Gen. sp.	Gen. sp.	Gen. sp. 1, Gen. sp. 2	Lepetodrilus sp. Peltospira sp.	Gen. sp.
Skeneidae		Protolira* valvatoides*			Protolira valvatoides Protolira sp.	
Conidae	Phymorhynchus moskalevi	Phymorhynchus sp. 1 Phymorhynchus sp. 2	Phymorhynchus moskalevi*	Phymorhynchus sp.	Phymorhynchus sp.	
Cerithiacea	Gen. sp.					
BIVALVIA						
Mytilidae	Bathymodiolus sp.	Bathymodiolus puteoserpentis*		Bathymodiolus sp.	Bathymodiolus sp.	Bathymodiolus sp.
Vesicomyiodae	Ectenagena sp.	Gen. sp.	Gen. sp.?			
ACARI						
Halacaridae		Agauopsis auzendei*	Gen. sp.		Halacarellus alvinus* Copidognathus alvinus*	Halacarus prolongatus*
PANTOPODA		Sericosura mitrata Ammothea sp.			Sericosura heteroscela* Gen. sp.?	Sericosura heteroscela
CIRRIPEDIA						
Balanomorpha						
OSTRACODA		Poseidonamicus sp.				

Table 3   Continued

	Logatchev ("14-45")	Snake Pit	TAG	Broken Spur	Lucky Strike	Menez Gwen
COPEPODA						
Siphonostomatoida		*Stygiopontius pectinatus** *Stygiopontius cladarus** *Stygiopontius serratus** *Stygiopontius teres** *S. regius** *S. bulbisetiger** *S. mirus** *S. latulus** *Aphotopontius forcipatus** *Rimipontius mediospinifer**	*Stygiopontius pectinatus** Gen. sp.		*Aphotopontius atlanteus**	
Harpacticoida		Gen. sp. 1; Gen. sp. 2; Gen. sp. 3; Gen. sp. 4	Gen. sp.	Gen. sp.		
AMPHIPODA						
Stegocephalidae		*Euandania* aff. *ingens* *Steleuthera ecoprophycea**				
Amphilochidae					*Gitanopsis alvina** *Bouvierella curtirama**	
Eusiridae					*Luckia striki**	
Lysianassidae s. l.				*Hirondella brevicaudata*		
TANAIDACEA		*Typhlotanais* sp.				
ISOPODA			Gen. sp.		Gen. sp.	
DECAPODA						

Table 3 Continued

	Logatchev ("14–45")	Snake Pit	TAG	Broken Spur	Lucky Strike	Menez Gwen
MACRURA, Natantia Alvinocarididae =	*Alvinocaris* aff. *muricola*	***A. markensis****	***A. markensis***	***A. markensis?***		
Bresiliidae ex parte	*A.* aff. *longirostris* ***Rimicaris exoculata***	***Chorocaris chacei*** *Rimicaris exoculata* *Iorania concordia*	***Chorocaris chacei**** ***Rimicaris* exoculata*** ***Iorania* concordia***	***Chorocaris chacei*** ***Rimicaris exoculata*** *Iorania concordia*	***Chorocaris chacei*** *Rimicaris* sp.	*Chorocaris* sp.
Mirocarididae	***Mirocaris keldyshi***		***Mirocaris* keldyshi***	***Mirocaris keldyshi*** *M. fortunata*	***Mirocaris fortunata****	
Nephropsidae		*Thimopides* sp.				
Pandalidae						*Plesionika* sp.1
BRACHYURA						
Bythograeidae	***Segonzacia* sp.**	***Segonzacia* mesatlantica***	***Segonzacia mesatlantica***	***Segonzacia mesatlantica***	***Segonzacia mesatlantica***	
Geryonidae						*Chaceon affinis*
ANOMURA						
Galatheidae	*Munidopsis* sp.	*Munidopsis crassa*	*Munidopsis* sp.	*Munidopsis* sp.		
ECHINODERMATA HOLOTHUROIDEA						
Chiridotidae		*Chiridota* aff. *laevis*	Gen. sp.	Gen. sp.		
OPHIUROIDEA						
Ophiuridae	***Ophiotenella acies***	***Ophiotenella* acies***	*Ophiotenella acies?*	***Ophiotenella* acies***		Gen. sp.
ECHINOIDEA						
Echinidae					*Echinus alexandri*	
PISCES						
Synaphobranchidae		Gen. sp.	Gen. sp.	*Haptenchelys texis?*		*Synaphobranchus* sp.
Macrouridae		*Nematonurus armatus*	Gen. sp.	Gen. sp.	Gen. sp.	
Gadidae					*Gaidopsarus* sp.	
Zoarcidae	***Pachycara thermophilum***	***Pachycara thermophilum****	Gen. sp.	***Pachycara thermophilum***		
Bythitidae					*Cataetyx laiceps*	
Chimaeridae					*Hydrolagus mirabilis*	

*Type locality; bold type, endemic to MAR; underlined, endemic to vents/cold seeps.

supplemented by unpublished preliminary identifications (sponges – Tabachnik; polychaetes – Detinova; bivalves – Krylova; gastropods – Sysoev and Kantor; ophiuroids – Litvinova; pisces – Parin). At present only 44 species have reliable identifications. Of these, at least nine species, *Halisiphonia arctica, Parasicyonis ingolfi, Sericosura mitrata, Munidopsis crassa, Nematonurus armatus, Cataetyx laticeps, Hydrolagus mirabilis, Haptenchelys texis* and *Candelabrum phrygium*, belong to the background fauna and were known long before the discovery of hydrothermal vents. Thirty-one species (including nine copepod species, commensals of shrimps), nine genera, nearly all of them monotypical (*Amathys, Iorania, Luckia, Mirocaris, Ophioctenella, Protolira, Rimicaris, Rimipontius* and *Segonzacia*) and one family (Mirocarididae) are endemic to MAR hydrothermal systems. However, some of them (non-specialized) may be anticipated to occur away from hydrothermal fields, where they have not, so far, found owing to less intensive investigations of the regular abyssal fauna.

Phymorhynchus moskalevi, Bathymodiolus puteoserpentis, Alvinocaris markensis, Archinome rosacea and *Pachycara thermophylum*, belong to genera known also from the East Pacific. *Alvinocaris, Chorocaris, Bathymodiolus, Pseudorimula* and *Phymorhynchus* occur also in the West Pacific. *Alvinocaris, Bathymodiolus* and *Branchipolynoe* are also known from cold seeps in the Atlantic. Thus, the distribution of hydrothermal genera indicates historic links between different geographic regions, namely the East Pacific, the West Pacific and the Atlantic, and between the fauna of hydrothermal vents and cold seeps.

The distribution of such genera as *Alvinocaris, Bathymodiolus* and *Phymorhynchus* (Figures 7–9) indicates clear faunistic links between the West and the East Pacific and the Atlantic. The pattern of distribution (circumtropical, low to moderate latitudes) resembles that known for many marine (including deep-sea) groups of Tethys origin (Menzies *et al.*, 1973; Mironov, 1983, 1985; Briggs, 1995). The history of the hydrothermal fauna is thus not entirely related to the history and configuration of the Mid-Ocean Ridge system (Martin and Hessler, 1990; Tunnicliffe and Fowler, 1996). All major hydrothermal groups, for example Vestimentifera, Vesicomyidae, Mytilidae, Alvinocarididae, Provannidae, Dirivultidae (Copepoda) are common not only for hot vent but also for cold seep communities with a number of genera (*Calyptogena, Ectenagena, Bathymodiolus, Lamellibrachia, Alvinocaris, Provanna*, etc.) and even species (*Branchipolynoe seepensis, ?Alvinocaris muricola*) shared (Warén and Bouchet, 1993; Vrijenhoek *et al.*, 1994; Craddock *et al.*, 1995; and original data). Although the life history and dispersal capabilities of most hydrothermal organisms remain unknown (Mullineaux and France, 1995), the occurrence of some hydrothermal and seep faunal elements far away from geochemical sites, for example on dead whale carcasses (Smith, *et al.*, 1989) and in organic-rich cargoes of wrecked ships (Dando *et al.*, 1992), indicate that larvae of hydrothermal vent animals are likely to be transported thousands of kilometres away from vents and may be able to colonize even small hot spots with favourable conditions. At least five species

(*Vesicomya gigas*, *Calyptogena pacifica*, the mussel *Idasola washingtonia* and two gastropod species of the family Pyroptelidae) are known from both hydrothermal vents and old whale skeletons (Smith *et al.*, 1989; Warén and Bouchet, 1993); one of them, *C. pacifica*, occurs also at cold seeps on the Californian slope (Smith *et al.*, 1989).

The number of species known from both the Atlantic and the Pacific is at present very small: only two copepods, commensals of shrimps, have been reported so far from the both oceans: *Stygiopontius pectinatus* – from the MAR and the Mariana Back-Arc Basin, and *Aphotopontius forcipatus* from the MAR and the north-east Pacific sites (Humes, 1996). Additionally, the polychaete *Archinome rosacea*, known from Galapagos and Guaymas sites, has been found at TAG (N. Detinova, personal communication, Table 2) These facts, although limited, suggest a link (within a recent geological period) between the faunas of the two oceans, as proposed earlier by Martin and Hessler (1990).

It may be assumed that at least some of the vent/seep groups were widely distributed throughout the Tethys Sea Basin, one of the major marine basins of the past, thriving in different reduced habitats with favourable conditions for prokaryotic/eukaryotic symbiosis. After the break-up of Tethys in the Miocene, four Tethys-derived regions appeared: the isolated Gulf of Mexico; the Caribbean; the Mediterranean; and the Indo-Pacific (Menzies *et al.*, 1973). The Pacific coast of Mexico also appears to contain faunal elements derived from Tethys (W.A. Newman, personal communication). These regions took on the role of centres of origin and distribution, whereas some older taxa retained the worldwide circumtropical distribution. Notably, widely distributed vent/seep genera (Figures 7, 8 and 9), presumably of Tethys origin, do not occur in the north-east Pacific which, thus, may have a younger fauna (of post-Miocene age). Clear faunistic links (on the generic level) exist between the faunas of the MAR and the Gulf of Mexico. The two faunas also share a species (*Branchipolynoe seepensis*), or maybe two, if the record of *Alvinocaris muricola* on the MAR (see p. 111) is confirmed.

Intensive radiation of hydrothermal shrimps seems to have occurred on the MAR, where most of the endemic genera and species are present (Figure 10). Less specialized to hydrothermal habitats, the genus *Alvinocaris* is also widely distributed in the Pacific, together with the slightly more specialized genera *Opaepele* and *Chorocaris*. Additionally, *Alvinocaris*, like *Bathymodiolus*, is known from cold-seep habitats. It appears that in general the non-vent fauna of the MAR is more closely linked to the slope fauna of the west Atlantic (Mironov, 1983, 1985; Zezina, 1985) than that of the east Atlantic. On the other hand, the distribution of the non-vent-specific *Phymorhynchus*, which is very common in hydrothermal communities (Figure 9), demonstrates the links of the MAR fauna with both sides of the Atlantic, presumably as a result of the Tethys-age distribution. Similarly, indications exist for the relations between the Guaymas Basin fauna and the fauna in the coastal vents and seeps off the western USA (Warén and Bouchet, 1993). Additionally, recent studies of the molecular

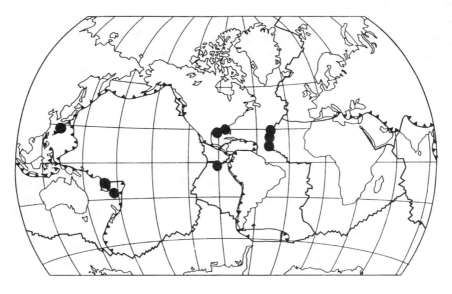

Figure 7   Known global distribution of the genus *Alvinocaris*.

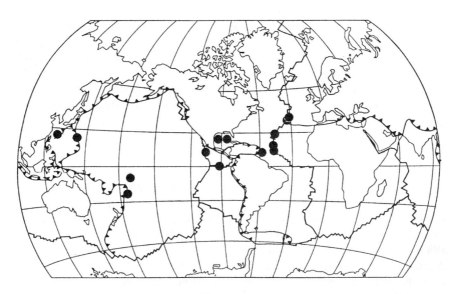

Figure 8   Known global distribution of the genus *Bathymodiolus* (after Von Cosel *et al.*, 1994, with additions).

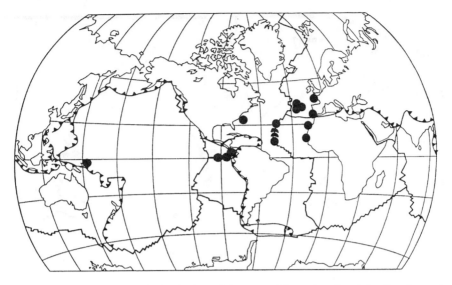

Figure 9   Known global distribution of the genus *Phymorhynchus* (after Bouchet and Warén, 1980; Sysoev and Kantor, 1995, with additions).

phylogeny of *Bathymodiolus* have demonstrated seep–vent links, with the vent group of species presumably deriving from those adapted to seeps (Craddock *et al.*, 1995).

We cannot thus exclude an ancestral role for seep fauna as a source of different hydrothermal animals. Cold-water seeps are widely distributed on passive margins along the continental shelf (in a broad range of depths) and at the base of active subduction zones (see Olu *et al.*, 1996 for recent review). This hypothesis is supported by the apparent stability of seep habitats and the seep ancestors that gave rise to hydrothermal forms, at least among some groups (e.g. Alvinocarididae and Mytilidae).

Comparison of the hydrothermal fauna from the six Atlantic sites is limited by the small set of reliable data. Many of the sites share some shrimp species (Figure 10). *R. exoculata* is known at present from four sites: from 14–45 in the south to Broken Spur in the north. *C. chacei* has been recorded at the Snake Pit, TAG and Broken Spur sites which share also some other hydrothermal species. *Mirocaris keldyshi* occurs at Broken Spur, TAG and 14–45, the latter population having some difference at the subspecies level from the other two (original data). *Iorania concordia* is known only from TAG and Broken Spur. TAG and Broken Spur also share the *Mirocaris keldyshi* species. *Alvinocaris markensis* is known from Snake Pit, TAG and Broken Spur, whereas at 14–45 two other species have been recorded: *A.* aff. *muricola* and *A.* aff. *longirostris*. Finally, Broken Spur shares the

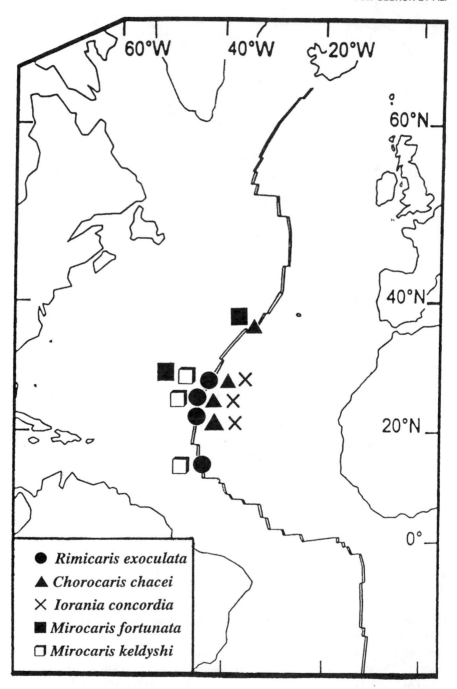

Figure 10 Presently known distribution of alvinocaridid and mirocarid species along the MAR.

species *Mirocaris* (=*Chorocaris*) *fortunata* with Lucky Strike (Martin and Christiansen, 1995).

With increasing investigation, species first described as endemic to one site, are often found elsewhere. Thus, *Phymorhynchus moskalevi* described from TAG (Sysoev and Kantor, 1995) was recently found at 14–45 (A. V. Sysoev, personal communication). The ophiuroid *Ophioctenella acies* originally described from Snake Pit and Broken Spur (Tyler *et al.*, 1995) was also recently found at 14–45 (N.M. Litvinova, personal communication). Thus, it seems there may be no major barriers to dispersal along the MAR, at least between the five deep-water sites investigated (except perhaps for Menez Gwen). It seems possible that the numerous microhabitats described for hydrothermal sites has led to greater diversity in groups such as shrimps that can exploit the varied trophic niches thus made available.

ACKNOWLEDGEMENTS

We thank many colleagues for help and assistance at sea and in the laboratory. Especial thanks are due to Dr Eve Southward for checking the taxonomy and the literature. Financial support came from INTAS (grant 94-592) and NERC (BRIDGE).

REFERENCES

Allen, C.E., Varney, M.S. and Tyler, P.A. (1996) Lipid composition of *Rimicaris exoculata* from the TAG hydrothermal field, 26°N MAR. *BRIDGE Newsletter* **10**, 18–20.
Anon (1995) Actually, there are many more "black smokers". *Priroda* **5**, 103–104. (In Russian).
Arzamastsev, I.S. and Preobrazhenskii, B.V. (1990) "Atlas of underwater landscapes of the Sea of Japan". Nauka, Moscow. (In Russian).
Avdonin, V.V. (1995) Analogies of black smokers from pyrite ore deposits. *Vestnik Moskovskogo Universiteta* **5**, 50–56. (In Russian).
Bartsch, I. (1996) Halacarid mites (Acari) from the Mid-Atlantic Ridge. *Cahiers de Biologie Marine* **37**(2), 159–167.
Batuyev, B.N., Krotov, A.G., Markov, V.F., Cherkashev, G.A., Krasnov, S.G. and Lisitsyn, Y.D. (1994) Massive sulphide deposits discovered and sampled at 14°45'N, Mid-Atlantic Ridge. *BRIDGE Newsletter* **6**, 6–10.
Bellan-Santini, D. and Thurston, M. (1996) Amphipoda of the hydrothermal vents along the Mid-Atlantic Ridge. *Journal of Natural History* **30**, 685–702.
Bogdanov, Y.A., Sagalevitch, A.M., Chernyaev, E.S., Ashadze, A.M., Gurvich, E.G., Lukashin, V.N., Ivanov, G.V. and Peresypkin,V.N. (1995a) Hydrothermal field on 14°45' on the Mid-Atlantic Ridge. *Doklady Akademii Nauk* **343**(3), 353–357. (In Russian).
Bogdanov, Y.A., Sagalevitch, A.M., Chernyaev, E.S., Ashadze, A.M., Gurvich, E.G.,

Lukashin, V.N., Ivanov, G.V. and Peresypkin, V.N. (1995b) A study of the hydrothermal field at 14°45' on the Mid-Atlantic Ridge using the "Mir" submersibles. *BRIDGE Newsletter* **9**, 9–13.

Bouchet, P. and Warén, A. (1980) Revision of the northeast Atlantic bathyal and abyssal Turridae (Mollusca, Gastropoda). *Journal of Molluscan Studies*, Supplement 8, 121 pp.

Briggs, J.C. (1995) "Global Biogeography". Elsevier Science B.V., Amsterdam, The Netherlands.

Casanova, B., Brunet, M. and Segonzac, M. (1993) L'impact d'une epibiose bacterienne sur la morphologie fonctionelle de crevettes associées à l'hydrothermalisme médio-Atlantique. *Cahiers de Biologie Marine* **34**, 573–588.

Child, C.A. and Segonzac, M. (1996) *Sericosura heteroscela* and *S. cyrtoma*, new species, and other Pycnogonida from Atlantic and Pacific hydrothermal vents, with notes on habitat and environment. *Proceedings of the Biological Society of Washington* **109**, 664–676.

Childress, J.J., Favuzzi, J.A, Sanders, N.K, and Alayse, A.M. (1991) Sulfide-driven autotrophic balance in the bacterial-symbiont-containing hydrothermal tubeworm, *Riftia pachyptila* Jones. *Biological Bulletin, Marine Biological Laboratory, Woods Hole* **180**, 135-153.

Choukroune, P., Francheteau, J., Auvray, B., Auzende, J.M., Brun, J.B., Sichler, B., Arthaud, F. and Lepine, C. (1988) Tectonics of an incipient oceanic rift. *Marine Geophysical Researches* **9**, 147–163.

Christoffersen, M.L. (1986). Phylogenetic relationships between Oplophoridae, Atyidae, Pasiphaeidae, Alvinocarididae Fam. N., Bresiliidae, Psalidopodidae and Disciadidae (Crustacea Caridea Atyoidea). *Boletim de Zoologia, Universidade de Sao Paulo* **10**, 273–281.

Christoffersen, M.L. (1990) A new super family classification of the Caridea (Crustacea: Pleocyemata) based on phytogenetic pattern. *Zeitschrift fur zoologische Systematic und Evolutionsforschung* **28**, 94–106.

Copley, J.T.P. and Tyler, P.A. (1995) Faunal distribution at Broken Spur Vent Field: Recovery Revisited. *BRIDGE Newsletter* **9**, 29–33.

Copley, J.T.P., Tyler, P.A., Sheader, M., Murton, B.J. and German, C.R. (1995) Non-vent fauna of the Reykjanes Ridge. *BRIDGE Newsletter* **8**, 43–48.

Copley, J.T.P., Tyler, P.A., Sheader, M., Murton, B.J. and German, C.R. (1996) Megafauna from sublitoral to abyssal depths along the Mid-Atlantic Ridge south of Iceland. *Oceanologica Acta* **19**(5), 549–559.

Craddock, C., Hoeh, W.R., Gustafson, R.G., Lutz, R.A., Hashimoto, J. and Vrijenhoek, R.J. (1995) Evolutionary relationships among deep-sea mytilids (Bivalvia: Mytilidae) from hydrothermal vents and cold-water methane/sulfide seeps. *Marine Biology* **121**, 477–85.

Crane, K., Johnson, L., Appelgate, B., Buck, R., Jones, C., Gurvich, E., Moskalev, L., Gebruk, A., Lukashin, V., Luk'yanov, S., Scherbinin, A., Serova, V. and Rudenko, M. (1992) Investigating the Reykjanes Ridge near 59°50'N with the Russian Mir submersibles. *EOS Transactions of the American Geophysical Union* **73**, 531.

Creasey, S., Rogers, A.D. and Tyler, P.A. (1996) Genetic comparison of two populations of the deep-sea vent shrimp *Rimicaris exoculata* (Decapoda: Bresiliidae) from the Mid-Atlantic Ridge. *Marine Biology* **125**, 473–482.

Dando, P.R., Austen, M.C., Burke, R.J., Kendall, M.A., Kennicutt, M.C., Judd, A.J., Moore, D.C. O'Hara, S.C.M., Schmaljohann, R. and Southward, A.J. (1991) Ecology of a North Sea pockmark with an active methane seep. *Marine Ecology Progress Series* **70**, 49–63.

Dando, P.R., Southward, A.J., Southward, E.C., Dixon, D.R., Crawford, A. and Crawford, M. (1992) Shipwrecked tube worms. *Nature* **356**, 667.

Dando, P.R., Bussmann, I., Niven, S.J., O'Hara, S.C.M., Schmaljohann, R. and Taylor, L.J. (1994) A methane seep area in the Skagerrak, the habitat of the pogonophore *Siboglinum poseidoni* and the bivalve mollusc *Thyasira sarsi*. *Marine Ecology Progress Series* **107**, 157–167.

Desbruyères, D. and Laubier, L. (1996) A new genus and species of ampharetid polychaete from deep-sea hydrothermal vent community in the Azores triple-junction area. *Proceedings of the Biological Society of Washington* **109**(2), 248–255.

Desbruyères, D., Gaill, F., Laubier, L., Prieur, D. and Rau, G. (1983) Unusual nutrition of "Pompeii" worm *Alvinella pompejana* (polychaetous annelid) from a hydrothermal vent environment: SEM, TEM ^{13}C and ^{15}N evidence. *Marine Biology* **75**, 201–205.

Desbruyères, D., Alayse, A.-M., Antoine, E., Barbier, G., Barriga, F., Biscoito, M., Briand, P., Brulport, J.-P., Comet, T., Cornec, L., Crassous, P., Dando, P., Fabri, M.C., Felbeck, H., Lallier, F., Fiala-Médioni, A., Gonçalves, J., Ménard, F., Kerdoncuff, J., Patching, J., Saldanha, L. and Saradin, P.-M. (1994) New information on the ecology of deep-sea vent communities in the Azores Triple Junction area: Preliminary results of the DIVA 2 Cruise (May 31–July 4, 1994). *InterRidge News* **3**(2), 18–19.

Dixon, D.R. and Dixon, L.R.J. (1996) Results of DNA analyses conducted on vent-shrimp postlarvae collected above the Broken Spur vent field during the CD95 cruise, August 1995. *BRIDGE Newsletter* **11**, 9–15.

Enright, J.T., Newman, W.A., Hessler, R.R. and J.A. McGowan (1981) Deep-ocean hydrothermal vent communities. *Nature* **289**, 219–221.

Fedorov, V.D. (1982) "Recommendation of the Method of Marine Landscape Investigations for Fishery needs". VNIRO, Moscow. (In Russian).

Fisher, C.R. (1996) Ecophysiology of primary production at deep-sea vents and seeps. *In* "Deep-sea and Shallow-water Extreme Habitats: Affinities and Adaptations" (F. Uiblein, J. Ott and M. Stachowitsch, eds), pp. 313–336. *Biosystematics and Ecology Series 11*, Vienna.

Fisher, C.R., Childress, J.J., Oremland, R.S. and Bidigare, R.R. (1987) The importance of methane and thiosulfate in the metabolism of the bacterial symbionts of two deep-sea mussels. *Marine Biology* **96**, 59–71.

Fülgel, H.J. and Langhof, I. (1983) A new hermaphroditic pogonophore from the Skagerrak. *Sarsia* **67**, 211–212.

Fornari, D., Van Dover, C.L., Shank, T., Lutz, R. and M. Olsson (1994) A versatile, low-cost temperature-sensing device for time-series measurements at deep-sea hydrothermal vents. *BRIDGE Newsletter* **6**, 40–47.

Fouquet, Y., Wafik, A., Cambon, P., Mevel, C., Meyer, G. and Gente. P. (1993) Tectonic setting and mineralogical and geochemical zonation in the Snake Pit sulfide deposit (Mid-Atlantic Ridge at 23°N). *Economic Geology* **88**, 2018–2036.

Fouquet, Y., Charlou, J.-L. Costa, I., Dnavl, J.-P., Radford-Knoery, J., Pelle, H., Ondreas, H., Lourenco, N., Segonzac, M. and Tivey, M.K. (1994) A detailed study of the Lucky Strike Hydrothermal Site and Discovery of a new Hydrothermal Site: Menez Gwen; Preliminary Results of the DIVA 1 Cruise (5–29 May, 1994). *InterRidge News* **3**(2), 14–17.

Fustec, A., Desbruyères, D. and Juniper, K. (1987) Deep-sea hydrothermal vent communities at 13°N on the East Pacific Rise: microdistribution and temporal variations. *Biological Oceanography* **4**(2), 121–164.

Gal'chenko, V.F., Galkin, S.V., Lein, A.Y., Moskalev, L.I. and Ivanov, M.V. (1988) Role of bacterial symbionts in nutrition of invertebrates from areas of active underwater hydrothermal systems. *Oceanology* **28**, 786–794.

Galkin, S.V. (1992) Bottom fauna of the hydrothermal vents in the Manus Basin. *Okeanologiya* **32**(6), 1102–1110. (In Russian, English translation in *Oceanology* **32**(6), 768–774).

Galkin, S.V. (1993) Deep sea landscapes. *Deep Sea Newsletter* **20**, 18–19.

Galkin, S.V. and Moskalev, L.I. (1990a) Investigation of the abyssal fauna of the North Atlantic with deep-water manned submersibles. *Okeanologiya* **30**(4), 682–689. (In Russian).

Galkin, S.V. and Moskalev, L.I. (1990b) Hydrothermal fauna of the Mid-Atlantic Ridge. *Okeanologiya* **30**(5), 842–847. (In Russian, English translation in *Oceanology* **30**(5), 624–627).

Galkin, S.V. and Moskalev, L.I. (1990c). Benthos of the Axial Seamount, Juan de Fuca Ridge. *In* "Geological Structure and Hydrothermal Deposits of the Juan de Fuca Ridge". (A.P. Lisitsyn, ed.), pp. 157–177. Nauka, Moscow, Russia. (In Russian).

Gebruk, A.V. (1996) Feeding strategies of hydrothermal vent shrimp. *Journal of Conference Abstracts* **1**(2), 791–792.

Gebruk, A.V., Pimenov, N.V. and Savvichev, A.S. (1992) Bacterial "farm" on the shrimp mouthparts. *Priroda (Moscow)* **6**, 37–39. (In Russian).

Gebruk, A.V., Pimenov, N.V. and Savvichev, A.S. (1993) Feeding specialization of bresiliid shrimps in the TAG site hydrothermal community. *Marine Ecology Progress Series* **98**, 247–253.

Gebruk, A.V., Lein, A.Y., Pimenov, N.V., Miller, Y.V., Galkin, S.V. and Ivanov, M.V. (1997) Trophic Structure of the Broken Spur hydrothermal community as revealed by stable isotope and C:H:N:S data. *BRIDGE Newsletter* **12**.

German, C.R., Parson, L.M. and HEAT Scientific Team (1996) Hydrothermal exploration near the Azores Triple Junction: tectonic control of venting at slow-spreading ridges? *Earth and Planetary Science Letters* **138**, 93–104.

Glaessner, M.F. (1969) Decapoda. *In* "Treatise on Invertebrate Paleontology". Pt *R, Arthropoda* **4**, 2, 400–535. Geological Society of America, Boulder, CO.

Grassle, J.F. (1986) The ecology of deep-sea hydrothermal vent communities. *Advances in Marine Biology* **23**, 301–362.

Hashimoto, J., Miura, T., Fujikura, K. and Osaka, J. (1993) Discovery of vestimentiferan tube-worms in the euphotic zone. *Zoological Science* **10**, 1063–1067.

Herring, P.J. (1996) Travelling shrimp. *BRIDGE Newsletter* **11**, 6–8.

Hessler, R.R. and. Kaharl, V.A. (1995) The deep-sea hydrothermal vent community: an overview. *In* "Seafloor Hydrothermal Systems". (S.E Humphris, R.A. Zierenberg, L.S. Mullineaux and R.E. Thomson, eds), pp. 47–71. American Geophysical Union, Washington, D.C., USA.

Hessler, R., Lonsdale. P. and Hawkins, J. (1988) Patterns on the ocean floor. *New Scientist* **177** (1605), 47–51.

Holthuis, L.B. (1993) "The Recent Genera of the Caridean and Stenopodidean Shrimps (Crustacea, Decapoda); with an Appendix on the Order Amphionidacea"Nationaal Natuurhistorisch Museum, Leiden.

Hovland, M. and Judd, A. G. (1988) "Seabed Pockmarks and Seepages: Impact on Geology, Biology and the Marine Environment". Graham & Trotman, London.

Hovland, M. and Thomsen E., (1989) Hydrocarbon-based communities in the North Sea? *Sarsia* **74**, 29–42.

Humes, A.G. (1996) Deep-sea Copepoda (Siphonostomatoida) from hydrothermal sites on the Mid-Atlantic Ridge at 23° and 37°N. *Bulletin of Marine Science* **58**(3), 609–653.

Humphris, S.E., Herzig, P.M. and Miller, D.J. (+22 authors) (1995) The internal structure of an active sea-floor massive sulphide deposit. *Nature* **377**(6551), 713–716.

Jannasch, H.W. (1995) Microbial Interactions with hydrothermal fluids. *In* "Seafloor

Hydrothermal Systems" (S.E Humphris, R.A. Zierenberg, L.S. Mullineaux and R.E. Thomson, eds), pp. 273–296. American Geophysical Union, Washington, D.C., USA.

Jannasch, H.W., Wirsen, C.O. and Molyneaux, S.J. (1991) Chemosynthetic microbial activities at the 23° and 26°N Mid-Atlantic Ridge vent sites. *RIDGE Events* **2**(2), (Nov. 1991), p.19.

Johnson, M.L., Shelton, P.M.J., Herring, P.J. and Gardner, S. (1995) Spectral responses from the dorsal organ of a juvenile *Rimicaris exoculata* from the TAG hydrothermal vent site. *BRIDGE Newsletter* **8**, 38–42.

Jones, M.L (ed.) (1985) "Hydrothermal Vents of the Eastern Pacific: An Overview". *Bulletin of the Biological Society of Washington* **6**.

Juniper, S.K., Tunnicliffe, V. and Desbruyères, D. (1990) Regional-scale features of northeast Pacific, East Pacific Rise, and Gulf of Aden vent communities. *In* "Gorda Ridge. A Seafloor Spreading Center in the United States Exclusive Economic Zone" (G.R. McMurray, ed.), pp. 265–278. Springer, New York.

Juniper, S.K., Jonasson, I.R., Tunnicliffe, V. and Southward, A.J. (1992) Alteration of hydrothermal chimney mineralization by a tube-building polychaete. *Geology* **20**, 895–898.

Kamenev, G.M., Fadeev, V.I., Selin, N.I., Tarasov, V.G. and Malakhov, V.V. (1993) Composition and distribution of macro- and meiobenthos around sublitoral hydrothermal vents in the Bay of Plenty, New Zealand. *New Zealand Journal of Marine and Freshwater Research* **27**, 407–418.

Karl, D.M. (1995) Ecology of Free-Living, Hydrothermal Vent Microbial Communities. *In* "The Microbiology of the Deep-Sea Hydrothermal Vents" (D.M. Karl, ed.), pp. 35–124. CRC Press, London.

Karson, J.A. and Brown, J.R. (1988) Geologic setting of the Snake Pit hydrothermal site: an active vent field on the Mid-Atlantic Ridge. *Marine Geophysical Research* **10**, 91–107.

Kennicutt M.C. II, Brooks, J.M., Bidigare, R.R. and Denoux, G. (1988) Gulf of Mexico hydrocarbon seep comunities – I. Regional distribution of hydrocarbon seepage and associated fauna. *Deep-Sea Research* **35**(9), 1639–1651.

Kennicutt, M.C. II, Brooks, J.M., Bidigare, R.R., McDonald, S.J., Adkison, D.L. and Macko, S.A. (1989) An upper slope "cold" community: Northern California. *Limnology and Oceanography* **34**(3), 635–640.

Kikuchi, T. and Ohta, S. (1995) Two caridean shrimps of the families Bresiliidae and Hippolytidae from a hydrothermal field on the Iheya Ridge, off Ryuku Islands, Japan. *Journal of Crustacean Biology* **15**, 771–785.

Kong, L., Ryan, W.B., Mayer, L., Detrik, R.S., Fox, P.J. and Manchester, K. (1985) Bare-rock drill sites, ODP Legs 106 and 109: evidence for hydrothermal activity at 23°N on the Mid-Atlantic Ridge. *EOS* **66**, 1106.

Krasnov, S.G., Cherkashev, G.A., Stepanova, T.V. and ten other authors (1995) Detailed geological studies of hydrothermal fields in the North Atlantic. *In* "Hydrothermal Vents and Processes" (L.M. Parson, C.L Walker and D.R Dixon, eds), pp. 43–64. Geological Society, London. *Special Publication* **87**.

Krasnov, S.G., Cherkashev, G.A., Poroshina, I.M., Fouquet, Y., Prieur, D. and Ashadze, A.M. (1996) 15°N, Mid-Atlantic Ridge – Logatchev Hydrothermal Field. *Journal of Conference Abstracts* **1**(2), 809–810.

Kuznetsov, A.P. (ed.) (1985) "Bottom Fauna of the Open-oceanic Elevations (Northern Atlantic)". *Trudy Instituta Okeanologii*, Moscow **120**, 192 pp. (In Russian).

Kuznetsov, A.P., Galkin, S.V. and Rass, T.S. (1987) New benthic community based on photo- and chemosynthetic food sources. *In* "Nutrition of Marine Invertebrates and its

Role in Formation of Communities" (A.P. Kuznetsov, ed.), pp. 6–15. Akademii Nauk SSSR, Moscow. (In Russian).

Laubier, L. (1986) "Des Oasis au Fond des Mers". Science et Decouvertes 3. Le Rocher.

Lisitsyn, A.P., Bogdanov, Y.A., Zonenshain, L.P., Kuz'min, M.N. and Sagalevitch, A.M. (1989) Hydrothermal manifestations on 26°N of the Mid-Atlantic Ridge (TAG hydrothermal field). *Izvestiya Akademii Nauk SSSR*, Ser. Geol. **12**, 3–20. (In Russian).

Lisitsyn, A.P., Cherkashov, G.A. and Shashkov, N.L. (1990) Study of a hydrothermal spring in the Atlantic Ocean from submersibles "Mir". *Doklady Akademii Nauk SSSR* **311**(6), 1462–1467.

Little, C.T.S., Herrington, R.J. and Zaykov, V.V. (1995) A Report on a Silurian-age hydrothermal vent community from the southern Urals, Russia. *BRIDGE Newsletter,* **9**, 5–8.

Little, C.T.S., Herrington, R.J., Maslennikov, V.V., Morris, N.J. and Zaykov, V.V. (1997) Silurian hydrothermal-vent community from the southern Urals, Russia. *Nature, London,* **385**, 146–148.

MacDonald, I.R., Boland, G.S., Baker, J.S., Brooks, J.M., Kennicutt, M.C. II and Bidigare, R.R. (1989) Gulf of Mexico hydrocarbon seep communities – II. Spatial distribution of seep organisms and hydrocarbons at Bush Hill. *Marine Biology* **101**, 235–247.

Malakhov, V.V., Obzhirov, A.I. and Tarasov, V.G. (1992) On the relation of pogonophoran genus *Siboglinum* to zones of high methane concentration. *Doklady Akademii Nauk* **325**(1), 195–197 (In Russian).

Martin, J.W. and Christiansen, J.C. (1995) A new species of the shrimp genus *Chorocaris* Martin and Hessler, 1990 (Crustacea: Decapoda: Bresiliidae) from hydrothermal vents along the Mid-Atlantic Ridge. *Proceedings of the Biological Society of Washington* **108**(2), 220–227.

Martin, J.W. and Hessler, R.R. (1990) *Chorocaris vandoverae*, a new genus and species of hydrothermal vent shrimp (Crustacea, Decapoda, Bresiliidae). *Contributions in Science, Los Angeles County Museum* **417**, 1–11.

Maslennikov, V.V., Zaykov, V.V. and Kuznetsov, A.P. (1995) Hydrothermal-like fauna in segments of rift valleys in the Urals Paleoocean. *BRIDGE Newsletter* **8**, 81.

Menzies, J.R., George, R.Y. and Rowe, G.T. (1973) "Abyssal Environment and Ecology of World Oceans". Wiley, New York.

Mevel, C., Auzende, J.-M., Cannat, M., Donval, J.-P., Dubois, J., Fouquet, Y., Gente, P., Grimaud, D., Karson, J., Segonzac, M. and Stievenard, M. (1989) La ride du Snake Pit (dorsale medio-Atlantique, 23°22'N): résultats préliminaires de la campagne HYDRO-SNAKE. *Comptes Rendus Hebdomadaires des Séances de l'Académie des Sciences, Paris*, **308**(II), 545–552.

Mironov, A.N. (1983) Accumulation effect in the distribution of sea urchins. *Zoologiches-kii Zhurnal* **62**(8), 1202–1208. (In Russian, English summary).

Mironov, A.N. (1985) Circumcontinental zonation in distribution of Atlantic echinoids. *Trudy Instituta Okeanologii* **120**, 70–95. (In Russian).

Mullineaux, L.S. and France, S.C. (1995) Dispersal mechanisms of deep-sea hydrothermal vent fauna. *In* "Seafloor Hydrothermal Systems" (S.E Humphris, R.A. Zierenberg, L.S. Mullineaux and R.E. Thomson, eds), pp. 408–424. American Geophysical Union, Washington, D.C., USA.

Murton, B.J., Klinkhammer, G. +11 authors (1994) Direct evidence for the distribution and occurrence of hydrothermal activity between 27–30°N on the Mid-Atlantic Ridge. *Earth and Planetary Science Letters* **125**, 119–128.

Murton, B.J., Van Dover, C.L. and Southward, E.C. (1995) Geological setting and ecology of the Broken Spur hydrothermal vent field at 29°10'N on the Mid-Atlantic Ridge. *In* "Hydrothermal Vents and Processes" (L.M. Parson, C.L. Walker and D.R. Dixon, eds), pp. 33–42. *Geological Society Special Publication* **87**.

Nelsen, T.A. (1988) New venting in TAG Field. *EOS Transactions of the American Geophysical Union* **69**(46), 1573.

Nelson, D.C. and Fisher. C.R. (1995) Chemoautotrophic and methanotrophic endosymbiotic bacteria at vents and seeps. *In* "Microbiology of Deep-sea Hydrothermal Vents" (D.M. Karl, ed.), pp. 125–167. CRC Press, Boca Raton, FL.

Nesbitt, R.W. (1995) The geology of the Broken Spur Hydrothermal Vent Site; a new look at an old field. *BRIDGE Newsletter* **8**, 30–34.

Nix, E.R., Fisher, C.R.,Vodenichar, J. and Scott, K.M. (1995) Physiological ecology of a mussel with methanoautotrophic endosymbionts at three hydrocarbon seep sites in the Gulf of Mexico. *Marine Biology* **122**, 605–617.

Nuckley, D.J., Jinks, R.N., Battelle, B.-A., Herzog, E.D., Kass, L., Renninger, G.H. and Chamberlain, S.C. (1996) Retinal anatomy of a new species of bresiliid shrimp from a Hydrothermal vent field on the Mid-Atlantic Ridge. *Biological Bulletin, Marine Biological Laboratory, Woods Hole* **190**, 98–110.

Olu, K., Duperet, A., Sibuet, M., Foucher, J.-P. and Fiala-Medioni, A. (1996) Structure and distribution of cold communities along the Peruvian active margin: relationship to geological and fluid patterns. *Marine Ecology Progress Series* **132**, 109–125.

O'Neil, P.J., Jinks, R.N., Herzog, E.D., Battelle, B.-A., Kass, L., Renninger, G.H., and Chamberlain, S.C. (1995) The morphlogy of the dorsal eye of the hydrothermal vent shrimp, *Rimicaris exoculata*. *Visual Neuroscience* **12**, 861–875.

Page, H.M., Fisher, C.R. and Childress, J.J. (1990) Role of filter-feeding in the nutritional biology of a deep-sea mussel with methanotrophic symbionts. *Marine Biology* **104**, 251–257.

Petrov, K.M. (1971) Sea-shore zone as a landscape system. *Izvestiya VGO* **103**(5), 391–396. (In Russian).

Pettersson, H. (1970) Around the World on the "Albatross". Gidrometeoizdat, Leningrad. (Russian translation of: Petersson, H. (1954) "La croisière aux abimes. Autour du monde avec l'expedition océanographique suedoise". Biblioteque de la Mer, Paris).

Pimenov, N.V., Savvichev, S.A., Gebruk, A.V., Moskalev, L.I., Lein, A.Y. and Ivanov, M.V. (1992) Trophical specialization of bresiliid shrimps in the TAG site hydrothermal community. *Doklady Akademii Nauk* **323**(3), 567–571 (In Russian).

Polz, M.F. and Cavanaugh, C.M. (1995a) The ecology of ectosymbiosis at the Mid-Atlantic Ridge hydrothermal vents. Symposium Abstracts "Deep-sea and Extreme Shallowwater Habitats: Affinities and Adaptations", October 6–8, Vienna 1995.

Polz, M.F. and Cavanaugh, C.M (1995b) Dominance of one bacterial species at a Mid-Atlantic Ridge hydrothermal vent site. *Proceedings of the National Academy of Sciences of USA* **92**, 7232–7236.

Polz, M.F. and Cavanaugh, C.M. (1996) The ecology of ectosymbiosis at a Mid-Atlantic Ridge hydrothermal vent site. *In* "Deep-sea and Extreme Shallow-water Habitats: Affinities and Adaptations" (F. Ublein, J. Ott and M. Stachowtisch, eds), pp. 337–352. *Biosystematics and Ecology Series* **11**.

Preobrazhenskii, B.V. (1980) Landscape as a characteristic of ecosystems. *In* "Methods of Complex Mapping of Shelf Ecosystems". Vladivostok, DVNC AN SSSR, 23–28. (In Russian).

Preobrazhenskii, B.V. (1984) Main purposes of marine landscape science. *Georgaphiya i prirodnye resursy* (Geography and natural resources) **1**, 15–22. (In Russian).

Renninger, G.H., Kass, L., Gleeson, R.A., Van Dover, C.L., Battelle, B.A., Jinks, R.N. and Herzog, E.D. (1995) Sulfide as a chemical stimulus for deep-sea hydrothermal vent shrimp. *Biological Bulletin* **189**, 69–76.

Rieley, G., Van Dover, C.L. Hedrick, D.B. and Eglinton, G. (1996) Trophic ecology of *Rimicaris exoculata* from TAG as revealed by relative abundances and isotopic compositions of individual fatty acids. *Journal of Conference Abstracts* **1**(2), 845–847.

Rona, P.A. and Scott, S.D. (1993) Special issue on sea floor hydrothermal mineralization: New perspectives. Preface. *Economic Geology* **88**, 1935–1976.

Rona, P.A., Klinkhammer, G., Nelsen, T.A., Trefry, J.H. and Elderfield, H. (1986) Black smokers, massive sulfides and biota at the Mid-Atlantic Ridge. *Nature* **312**, 33–37.

Rudnicki, M.D., German, C.R., Kirk, R.E., Sinha, M., Elderfield, H. and Riches, S. (1995) New instrument platform tested at Mid-Atlantic Ridge. *EOS* **76**(33), 329–330.

Russell, M.J. (1996) The generation at hot springs of sedimentary ore deposits, microbialites and life. *Ore Geology Reviews* **10**, 199–214.

Sagalevitch, A.M and Bogdanov, Y.A. (1995) First dives on the "Mir" submersibles on a new hydrothermal field in the Atlantic. *Ocean's 95, MTS/IEEE, Conference Proceedings, San Diego* **3**, 1511–1515.

Sagalevitch, A.M. and Moskalev, L.I. (1992) Chemobios on the Pacific Ocean seafloor. *Priroda* **5**, 33–40. (In Russian).

Sagalevitch, A.M., Torohov, P.V., Matveenkov, V.V., Galkin, S.V. and Moskalev, L.I. (1992) Hydrothermal manifestations of Piip's submarine volcano (Bering Sea). *Izvesiya Akademii Nauk, Seriya Geologicheskaya* **9**, 104–114. (In Russian).

Saldanha, L. and Biscoito, M. (1996) On the Fish Fauna from the Lucky Strike and Menez Gwen hydrothermal vent fields (Mid-Atlantic Ridge). *Journal of Conference Abstracts* **1**(2), 853.

Saldanha, L., Biscoito, M. and Desbruyères, D. (1996) The Azorean deep-sea hydrothermal ecosystem: Its recent discovery. *In* "Deep-sea and Shallow-water Extreme Habitats: Affinities and Adaptations" (F. Uiblein, J. Ott and M. Stachowitsch, eds), pp. 383–388. *Biosystematics and Ecology Series* 11.

Schmaljohann, R. and Flügel, H.J. (1987) Methane-oxidizing bacteria in Pogonophora. *Sarsia* **72**, 91–98.

Schmaljohann, R., Faber, E., Whiticar, M.J. and Dando, P.R. (1990) Co-existence of methane and sulphur-based endosymbioses between bacteria and invertebrates at a site in the Skagerrak. *Marine Ecology Progress Series* **61**, 119–124.

Segonzac, M. (1992) Les peuplements associés a l'hydrothermalisme océanique du Snake Pit (dorsale médio-atlantique; 23°N, 3480 m): composition et microdistribution de la mégafaune. *Comptes Rendus Hebdomadaires des Séances de l'Académie des Sciences de Paris.* Ser. III **314**(13), 593–600.

Segonzac, M. and Vervoort, W. (1995). First record of the genus *Candelabrum* (Cnidaria, Hydrozoa, Athecata) from the Mid-Atlantic Ridge: a description of a new species and a review of the genus. *Bulletin du Muséum National d'Histoire Naturelle, Sect. A, Zoologie* **17**, 31–63.

Segonzac, M., de Saint Laurent, M. and Casanova, B. (1993) L'énigme du comportement trophique des crevettes Alvinocarididae des sites hydrothermaux de la dorsale médio-atlantique. *Cahiers de Biologie Marine* **34**, 535–571.

Smith, C.R., Kukert, H., Wheatcroft, R.A., Jumars, P.A. and Deming, J.W. (1989) Vent fauna on whale remains. *Nature* **341** (6237), 27–28.

Stephenson T.A and Stephenson, A. (1972) "Life Between Tidemarks on Rocky Shores". Freeman, San Francisco.

Sudarikov, S.M. and Galkin, S.V. (1994) Hydrothermal ecosystems and biogeochemistry of the Mid-Atlantic Ridge. *Abstracts Seventh Deep-Sea Symposium*, Institute of Marine Biology, Crete, p. 36.

Sudarikov, S.M. and Galkin, S.V. (1995) Geochemistry of the Snake Pit vent field and its implications for vent and non-vent fauna. *In* "Hydrothermal Vents and Processes" (L.M. Parson, C.L. Walker, and D.R. Dixon, eds), pp. 319–327. Geological Society, London. Special Publication **87**.

Sysoev, A.V. and Kantor, Y.I. (1995) Two new species of *Phymorhynchus* (Gastropoda, Conoidea, Conidae) from the hydrothermal vents. *Ruthenica* **5**(1), 17–26.

Tarasov, V.G., Propp, M.V., Propp, L.N., Zhirmunsky, A.V., Namsaraev, B.B., Gorlenko, V.M. and Starynin, D.A. (1990) Shallow-water gasohydrothermal vents of Ushishir Volcano and the ecosystem of Kraternaya Bight (The Kuril Islands). *Marine Ecology* **11**, 1–23.

Tarasov, V.G., Sorokin, Y.I., Propp, M.V., Shul'kin, V.M., Namsaraev, B.B., Bonch-Osmolovskaya, E.A., Starynin, D.A., Kamenev, G.M., Fadeev, V.I.., Malakhov, V.V., Kosmynin, V.N. and Gebruk A.V. (1993) Structural and functional characteristics of marine ecosystems in zones of shallow-water gasohydrothermal activity in the Western Pacific Ocean. *Izvestiya Academii Nauk, Seriya Biologicheskaya* **6**, 914–926. (In Russian).

Thompson, G., Humphris, S.E., Shroeder, B., Sulanovska, M. and Rona, P.A. (1988) Hydrothermal mineralization on the Mid-Atlanitc Ridge. *Canadian Mineralogist* **26**, 697–711.

Tunnicliffe, V. (1988) Biogeography and evolution of hydrothermal vent fauna in the eastern Pacific Ocean. *Proceedings of the Royal Society of London* **B 233**, 347–366.

Tunnicliffe, V. (1991) The biology of hydrothermal vents: ecology and evolution. *Oceanography and Marine Biology Annual Review* **29**, 319–407.

Tunnicliffe, V. and Fowler, C.M.R. (1996) Influence of sea-floor spreading on the global hydrothermal vent fauna. *Nature* **379**, 531–533.

Tyler, P.A., Paterson, G.J.L., Sibuet, M., Guille, A. Murton, B.J. and Segonzac, M. (1995) A new genus of ophiuroid (Echinodermata: Ophiuroidea) from hydrothermal mounds along the Mid-Atlantic Ridge. *Journal of the Marine Biological Association of the UK* **75**, 977– 986.

Van Dover, C.L. (1995) Ecology of Mid-Alantic Ridge hydrothermal vents. *In* "Hydrothermal Vents and Processes" (L.M. Parson, C.L. Walker and D.R. Dixon, eds), pp. 257–294. Geological Society, London. *Special Publication* **87**.

Van Dover, C.L., Fry, B., Grassle, J.F., Humphris, S. and Rona, P.A. (1988). Feeding biology of the shrimp *Rimicaris exoculata* at hydrothermal vents on the Mid-Atlantic Ridge. *Marine Biology* **98**, 209–216.

Van Dover, C.L., Szuts, E.Z., Chamberlain, S.C. and Cann, J.R. (1989) A novel "eye" in "eyeless" shrimp from hydrothermal vents of the Mid-Atlantic Ridge. *Nature* **337**, 458–460.

Van Dover, C.L., Desbruyères, D., Segonzac, M., Comtet, T., Saldanha, L., Fiala-Médioni, A. and Langmuir, C. (1996). Biology of the Lucky Strike hydrothermal field. *Deep-Sea Research* **43**, 1509–1529.

Vereshchaka, A.L. (1996) A new genus and species of caridean shrimp (Crustacea, Decapoda, Alvinocarididae) from North Atlantic hydrothermal vents. *Journal of the Marine Biological Association of the United Kingdom* **76**, 951–961.

Vereshchaka, A.L. (1997) A new family for a deep-sea caridean shrimp from North Atlantic hydrothermal vents. *Journal of the Marine Biological Association of the United Kingdom* **77**, 425–438.

Vinogradov, G.M. (1995) Amphipods from hydrothermal vents and seep fields of the Ocean. *Okeanologiya* **35**(1), 75–81. (In Russian).

Vinogradov, M.E. and Vereshchaka, A.L. (1995) The micro-scale distribution of the hydrothermal near-bottom shrimp fauna. *Deep-Sea Newsletter* **23**, 18–21.

Vinogradov, M.E., Vereshchaka, A.L. and Shushkina, E.A. (1996) Vertical structure of the zooplankton communities in the oligotrophic areas of the northern Atlantic, and influence of the hydrothermal vent. *Okeanologiya* **36**(1), 71–79.

Von Cosel, R., Metivier, B. and Hashimoto, J. (1994). Three new species of *Bathymodiolus* (Bivalvia: Mytilidae) from hydrothermal vents in the Lau Basin and North Fiji Basin,Western Pacific, and the Snake Pit Area, Mid-Atlantic Ridge. *The Veliger* **37**(4), 374–392.

Vrijenhoek, R.C., Schutz, S.J., Gustafson, R.G. and Lutz, R.A. (1994) Cryptic species of deep-sea clams (Mollusca: Bivalvia: Vesicomyidae) from hydrothermal vent and cool-water seep environments. *Deep-Sea Research* **41**, 1171–1189.

Warén, A. and Bouchet, A. (1993) New records, species, genera, and a new family of gastropods from hydrothermal vents and hydrocarbon seeps. *Zoologica Scripta* **22**(1), 1–90.

Williams, A.B. 1988. New marine decapod crustaceans from waters influenced by hydrothermal discharge, brine and hydrocarbon seepage. *Fisheries Bulletin, US* **86**, 263–287.

Williams, A.B. and Chace, F.A. (1982) A new caridean shrimp of the family Bresiliidae from thermal vents of the Galapagos Rift. *Journal of Crustacean Biology* **2**, 136–147.

Williams, A.B. and Dobbs, F.C. (1995) A new genus and species of caridean shrimp (Crustacea: Decapoda: Bresiliidae) from hydrothermal vents on Loihi Seamount, Hawaii. *Proceedings of the Biological Sciety of Washington* **108**(2), 228–237.

Williams, A.B. and Rona, P.A. (1986) Two new caridean shrimps (Bresiliidae) from a hydrothermal field on the Mid-Atlantic Ridge. *Journal of Crustacean Biology* **6**(3), 446–462.

Wirsen, C.O., Jannasch, H.W. and Molyneaux, S.J. (1993) Chemosynthetic microbial activity at Mid-Atlantic Ridge hydrothermal vent sites. *Journal of Geophysical Research,* **98**(B6), 9693–9703.

Zarenkov, N.A. (1983) "Arthropoda. Crustaceans". Part II. Izdatel'stvo Moskovskogo Universiteta, Moscow. (In Russian).

Zezina, O.N. (1985) "Recent Brachiopods and Problems of the Ocean Bathyal Zone". Nauka, Moscow. (In Russian).

Zhukova N.V., Kharlamenko, V.I. and Gebruk, A.V. (1991) Fatty acids of *Axinopsida orbiculata* – potential for detection of symbiosis with chemoautotrophic bacteria. *In* "Shallow-water Vents and Ecosystem of Kraternaya Bight (Volcano Ushishir, Kuril Islands)", vol. 1, Part II, pp. 63–78. DVO RAN, Vladivostok. (In Russian).

Zonenshain, L.P., Kuz'min, M.I., Lisitsyn, A.P., Bogdanov, Y.A. and Baranov, B.V. (1989) Tectonics of the Mid-Atlantic rift valley between the TAG and MARK areas (26–24° N): evidence for vertical tectonism. *Tectonophysics* **159**, 1–23.

ADDITIONAL NOTE

New observations of vent shrimp distribution were made during a collaborative cruise of "Atlantis" and "Alvin" (Rutgers University and BRIDGE) in July 1997. The ranges of many species were extended and an additional species of *Alvinocaris* was found at TAG. The new vent field "Rainbow" has at least four species, while Broken Spur has six, and five occur at TAG, Snakepit and Logatchev. The smallest number of shrimp species occurs at the shallow sites, Menez Gwen (two) and Lucky Strike (three).

Biology of the Nazca and Sala y Gómez Submarine Ridges, an Outpost of the Indo-West Pacific Fauna in the Eastern Pacific Ocean: Composition and Distribution of the Fauna, its Communities and History

N.V. Parin, A.N. Mironov and K.N. Nesis

P.P. Shirshov Institute of Oceanology, Russian Academy of Sciences, Nakhimovsky Prospekt 36, Moscow 117851, Russia

ADVANCES IN MARINE BIOLOGY VOL. 32
ISBN 0-12-026132-4

ABSTRACT

Study of the fauna of the submarine ridges in the south-eastern Pacific began in
the 1950s, but the most detailed investigations were made in 1973–1987 during
cruises of several Russian research vessels, notably the "Ikhtiandr"", "Professor
Mesyatzev" and "Professor Shtokman". At 22 seamounts of the Sala y Gómez
and Nazca ridges, 177 genera and 192 species of benthopelagic and benthic
invertebrates and 128 genera and 171 species of fishes were identified. Seven
genera and 150 species were described for the first time: four and 74 among
invertebrates, three and 76 among fishes.

Bottom invertebrate communities of the seamount summits are characterized
by strong dominance of a few species. At depths less than 400 m the spiny lobster
Projasus bahamondei is dominant to the east of 83°W, while sea urchins
predominate to the west. At greater depths sponges, gorgonarians, starfishes or
shrimps are most abundant, in various combinations. Horse mackerel, *Trachurus
symmetricus murphyi* (a temporary visitor from Chilean waters), usually
dominates benthopelagic fish communities above the seamounts eastward of
85°W. Other abundant species are *Emmelichthys cyanescens*, *E. elongatus*,
Decapterus muroadsi, *Zenopsis oblongus*, *Epigonus elegans* and *Pentaceros
quinquespinis*, which all form the basis of. a commercial fishery. Among bottom
fishes *Caelorinchus immaculatus* (Nazca Ridge) and *Pterygotrigla picta* were
noted dominants. The greatest diversity of fishes was observed at depths shallower
than 500–600 m; the communities of soft-bottom and rocky biotopes differ
significantly.

The fauna of benthic and benthopelagic invertebrates and fishes of the area is
much more closely related to Indo-West Pacific than to the Eastern Pacific fauna
and is characterized by very high degree of endemism at species level (51% among
identified bottom invertebrates, 44% among fishes). The seamounts of the Sala y
Gómez Ridge and transitional Sala y Gómez/Nazca area westward of 83– 84°W
should be considered as a separate Nazcaplatensis Province of the Indo-West
Pacific Region. Based on fish populations, the seamounts of the Nazca Ridge
proper may be included in the same province, as a faunistically impoverished
portion. On the basis of benthic invertebrates there is a link to the Eastern Pacific
subcontinental Region.

The faunistic composition of the main fauna of the Sala y Gómez and Nazca
ridges can be explained by two main processes: eastward dispersal of the western
Pacific fauna and active speciation *in situ*.

1. INTRODUCTION

The study of the biota of underwater ridges or "mountains in the ocean" has received much attention during the past 10 years or so. Collective monographs and reviews have been devoted to biological investigations of seamounts (Uchida *et al.*, 1986; Keating *et al.*, 1987; Wilson and Kaufmann, 1987; Mironov and Rudjakov, 1990; Parin and Becker, 1990; Parin, 1993; Kuznetsov and Mironov, 1994; Rogers, 1994). A considerable number of research contributions have been published (e.g. Parin *et al.*, 1985; Rudjakov and Tseitlin, 1985; Tseitlin, 1985; Wilson *et al.*, 1985; Boehlert, 1988; Genin *et al.*, 1988; Rudjakov and Zaikin, 1990; Wishner *et al.*, 1990, 1995; Levin *et al.*, 1991). This interest in the biota of seamounts has several causes. One is that unusually dense concentrations of valuable fishes and invertebrates of great commercial importance are found on some seamounts. This richness is thought to be related in part to the occurrence of baroclinic perturbations above and around underwater rises (e.g. Hubbs, 1959; Borets and Kulikov, 1986; Uchida *et al.*, 1986; Rogers, 1994). Some seamounts are rare examples of relatively closed ecosystems in the open ocean whose populations have developed sophisticated adaptations to withstand the risk of propagules being lost by the currents (Rudjakov and Tseitlin, 1985; Parker and Tunnicliffe, 1994). Another reason for interest in seamounts, especially to biogeographers, is the great diversity of seamounts in summit depths, geomorphological features and degree of geographic isolation. The endemism of the fauna of some seamounts or underwater ridges is very high in comparison with other marine areas of the same size (Newman and Foster, 1983; Mironov, 1985b, 1994; Newman, 1986; Parin and Nesis, 1986; Parin, 1990a, 1991a; Sazonov and Iwamoto, 1992). Study of the biota of seamounts, of its sources and ways of dispersion, throws light on a much disputed problem in marine biogeography: the relative roles of vicariance and dispersal (Rotondo *et al.*, 1981; Springer, 1982; Newman and Foster, 1983; Nesis, 1990, 1993a,b).

Interdisciplinary oceanological and biological investigations of seamounts were first conducted in the Atlantic Ocean on the Vema and Great Meteor seamounts (Simpson and Heydorn, 1965; Berrisford, 1969; Thiel, 1970). Some 20 years later the number of seamounts yielding biological data increased to approximately 100 (Wilson and Kaufmann, 1987) and today the number is nearly 200. The most intensively studied ones were in the North Pacific (Kyushu–Palau Ridge, Emperor Seamount Chain and North Hawaiian Ridge, Mid-Pacific Mountains, Cobb Seamount and some seamounts in the Gulf of Alaska) and eastern Atlantic (Atlantis–Great Meteor Chain, Josephine Seamount, Valdivia Seamount, etc.) (Wilson and Kaufmann, 1987).

Russian expeditions to seamounts began in the late 1960s. The first results of these expeditions were summarized by Andriashev (1979). Later biological collections were made on seamounts of most areas of the World Ocean: the northern and southern Atlantic, northern and southern Pacific, western Indian and

Southern Ocean (Pakhorukov, 1976, 1981; Zevina *et al.*, 1979; Kuznetsov and Mironov, 1981; Fedorov, 1982; Kuznetsov, 1985; Boldyrev, 1986; Borets, 1986; Borets and Kulikov, 1986; Darnitsky *et al.*, 1986; Pshenichny *et al.*, 1986; Sokolov *et al.*, 1986; Boldyrev *et al.*, 1987; Parin and Mironov, 1987; Shcherbachev *et al.*, 1989; Trofimov *et al.*, 1989a,b; Mironov and Rudjakov, 1990; Parin, 1990a, 1991a; Parin and Becker, 1990; Nesis, 1993a,b, 1994; Parin, 1993; Parin *et al.*, 1993).

The seamounts of the intraplate Nazca and Sala y Gómez ridges, south-eastern Pacific Ocean, are probably the most profitable area for studies of the seamount fauna and its genesis. These seamounts form a very long and narrow chain stretching from west to east and then northeast. The area includes seamounts differing greatly in form, summit depth, degree of geomorphological isolation and oceanographic conditions. These mounts are located on the Nazca Plate, which is moving eastward of the East Pacific Ridge (EPR) while the huge Pacific Plate, populated predominantly by the tropical Indo-West Pacific fauna, has moved northwestward from the EPR. The spreading rate of the EPR at its junction with the Sala y Gómez Ridge is the world's greatest, about 180 mm year^{-1} (Krasnov, 1995). Thus, if a spreading zone with such a large spreading rate formed a substantial barrier to distribution of the bottom and near-bottom fauna, we could expect that the faunas of the two diverging plates, the Pacific and the Nazca Plate, would be different. The first few studies have shown that a large part of the ridges on the Nazca Plate is populated by a highly endemic fauna with well-expressed Indo-West Pacific affinities; only in the easternmost section of the ridges are there some species with affinities to the fauna of western South America (Parin *et al.*, 1980; Parin and Golovan', 1982; Parin *et al.*, 1988a). Detailed investigations support this conclusion (Mironov and Detinova, 1990; Nesis, 1990; Parin, 1990a, 1991a). Thus, this chain of seamounts is populated by descendants of the fauna of the geographically very distant Indo-West Pacific Region, inhabiting the Pacific Plate and other plates to the west and southwest of it, and not by that of the neighbouring Eastern Pacific Region. The endemism of the fauna of the Nazca and Sala y Gómez ridges is unprecedently high for a deep water area (Parin, 1991a).

Although exploration of the Nazca and Sala y Gómez ridges did not discover such huge concentrations of commercial fishes as were found, for example, on the seamounts of the Emperor Ridge (Borets and Kulikov, 1986; Uchida *et al.*, 1986), there were abundant populations of some valuable commercial crustaceans (small spiny lobsters, crabs and shrimps) and of fishes (jack mackerels, redbaits and blue-mouth perches) (Fedorov and Chistikov, 1985; Parin *et al.*, 1988a; Fedorov, 1990; Mironov *et al.*, 1990; Prosvirov, 1990; Rudjakov *et al.*, 1990). It appeared that the enhanced biological productivity predicted for some seamounts was rather well expressed there (Amarov and Korostin, 1981; Kashkin, 1984; Gevorkyan *et al.*, 1986).

The fauna of oceanic seamounts of the Nazca Ridge was first investigated in the late 1950s by the US *Downing* Expedition (Hubbs, 1959; Zullo and Newman,

1964; Allison *et al.*, 1967). Bottom animals were collected here for the first time by the US R/V "Horizon" on 26 January, 1958 with a biological dredge, sample No. HD-1973. The vessel participated in the 1957–1958 *Downing* cruise during the International Geophysical Year (Allison *et al.*, 1967). The sample No. HD-73 was taken at 25°44'S, 85°25'W, depths 210–227 m. The sampled guyot was called "Shoal Guyot" in earlier publications (Zullo and Newman, 1964) and "Bolshaya Banka" ("Great Bank") in subsequent Russian publications (Parin *et al.*, 1980, 1988a; Parin, 1982a, 1990a, 1991a; Mironov and Detinova, 1990). Some dead corals were dredged the same day a few kilometres to the north-east of this guyot, in the sample No. HD-72, at 25°31'S, 85°14'W, depths 870–950 m. Some small collections were made much later by Japanese scientists (R/V "Ibiku-Maru", 1983; see Okutani and Kuroiwa, 1985). However, the most detailed investigations of the bottom, near-bottom, benthopelagic and pelagic fauna in the area of the Nazca and Sala y Gómez ridges have been those carried out by expeditions of the former USSR.

The USSR studies began during November, 1973 (R/V "Poseidon") and June–July, 1975 (R/V "Astronom") but the first large-scale surveys were conducted by the R/V "Gerakl" during October, 1975–January, 1976 (Grossman, 1978). In 1979–1983 more detailed fisheries studies were undertaken from the R/V "Ikhtiandr" and "Odissey" supported by the manned submersible "Sever 2". The fifth (August–November, 1979) and sixth (June–December, 1980) cruises of the R/V "Ikhtiandr" provided particularly interesting results (Parin *et al.*, 1980; Parin and Golovan', 1982; Fedorov and Chistikov, 1985; Gevorkyan *et al.*, 1986; Parin, 1990a). Some collections were made also from the R/V "Akademik Kurchatov" (1982) and "Professor Mesyatzev" (1983–1985) and some other vessels. By far the richest samples of marine invertebrates and fishes were, however, collected during the 18th cruise of the R/V "Professor Shtokman", in April–May, 1987 (Parin *et al.*, 1988a, Mironov and Rudjakov, 1990; Parin and Becker, 1990). Details of this cruise, its route and the sampling gear used were presented in Parin *et al.* (1988a); Parin (1990a) and Rudjakov and Zaikin (1990).

The expeditions on the R/Vs "Poseidon", "Astronom" and "Gerakl" were organized by the Pacific Research Institute of Fisheries and Oceanography (TINRO, Vladivostok), those on the R/V "Ikhtiandr", "Odissey" and "Professor Mesyatzev" by the All-USSR (now Russian Federal) Institute of Marine Fisheries and Oceanography (VNIRO, Moscow), and the Sevastopol Design Office of Commercial Fisheries, and those on the R/V "Akademik Kurchatov" and "Professor Shtokman" by the P.P. Shirshov Institute of Oceanology of the USSR (now Russian) Academy of Sciences (IO RAS, Moscow).

It should be noted that all the Russian investigations of the seamounts of the Nazca and Sala y Gómez ridges were limited by the 200 nautical mile EEZ off Peru and Chile. Thus, our investigations stopped 200 nautical miles from Easter, Sala y Gómez, San Felix and San Ambrosio Is and from the shore of South America, so that many interesting seamounts remain unstudied.

2. GEOLOGY OF THE SEAMOUNTS, AND HYDROLOGY AND PLANKTON OF THE OVERLYING WATERS

2.1. Geographic Position and Geology of the Nazca and Sala y Gómez Ridges

The Nazca and Sala y Gómez ridges are aseismic intraplate block–volcanic ridges located on the Nazca Plate in the Southeastern Pacific; they divide the region into the Peru and Chile basins. The Sala y Gómez Ridge is sublatitudinal and stretches approximately along 25°S, while the Nazca Ridge stretches north-eastward from approximately 27°S, 87°W to 15°S, 76°W, where it is suddenly chopped off by the slope of the Peru–Chile Trench. Here, off the City of Nazca (Peru), this trench is divided by a saddle of 4950 m depth into the Peru and Chile trenches (Menard, 1964; Udintsev, 1972).

The length of the Nazca Ridge is approximately 1100 km and with the inclusion of the transitional Nazca/Sala y Gómez area it reaches approximately 1600 km. It is limited by the 4000 m isobath and its width is approximately 100–150 km. The north-western slope of this ridge is more gentle than the steep south-eastern one, while the north-eastern part of the ridge is deeper than the central and south-western parts. The central region, between 20 and 24°S, is the highest, with a rather flat plateau at depths of 2000–2500 m. The seamounts arise from this plateau. The summits are deeper to the north-east. The northernmost and easternmost summit, Nachalnaya (Initial) Seamount (Figure 1, No. 21) is the deepest, at approx. 850 m and the rim of the summit is at 950 m. Larger and shallower (240–340 m) seamounts follow: Zvezda (Star), Professor Mesyatzev (or Nazca Guyot), Soldatov, Ekliptika (Ecliptic) and Ikhtiandr (Ichthyandr, or Southern Tropic, located at the Tropic of Capricorn) seamounts (Figure 1, Nos 17–20, 22). All these seamounts are located at 19°30'–23°30'S, 80°00'-83°30'W and are separated from the deep north-eastern part of the ridge by a transverse valley, with a maximum depth of 3700 m (Manimerickx *et al.*, 1975; Newman and Foster, 1983; Fedorov and Ivanov, 1984; Gevorkyan *et al.*, 1986, 1987; Parin *et al.*, 1988a).

Next, a group of rather shallow guyots is located in the area where the Nazca and Sala y Gómez ridges intersect, the transitional Nazca/Sala y Gómez area. These are: Albert and Kommunar (Communard) seamounts (Figure 1, Nos 13 and 16) on the continuation of the Nazca Ridge axis, Yuzhnaya (South, not shown on Figure 1), Novaya (New), Bolshaya (Great Bank, or Shoal Guyot), Dorofeyev, Dlinnaya (Long), Baral and Zapadnaya (Western) seamounts (Figure 1, Nos 9–12, 14–15) on the continuation of the Sala y Gómez Ridge axis (Mammerickx *et al.*, 1975; Fedorov and Ivanov, 1984; Fedorov, 1985; Gevorkyan *et al.*, 1986, 1987; Parin *et al.*, 1988a). To the south of these seamounts there are some others, including the large Merriam Ridge seamount complex with the Professor Shtokman Seamount, not studied by the USSR expeditions considered in this chapter. The summit plateaux of these seamounts are at 160–380 m below the

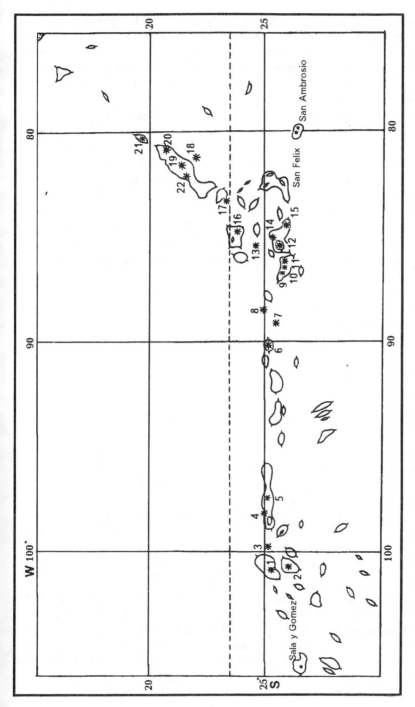

Figure 1   Seamounts of the Sala y Gómez and Nazca ridges where bottom fauna samples were obtained. 1, Igolnaya (Needle); 2, Utes (Rock); 3, Ichthyologists; 4, Stolbovaya (Pillar); 5, Kupol (Cupole); 6, Pervomayskaya (Mayday); 7, Zhemchuzhnaya (Pearl); 8, Yantarnaya (Amber); 9, Zapadnaya (Western); 10, Baral; 11, Dlinnaya (Long); 12, Bolshaya (Great); 13, Kommunar (Communard); 14, Novaya (New); 15, Dorofeyev; 16, Albert; 17, Ikhtiandr (Ichthyandr); 18, Ekliptika (Ecliptic); 19, Professor Mesyatzev; 20, Zvezda (Star); 21, Nachalnaya (Initial); 22, Soldatov (after Parin *et al.* 1988a, modified).

surface and their relative heights reach 3700 m. This group is separated from the Nazca Ridge by a valley 3800 m deep and divided from the Sala y Gómez Ridge by a fracture of the Easter Fracture Zone more than 3650 m deep. Some groups of seamounts are also separated by depressions where the depths are as much as 3800–4000 m (Mammerickx *et al.*, 1975; Fedorov and Ivanov, 1984; Fedorov, 1985; Gevorkyan *et al.*, 1986, 1987).

On the Sala y Gómez Ridge there are some 15 guyots located at 25–26°S, between 88 and 103°W. Their summits are mostly at 220–500 m below the sea surface, but in the central part of this ridge, at 90–97°W there are no shallow seamounts (Newman and Foster, 1983). These particular seamounts have no established names and we use the "expedition names" introduced during the 18th cruise of the R/V "Professor Shtokman", not yet inserted on charts. Three seamounts on the eastern part of the Sala y Gómez Ridge were investigated during this cruise (at 25°S, 88°30'–89°30'W): Yantarnaya (Amber), Zhemchuzhnaya (Pearl) and Pervomayskaya (Mayday) (Figure 1, Nos 6–8), as were five in the western part (25–26°S, 97–101°W): Kupol (Cupole), Stolbovaya (Pillar), Ichthyologists, Igolnaya (Needle) and Utes (Rock) (Figure 1, Nos 1–5) (see Mironov and Detinova (1990) and Rudjakov and Zaikin (1990)).

Both ridges have a block structure but the Nazca Ridge is predominantly a tectonic construction and the Sala y Gómez Ridge is predominantly a volcanic one. The latter is younger than the Nazca Ridge and, from morphometrical and geomorphological features, resembles the Walvis Ridge in the south-eastern Atlantic (Zakharov *et al.*, 1984).

2.2. Geomorphology and Geological History of the Seamounts of the Nazca Plate

The seamounts of the Nazca and Sala y Gómez ridges are drowned coral atolls with almost intact atoll structures, including fringing and barrier reefs, whose thickness is not less than 250 m and may reach 1200 m (Fedorov and Ivanov, 1981; Gevorkyan *et al.*, 1987). The fossil corals and gastropods dredged from some seamounts of the Nazca Ridge (e.g. Professor Mesyatzev Seamount) have been dated as Miocene (Menard, 1964; Gevorkyan *et al.*, 1987). Studies from the manned submersible "Sever 2" confirmed the existence of drowned shorelines (2–7 lines, each up to 3 m in height), a characteristic of these seamounts, providing evidence that during post-Miocene time they were to 300–375 m higher than now and that the summits of most of the seamounts would then have been emergent islands. Drowning of these islands was approximately simultaneous. Typical atoll geomorphology with a volcanic base and traces of eruptions has been found on the summits of the Zvezda, Professor Mesyatzev, Ekliptika, Bolshaya, Yuzhnaya, Albert and Baral seamounts. Fringing reefs were found on all seamounts except Professor Mesyatzev, where a typical barrier reef was discovered showing that in addition to small volcanic peaks, as in the modern Sala y Gómez Is, a whole summit has sometimes emerged. Both subaerial and submarine relief forms were discovered on the summits of shallow guyots (Fedorov and Ivanov, 1981, 1984; Fedorov, 1985, 1987; Ivanov, 1988). There is still dispute over whether the submersion of the summits was the result of a large global rise in sea level or by a lowering of the ridges (Udintsev, 1972; Fedorov, 198; Gevorkyan *et al.*, 1987), but differential tectonic vertical movements are characteristic of different blocks of seamounts (Fedorov, 1985; Ivanov, 1988). It may be supposed that the extinction of corals on the summits was caused by the general cooling that occurred during the Pliocene and Pleistocene periods, combined with tectonic submersion of the blocks of ridges.

Three groups of seamounts can be distinguished by the morphology of their summits: (i) seamounts with flat sub-horizontal summits (Zvezda, Professor Mesyatzev, Ekliptika); (ii) seamounts with convex summits sloping from the centre to the rim (Nachalnaya, Albert, Kommunar, Bolshaya, Yuzhnaya, Dlinnaya, Baral); and (iii) acute-topped seamounts (Ikhtiandr, Stolbovaya and some unnamed). The last group probably represents tectonically disrupted coral constructions of the first two types (Gevorkyan *et al.*, 1987).

The summits of the guyots are more or less smooth. This smoothing was caused mostly by abrasion. The very thin, loose sediments on all the seamounts are biogenic sands derived from local mollusc shells, sea urchin tests, bottom foraminiferans, etc., with some pelagic pteropod and foraminiferal material. Sand ripples occur on the slopes at depths down to 1000 m and the currents on the subunits are strong, so that sediment accumulates only in depressions. Sand and pebble avalanches (turbidites) are common on all slopes (Fedorov and Ivanov, 1984; Fedorov, 1985; Gevorkyan *et al.*, 1987; Ivanov, 1988).

It may be supposed that both the Nazca and Sala y Gómez ridges are rather old, much older than the South Pacific Rise (Zakharov *et al.*, 1984; Gevorkyan *et al.*, 1987). The seamounts on these ridges indicate both general sinking of the ridges themselves and intense differential vertical movements of separate blocks. The modern geomorphology of guyots is the result of complex processes, including local volcanism, crustal tectonic movements, coral morpholithogenesis, sea-level changes and wave action, weakly modified by pelagic sedimentation and underwater erosion (Fedorov and Ivanov, 1984).

2.3. Hydrology of the Overlying Waters

The Sala y Gómez Ridge and the transitional Nazca/Sala y Gómez area are located in the Southern Pacific subtropical anticyclonic gyre formed by the east-going South Pacific Current, north-going Peru Current and west-going South Equatorial Current. The areas influenced by the South Pacific Current and South Equatorial Current are divided by the Southern Subtropical Convergence, which is situated south of the Sala y Gómez Ridge, approximately along 30°S. The eastern part of the chain of ridges is influenced by the Oceanic Branch of the Peru Current and the Peru Countercurrent between the Oceanic and Coastal branches of the Peru Current. The Peru Countercurrent is a continuation of the Equatorial Undercurrent (Cromwell Current) which extends southward approximately along 80°W. It is a surface countercurrent during the southern summer and a subsurface undercurrent during other seasons (e.g. Wyrtki, 1966; Berman, 1969; Burkov, 1972, 1980; Stepanov, 1974; Amarov and Korostin, 1981; Gar'kusha and Kotelnikov, 1981; Sheremet and Zalesinsky, 1981; Gevorkyan *et al.*, 1986).

The Southern Tropical Front lies along the southern boundary of the South Equatorial Current and the western boundary of the Peru Current and crosses the studied area over the northern part of the Nazca Ridge (Stepanov, 1974).

Two surface water masses are found in the area studied: the Eastern Subtropical Surface Water (ESSW) and the South Pacific Subtropical Surface Water (SPSSW). The ESSW (average temperature 18.8°C, salinity 34.8‰) is the dominant water mass in the upper 100–120 m layer over the transitional Nazca/Sala y Gómez area, whereas the SPSSW (average temperature 21°C, salinity 35.7‰) is dominant in other areas, as deep as 80–150 m over the Nazca and to 200–250 m over the western (westward of 94°W) part of the Sala y Gómez Ridge; the SPSSW is dominant in the area of 98–100°W.

In the subsurface zone the Subantarctic Subsurface Water (SASW), with an average temperature of 12.5°C, salinity 34.0‰, occupies the area; its thickness diminishing north-eastward over the Nazca Ridge from 150–260 m over the southern to 120–170 m over the northern segment. It is relatively constant in thickness (80–200 m) over the Sala y Gómez Ridge.

In the intermediate zone two water masses can be distinguished: Tropical Intermediate Water (TIW) with an oxygen-minimum layer ($0.15–0.20$ ml O_2 l^{-1}) and Antarctic Intermediate Water (AIW), which is well oxygenated. The TIW (average temperature 10.8°C, salinity 34.9‰) occurs over the central part of the Nazca Ridge at depths of about 130–900 m, its thickness diminishes over the Nazca Ridge toward the southwest and is further reduced (< 50%) at 24°S. The AIW (average temperature 5.4°C, salinity 34.1‰) is dominant over the Sala y Gómez Ridge at depths from 380 m to approximately 1200 m, particularly westward of 92°W. The AIW component is reduced over the Nazca Ridge to the northeast and falls to less than 50% in the central part.

Depths greater than 1250 m are occupied by Central Pacific Deep Water (CPDW) with an average temperature of 0.5°C and salinity of 34.6‰ (Kudryavtsev, 1981).

The oxygen-minimum layer is associated with the TIW and located at 250–300 m or 250–350 m over the southern and central parts of the Nazca Ridge and 150–1000 m in its northern part. From time to time water deficient in oxygen occupies the near-bottom layer on the summits of some guyots (e.g. Professor Mesyatzev) while the shallower summits (e.g. Bolshaya and Ekliptika) lie above the oxygen minimum layer (Gevorkyan *et al.*, 1986).

The Peru Countercurrent determines the spread of the oxygen-deficient water in the area over the Nazca Ridge. Changes in the strength of advection of the TIW, under the influence of this countercurrent, can cause great changes in the oxygen concentration in the near-bottom water over the guyots whose summits lie in this layer; for example from 0.14 to 0.90 ml l^{-1} on Professor Mesyatzev Seamount (Gevorkyan *et al.*, 1986). Hydrological conditions, including the oxygen concentration over the summits of seamounts, may change greatly in connection with El Niño–Southern Oscillation (ENSO) events, although the area studied lies outside the region of maximum changes (Wyrtki, 1966; Amarov and Kuznetsov, 1986). The water in the area of the Sala y Gómez Ridge is well oxygenated (Wyrtki, 1966; Stepanov, 1974).

Quasi-stationary circulations of the "Taylor column" type (mostly anti-cyclonic) have been reported over some of the shallower seamounts (Professor Mesyatzev, Bolshaya and Ekliptika). Here microscale upwellings were associated with enhanced biological productivity and the advection of relatively deep-water zooplankton into surface waters (Grossman, 1978; Amarov and Korostin, 1981; Kolodnitsky and Kudryavtsev, 1982; Gevorkyan *et al.*, 1986). No plankton enrichment was found along either side of a frontal division northward of the Nazca Ridge during February, 1982 (Vinogradov *et al.*, 1984).

2.4. Plankton

The zooplankton in the area of the Nazca and Sala y Gómez ridges was studied by Grossman (1978) and in the area near the the Nazca Ridge by Vinogradov *et al.* (1984) and Kashkin and Kashkina (1984). There are observations in the literature on copepods (Heinrich, 1990), pelagic amphipods (Vinogradov, 1990), euphausiids, mysids, pelagic shrimps and decapod larvae (Vereshchaka, 1990a,b,c).

Macro- and mesoplankton biomass during October, 1975–January, 1976 (spring and summer of southern hemisphere) over the Nazca and Sala y Gómez ridges comprised 5.0 and 2.4 mg m^{-3}, respectively. In both areas the biomass was greater immediately over the western slopes of seamounts, probably because of local upwelling. Over the Sala y Gómez Ridge the biomass of mesoplankton decreased toward the east, from 450 to 300 mg m^{-3}, and over the Nazca Ridge it decreased northeastward from 700 to 550 mg m^{-3}. There was a maximum over the central part where a huge concentration of salps provided a biomass of 15.1 g m^{-3}. Positions of enhanced values of macro- and mesoplankton biomass coincide on both ridges (Grossman, 1978). During April–early May, 1987 the abundance of adult copepods and copepodite stages decreased from Ekliptika westward to Baral and then increased again over the Yantarnaya and Zhemchuzhnaya seamounts (Heinrich, 1990), which agrees with the general trend of distribution of the macro- and mesoplankton biomass in 1976.

On the section immediately to the north of the Nazca Ridge (19°20'–19°45'S,

81°45'–81°25'W) during late February, 1981, the biomass of mesoplankton was 4.4–5.2 g under 1 m^2 (Evseenko and Gorbunova, 1984). The section crossed a hydrologic front with a temperature change up to 0.5°C per 2 nautical miles. The water on both sides of this front was oligotrophic, of low productivity (primary production not more than 0.1 g C m^{-2} day^{-1}), and the plankton was in the same successional state. But the ichthyoplankton composition of the warm- and cold-water sides of the frontal zone was different (Evseenko and Gorbunova, 1984). The average biomass of macroplankton in the centre of the section was 7.6 mg m^{-3}, with fishes, euphausiids and shrimps dominating. The biomass of mesoplankton was as usual for Central Water but macroplankton biomass was comparatively high, possibly as a result of the proximity of the rich waters of the Peru Current, eastward of the area studied (Kashkin and Kashkina, 1984; Vinogradov *et al.*, 1984). The biomass of euphausiids and decapod larvae in the 0–300-m layer during April–early May, 1987 was also enhanced over the Nazca Ridge (Ekliptika) and diminished towards the western part of the Sala y Gómez Ridge (Vereshchaka, 1990a,c).

3. MATERIAL

3.1. Bottom Invertebrates

Bottom, near-bottom and benthopelagic fauna were collected in the area of the Nazca and Sala y Gómez ridges by expeditions of R/V "Ikhtiandr" (1979 and 1980) and "Professor Shtokman" (1987). Bottom invertebrates were obtained on 22 seamounts at depths from 162 to 1900 m (Mironov and Detinova, 1990). One hundred and thirty samples of bottom invertebrates were collected; of these 54 samples were taken with shrimp otter-trawls, 39 with Sigsbee (Agassiz) trawls, 24 with baited traps, eight by bottom grabs and five by geological dredges.

Most stations were located on the flat tops of the guyots. There is little information about the populations of the steep and rough-bottom seamount slopes. Samples taken with different gear provided different faunal compositions. The otter-trawl hauls produced abundant mega- and macrofauna, and some branches of gorgonians reached 170 cm in height, but smaller animals, less than 10 cm, were relatively rare. Such animals were caught mostly by the Sigsbee trawls and dredges. Baited traps caught mostly large shrimps, brachyuran crabs and spiny lobsters. Bottom grabs were ineffective because of the predominance of hard substrata with only a thin covering of soft sediments on the tops of seamounts.

3.2. Bottom and Near-Bottom Fishes and Cephalopods

Most of the important observations and collections of bottom and near-bottom fishes and cephalopods were made from 1979 to 1987 during expeditions of the

Table 1   Main fish-collecting USSR expeditions to the Nazca–Sala y Gómez area.

Vessel, cruise No.	Year, month	Fishing gear, number of operations (in parentheses)
"Ikhtiandr", cr. 5	1979, VIII–X	BOT (50)
"Ikhtiandr", cr. 6	1980, VIII–X	BOT (About 100)
"Odyssey", cr. 2	1982, XII–1983, I	BOT (24), BT (5)
"Professor Mesyatzev", cr. 13	1983, IX–XI	BOT (85)
"Professor Mesyatzev", cr. 15	1984, VIII–X	BOT (23)
"Professor Shtokman", cr. 18	1987, IV–V	BOT (35), OMT (14), ST (32), BT (28), BLL (14), VLL (18)

*BOT, bottom otter-trawl, OMT, off-bottom midwater trawl (TIKSA); ST, Sigsbee trawl; BT, bottom baited trap; BLL, bottom long-line; VLL, vertical long-line.

"Ikhtiandr", "Odissey", "Professor Mesyatzev" and "Professor Shtokman" (Table 1). A few small samples obtained from other cruises ("Poseidon", "Astronom", "Darwin", "Zvezda", "Kommunar" and others) were also of considerable importance, especially when they contained specimens of otherwise rare species. Altogether more than 350 fish samples were collected, mainly by bottom trawls, from 22 seamounts of the Nazca and Sala y Gómez ridges, depths from 160 to 1600–2000 m. As in the case of bottom invertebrates, most samples were taken on the flat tops of guyots at depths of 160–580 m.

During the 18th cruise of the "Professor Shtokman" the bottom and near-bottom fish and cephalopods were collected by otter-trawls, Sigsbee trawls, horizontal and vertical bottom-set long-lines and baited traps (Table 1). The otter- trawls used were the one-wire type shrimp trawls with ground-ropes 19.4 and 29.0 m respectively, headrope 23.3 and 38.0 m, horizontal opening 9–10 and 12–13 m and vertical opening 1.5–2 and 2–2.5 m; a bottom-pelagic trawl with footrope 29.0 m, horizontal and vertical openings 12 and 8 m respectively was also deployed.

Off-bottom and pelagic fish were caught by an Isaacs–Kidd midwater trawl with a Samyshev–Aseyev net (TIKSA) and by pelagic long-line. The TIKSA was equipped with a pressure sensor DDV-200 that reported data to an on-board computer EC-1010 so that constant control of the trawl path was possible. With this equipment it was possible to tow the trawl some metres above the flat summits of guyots, for specific sampling of off-bottom animals. The Samyshev–Aseyev net is 17 m long with a mouth area 5.5 m^2 and 5 mm mesh through its whole length, having a 5 m long gauze net with mesh 1.3 mm in the cod-end. For catching pelagic animals the usual versions of the TIKSA or IKMT without the sensor were used. Details of these devices are given by Parin *et al.* (1988a) and Rudjakov and Zaikin (1990).

In general during this expedition 46 tows were made by TIKSA/IKMT, 36 with bottom otter-trawls, three with pelagic otter-trawl, 32 tows by Sigsbee trawl, 30 placements of bottom traps, 14 of bottom long-lines, two of pelagic long-line and 18 of vertical bottom-set long-line.

Collections of fish and cephalopods gathered during the expeditions of the "Gerakl", "Professor Mesyatzev", "Ikhtiandr", "Odissey" and small samples obtained by "Poseidon", "Zvezda", "Kommunar", "Darwin", "Ocher" were made by various large-sized commercial otter-trawls, either bottom trawls or (more commonly) bottom versions of midwater trawls. Large commercial midwater otter-trawls were also used.

Fishes and cephalopods caught by bottom trawls and by the TIKSA towing over the bottom may be either benthic, near-bottom or off-bottom ones while those caught by Sigsbee trawl or trap were caught definitely at the bottom. But even the latter may not necessarily be bottom inhabitants, they may descend to the bottom during active vertical migration or passive horizontal advection or both (Kashkin, 1984; Nesis, 1990, 1993b, 1994; Parin, 1990a; Rudjakov and Zaikin, 1990; Heinrich et al., 1993).

Lists of seamounts where fishes and invertebrates were collected during the 18th cruise of the R/V "Professor Shtokman", with station numbers, coordinates and depths, were given in Parin et al. (1988a) and Rudjakov and Zaikin (1990). Data on collections made by other expeditions were listed in Parin et al. (1980) and Parin (1982a).

4.   LISTS OF TAXA

4.1.   Bottom and Near-bottom Invertebrates

The list of taxa named in sample No. HD-73 from "Shoal Guyot" (Allison et al., 1967) includes 21 genera and nine species. Among these, four genera of corals are fossil, four molluscan genera were probably dead and others were living (Table 2). Species identifications were made for Cirripedia (Zullo and Newman, 1964) and Echinoidea (Allison et al., 1967). Later one decapod species was added (Garth, 1992). The fauna in sample HD-72 sample consisted of fossil corals Stylophora pistillata (Esper, 1797) and Plesiastrea (Allison et al., 1967).

Lists of the bottom invertebrates (to generic or species level), collected by the Russian expeditions to the Nazca and Sala y Gómez ridges are presented in Table 3. Details of identifications may be found in the following publications: Spongia Hexactinellida: Tabachnick (1990), Cirripedia: Zevina (1983, 1990), Tanaidacea: Kudinova-Pasternak (1990), Amphipoda: Vinogradov and Vinogradov (1991), Decapoda Caridea: Burukovsky (1990), Decapoda Stenopodidea, Brachyura and Anomura: Zarenkov (1990), Gastropoda Turridae: Sysoev and Ivanov (1985) and Sysoev (1990), Bivalvia Septibranchia: Krylova (1991, 1994, 1995), Cephalopoda (including near-bottom species): Nesis (1990), Echinodermata Echinoidea: Mironov and Sagaidachny (1984), Markov (1988, 1989) and Mironov (1989), Asteroidea: Galkin (1993), Ophiuroidea subfam. Ophiohelinae: Litvinova (1992;

Table 2   Flora and fauna collected by R/V "Horizon" in 1958; sample HD-73. Taxa listed were either live specimens or interpreted as recent unless indicated otherwise (from Allison *et al.*, 1967, with addition).

ALGAE	Unidentified coralline algae (?fossil)
FORAMINIFERA	*Homotrema*
	Sedimentary gut contents of echinoids *Scrippsechinus*
	fisheri consisted mostly of planktonic foraminifera
CORALS	*Stylophora* (?fossil) (HD-72 specimens identified as *S. pistillata*
	(Esper, 1797))
	Pocillopora (fossil)
	Leptoseris (fossil)
	Porites (fossil)
BRYOZOA	Unidentified
ANNELIDA	Encrusting calcareous tubes, unidentified (?recent)
MOLLUSCA	*Turbonilla (Strioturbonilla)*
	Glycymeris (?living)
	Xenophora (?living)
	Ctena (?living)
	Colus
	Acesta
	Arcopsis (?living)
	Pteropoda in the gut contents of *Scrippsechinus fisheri*
CIRRIPEDIA	*Megalasma (Megalasma) elegans* Newman, 1964
	Heteralepas mystacophora Newman, 1964
	Verruca (Verruca) scrippsae Zullo, 1964
	Balanus (Solidobalanus) nascanus Zullo, 1964
DECAPODA	*Panthenope (Platylambrus) mironovi* (Zarenkov, 1990)
ECHINOIDEA	*Salenia scrippsae* Zullo and Allison, 1964
	Stereocidaris nascaensis Allison, Durham and Mintz, 1967
	Echinocyamus incertus Clark, 1914 (gut contents of
	Scrippsechimus fisheri)
	Scrippsechinus fisheri Allison, Durham and Mintz, 1967
PISCES	*Pterygotrigla picta* (Günther, 1880)

N.M. Litvinova, unpublished information), Brachiopoda: Zezina (1990). Mironov and Detinova (1990) presented also some identifications in other groups: Polychaeta (N.N Detinova), Isopoda (B.V. Mezhov), Loricata (B.I. Sirenko), Ophiuroidea (N.M. Litvinova) and Echinoidea (A.N. Mironov). We have also used generic and species identifications of hermit crabs, family Parapaguridae, by D.G. Zhadan and gorgonarian and antipatharian corals by F.A. Pasternak.

In total, we have at our disposal specimens of 177 genera of bottom invertebrates recorded on the seamounts of the Nazca and Sala y Gómez ridges (Table 3). For 143 genera recorded on these ridges the number of species is known in the area studied. These 143 genera are represented on the ridges by 192 species, of which 136 were identified to species level, nine are designated as "n. sp." without a species name, and 47 as "sp." (Table 3). One hundred and seventeen

Table 3 List of genera and species of bottom algae and invertebrates recorded from the Sala y Gómez and Nazca ridges and their distribution on the seamounts in the area. Seamount numbers as on Figure 1. Seamount numbers in parentheses are records deeper than 800 m. Endemic species are indicated by asterisk. Only genera are numerated in the left column. "Sp" or "sp.1 and 2" is indicated when the number of species in a genus in the area studied is known. When the number of species is not reported, only the generic name is indicated.

Species	Seamounts

ALGAE
  CYANOPHYCOPHYTA
  1. *Rivularia*
  2. *Scytonema*
  3. *Entophysalis*

  RHODOPHYCOTA
  4. *Melobesia*
  5. *Lithothamnion*
  6. *Lithophyllum*
  7. *Archeolithothamnium*
  8. *Peyssonelia*

  CHLOROPHYCOTA
  Four genera, probably drifted

ANIMALIA

SPONGIA
HEXACTINELLIDA
  1. *Aphorme horrida* Schulze, 1899   18
  2. *Euplectella* sp.   3
  3. *Eurete (Pararete) farreopsis* Carter, 1877   8
  4. **Pheronema naskaniensis* Tabachnick, 1990   4,5,6,7,8,9,10,(11),12,(21)
  5. **Pseudoplectella dentatum* Tabachnick, 1990   3
  6. *Regadrella peru Tabachnick, 1990   8
  7. *Shulzeviella gigas spinosum* Tabachnick, 1990   (21)

DEMOSPONGIAE
  8. *Radiella* sp.   7,8

OCTOCORALLIA

GORGONARIA
  9. *Acanthogorgia*
HEXACORALLIA   10
ANTIPATHARIA
  10. *Bathypathes*   3
  11. *Cirripathes*   1,12,14,19
  12. *Curtipathes*   3

Table 3   Continued

Species	Seamounts
POLYCHAETA	
13. *Anaitides*	1,2,3,5,12
14. *Chloeia*	1,3,5,8,10,12
15. *Crysopetalum*	1,3,5,8,10
16. *Eunice*	3,5,12,15,18,19,20,(21)
17. *Lanice*	4,12,14
18. *Leocrates*	4,5,7,8,10,(21)
19. *Nothria*	1,2,3,5,7,8,10,12,14,15,18
20. *Phalacrostemna*	1,3,5,12
21. *Pholoe*	1,2,3,7,8,12,15
22. Polynoidae gen.sp.n. (from Hexactinellida)	5,7,10
CIRRIPEDIA	
23. *Altiverruca mollae* Zevina, 1990	3,7,8
24. *Balanus (Solidobalanus) nascanus* Zullo, 1964	1,12
25. *Heteralepas mysthacophora* Newman, 1964	10,12,13,15,18
26. *Litoscalpellum piliferum* Zevina, 1983	18,19
27. *Megalasma (Glyptelasma) caudata* Zevina, 1990	1,3,5
Megalasma (Megalasma) elegans Newman, 1964	3,12
28. *Metaverruca pallida* Zevina, 1990	1,2,5,11
Metaverruca tarasovi (Zevina, 1971)	12,15
29. *Oxynaspis michi* Zevina, 1983	12
30. *Paralepas ichtiandri* Zevina, 1983	11
Paralepas nascai Zevina, 1990	1,3,5,10,12
31. *Poecilasma crassa* (Gray, 1848)	13,14,17,18,20,22
Poecilasma kaempferi litum Pilsbry, 1907	12,13,14,17,18,19,20,22
32. *Verruca scrippsae* Zullo, 1964	1,12
TANAIDACEA	
33. *Apseudes diversus* (Lang, 1968)	17
34. *Collettea minima* (Hansen, 1913)	3
35. *Cryptocopoides arctica* (Hansen, 1913)	3,5
36. *Leptognathia breviremis* (Lilljeborg, 1864)	3,5
37. *Parafilitanais similis* Kudinova-Pasternak, 1990	3
38. *Paranathura insignis* Hansen, 1913	3
39. *Pseudobathytanais shtokmani* Kudinova-Pasternak, 1990	5
40. *Typhlotanais inermis* Hansen, 1913	3
Typhlotanais spinicauda Hansen, 1913	3
ISOPODA	
41. *Abyssoniscus* sp.	17
42. *Aega*	3,12
43. *Austroniscus* sp.1 and 2	8,14,17,19
44. *Austrosignum*	19
45. *Baenectes*	5
46. *Bathygnathia*	(14)
47. *Betamorpha*	8

Table 3   Continued

Species	Seamounts
48. *Cirolana* sp.1, 2 and 3	1,5,12,14,17
49. *Deutroiton*	17
50. *Disconectes* sp.	3,17
51. *Echinimunna*	17
52. *Eurycope*	3,(10),17
53. *Gnathia*	1,3,5,7,8,11,16,17
54. *Ilyarachna* sp.1 and 2	3,8,10,14,17,18
55. *Jaeropsis*	12
56. *Mesosignum*	17
57. *Microthambema*	5
58. *Munnogonium*	3
59. *Panetela*	5
60. *Paramunna*	14,17
61. *Pleurocope*	1
62. *Pleurogonium*	18
63. *Stenetrium*	5
AMPHIPODA	
64. *Cyclocaris tahitensis* Stebbing, 1888	8
65. *Orchomene montana* Vinogradov and Vinogradov	8
DECAPODA	
STENOPODIDEA	
66. **Spongicola parvispina* Zarenkov, 1990	5
CARIDEA	
67. *Acanthephyra eximia* S. Smith, 1884	8,15
68. **Alpheus romensky* Burukovsky, 1990	7
69. *Aristeomorpha foliacea* (Risso, 1827)	5
70. *Benthesicymus investigatoris* Alcock and	3
Anderson, 1899	
71. **Glyphocrangon wagini* Burukovsky, 1990	3,7,8
72. *Hadropenaeus lucasii* Bate, 1881	2,5,6,7,10,12,14
73. *Heterocarpus fenneri* (Crosnier, 1986)	5,7
Heterocarpus laevigatus Bate, 1888	4,5,6,7,8,13,17
Heterocarpus sibogae de Man, 1917	2,4,5,6,7,8,10,11,14
74. *Hymenopenaeus halli* Bruce, 1966	3
75. **Metapenaeopsis stokmani* Burukovsky, 1990	12
76. **Nematocarcinus pseudocurso* Burukovsky, 1990	3,7,8,(22)
Nematocarcinus undulatipes Bate, 1888	3,17
77. *Oplophorus spinosus* Brulle, 1839	8
78. **Pandalina nana* Burukovsky, 1990	5,8,14,17
79. *Pasiphaea americana* Faxon, 1893	8
Pasiphaea flagellata Rathbun, 1906	11
80. *Periclimenes alcocki* Kemp, 1922	5
81. *Plesionika edwardsii* (Brandt, 1851)	2,3,11,14
Plesionika ensis (A. Milne-Edw., 1881)	5,7,8,10
Plesionika martia (A. Milne-Edw., 1883)	3,5,8
Plesionika ocellus (Bate, 1888)	5

Table 3   Continued

Species	Seamounts
Plesionika aff. *williamsi* Forest, 1963	10
82. *Pontocaris rathbuni* (de Man, 1918)	5
83. *Pontophilus gracilis juncaeus* Bate, 1888	3,7,8
Pontophilus nikiforovi Burukovsky, 1990	5,12
84. *Processa pygmaea* Burukovsky, 1990	2,5
85. *Sicyonia nasica* Burukovsky, 1990	5
86. *Stylodactylus pubescens* Burukovsky, 1990	5,6
BRACHYURA	
87. *Cancer balssii* Zarenkov, 1990	10,12,14
88. *Chaceon chilensis* (Chirino-Galvez and Manning, 1989)	7,14,18
89. *Cyrtomaja danieli* Zarenkov, 1990	3,6,8,10
Cyrtomaja platypes Yokoya, 1933	2,5
90. *Dromia dehaani* Rathbun, 1923	12
91. *Ebalia sculpta* Zarenkov, 1990	2,12
92. *Heterocrypta epibranchialis* Zarenkov, 1990	1,2
93. *Homologenus orientalis* Zarenkov, 1990	3
94. *Latreillia phalangium* de Haan, 1835	12
95. *Mursia aspera* Alcock, 1899	3,5,12,18
Mursia gaudichaudii H. Milne-Edwards, 1834	18
96. *Paramola japonica* Parisi, 1915	19,15,12,13,14,15,18
97. *Parthenope (Platylambrus) mironovi* (Zarenkov, 1990)	3,5,12
98. *Progeryon mararae* Guinot and Richer de Forges, 1981	5
99. *Projasus bahamondei* George, 1976	16,17,18,19,20
100. *Randallia nana* Zarenkov, 1990	5
ANOMURA	
101. *Bivalvopagurus sinensis* (de Saint-Laurent, 1972)	3,7,8
102. *Porcellanopagurus foresti* Zarenkov, 1990	12
103. *Sympagurus affinis* (Henderson, 1888)	6,7,8
Sympagurus africanus subsp.nov.	18,19,20
Sympagurus boletifer (de Saint-Laurent, 1972)	3,5,12
Sympagurus rectichela (Zarenkov, 1990)	3,6,8,10,12
Sympagurus ruticheles (A. Milne-Edwards, 1891)	2,5
Sympagurus wallisi Lemaitre, 1994	5
MOLLUSCA	
LORICATA	
104. *Leptochiton* sp.	1,(10),12,15,16,18,19,20
GASTROPODA	
105. *Profundiconus smirna* (Bartsch and Rehder, 1943) (=*Conus profundorum* Kuroda, 1956)	2,5,12,14
106. *Trochus* sp.	19
(TURRIDAE)	
107. *Benthomangelia brevis* Sysoev and Ivanov, 1985	(14)
108. *Daphnella (Eubela) ichthyandri* Sysoev and Ivanov, 1985	3,17

Table 3 Continued

Species	Seamounts
109. *Famelica nitida* Sysoev, 1990	4
110. *Gemmula (Ptychosyrinx) bisinuata* (Sysoev and Ivanov, 1985)	3,5,8,11,13
Gemmula (P.) naskensis (Sysoev and Ivanov, 1985)	15,16,18
111. *Gymnobela altispira* Sysoev and Ivanov, 1985	(14)
Gymnobela brachypleura Sysoev, 1990	3
Gymnobela brunnistriata Sysoev, 1990	5
Gymnobela chisticovi Sysoev and Ivanov, 1985	(14)
Gymnobela crassilirata Sysoev, 1990	5
Gymnobela eugeni Sysoev and Ivanov, 1985	3,12,17
Gymnobela gracilis Sysoev, 1990	5
Gymnobela granulisculpturata Sysoév, 1990	5
Gymnobela laticaudata Sysoev, 1990	2
Gymnobela nivea Sysoev, 1990	8
Gymnobela rotundata Sysoev, 1990	10
Gymnobela turrispina Sysoev, 1990	3
112. *Kuroshiodaphne phaeacme* Sysoev, 1990	5
113. *Leucosyrinx turridus* Sysoev, 1990	2
114. *Mitromorpha (Mitrolumna) maculata* Sysoev, 1990	5
115. *Naskia axiplicata* Sysoev and Ivanov, 1985	11
116. *Pleurotomella (Anomalotomella) minuta* Sysoev and Ivanov, 1985	3,17
117. *Pyrgocythara nodulosa* Sysoev and Ivanov, 1985	12
118. *Splendrillia basilirata* Sysoev, 1990	5
Splendrillia obscura Sysoev, 1990	10
119. *Xanthodaphne tropica* Sysoev and Ivanov, 1985	(14), 17
BIVALVIA	
SEPTIBRANCHIA	
120. *Cribrosoconcha alephtinae* Krylova, 1991	(22)
Cribrosoconcha elegans Krylova, 1991	5,7,8
121. *Cuspidaria macrorhynchus* E. A. Smith, 1895	2,3,5,7,8
Cuspidaria n.sp	8
Cuspidaria sp.	5
122. *Multitentacula (Multitentacula) admirabilis* Krylova, 1995	5
123. *Rhinoclama (Austroneaera) similis* Krylova, 1994	5,8,10
CEPHALOPODA	
124. *Danoctopus* sp.cf. *hoylei* (Berry, 1909)	5
125. *Heteroteuthis* sp.n. aff. *dispar* (Rüppell, 1845)	3,7,8
126. *Iridoteuthis* sp.n. aff. *maoria* Dell, 1959	2,3,5
127. *Nototodarus hawaiiensis* (Berry, 1912)	5,10,12
128. *Scaeurgus unicirrhus patagiatus* Berry, 1913	1,3,5,11,12
129. *Sepioloidea* sp.cf. *pacifica* (Kirk, 1882)	5
130. *Stoloteuthis* sp.n.aff. *leucoptera* (Verrill, 1878)	20

Table 3   Continued

Species	Seamounts
BRACHIOPODA	
131. *Dallithyris murrayi* Muir-Wood, 1959	2,3,5,6,7,8,12
132. *Pelagodiscus atlanticus* (King, 1868)	12
133. *Platidia anomoides* (Scacchi and Philippi, 1884)	1,2,7,12
134. **Septicollarina oceanica* Zezina, 1990	2,5
ECHINODERMATA	
ECHINOIDEA	
135. **Aspidodiadema* n.sp.	3,4,5,6,8,10,11,13
Aspidodiadema sp.	(14,18,19,22)
136. *Caenopedina* n. sp.	1,3,5,8,12
137. *Centrostephanus sylviae* Fell, 1975	8,12,18
138. *Clypeaster isolatus* Serafy, 1971	1,5,12
139. **Coelopleurus* n.sp	12
Coelopleurus sp.	5
140. *Echinocyamus incertus* H. L. Clark, 1914	1,2,3,5,7,8,10,11,12,(14), 16,17
141. *Echinus* sp.	8
142. *Habrocidaris argentea* Agassiz and Clark, 1907	3,8
143. **Homolampas* n.sp.	8
144. *Lampechinus nitidus nascaensis* Markov, 1989	12,17
145. **Podocidaris* n.sp.	
146. **Salenia scrippsae* Zullo and Allison, 1964	2,3,5,9,(10),12,13,16,17
147. **Salenocidaris nudispina* Markov, 1988	3,5,12
148. *Scrippsechinus fisheri* Allison, Durham and Mintz, 1967	8,17
149. **Stereocidaris nascaensis* Allison, Durham and Mintz, 1967	1,2,3,5,10,12 1,3,5,12,13,15
150. *Trigonocidaris albida* Agassiz, 1869	2,3,5,12
151. *Tromikosoma hispidum* Agassiz, 1898	(18,22)
ASTEROIDEA	
152. *Astropecten* sp.1	5
Astropecten sp.2	12
Astropecten sp.3 aff. *ornatissimus* Fisher, 1906	3,5,6,8,10
153. *Brisinga eucoryne* Fisher, 1919	6,10,19
154. **Brisingella* sp.n.	8
155. *Ceramaster* sp.	2,12
156. *Coronaster* sp.	10
157. Goniasteridae g.sp.	10,12
158. *Henricia* spp.	3,12,18,19
159. *Hydrasterias* sp.	8,9,10,12
160. *Luidia* sp.	12
161. *Marginaster* sp.	3,17
162. *Pectinaster* sp.	3,7,8,10,11
163. *Plinthaster* sp.1	5

Table 3   Continued

Species	Seamounts
Plinthaster sp.aff. *ceramoidea* (Fisher, 1919)	3,6,8
164. *Pseudarchaster* sp.aff. *pusillus* Fisher, 1904	8
165. *Pteraster* sp.	17
166. *Sclerasterias* sp.	1,10,12,16
167. *Tamaria* sp.	3,5,12,13,16
168. *Zoroaster* sp.	8
OPHIUROIDEA	
169. *Amphiophiura* sp.	(22)
170. *Asteroschema* sp.	3
171. *Astrophiura* sp.	6
172. *Ophioleuce brevispina* (H. L. Clark, 1911)	3,7
173. *Ophiomyces* sp.	5,7,10,12
174. *Ophiophyces* sp.	(21)
175. *Ophiotholia multispina* Koehler, 1904	3
176. *Ophiura* spp. 1–5	3,5,7,8,9,10
177. *Ophiurases* sp.	10,12

genera (82% of all) are represented on the seamounts of these ridges by one species and only four by more than three species each. These are shrimps *Plesionika* (five species), hermit crabs *Sympagurus* (six), gastropods *Gymnobela* (11) and brittle stars *Ophiura* (five).

Most species not included in Table 3 are hydroids, gorgonarians, actiniarians, solitary corals, non-turrid gastropods, bivalves and bryozoans. The number of holothurians and comatulid crinoids was very low and stalked crinoids were never recorded.

Most material, including type specimens of newly described species, is kept in the collections of the P.P. Shirshov Institute of Oceanology (Moscow), Zoological Museum of the Moscow State University and the Zoological Institute (St Petersburg).

4.2.   Bottom and Near-bottom Fishes

Fishes were collected from 22 seamounts of the Nazca and Sala y Gómez ridges. Annotated lists of all seamount-associated benthic, near-bottom and neritopelagic fishes with their distributional data and depth ranges in the area were presented by Parin (1990a, 1991a). A somewhat modified version of this list corrected on the base of subsequent publications (Anderson and Randall, 1991; Becker, 1992; Fricke, 1992; Kotlyar, 1991; McKosker and Parin, 1995; Markle and Olney, 1990; Parin, 1991b, 1992, 1994; Parin and Kobyliansky, 1993; Sazonov and Iwamoto, 1992; Williams and Machida, 1992) is presented in Table 4.

Description of fishes and data on their distribution were published in the following papers: Amaoka and Parin (1990), Amaoka *et al.* (1997), Anderson and Johnson (1984), Anderson *et al.* (1990), Anderson and Randall (1991), Becker (1992), Belyanina (1989, 1990), Borodulina (1981), Dolganov (1984), Fricke (1992), Golovan' and Pakhorukov (1986), Heemstra and Anderson (1983), Hensley and Suzumoto (1990), Karmovskaya (1990), Kotlyar (1982a,b, 1988b, 1990, 1991), Kotlyar and Parin (1986), Le Danois (1984), Mandritsa (1992), Markle and Olney (1990), McCosker and Parin (1995), Parin (1982a,b, 1983, 1984, 1985, 1987, 1989, 1990a,b, 1991, 1992, 1994), Parin and Abramov (1986), Parin and Borodulina (1986, 1990), Parin and Karmovskaya (1985), Parin and Kobyliansky (1993), Parin and Kotlyar (1985, 1988, 1989), Parin and Sazonov (1990), Parin and Shcherbachev (1982), Parin *et al.* (1980, 1990a,b), Paulin (1991), Sazonov (1989), Sazonov and Iwamoto (1992), Sazonov and Shcherbachev (1982), Svetovidov (1986), Williams and Machida (1992).

The full list of fishes associated with seamounts in the area studied includes 171 species from 64 families. Elasmobranchs are represented by 13 species of nine genera and five families (10 sharks, including eight in the family Squalidae, two skates, one chimaera). Among teleostean fishes the most species-rich families are Macrouridae (25 species), Moridae (12), Alepocephalidae (seven), Scorpaenidae and Gempylidae (six in each), Myctophidae, Emmelichthyidae, Percophidae (five in each), Congridae, Chlorophthalmidae, Serranidae, Draconettidae and Bothidae (four). The most speciose genera are *Physiculus* and *Hymenocephalus* (five species each), *Etmopterus*, *Caelorinchus*, *Ventrifossa*, *Centrodraco* (four each), *Aldrovandia*, *Gnathophis*, *Diaphus* and *Epigonus* (three). Thirty-four families are represented by one species each.

At least 30 fish species (Table 4), collected only at depths between 600 and 3000 m belong to the true deep-water ichthyofauna, which is poorly represented in the collections. However, the exclusion of these species from consideration does not greatly change the pattern of species composition.

Most of the fishes collected, including type specimens of newly described species, are kept in the collections of the same institutions as for invertebrates: P.P. Shirshov Institute of Oceanology of the Russian Academy of Sciences (Moscow), Zoological Museum of the Moscow State University and Zoological Institute of the Russian Academy of Sciences (St Petersburg).

5. COMMUNITIES

5.1. Benthic Communities

5.1.1. *Distribution Patterns*

As a rule, only one to three species were dominant in most trawl catches taken in the area of the Nazca and Sala y Gómez ridges. According to the data obtained

Table 4   List of benthic and benthopelagic fish species recorded from the Nazca and
Sala y Gómez ridges and their distribution on seamounts in the area. Endemics and
probable endemics for the area are indicated with an asterisk. Seamount numbers in
brackets – deepwater records or records of a deep-water species near this seamount.

Family and species	Seamount numbers	Depth range in the area
Hexanchidae	19	300–320
Hexanchus griseus (Bonnaterre)		
Echinorhinidae		
Echinorhinus cookei Pietschman	19, 20	345–540
Squalidae		
Centroscymnus owstoni Garman	(5,7,8,13,14,21)	460–800
Etmopterus litvinovi Parin and Kotlyar	5,7,10,12,14,21	630–1100
E. *lucifer* Jordan and Snyder	2,3,5,6,7,8,10,13	230–800
E. pycnolepis Kotlyar	3,8,10,12	410–760
E. *pusillus* (Lowe)	3,5,8,11	260–770
Mollisquama parini Dolganov	19	320
Somniosus rostratus (Risso)	(21)	800
Squalus mitsukurii Jordan and Snyder	2,3,5,6,8,10,11,12,13,14, 15,16	160–545
Torpedinidae		
Torpedo microdiscus Parin and Kotlyar	11,12	180–280
T. semipelagica Parin and Kotlyar	12	165–290
Chimaeridae		
Hydrolagus sp.	(21)	800
Halosauridae		
Aldrovandia affinis (Günther)	(17)	1220–1230
A. *oleosa* Sulak	(21)	980
A. *phalacra* (Vaillant)	(17,21,22)	980–1230
Notacanthidae		
Notacanthus chemnitzi Bloch	18,19	230–1200
Ophichthidae		
Muraenichthys profundorum McCosker and Parin	19	310
Synaphobranchidae		
Ilyophis blachei Saldanha and Merritt	3	775
Synaphobranchus affinis Günther	3,5,6,7,8,10,13,14,15,17,18	290–750
Simenchelys parasiticus Gill	3	750–800
Nettastomatidae		
Facciolella castlei Parin and Karmovskaya	18	230–500
Nettastoma falcinaris Parin and Karmovskaya	3,6,10,12	420–1100

Table 4 Continued

Family and species	Seamount numbers	Depth range in the area
Congridae		
Gnathophis andriashevi Karmovskaya	5	270–330
G. parini Karmovskaya	7,8	540–560
G. smithi Karmovskaya	12	160–260
Bassanago nielseni (Karmovskaya)	12,13,15,17,18	160–340
Argentinidae		
Glossanodon danieli Parin and Shcherbachev	3,5,10,11,18,19	230–490
G. nazca Parin and Shcherbachev	3,5,12,17,19	160–350
Platytroctidae		
Sagamichthys abei Parr	(21)	980
Alepocephalidae		
Alepocephalus sp.	(12)	1050–1080
Bathytroctes oligolepis (Krefft)	(12)	1050–1080
Conocara fiolenti Sazonov and Ivanov	(13)	1050–1230
Photostylus pycnopterus Beebe	(3)	700–1000
Rouleina attrita (Vaillant)	(12)	1050–1080
R. maderensis Maul	(12,20,21)	1070–1100
Talismania bussingi Sazonov	20	600
Photichthyidae		
Polymetme andriashevi Parin and Borodulina	6,7,8	535–600
Sternoptychidae		
Argyripnus electronus Parin	5,7,8	545–600
Maurolicus rudjakovi Parin and Kobyliansky	7,10,11,12,13,15,18,19	300–400
Polyipnus inermis Borodulina	1,6,7,8,10	420–590
Aulopidae		
Hime macrops Parin and Kotlyar	12	160–170
Chlorophthalmidae		
Bathypterois atricolor Alcock	(21)	980
Bathytyphlops marionae Mead	(13)	1220–1230
Chlorophthalmus ichthyandri Kotlyar and Parin	3,6,7,8,10,12	285–590
C. zvezdae Kotlyar and Parin	15	?
Synodontidae		
Bathysaurus ferox Günther	(12,15)	980–1230
Synodus cf. *doaki* Russel and Gressey larva	5	–

Table 4   Continued

Family and species	Seamount numbers	Depth range in the area
Neoscopelidae		
Neoscopelus macrolepidotus Johnson	3	730–800
Myctophidae		
Diaphus adenomus Gilbert	10,11,12, 13,16	225–320
*D. confusus Becker	8	545–560
*D. parini Becker	11,13	305–320
Idiolychnus urolampus (Gilbert and Cramer)	1,2,3,5	270–360
Lampadena urophaos Paxton	3,5,6,8,19	310–800
Moridae		
Antimora rostrata Günther	(12,15)	1070–1800
*Gadella obscurus (Parin)	3,5,12,13,22	175–330
Halargyreus johnsoni Günther	(12,15)	1220–1230
Laemonema rhodochir Gilbert	3,7,10,15	280–600
*L. yuvto Parin and Sazonov	6	545–600
Lepidion guentheri (Giglioli)	(21)	750–800
*Physiculus hexacytus Parin	5,10,12,13,14,15,17, 18,19,20	240–420
*P. longicavis Parin	5,12	160–285
*P. parini Paulin	3,5	330–360
*P. sazonovi Paulin	5	260–270
Physiculus sp.	5	260–270
*Tripterophycis svetovidovi Sazonov and Shcherbachev	6,7,10	385–610
Gadidae		
*Gaidropsarus parini Svetovidov	17	300
Merlucciidae		
Macruronus magellanicus Lönnberg	18	230
Macrouridae		
Cetonurus crassiceps (Günther)	(12,21)	940–1100
*Caelorinchus immaculatus Sazonov and Iwamoto	7,8,20,21	505–730
*C. multifasciatus Savonov and Iwamoto	3	350–490
*C. nazcaensis Sazonov and Iwamoto	18,19,22	230–530
*C. spilonotus Sazonov and Iwamoto	6,7,8,10	350–600
Coryphaenoides paradoxus (Smith and Radcliffe)	(12,20,21)	980–1230
Gadomus cf. *melanopterus* Gilbert	(12,15,16)	1050–1230
Hymenocephalus cf *aterrimus* (Gilbert)	(12,16)	780–1100
H. gracilis Gilbert and Hubbs	10,12,13	370–410
*H. neglectissimus Sazonov and Iwamoto	7,8	525–600
*H. semipellucidus Sazonov and Iwamoto	3,6,7,8,10	550–800

Table 4　Continued

Family and species	Seamount numbers	Depth range in the area
H. striatulus Gilbert	3,6,7,8,10	450–800
*Kuronezumia pallida Sazonov and Iwamoto	3,8	540–800
Macrouroides inflaticeps Smith and Radcliffe	(12)	1600–2000
Malacocephalus laevis (Lowe)	5,6,7,8,10,11	385–590
Mataeocephalus acipenserinus (Gilbert and Cramer)	3	730–800
Nezumia convergens (Garman)	(15)	1050–1080
N. propinqua (Gilbert and Cramer)	3,7,8	570–800
*Pseudocetonurus septifer Sazonov and Shcherbachev	(11,21,22)	780–950
Squalogadus modificatus Gilbert and Hubbs	(15)	1050–1080
Trachonurus villosus (Günther)	(12,15)	1050–1100
Ventrifossa johnboborum Iwamoto	3,8	750–800
*V. macrodon Sazonov and Iwamoto	3,6,7,8	540–800
*V. obtusirostris Sazonov and Iwamoto	3	750–800
*V. teres Sazonov and Iwamoto Ophidiidae	3,6,7,8	345–700
Dicrolene nigra Garman Carapidae	(21)	940–960
Eurypleuron cinereum (Smith)	5,12	160–330
*Pyramodon parini Markle and Olney Lophiidae	5,6,10,12	185–330
Lophiodes mutilus (Alcock) Chaunacidae	7	563–590
*Chaunax latipunctatus Le Danois Ogcocephalidae	3,6,7,8	545–790
*Dibranchus sp. nova Bradbury Ateleopodidae	3	750–800
Ateleopus sp. Monocentridae	1	280–300
Monocentris reedi Schultz Berycidae	12	165–235
Beryx splendens Lowe Polymixiidae	3,8,12,15,18,19	230–770
*Polymixia salagomezensis Kotlyar	5	330
*P. yuri Kotlyar	12	180–250

Table 4   Continued

Family and species	Seamount numbers	Depth range in the area
Zeidae		
Cyttomimus stelgis Gilbert	3,10,?19	330–410
Stethopristes eos Gilbert	10	380–530
*Zenopsis oblongus Parin	5,11,12,13,15,16,17,18	180–330
Oreosomatidae		
Neocyttus rhomboidalis Gilbert	12	500–700
Zeniontidae		
Zenion hololepis Goode and Bean	3,5	325–490
Caproidae		
*Antigonia aurorosea Parin and Borodulina	2,5	240–310
A. capros Lowe	1,2,5,12	210–320
Macrorhamposidae		
Macrorhamposus scolopax (Linnaeus)	12	145–260
Notopogon fernandezianus (Dolfin)	3,5,12,15	210–530
Scorpaenidae		
Helicolenus lengerichi Norman	10,11,12,13,14,15,16, 18,19,20	215–530
*Phenacoscorpius eschmeyeri Parin and Mandritsa	5	500
*Plectrogenium barsukovi Mandritsa	10,11,14	290–380
Scorpaena uncinata De Buen	12	160–240
Setarches guentheri Johnson	3,13	320–400
Trachyscorpia cristullata (Goode and Bean)	12	360–400
Triglidae		
Pterygotrigla picta (Günther)	1,2,3,5,11,12,16	160–350
Platycephalidae		
Bembradium roseum Gilbert	4,6,10	380–540
Hoplichthydae		
Hoplichthys citrinus Gilbert	3,5	230–490
Percichthydae		
Sphyraenops bairdianus Poey	5	240–275
Serranidae		
*Anatolanthias apiomycter Anderson, Parin and Randall	12	160–170
Caprodon longimanus (Güenther)	12	160–235
*Plectranthias exsul Heemstra and Anderson	12	160–225
*P. parini Anderson and Randall	5	260–270

Table 4   Continued

Family and species	Seamount numbers	Depth range in the area
Callanthiidae		
Callanthias parini Anderson and Johnson	2,18,19	230–350
Grammatonotus laysanus Gilbert	1,5,12	240–330
Lutjanidae		
Symphysanodon maunaloae Anderson	1,5	240–300
Priacanthidae		
Cookeolus japonicus (Cuvier)	5	240–300
Epigonidae		
Epigonus atherinoides (Gilbert)	8	540
E. elegans Parin and Abramov	10,13,14,15,18,19	230–410
E. notacanthus Parin and Abramov	10,12,18,19	230–410
Carangidae		
Decapterus muroadsi (Schlegel)	12,14	90–270
Seriola lalandi (Valenciennes)	18	150–410
Emmelichthyidae		
Emmelichthys cyanescens (Guichenot)	12,15,18,19	145–330
E. elongatus Kotlyar	2,5,12	150–245
Erythrocles scintillans (Jordan and Thompson)	2	240–275
Plagiogeneion geminatum Parin	5	200
P. unispina Parin	18	280–310
Pentacerotidae		
Pentaceros quinquespinis Parin and Kotlyar	5,12	165–280
Cheilodactylidae		
Acantholatris gayi (Kner)	12	160–220
Mugiloididae		
Parapercis dockinsi McCosker	5,12	180–290
Percophidae		
Chrionema chryseros Gilbert	3	350–490
C. pallidum Parin	1,3,5	230–350
Dactylopsaron dimorphicum Parin and Belyanina	5,10	240–345
Enigmapercis acutirostris Parin	5	470–485
Osopsaron karlik Parin	1,2,5,10,11,14	260–450
Draconettidae		
Centrodraco gegonipa (Parin)	3,5	260–480
C. nanus (Parin)	?	?
C. nakaboi Fricke	6	545–600
C. striatus Parin	10,11,18,19	230–410

Table 4   Continued

Family and species	Seamount numbers	Depth range in the area
Schindleriidae		
Schindleria praematurus (Schindler)	5,12	200–220
Scombrolabracidae		
Scombrolabrax heterolepis Roule	19,21	330–350
Gempylidae		
Lepidocybium flavobrunneum (Smith)	19	300
Nesiarchus nasutus Johnson	19	150–210
Promethichthys prometheus (Cuvier)	5	290
*Rexea antefurcata Parin	3,5,6,7,8,11,12,18	160–770
*R. brevilineata Parin	10,11,12,18,20	180–400
Ruvettus pretiosus Cocco	18,19,20	320–630
Trichiuridae		
Aphanopus capricornis Parin	(21)	900
Assurger anzac (Alexander)	8,16	200–230
Benthodesmus elongatus (Clarke)	6,8,10,16	260–575
Centrolophidae		
Seriolella labirinthica (McAllister and Randall)	5	240–275
Ariommidae		
Ariomma luridum Jordan and Snyder	2,5	280–310
Samaridae		
Samariscus cf. nielseni Quero, Hansley and Mauge	4	470
Bothidae		
*Arnoglossus multirastris Parin	3,5,10,11,12,13,15	160–490
*Chascanopsetta megagnatha Amaoka and Parin	7,8,10	400–590
Engyproposon cf. regani Hansley and Suzumoto larva	5	—
*Parabothus amaokai Parin	12	150–220

in 1979 and 1980 by "Ikhtiandr" (Fedorov and Chistikov, 1985), spiny lobsters
Projasus bahamondei were dominant by weight in the bottom communities on the
summits of the Zvezda, Professor Mesyatzev and Ekliptika seamounts. These
spiny lobsters were noted also on the summits of the Ikhtiandr and Albert
seamounts, where they were less abundant. A different picture was found on the
summit of the Bolshaya Seamount where the bulk of the catch consisted of the
irregular echinoid *Scrippsechinus fisheri* and the regular echinoid *Stereocidaris
nascaensis* (Mironov, 1985a). The trawl catches obtained during the 18th cruise

of "Professor Shtokman" may be grouped as follows, according to the dominant animals (Mironov and Detinova, 1990) (Figure 2):

- In seven out of the 37 most successful hauls the spiny lobster *Projasus bahamondei* strongly dominated by weight. These catches were obtained on the tops or upper slopes (depths 227–420 m) of the Ekliptika, Professor Mesyatzev and Zvezda seamounts (Nos 18, 19 and 20 on Figure 1). All these seamounts are located to the east of 82°W. In the central parts of the guyot tops, in addition to the spiny lobsters, solitary Zoantharia were also abundant. In catches taken at the peripheries of the tops, places with more complex relief, the sea star *Henricia* and sponges were common.

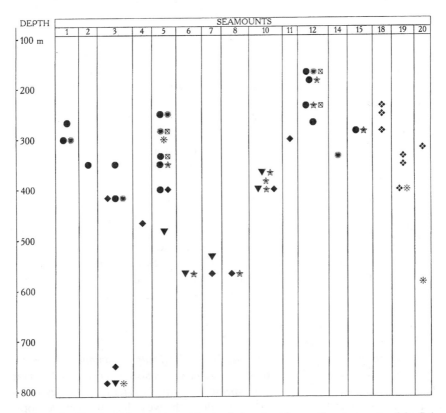

Figure 2   Dominant bottom invertebrates in trawl catches on the seamounts of the Sala y Gómez and Nazca ridges. The names of the numbered seamounts on this and following figures are given in Figure 1 (after Mironov and Detinova, 1990). ● *Stereocidaris, etc.*; ✳ *Aspidodiadema, etc.*; ◆ Decapoda; ❖ *Projasus*; ▼ Spongia; ★ Asteroidea; ✳ Gorgonaria; ⊠ encrusting red algae.

•   In 16 catches sea urchins predominated. On flat bottoms the most common sea
    urchin species were *Scrippsechinus fisheri* and *Stereocidaris nascaen-
    sis*; on irregular bottoms the genera *Aspidodiadema*, *Centrostephanus* or
    Coelopleurus predominated. These catches were obtained on the tops and
    upper slopes (depths 162–400 m) of the seamounts Igolnaya, Utes, Ich-
    thyologists, Kupol, Bolshaya, Novaya and Dorofeyev (Figure 1, Nos 1–3, 5,
    12, 14–15), all located to the west of 82°W. Trawl hauls on the Bolshaya and
    Dorofeyev seamounts also brought up many sea stars, hauls on the Kupol
    Seamount produced sea stars and shrimps, and hauls on the Ichthyologists
    Seamount produced shrimps.

    Crustose red algae (Rhodophyceae) were found on the tops of the Igolnaya,
Utes, Kupol and Bolshaya seamounts, on stones, pebbles and shells. These
algae include representatives of the families Squamariaceae and Corallinaceae
and genera *Melobesia*, *Lithothamnion*, *Lithophyllum*, *Archeolithothamnium*
and *Peyssonelia* (Vozzhinskaya, 1990). Crustose red algae were noted on Utes
Seamount at depths down to 330 m (Mironov and Detinova, 1990). Earlier such
algae had been reported only to 268 m (Littler *et al.*, 1985).

•   In 14 catches sponges, gorgonarians, starfishes or shrimps, in various
    combinations, predominated. Some catches were obtained on the summits of
    the seamounts Stolbovaya, Pervomayskaya, Zhemchuzhnaya, Yantarnaya,
    Baral and Dlinnaya. The first four seamounts (Nos 4, 6, 7 and 8 on Figure 1)
    had deep summits (deeper than 400 m). Other catches of this type were
    obtained on the slopes of Ichthyologists, Kupol and Zvezda seamounts, also
    rather deep. All catches of this type came from depths of 300–800 m.

These three groups of catches differ in their horizontal and vertical
distributions. If we compare rather shallow-water catches, obtained at depths of
162–400 m, we see that about longitude 82°W a rather abrupt change was recorded
from one group of widely distributed communities dominated by spiny lobsters,
to another dominated by sea urchins. Such differences were not found at depths
over 450 m. The communities also changed with increased depth: from dominance
by spiny lobsters or sea urchins to dominance by sponges, gorgonarians, starfishes
and shrimps, at depths of 350–400 m.

The composition of baited trap catches also demonstrates a change with the
geographic, position of a seamount (Mironov and Detinova, 1990) (Figure 3). East
of 84°W the traps caught many spiny lobsters *Projasus bahamondei* and crabs
Chaceon chilensis while west of 84°W catches consisted mostly of the crab
Paramola japonica and five shrimp species (Figure 3). The best catches in the
eastern area were obtained on summits while in the west they were on the slopes
of seamounts.

Galkin (1993) has shown an association of certain starfish species with the
predominant animals in trawl catches. In particular, starfishes of the genus
Henricia were definitely associated with spiny lobsters (probability $P > 0.999$)

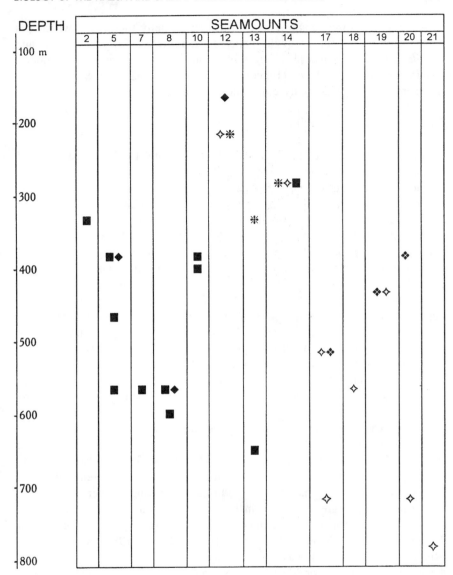

Figure 3 Dominant bottom invertebrates in baited trap catches obtained on the seamounts of the Sala y Gómez and Nazca ridge (after Mironov and Detinova, 1990). ■ shrimps; ◆ other crustacea; ❖ *projasus*; ✧ *Geryon*; ✴ Paramola.

while the genera *Tamaria* ($P > 0.99$) and *Luidia* and two species of *Astropecten* (spp. 1 and 2) were found only in communities dominated by sea urchins. Seston-feeding starfishes of the family Brisingidae were associated with other seston-feeders (Spongia, Gorgonaria) and with shrimps but not with sea urchins

($P = 0.95$). *Sclerasterias* and *Hydrasterias* were usually found in trawl catches with many fishes but not with spiny lobsters ($P > 0.95$). *Plinthaster* and *Pseudarchaster* were recorded only with sponges, shrimps and sea anemones. Galkin concluded that the bottom communities of the Nazca and Sala y Gómez ridges are well defined and have distinct boundaries.

5.1.2. *Population Stability*

Significant changes were noted between 1979–1980 and 1987 in the structure of the bottom communities on the Bolshaya and Ekliptika seamounts (Mironov and Detinova, 1990). During that time the antipatharian *Cirripathes* and cirripede *Heteralepas mystacophora* disappeared from the sea urchin community on the Bolshaya Seamount and the crab *Paramola japonica* disappeared from the spiny lobster community on the Ekliptika Seamount. Colonies of *H. mystacophora* are attached to *Cirripathes*, looking like large globular floats hanging on long twigs. All three forms were common in 1970–1980 but the bushes of fragile antipatharians were destroyed by the bottom otter-trawl commonly used in these years, and *H. mystacophora* were lost with their substratum animals. It is important to note that commercial bottom trawling on seamounts is only possible along paths free from obstacles, which are very common there and may seriously damage gear. Thus, almost all trawl tows follow the same routes while other, neighbouring areas with hard bottom and obstacles may remain unfished. Sea urchins and spiny lobsters were found to be more stable constituents of communities and retained their leading role in spite of being caught with large bottom trawls. Populations of the sea urchin *Stereocidaris nascaensis* on the Bolshaya Seamount declined following the destruction of *Heteralepas mystacophora* which, in 1980, formed the major part of the gut contents of these animals (Mironov, 1985a; Sokolova, 1987). One reason for the relative stability of *Stereocidaris nascaensis* populations on the Bolshaya Seamount may be its highly adaptable feeding habits, discussed in the following chapter.

To evaluate the degree of stability in the size-weight structure of the populations of spiny lobster *Projasus bahamondei*, comparisons were made between sets of data obtained during September–October, 1980, February, 1982 and April, 1987 (Rudjakov *et al.*, 1990). It was found that populations inhabiting the summits of the Zvezda and Professor Mesyatzev seamounts consisted of animals of two size–weight groups, one approximately twice the weight of the other and 1.2–1.4 times as long. These two groups were treated as two subsequent stages in the moulting cycle. On the summit of Ekliptika Seamount the differentiation into two groups was less apparent. In 1980 large lobsters predominated on the Zvezda and Professor Mesyatzev seamounts, whereas in 1982 and 1987 small ones predominated. On the Ekliptika Seamount the size distribution was almost constant and small lobsters predominated in all years (Figure 4). Rudjakov *et al.* suggested two factors which may influence these temporal differences, natural

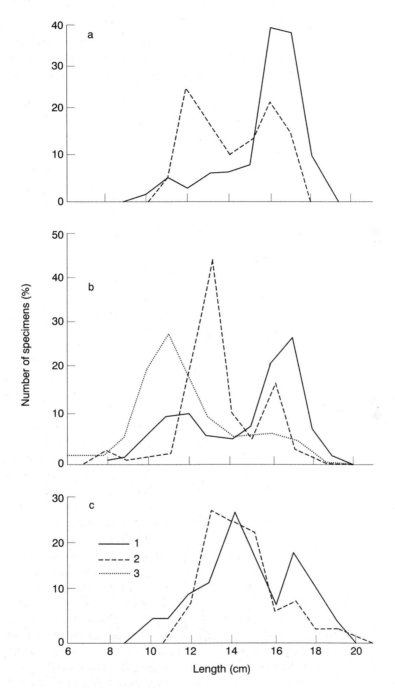

Figure 4   Body length frequency distribution of the spiny lobster *Projasus bahamondei* as percent, on the seamounts Zvezda (a), Professor Mesyatzcv (b) and Ekliptika (2c). 1, September–October, 1980; 2, April, 1987; 3, February, 1982 (after Rudjakov *et al.*, 1990).

fecundity fluctuations (probably determined by the success of larval settlement) and seasonality of reproduction and growth. Seasonality in lobster reproduction was shown by Prosvirov (1990). It is possible that during February to April (the sampling period in 1982 and 1987) most lobsters have not yet reached the age of the animals in the dominant size group in October, 1980 (Rudjakov *et al.*, 1990). We cannot exclude the probable effect of lobster overfishing with otter-trawls. In this study no evidence was found of genetic isolation of lobster populations on individual seamounts and the authors concluded that the pelagic larval stage in *P. bahamondei* is long enough to overcome the distances between neighbouring seamounts. This accords with existing data on the larval biology of related lobster species (Phillips and McMillan, 1986) and on the very wide dispersal of lobster larvae (Lutjeharms and Heydorn, 1981; Phillips, 1982).

5.1.3. *Trophic Structure of the Benthic Communities in its Relation to Environmental Factors*

The communities of bottom invertebrates classified according to dominant taxa (see above) differ also in their trophic structure. The first group is characterized by carnivores, such as spiny lobsters *Projasus bahamondei* (Fedorov and Chistikov, 1985; Mironov and Detinova, 1990), the second group by omnivores (sea urchins such as *Stereocidaris nascaensis*, *Aspidodiadema*, *Centrostephanus*, *Coelopleurus*, and starfishes) and detritus-feeders such as the sea urchin *Scrippsechinus fisheri* (Mironov, 1985a; Sokolova, 1987, 1990, 1994; Mironov and Detinova, 1990). The third group of communities can be divided into two subgroups: suspension-feeders are most common in one (hyaline sponges, gorgonarians, sea stars *Brisinga eucoryne*), while carnivores and omnivores prevail in the other, mostly shrimps and sea stars, family Goniasteridae and genus *Pectinaster* (Mironov and Detinova, 1900; Galkin, 1993). Details of the type of feeding in the last group of animals require further investigation.

Fedorov and Chistikov (1985), have shown that carnivorous animals form up to 99.9% of the biomass in the bottom communities on the summit of Professor Mesyatzev Seamount. To account for such a strong predominance of benthic consumers the following sequence of cause-and-effect features was proposed: the formation of quasi-stationary eddies over the summit, enhancement of productivity in these eddies, a high abundance of macroplankton and mesopelagic fish in the deep scattering layer (DSL), the concentration of macroplankton and pelagic fish in the vicinity of the bottom during the daytime descent of the DSL on to the flat top of the seamount, consumption of the DSL animals by demersal and benthopelagic fish, and consumption by the spiny lobsters of the remains of fish food. Knowledge of the hydrodynamics and biology of the waters surrounding the Professor Mesyatzev Seamount seems to support the ideas of Fedorov and Chistikov (1985). Current–topography interactions, including formation of eddies, local upwelling and closed circulation patterns called Taylor columns, are well known in various areas of the World Ocean and have a profound impact on seamount ecosystems (Darnitsky, 1979; Darnitsky et al., 1986; Boehlert and Genin, 1987; Boehlert, 1988). Over the Professor Mesyatzev Seamount, stationary cyclonic and anticyclonic eddies identified as Taylor columns have been reported and the biomass of plankton near the western slopes of the mount was found to be two to three times higher than at a distance from the seamount (Grossman, 1978; Amarov and Korostin, 1981; Kolodnitsky and Kudryavtsev, 1982; Fedorov and Chistikov, 1985; Gevorkyan et al., 1986). The "settling" of the DSL on the summit has been observed on the Professor Mesyatzev Seamount and on another south-east Pacific seamount with a minimal depth of 101 m (Kashkin, 1984). Finally, the lower limit of the distribution of P. bahamondei on this seamount coincided with the daytime position of the DSL and the stomachs of these spiny lobsters contained the remnants of DSL animals, such as squids, shrimps and gonostomatid fishes (Fedorov and Chistikov, 1985). Thus, the prevalence of carnivores on the summit of Professor Mesyatzev Seamount is likely to be the consequence of the enhanced productivity of surrounding waters (Fedorov and Chistikov, 1985; Fedorov, 1990).

This hypothesis was also supported by the known scarcity of carnivores on seamounts surrounded by waters with low plankton biomass. In the area of the Nazca Ridge the biomass of plankton diminishes towards the south-west, from the Zvezda to the Bolshaya Seamount (Gevorkyan et al., 1986). The survey was made in late winter (August, 1980) when the absolute figures of biomass were relatively low (Figure 5). The number of spiny lobsters diminished from four specimens per 10 m² on the Professor Mesyatzev Seamount and an even higher number on Zvezda Seamount to 0.2 specimens per 10 m² on the Albert Seamount and total absence on Bolshaya Seamount (Fedorov and Chistikov, 1985: Figure 3).

The sequence hypothesized by Fedorov and Chistikov (1985) for the spiny lobsters, was proposed by Mironov (1985a) to explain the strong domination of omnivorous sea urchins on some seamounts. He also suggested that the corpses and faeces of benthopelagic fishes are important food objects for these echinoids because the remnants of fishes and homogeneous faecal pellets were found in the guts of the sea urchins. Data for an area off California show that midwater-fish faecal matter may be a very important source of organic matter for benthic animals (Robison and Bailey, 1981).

However, there is not enough evidence that orographically determined eddies over seamounts remain stationary long enough for the successive changes to be completed, from upwelling to the growth of secondary and tertiary consumers, such as macroplankton and micronekton. If this succession is to be completed, the concentrations of macroplankton and micronekton must remain over the seamount summit and not far downcurrent (Parin et al., 1985). The descent of the DSL on to a seamount summit probably ensures the food of bottom and near-bottom fishes and benthic invertebrates as advected macroplankton and micronekton without enhancement of productivity by quasi-stationary eddies (Isaacs and Schwatzlose,

1965; Kashkin, 1984; Parin *et al.*, 1985; Genin *et al.*, 1988; Parin, 1988; Nesis, 1993b).

Echinoid dominance has been recorded on six seamounts, located in the Pacific, Indian and Atlantic Oceans (Mironov, 1985a). On five of them, including the Bolshaya Seamount of the Nazca Ridge, planktonic invertebrates and fish remnants formed the major part of the echinoid gut contents. Fish remnants predominated in the gut of the sea urchin *Centrostephanus nitidus* from the summit of the Error Seamount (north-western Indian Ocean), located in an eutrophic area. Mironov (1985a) suggested that the large fish concentrations over the summits have a strong impact on the trophic structure of the communities of bottom animals, and that the prevalence of necrophages in these communities may indicate the existence of fish concentrations.

Sokolova (1987, 1990, 1994) investigated in great detail the gut contents of sea urchins on the seamounts. She did not confirm the presence of fish faeces in the food of echinoids. The gut contents of *Stereocidaris nascaensis* from the Bolshaya Seamount was studied in the collections made in 1979, 1980, 1984 and 1987. During this time the main foods consumed continually changed, from the remnants of dead pelagic invertebrates and fish to bottom invertebrates captured alive. This change and observations of large differences in the gut contents between specimens from different seamounts on the Sala y Gómez Ridge demonstrate great plasticity in the feeding of *S. nascaensis*. A necrophagous scavenger, *Coelopleurus* sp. 2 from the Kupol Seamount fed much more on pelagic animals than did *S. nascaensis*. Seventy-five per cent of its food had a midwater origin (planktonic crustaceans, chaetognaths and fish). The highest proportion of pelagic food items (82–100%) was recorded in sea urchins from the Error Seamount in the Indian Ocean. Sokolova (1994) studied sea urchins from seamounts in eutrophic, oligotrophic and intermediate areas of the oceans and reported a high proportion of the "rain of corpses" in their diet everywhere. In her opinion the summits of seamounts are habitats where feeding conditions allow the macrobenthos to be self-sufficient. Productivity of the overlying water is considered more important than geographical position.

Thus, the role of carnivorous bottom animals on the summits of seamounts of the Nazca and Sala y Gómez ridges (depths 162–400 m) diminishes from almost complete dominance in the east to a prevalence of omnivorous and detritivorous animals in the west. However, all along the ridges the rain of remnants of planktonic animals and fish is very important in the diet of macrobenthic animals.

Figure 5 Biomass of plankton (B), and relative abundance of carnivorous bottom invertebrates (A), over and on the summits of seamounts of the Sala y Gómez and Nazca ridges. Seamount numbers and names as in Figure 1 (after Fedorov and Chistikov, 1985).

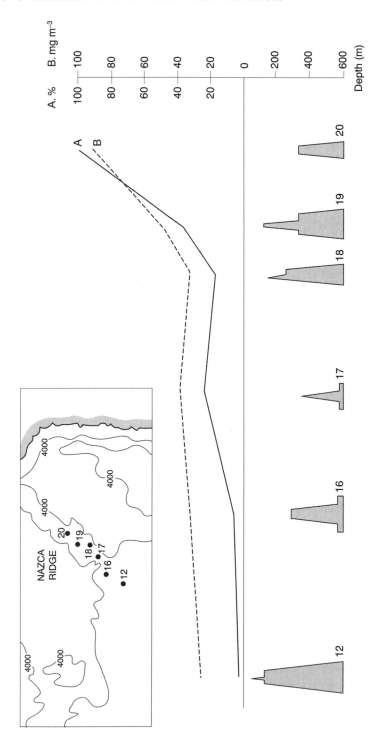

The prevalence of carnivores, omnivores and scavengers in bottom communities in depths of 162–400 m may be considered an indication of a possible local increase in the productivity of the overlying waters.

Another type of dependence on environmental factors is demonstrated by suspension-feeders (sestonophages). A predominance of suspension-feeders was observed in certain circumstances on the seamounts of the Nazca and Sala y Gómez ridges. First, places where there are small-scale changes of habitat, as when flat-bottom areas with loose sediments join areas of complex relief and hard substrata; second, areas of changing depth and increasing of the depth with areas with loose sediments (summits or gentle slopes). A manned submersible was used to investigate areas with complex relief and hard substrata on Professor Mesyatzev Seamount (Fedorov and Chistikov, 1985). At the transition from the flat top of the mount to the slope a community dominated by spiny lobsters changed into one with spiny lobsters and antipatharians (two to three colonies per 10 m^2). The same change was observed in trawl catches.

The most important factors responsible for a prevalence of suspension-feeding animals are usually considered to be rapid currents and high turbulence, increasing the food particle concentration in the water (Simpson and Heydorn, 1965; Grasshoff, 1972; Mironov and Pasternak, 1981; Grigg, 1984; Genin *et al.*, 1986; Moskalev and Galkin, 1986; Boehlert and Genin, 1987; Grigg *et al.*, 1987; Mironov and Sagalevich, 1987; Kaufmann *et al.*, 1989; Rogers, 1994).

Domination by suspension-feeders (up to 99% of trawl catches by weight) was observed on some flat areas with loose sediment in 365–600 m depth, on the summits of the Baral, Zhemchuzhnaya and Pervomayskaya Seamounts and on the slopes of the Baral and Kupol Seamounts. Here the hexactinellid sponge *Pheronema nascaniensis* and the sea star *Brisinga eucoryne* prevailed. The sponges had reduced oscula and atrial cavities and greatly enlarged basal tufts in the dead lower part of body, which were regarded as adaptations to life on soft substrata (Tabachnick, 1991).

Galkin (1993) stated that the depth (300–360 m) of the changeover from a sea urchin community to suspension-feeders and crustaceans coincided with a change of sediment from calcareous–foraminiferous–detritic sand to pteropod ooze. We suppose that the predominance of suspension-feeders on loose sediments is linked to the decline of available food as depth increases. Suspension-feeders also predominate on deep-water summits of guyots of the Mid-Pacific Mountains covered with loose sediments (Mironov and Pasternak, 1981). The development of large populations of suspension-feeders in places where the concentrations of plankton and organic particles in the water are low may be explained as the clearance of a large volume of near-bottom water by the animals.

The vertical distribution of bottom animals on the Nazca and Sala y Gómez ridges is certainly influenced by the oxygen-deficient layer. The lowest oxygen concentration was usually observed at 250–350 m; the summits of some guyots are in the oxygen-minimum zone or may be temporarily exposed to it during its vertical displacements. For example, at the summit of the Professor Mesyatzev Seamount the core of Tropical Intermediate Water has its salinity maximum and oxygen minimum (0.15–0.20 ml l^{-1}) at 250–340 m; that is, level with the summit. When the subsurface Peru Undercurrent is strong the concentration of oxygen in the near-bottom water at the summit may drop to 0.14 ml l^{-1} (Kudryavtsev, 1981; Gevorkyan et al., 1986). When the undercurrent is weak the summit may be washed by the deeper, oxygen-rich, Antarctic Intermediate Water (AIW). This change influences the distribution of bottom animals. When the near-bottom oxygen concentration falls to 0.15–0.20 ml l^{-1} the lobsters and the fish (blue-mouths *Helicolenus lengerichi*) migrate from the summit to the upper slopes. When it rises to 0.4–0.6 ml l^{-1} the lobsters return to the summit but fishes do not. Only when the AIW occupies the whole summit and the oxygen concentration rises to 2.5 ml l^{-1} do both the lobsters and blue-mouths settle on the summit; other fishes concentrate here from time to time (Golovan' et al., 1982). Studies at Volcano 7 in the Eastern Tropical Pacific have also shown a great influence of the oxygen minimum zone on the distribution of bottom fauna and the structure of bottom communities (Wishner et al., 1990, 1995; Levin et al., 1991).

5.2. Benthopelagic Communities

5.2.1. *Distribution of Fish Communities*

The data on depth distribution of benthic and benthopelagic fish tabulated in Parin (1990a, 1991a) show that almost all the fish inhabiting the seamounts are characteristic of intermediate depths (200–1500 m) but some with other bathymetric ranges are included.

Analysis of the vertical distribution of fish is seriously hindered by the non-uniform depth distribution of trawling stations because most trawl hauls were made on flat tops of seamounts, and thus the conclusions about bathymetric ranges are tentative. Nevertheless, vertical zonation of fish distribution is rather well marked, in spite of the presence of some very eurybathic species such as *Notacanthus chemnitzi* (230–1200) *Nettastoma falcinaris* (420–1100 m) or *Rexea antefurcata* (160–770 m). The number of fish species recorded for each 100 m depth range and their upper and/or lower limits are shown in Figure 6. It can be seen that these limits are most crowded in the 200–300 and 700–800-m ranges. These ranges coincide approximately with the limits, in this area, of the sublittoral and upper bathyal and of the upper and lower bathyal zones (= epizonal/upper mesozonal and upper/lower mesozonal zones according to Parin (1988)). There is also noticeable crowding of vertical range limits between 500 and 600 m. This may indicate some lack of uniformity in the upper bathyal fish fauna.

Five fish species occurred only in depths of less than 200–220 m. They belong to the genera *Hime, Anatolanthias, Plagiogeneion, Schindleria* and *Parabothus.* Four of them were found on the Bolshaya Seamount, the only seamount on these ridges with a summit shallower than 200 m; the fifth species (*Bassanago geminatus*) was caught over the Ekliptika Seamount well above the bottom. Twenty-three more species were caught in trawls taken partly in this zone; they were representatives of the genera *Squalus, Torpedo, Bassanago, Glossanodon, Physiculus, Eurypleuron, Monocentris, Polymixia, Zenopsis, Scorpaena, Pterygotrigla, Plectranthias, Emmelichthys, Rexea* and *Arnoglossus.*

The species diversity of the fish fauna was greatest in the depth range from 200 to 500–600 m. Here 138 species were found, 26 of which were not found deeper than 300 m; 83 did not go beyond the limits of the upper bathyal zone. Relatively shallow-living species belong to the genera including *Torpedo, Macrorhamposus, Monocentris, Caprodon, Plectranthias, Decapterus, Seriola, Emmelichthys, Pentaceros, Acantholatris* and *Arnoglossus.* In greater depths the typical genera are: *Squalus, Glossanodon, Maurolicus, Chlorophthalmus, Gadella, Physiculus, Caelorinchus* (*C. nazcaensis, C. spilonotus*), *Hymenocephalus, Malacocephalus, Pyramodon, Beryx, Polymixia, Cyttomimus, Zenion, Antigonia, Helicolenus, Scorpaena, Pterygotrigla, Bembradium, Callanthias, Epigonus, Parapercis, Chrionema, Centrodraco, Rexea, Benthodesmus* and *Chascanopsetta.*

Collections from depths between 600–1000 m included 43 fish species, the most typical of which are species of the genera *Etmopterus, Centroscymnus, Neoscopelus, Rouleina, Lepidion, Kuronezumia, Mataeocephalus, Nezumia, Pseudocetonurus* and *Neocyttus.*

Only 21 species were recorded deeper than 1000 m, belonging to genera such as *Aldrovandia, Notacanthus, Nettastoma, Alepocephalus, Rouleina, Bathypterois, Bathytyphlops, Benthosaurus, Antimora, Halargyreus, Cetonurus, Gadomus, Macrouroides, Squalogadus* and *Trachonurus.*

Judging from underwater observations (Golovan' and Pakhorukov, 1987) and differences in the catch composition of bottom, near-bottom and midwater trawls, the "mountain" fishes of the south-eastern Pacific vary greatly in the relative strength of their connection with the bottom. They include true *benthic* species, almost constantly in contact with the substratum (representatives of the genera *Aldrovandia, Synaphobranchus, Lophiodes, Dibranchus, Scorpaena, Centrodraco, Arnoglossus* and others), and purely *pelagic* species of neritic type unlikely to make contact with hard bottom (*Seriola, Decapterus, Emmelichthys* and *Ariomma*), but the overwhelming majority (approximately 68%) occupy intermediate positions between these two extremes. These species may be termed *benthopelagic* in a wide sense (Parin, 1988). Some benthopelagic fishes spending part of their time at the bottom, are termed *near-bottom* species (*Squalus, Chlorophthalmus, Physiculus, Caelorinchus, Callanthias,* and *Osopsaron*) and others live all the time in the near-bottom water layer, termed *off-bottom*

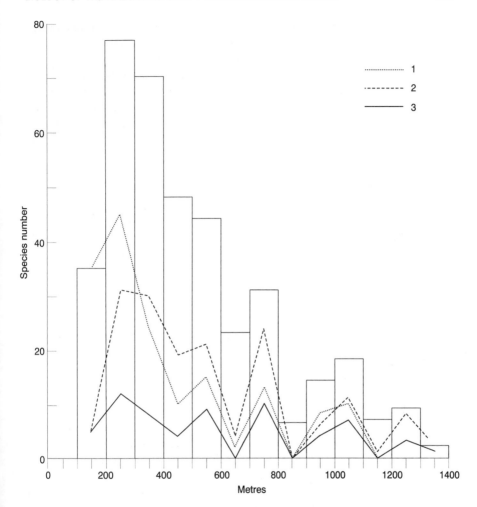

Figure 6   Number of fish species collected at the Nazca and Sala y Gómez ridges in each 100 m depth range. Columns show total number of species. Two lines show number of species in which the particular depth range is (1) the upper limit; (2) the lower limit. The third line (3) shows species restricted to this particular depth range.

species (*Glossanodon, Polymetme, Argyripnus, Diaphus adenomus* and *Hymeno-cephalus*); the third grouping, *off-bottom pelagic* species, can swim far away from the bottom, rising in midwater during diel vertical migrations (*Maurolicus, Idiolychnus, Macrorhamposus, Beryx, Zenopsis, Epigonus* and *Rexea*). Thus the benthopelagic group of fishes is not homogeneous and includes elements differing widely in their habits. Besides this, the habits of many species remain unknown and the placing of fishes with unknown ecology and behaviour in the various groups is tentative, based on general considerations and analogies.

A lack of homogeneity in the benthopelagic group is also seen for macroplanktonic crustaceans (mysids, euphausiids and decapods) (Vereshchaka, 1995) and macroplanktonic and micronektonic cephalopods (Nesis, 1983b). Vereshchaka (1995) distinguished three subgroups among benthopelagic crustaceans, of which the *hypo-benthopelagic* subgroup is approximately equivalent to the near-bottom fishes, *epi-benthopelagic* to off-bottom pelagic, and *amphibenthopelagic* to the off-bottom fishes.

Benthopelagic fishes predominate in number of species in almost all depth zones accounting for 54–78% (average 74%) of species (Table 5). The most important role is played by near-bottom fishes, composing from a third to almost a half of all species. Benthic and off-bottom species constitute from a quarter to a third of the species list in various depth ranges.

Between-seamount differences in depth, size and relief of the summits, sediment type and oceanographic features of surrounding water masses (including the biomass of plankton), determine notable biocoenotic differences between individual seamounts. This is the cause of differences in species lists for individual seamounts. For example, many species were found only on Bolshaya Seamount or on this and the neighbouring Dlinnaya and Novaya seamounts, including *Torpedo microdiscus*, *T. semipelagica*, *Monocentrus reedi*, *Macrorhamposus scolopax*, *Acantholatris gayi* and *Parabothus amaokai*. This indicates peculiar environmental conditions on this guyot, the largest and shallowest in the area. Conversely, on Bolshaya Seamount some of the species were common only to the westernmost seamounts of the Sala y Gómez Ridge: Kupol, Igolnaya and Ichthyologists seamounts (*Glossanodon nazca*, *Physiculus longicavis*, *Gadella obscurus*, *Antigonia capros*, *Grammatonotus laysanus*, *Emmelichthys elongatus*, *Pentaceros quinquespinis* and *Parapercis dockinsi*). There are probably some habitat characteristics in common between these western mountains and Bolshaya Seamount, although they are not identical because a rather large group of fish species (*Idiolychnus urolampus*, *Antigonia aurorosea*, *Hoplichthys citrinus*, *Cookeolus boops*, *Chrionema pallidum* and *Centrodraco gegonipa*) was recorded only on the western seamounts. The seamounts of the central part of the Sala y Gómez Ridge (Yantarnaya, Zhemchuzhnaya and Pervomayskaya) are faunistically rather peculiar because of the loose sediments on their flat summit, where suspension-feeding benthos is dominant (see Section 5.1.3). Eight species were found here which were absent in other areas (*Gnathophis parini*, *Polymetme andriashevi*, *Argyripnus electronus*, *Diaphus confusus*, *Laemonema yuvto*, *Hymenocephalus neglectissimus*, *Epigonus atherinoides* and *Centrodraco nakaboi*). In addition, there were six species in common with seamounts in the transitional Nazca/Sala y Gómez area, especially with Baral and Bolshaya seamounts (*Caelorinchus immaculatus*, *C. spilonotus*, *Malacocephalus laevis*, *Stethopristes eos*, *Assurger anzac* and *Chascanopsetta megagnatha*), and nine

Table 5   Ratio of main biotopic groups of fishes by bathymetric zones on the Nazca and Sala y Gomez ridges (per cent of total number of species known from each zone). Species found in several zones are considered in each of them.

Biotopic group	Bathymetric zones, m				
	< 200	200– 600	600– 1000	> 1000	All
A. Benthic	38	29	22	24	24
B. Benthopelagic	(54)	(68)	(78)	(76)	(74)
a. Near-bottom	33	40	44	47	46
b. Off-bottom	19	19	30	29	22
c. Migratory	2	9	4	—	6
C. Pelagic over seamounts	8	3	—	—	2

species in common with Ichthyologists and Kupol seamonts and, very rarely, with other western seamounts of the Sala y Gómez Ridge (*Polyipnus inermis, Hymenocephalus semipellucidus, H. striatulus, Kuronezumia pallida, Nezumia propinqua, Ventrifossa johnboborum, V. macrodon, V. teres* and *Chaunax latipunctatus*). There is a distinct group of fishes inhabiting only the Nazca Ridge and the transitional Nazca/Sala y Gómez area (*Bassanago nielseni, Helicolenus lengerichi, Epigonus elegans, E. notacanthus, Emmelichthys cyanescens, Centrodraco striatus* and *Rexea brevilineata*). Even among the fishes distributed mainly on the eastern seamounts (Nazca and Nazca/Sala y Gómez transitional area) some species are found on seamounts of the western (Sala y Gómez Ridge) area: *Beryx splendens* on Ichthyologists Seamount, *Zenopsis oblongus* on Kupol Seamount and *Callanthias parini* on Utes Seamount.

5.2.2. *Stability of Fish Communities*

Consecutive surveys of three seamounts – Professor Mesyatzev, Ekliptika and Bolshaya – undertaken in 1979–1987 showed the absence of obvious and, more importantly, clearly directed trends in the composition of their fish communities. Lists of fish species obtained during different cruises varied mostly in content of "rare" species while variations in the proportion of common, commercial species were due to the use of different trawl types (otter-trawls of various sizes, either bottom trawls, or midwater in bottom version) and variation in the time of day. The latter is very important because many fishes (*Beryx splendens, Macrorhamposus scolopax, Emmelichthys* spp., *Decapterus muroadsi*) stay near the bottom only in the daytime.

Commercial species formed 5% or more of the total catch in at least one survey; their species composition is shown for three seamounts in Tables 6–8. The catches of the Peruvian horse mackerel *Trachurus symmetricusmurphyi* were excluded because this pseudo-neritic species may form concentrations during some years over all three seamounts.

On the Professor Mesyatzev Seamount, the greatest stability was seen in the concentrations of the benthopelagic cardinal fish *Epigonus elegans*. The blue-mouth *Helicolenus lengerichi* predominated among benthic fishes and the alfonsino *Beryx splendens* was also important (Table 6). However, the redbait *Emmelichthys cyanescens*, which formed a substantial part of catches in 1979, was not recorded later on the Professor Mesyatzev Seamount. This species, the most pelagic among the commercial fishes, probably only migrates into this area in some years. The dominant fishes on the Ekliptika Seamount (Table 7) vary little from those on the Professor Mesyatzev Seamount except for the absence of *Glossanodon danieli*, *Callanthias parini* and *Ruvettus pretiosus* and the presence of *Zenopsis oblongus*. The apparent decrease on the part of *B. splendens*, increase of *H. lengerichi* and disappearance of *E. cyanescens* from the catches may be because during the 1987 survey (18th cruise of the R/V "Professor Shtokman") only a small bottom otter-trawl was used. The proportions of different species on the Bolshaya Seamount (Table 8) varied substantially but there were again no clear trends of increase or decrease in any of them. Morphometric analysis of two common species, *Zenopsis oblongus* and *Helicolenus lengerichi*, indicates possible separation of their populations on different seamounts of the Nazca Ridge and the Nazca/Sala y Gómez transitional area (Golovan' and Pakhorukov, 1987; Kotlyar, 1988a; Pakhorukov and Golovan', 1987, Parin *et al.*, 1988b).

5.2.3. *Trophic Structure of Fish Communities*

The feeding habits of fishes associated with underwater rises of the Nazca and Sala y Gómez ridges was studied by Parin *et al.* (1990b). More than 950 non-empty stomachs were investigated of 58 fish species, including 12 bottom, 21 near-bottom, 10 off-bottom, 10 migrating off-bottom pelagic and five true pelagic species. Data are summarized in Table 9 indicating the proportions of pelagic, near-bottom and benthic food items in the food spectra of various fishes.

Fishes in the benthic group had various feeding habits. *Notacanthus chemnitzi*, *Centrodraco striatus*, *C. nakaboi*, *Hoplichthys citrinus*, *Arnoglossus multistriatus* and *Parabothus amaokai* consumed only benthic and near-benthic animals and detritus was found in the stomachs of some. Others (*Torpedo semipelagica*, *Helicolenus lengerichi*, *Pterygotrigla picta* and *Chrionema pallidum*) consumed not only bottom and near-bottom but also truly pelagic animals. Predominance of pelagic animals, macroplanktonic and micronektonic tunicates, euphausiids, shrimp and fish, is particularly characteristic of the food of the blue-mouth (*H.*

Table 6 Species composition of trawl catches of commercial fishes on Professor Mesyatzev Seamount in different years and seasons as per cent of total catch by weight.

	Month, year						
Species	IX 1979	IX 1980	IX 1983	XI 1983	VIII 1984	IV 1985	IV 1987
Glossanodon danieli	+	+	+	+	+	+	5
Caelorinchus nazcaensis	10	−	+	+	+	+	+
Beryx splendens	40	+	5	−	+	10	−
Helicolenus lengerichi	20	+	+	+	20	15	65
Callanthias parini	+	+	+	+	+	10	+
Epigonus elegans	15	100	95	100	55	65	30
Emmelichthys cyanescens	15	−	−	−	−	−	−
Ruvettus pretiosus	+	+	−	+	25	+	−

+ Present in small quantities; − absent.

Table 7 Species composition of trawl catches of commercial fishes on Ekliptika Seamont in different years and seasons (per cent of total catch by weight).

	Month, year				
Species	IX 1983	XI 1983	III 1984	IV 1985	IV 1987
Caelorinchus nazcaenis	+	+	+	+	10
Beryx splendens	70	50	5	30	−
Zenopsis oblongus	+	+	10	+	+
Helicolenus lengerichi	15	20	15	+	80
Epigonus elegans	+	+	5	+	10
Emmelychthys cyanescens	15	30	65	70	−

+ Present in small quantities; − absent.

lengerichi), the commonest fish in the benthic community on the Nazca Ridge and transitional Nazca/Sala y Gómez area.

The near-bottom fishes also have varied diets. Benthos and inhabitants of the near-bottom layer comprise most of the food of the eels *Facciolella*, *Nettastoma*, *Gnathophis* and *Pseudoxenomystax* and an important part of the food of morids, genus *Physiculus*, and rattails, genera *Caelorinchus* (particularly large specimens of *C. nazcaensis*), *Malacocephalus* and *Ventrifossa*. The latter probably any

Table 8 Species composition of trawl catches of commercial fishes on Bolshaya Seamount in different years and seasons (per cent of total catch by weight).

Species	Month, year						
	IX-X 1979	IX 1983	XI 1983	VIII 1984	X 1984	IV 1985	IV 1987
Squalus mitsukurii	+	+	15	+	−	5	5
Monocentris reedi	−	+	+	−	+	−	5
Zenopsis oblongus	10	5	5	5	+	−	+
Macrorhamposus scolopax	5	5	30	5	75	15	15
Pterygotrigla picta	20	15	5	+	+	5	40
Carpodon longimanus	+	5	+	−	−	−	+
Decapterus muroadsi	−	−	−	−	5	−	−
Emmelichthys cyanescens	1	35	−	80	+	+	−
Emmelichthys elongatus	15	−	+	10	5	70	−
Pentaceros quinquespinis	40	35	45	+	10	5	35
Rexea antefurcata	+	+	+	−	5	−	+

+ Present in small quantities; − absent.

available food, meso- and macroplankton, micronekton, nektobenthos, benthos and detritus, having, like many other investigated rattails, very diverse stomach contents. Other species of the near-bottom group consume exclusively pelagic food (*Macrorhamposus*, *Callanthias* and *Symphysanodon*). The boarfish *Pentaceros quinquespinis* probably feeds both near the bottom and in midwater, on mainly pelagic food. Golovan' and Pakhorukov (1987) reported that boarfish may form concentrations extending some dozens of metres vertically. This may also be characteristic for an eel *Synaphobranchus affinis* in whose stomach micronektonic objects predominate.

Fishes of three other groups, off-bottom, off-bottom pelagic and truly pelagic, eat mainly pelagic organisms: small fishes, plankton advected from surrounding areas, macroplankton and nekton, either oceanic or seamount-associated. Near-bottom and even benthic animals may be found in the stomachs of many fish of these groups, captured when the fish swim near the bottom; but this food type does not seem to be important.

The food of 22 out of 58 studied species (Table 9) contained pelagic items only. Eighteen species should be treated as plankton-feeders and four (*Torpedo semipelagica*, *Seriola lalandi*, *Ruvettus pretiosus* and *Rexea brevilineata*) as predators and fish-eaters. Nineteen more species prefer pelagic items, judging by their occurrence in the food; one predator is included (*Rexea antefurcata*). Six fishes eat near-bottom animals only, and bottom animals predominate in the food of five others. According to our data, only two species are obligatory benthos-feeders – *Notacanthus chemnitzi* and *Centrodraco striatus* – and in four more species this diet predominates. In all, 71% of the fish community consume pelagic food, 19% near-bottom food, and only 10% benthic food. There are no obligatory benthos-eating species among the 16 most common fishes (marked by an asterisk in the Table 9) and only one is predominantly benthos-eating (*Pterygotrigla picta*).

The food of fish communities of separate seamounts from different parts of the Nazca Sala y Gómez ridges (Table 10) was examined, and including Ekliptika, Bolshaya, and the complex of the Yantarnaya and Zhemchuzhnaya seamounts, which are reasonably well studied. These mountains have different landscapes and faunal compositions. Off-bottom pelagic and near-bottom fishes predominate on the Ekliptika Seamount (68% of total), near-bottom and benthic predominate on the Bolshaya Seamount (67%), and near-bottom and off-bottom fishes predominate on the Yantarnaya and Zhemchuzhnaya seamounts (76%), but consumers of pelagic food are clearly the most important on all (the feeding types of fishes not studied by us was determined by analogy with related studied species).

Most of the food of the "mountain" fishes consists of pelagic organisms that make vertical migrations, usually in the deep scattering layers. The exceptions are non-migrating or weakly migrating copepods of the family Candaciidae and euphausiids of the genus *Stylocheiron*; these are present in the stomachs of some off-bottom pelagic and pelagic fishes and are specially characteristic of *Maurolicus rudjakovi* and *Emmelichthys elongatus*. Migrating interzonal plankton and small nekton organisms (copepods of the family Metridiidae and Aetideidae, euphausiids *Thysanopoda*, shrimps (Sergestidae) and micronektonic fishes (Myctophidae)) are important in the diet of both pelagic fishes and fishes belonging to the benthic and near-bottom complexes. Predatory fishes living on underwater rises (*Squalus mitsukurii, Zenopsis oblongus, Helicolenus lengerichi* and *Pentaceros quinquespinis*) consume more interzonal animals (lantern-fishes and shrimps) than regular inhabitants of the rises. Small swarming fishes living near the bottom, in particular small percophids, are eaten only by the flatfish *Arnoglossus multirastris* and the gurnard *Pterygotrigla picta*.

Interzonal food animals come in contact with the "mountain" fishes only at a definite time of day. They appear around the seamount slopes during the evening ascent, remain over the summits through the night and sink on to the summit and slopes of the rises during the morning descent. The last event is, of course, the most important for the fishes. The result is that the highest feeding intensity of the off-bottom pelagic and pelagic fishes, pearlsides (*Maurolicus*), alfonsins (*Beryx*), redbaits (*Emmelichthys*) and horse mackerels (*Trachurus*), is in the dark period.

Table 9 Relations of main groups of food objects in the stomachs of fishes from seamounts; high biomass species are marked by an asterisk. The sum of occurrences may be >100%.

Groups and species of fish	% Occurrence of food objects:			Predominant type of food
	Pelagic	Near-bottom	Benthic	
Benthic fishes				
Torpedo semipelagica	100	–	–	Nekton
Notacanthus chemnitzi	–	–	100	Benthos
Pyramodon parini	–	20	20	Benthoplankton
*Helicolenus lengerichi**	98	3	2	Macroplankton, micronekton
*Pterygotrigla picta**	28	61	39	Benthoplankton, benthos
Hoplichthys citrinus	–	50	66	Benthos, benthoplankton
Chrionema pallidum	50	75	–	Benthoplankton, plankton
Centrodraco nakaboi	–	100	–	Benthos (?)
C. striatus	–	–	100	Benthos
Arnoglossus multirastris	2	80	4	Benthoplankton
Parabothus amaokai	–	100	–	Benthoplankton
Chascanopsetta megagnatha	–	100	–	Benthoplankton
Near-bottom fishes				
*Squalus mitsukurii**	37	26	5	Micronekton, nektobenthos
*Synaphobranchus affinis**	80	–	20	Micronekton
Facciolella castlei	–	100	–	Benthoplankton
Nettastoma falcinaris	–	100	–	Benthoplankton
Gnathophis smithi	–	43	78	Benthos, benthoplankton
Bassanago nielseni	–	66	66	Benthoplankton, benthos
*Chlorophthalmus ichthyandri**	88	4	–	Mesoplankton
Physiculus hexacytus	33	35	23	Benthoplankton, plankton, benthos
Ph. longicavis	10	50	10	Benthoplankton
*Caelorinchus nazcaensis**	100	10	40	Plankton, benthos
C. immaculatus	53	80	58	Benthoplankton, benthos, plankton

Table 9   Continued

Groups and species of fish	% Occurrence of food objects:			Predominant type of food
	Pelagic	Near-bottom	Benthic	
Malacocephalus laevis	40	80	—	Benthoplankton
Ventrifossa macrodon	67	50	33	Plankton, benthoplankton, benthos
Monocentris reedi	60	10	10	Plankton
Antigonia capros	100	50	—	Plankton, benthoplankton
*Macrorhamposus scolopax**	100	—	—	Mesoplankton
Carpodon longimanus	+	—	—	Micronekton
Symphysanodon maunaloae	100	—	—	Plankton
Callanthias parini	100	—	—	Plankton
Dactylopsaron dimorphicum	100	—	—	Mesoplankton
*Pentaceros quinquespinis**	56	—	28	Macroplankton, benthos
Off-bottom fishes				
Glossanodon danieli	100	—	—	Plankton
Polyipnus inermis	100	—	—	Mesoplankton
Argyripnus electronus	100	—	—	Mesoplankton
Polymetme andriashevi	66	—	—	Plankton
Neoscopelus macrolepidotus	100	—	—	Plankton (?)
Hymenocephalus striatulus	82	23	—	Plankton, benthoplankton
H. semipellucidus	50	20	—	Plankton, benthoplankton
Zenion hololepis	100	—	—	Mesoplankton
Cyttomimus stelgis	80	10	—	Mesoplankton
Benthodesmus elongatus	85	15	—	Macroplankton
Off-bottom pelagic fishes				
*Maurolicus rudjakovi**	100	—	—	Mesoplankton
Diaphus adenomus	85	—	—	Plankton, micronekton

Table 9 Continued

Groups and species of fish	% Occurrence of food objects:			Predominant type of food
	Pelagic	Near-bottom	Benthic	
Idiolychnus urolampus	70	35	–	Plankton, benthoplankton
*Beryx splendens**	100	4	–	Macroplankton, micronekton
*Zenopsis oblongus**	100	–	–	Macroplankton, micronekton
*Epigonus elegans**	99	–	–	Mesoplankton
E. notacanthus	100	–	–	Mesoplankton
Ruvettus pretiosus	+	–	–	Nekton
Rexea antefurcata	64	43	7	Nekton, benthonekton
R. brevilineata	33	–	–	Nekton
Pelagic fishes				
*Decapterus muroadsi**	+	–	–	Mesoplankton
Seriola lalandi	+	–	–	Nekton
*Trachurus symmetricus**	100	–	1	Plankton, micronekton
*Emmelichthys cyanescens**	99	2	2	Mesoplankton
*E. elongatus**	100	–	–	Mesoplankton

Table 10 Fish species composition and main food sources of fishes on three seamounts of the Nazca and Sala y Gómez ridges.

Seamount	No. of fish species	Ecological grouping %					Main food source, %		
		Benthic	Near bottom	Off bottom	Off-bottom pelagic	Pelagic	Plankton and nekton	Near-bottom animals	Benthos
Ekliptika	22	9	32	5	36	18	77	14	9
Bolshaya	49	30	37	5	19	9	60	28	12
Yantarnaya and Zhemchuzhnaya	31	10	48	26	16	0	68	22	10

when they meet the interzonal food animals in midwater. A similar diel rhythm of feeding, in the same or related fishes, has been reported in other oceanic areas (Parin, 1988). On the other hand, the benthic and near-bottom fishes feed mostly during the morning and day time (pre-midday hours), when interzonal animals are migrating down to meet the bottom.

Thus the data obtained show very clearly the most significant characteristic of the feeding of seamount-inhabiting fishes: their main resources are plankton and nekton, predominantly animals undergoing vertical migrations in the DSLs.

5.2.4. *Size and Age Structure of Fish Communities*

Preliminary data on the age and growth of 60 fish species from the Nazca and Sala y Gómez ridges were presented by Kotlyar and Parin (1990). Figure 7 shows the predominance of small and intermediate-sized fishes: almost a third (30%) of species are not larger than 150 mm long, and three-quarters are not larger than 300 mm; only three (5% of total) have maximum body length exceeding 600 mm. According to the data available such a skewed size distribution is characteristic for all fishes of the seamounts studied. For example, in the list of teleostean fishes (without sharks, skates and chimaeras) collected on the 18th cruise of the R/V "Professor Shtokman" at depths of less than 540 m there are 90 seamount-associated species. Twelve per cent of them have a maximum body length < 75 mm, 30%, 75–150 mm and 34%, 150–300 m (together 76% of the small- and middle-sized species) and only 6% are larger than 600 m. The list of smallest fishes includes the near-bottom *Schindleria praematurus* and dwarf percophids of the genera *Dactylopsaron*, *Enigmapercis* and *Osopsaron* with body length only 20–30 mm and off-bottom species of the genera *Argyripnus*, *Polyipnus*, *Cyttomimus* and *Zenion* with lengths of 50–90 mm (Kotlyar and Parin, 1990; Parin, 1990a). The largest fishes on the ridges are the sharks *Echinorhinus*, *Hexanchus*, *Somniosus*, *Centroscymnus* (1.0–2.3 m) and the teleostean fishes *Pterygotrigla*, *Synaphobranchus*, *Rexea* (*R. antefurcata*), *Lepidion*, *Benthodesmus* and *Coryphaenoides* (55–100 cm) (Golovan' and Pakhorukov, 1986; Kotlyar and Parin, 1990; Litvinov, 1990; Parin, 1990a; Sazonov and Iwamoto, 1992).

Species with a short life cycle predominate among the fish studied (Figure 7). For 37 species (62%) the maximum age is 2–5 years, for 18 (30%) it is 6–9 years and for only five species (8%) is it 10 years or more.

Among the fishes whose age could be determined, there are species of the benthic, near-bottom, off-bottom, migrating off-bottom pelagic, and truly pelagic complexes (Kotlyar and Parin, 1990). Benthic fishes (14 spp.) reach the age of 2–9 years, live on average 5.0 years; near-bottom ones (22 spp.) have a similar life cycle: reaching 2–13 years, on average 5.1 years. The off-bottom species (13 spp.) have a shorter life span (2–9 years, on average 3.7 years) and the off-bottom pelagic and pelagic fishes (10 spp.) have longer life cycles (4–15 years, on average 9.6 years).

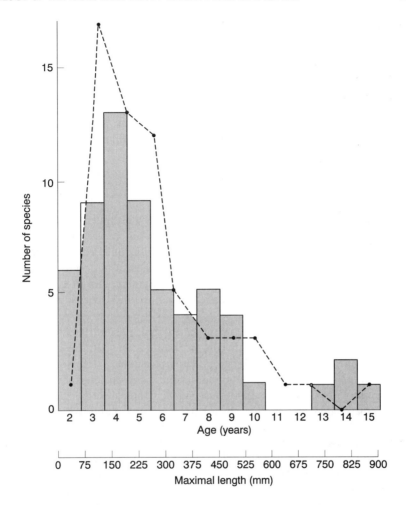

Figure 7 Age (columns) and size (line) distribution of fish species of the Nazca and Sala y Gómez ridges (after Kotlyar and Parin, 1990).

Fishes grouped according to their feeding habits show the following pattern. The benthos-feeders (seven spp.) live 2–7 years on average 4.0 years; the plankton-eaters (25 spp.) live 2–8 years, on average 4.6 years; the bentho-plankton eaters (11 spp.) live 3–13 years, on average 5.8 years. Fishes consuming both invertebrates (planktonic or benthic) and fish, that is facultative predators (10 spp.) have a life span of 4–13 years, on average 7.0 years; and obligatory predators (six spp.) have maximum longevity of 4–15 years, on average 8.2 years.

5.3. Pelagic Communities

More than 330 species of pelagic fishes were recorded in the open areas of the
Southeastern Pacific over the Nazca and Sala y Gómez ridges (Parin *et al.*, 1990a).
Of all species, 75.2% are rather deep-living, mostly mesopelagic, 17.7% are
epipelagic and 7.1% are interzonal. The following biogeographic groups were
distinguished: eastern Pacific tropical; widely tropical and equatorial–central
(absent in the tropical eastern Pacific); central; peripheric–central; southern
transitional; and notalian.

The Eastern Pacific tropical species are distributed primarily north of 7–8°S and
only half of these reach 18°S. They are distributed to the south with the Peru
Oceanic Countercurrent and are rarely found westward of 80°W, between 20 and
25°S. The central species are distributed mostly in the western and central parts
of the south-eastern Pacific west of 81–83°W between 20 and 30°S and they are
almost absent in the Peru Current. The distribution of equatorial–central and
widely tropical species almost coincides with that of the central ones. The
peripheral–central species are rare north of 20–25°S, their main area of distribution
lies south of 28–29°S. Southern transitional and notalian species are distributed
mostly to the south of 30°S.

Thus, over the eastern part of the ridges studied, Eastern Pacific fish species
predominate, while over the western part the central and equatorial–central groups
are mixed with peripherL–central species. The easternmost part of this area belongs
to the Eastern Pacific Tropical Region and its central and western parts belong to the
South Central Province of the Indo-West Pacific Tropical Region. The boundary
between these areas runs south-eastward and is located at approximately 81°W at
20°S and at 80°W at 23°S (Parin and Nesis, 1986; Parin *et al.*, 1990a).

It should also be noted that three species of neritic fish, horse mackerel
(*Trachurus symmetricus murphyi*), mackerel (*Scomber japonicus*), and pilchard
(*Sardinops sagax*), all of them very abundant in the Peruvian and Chilean waters,
penetrate sporadically from the east into the eastern part of the area under
consideration. One of these, the horse mackerel, from time to time forms huge
commercial concentrations at seamounts of the Nazca Ridge and the Nazca/Sala
y Gómez transitional area (from Zvezda to Baral Seamounts) both in midwater and
just above the bottom at depths of 150–400 m (Golovan' and Pakhorukov, 1987;
Parin, 1990a; Konchina, 1990, 1992).

In the area over the Nazca and Sala y Gómez ridges approximately 50 species
of pelagic cephalopods were found (K.N. Nesis, unpublished). Epipelagic oceanic
ommastrephid squids, observed by night at drifting stations and caught by jigs,
were numerous only over the easternmost part of the Nazca Ridge; catches
declined to single individuals at 83– 89°W and almost to zero west of Yantarnaya
Seamount (98°30'W). This trend corresponds to the distribution of plankton
biomass and the abundance of macroplankton (Vereshchaka, 1990a,c). However,
the average number of species and abundance of squids, primarily larval and

juvenile specimens, in the catches of midwater trawls (TIKSA and IKMT) was moderate at 80–84°W, decreased rapidly at 84–88°W but then clearly increased westward; at 95-101°W their abundance averaged twice that at 80–84°W. A similar distribution was reported also for mysidaceans (Vereshchaka, 1990b).

Among pelagic cephalopods, equatorial, narrow tropical, central, subtropical, peripheric, south–subtropical–notalian and notalian species as well as those with western (Indo-West Pacific, Atlanto-Indo-West Pacific) and eastern (Eastern Pacific) distribution were distinguished. According to the distribution of northern (equatorial and narrow tropical) and southern (peripheral, south–subtropical–notalian and notalian) species the area over the ridges should be treated as a region of mixing of the south–central subzone of the tropical zone and southern subtropical zone (Nesis, 1985; Parin and Nesis, 1986). The eastern species are not found west of the Ekliptika and Ikhtiandr seamounts and the western species rarely reach the area east of the Professor Mesyatzev Seamount. Over the Nachalnaya and Zvezda seamounts (80–81°W) the cephalopods are predominantly eastern species; the region over the Professor Mesyatzev, Soldatov, Ikhtiandr and Ekliptika seamounts (81–83°W) is an area of mixing of eastern and western species. The area at 84–90°W (Dorofeyev to Zhemchuzhnaya seamounts) is a zone of impoverished western fauna and the area westward of 97°W is dominated by the Western Indo-West Pacific pelagic fauna, fully expressed. Results from different cruises have shown that the boundaries of northern and southern, western and eastern species may alter from year to year and, probably, seasonally (K.N. Nesis, unpublished). Differences in the distribution of western and eastern species over the ridges were also noted for hyperiid amphipods (Vinogradov, 1990).

The area occupied by the eastern Pacific pelagic fauna is delimited by the zone of direct influence of the Peru Current system and is thus probably climatically determined. Over the ridges, the western and central areas are dominated by the Indo-West Pacific pelagic fauna, whereas the Eastern Pacific fauna predominates in the eastern area. The boundary between these two faunas is determined by the position of hydrological boundaries and probably changes seasonally and from year to year. The predominance of central and southern subtropical species over the ridges and the rarity of equatorial species also demonstrates the importance of climatic factors in the determination of the composition of the present seamount fauna and probably such factors were even more important during the cold epochs of the Pliocene.

6. BIOGEOGRAPHIC BOUNDARIES

6.1. Bottom and Near-bottom Invertebrate Fauna

The position of biogeographic boundaries in the area has been examined by two methods: by the method of latitudinal biogeographic transect introduced by A.N.

Mironov (1985b) and by determination of the homogeneity or distinctness of the fauna between different seamounts using indices of taxonomic similarity (Galkin, 1993). Here we attempt to redefine the position of biogeographic boundaries using both methods, with previous and additional data.

Using the method of latitudinal transect, Mironov and Detinova (1990) have divided the area studied into five geomorphologically distinct subareas (Figure 8). Four subareas, including the seamounts Nos 1–3, 4–5, 6–8 and 9–17 on Figure 1, are located approximately at the same latitude (25°S) but on different longitudes and the fifth subarea (seamounts Nos 18–21) differs from the others in both latitude and longitude, being the most eastern and northern. The distribution of 55 genera which were recorded at five or more stations is considered. Records deeper than 1000 m and records of juvenile specimens are omitted. Juvenile specimens of many species were found to be more widely distributed than adults on the Nazca and Sala y Gómez ridges, and it is possible that a general characteristic of these seamounts is the formation of pseudo-populations near range boundaries due to the constant advection of pelagic larvae (rarely adults) with currents from neighbouring seamounts where self-sufficient populations occur (Mironov, 1981; Mironov and Detinova, 1990). For the present chapter the same study has been repeated with two differences: the vertical zone of investigation was narrowed by exclusion of records deeper than 800 m (instead of 1000 m) and the number of genera included (recorded at five or more stations) has increased from 55 to 60.

Generic ranges boundaries are crowded around seamounts Nos 9–17, located between 83° and 87°W. This subarea is an eastern boundary of distribution for 41 genera and a western boundary for five genera. In total, 77% of the genera under consideration have distributional boundaries here.

For seven seamounts (Ekliptika, Professor Mesyatzev, Bolshaya, Baral, Yantarnaya, Kupol and Ichthyologists) the faunal similarity was calculated for species of sea stars (Galkin, 1993) using the Jaccard coefficient. The resulting dendrogram showed that the fauna of Ekliptika and Professor Mesyatzev seamounts is markedly different from that of other seamounts, in accordance with previous results. Among five other seamounts the Bolshaya Seamount is more distant from the rest. Galkin (1993) noted a change of species composition around the depths of 300–400 m.

Figure 8 Latitudinal faunistic transect along the Nazca and Sala y Gómez ridges at depths from 162 to 800 m. The area studied was divided into five geomorphologically distinct sub-areas – groups of seamounts whose names and numbers are indicated in the caption to Figure 1. The ranges of 60 recorded genera (horizontal lines) are regarded as uninterrupted between the westernmost and easternmost seamounts, where they were recorded. The distribution shows only those genera recorded from at least five stations. The genus numbers are from Table 3. A zone of change of generic range limits is located in the area of seamounts 9–17 and is treated as a biogeographic boundary (after Mironov and Detinova, 1990, with additions).

```
---------------------------------------------------------------------
genus |                    seamount groups                          |
number|------------------------------------------------------------ |
      | 1 - 3  |  4 - 5  |   6 - 8   |  9 - 17  |  18 - 20           |
------|--------|---------|-----------|----------|-------------------|
 23.  | ————————————————————————————— |          |                 |
 71.  | ————————————————————————————— |          |                 |
121.  | ————————————————————————————— |          |                 |
163.  | ————————————————————————————— |          |                 |
 13.  | ——————————————————————————————————————— |                 |
 14.  | ——————————————————————————————————————— |                 |
 15.  | ——————————————————————————————————————— |                 |
 20.  | ——————————————————————————————————————— |                 |
 21.  | ——————————————————————————————————————— |                 |
 27.  | ——————————————————————————————————————— |                 |
 30.  | ——————————————————————————————————————— |                 |
 48.  | ——————————————————————————————————————— |                 |
 53.  | ——————————————————————————————————————— |                 |
 72.  | ——————————————————————————————————————— |                 |
 73.  | ——————————————————————————————————————— |                 |
 76.  | ——————————————————————————————————————— |                 |
 81.  | ——————————————————————————————————————— |                 |
 83.  | ——————————————————————————————————————— |                 |
 89.  | ——————————————————————————————————————— |                 |
 97.  | ——————————————————————————————————————— |                 |
105.  | ——————————————————————————————————————— |                 |
111.  | ——————————————————————————————————————— |                 |
131.  | ——————————————————————————————————————— |                 |
133.  | ——————————————————————————————————————— |                 |
135.  | ——————————————————————————————————————— |                 |
136.  | ——————————————————————————————————————— |                 |
140.  | ——————————————————————————————————————— |                 |
145.  | ——————————————————————————————————————— |                 |
146.  | ——————————————————————————————————————— |                 |
148.  | ——————————————————————————————————————— |                 |
149.  | ——————————————————————————————————————— |                 |
150.  | ——————————————————————————————————————— |                 |
152.  | ——————————————————————————————————————— |                 |
162.  | ——————————————————————————————————————— |                 |
166.  | ——————————————————————————————————————— |                 |
167.  | ——————————————————————————————————————— |                 |
176.  | ——————————————————————————————————————— |                 |
  4.  |        | ———————————————————————————— |                 |
 17.  |        | ———————————————————————————— |                 |
 18.  |        | ———————————————————————————— |                 |
 22.  |        | ———————————————————————————— |                 |
 78.  |        | ———————————————————————————— |                 |
139.  |        | ———————————————————————————— |                 |
173.  |        | ———————————————————————————— |                 |
159.  |        |         | ———————————————————— |                 |
 88.  |        |         | ——————————————————————————————————————|
 25.  |        |         |           | ————————————————————————————|
 31.  |        |         |           | ————————————————————————————|
 43.  |        |         |           | ————————————————————————————|
 96.  |        |         |           | ————————————————————————————|
 99.  |        |         |           | ————————————————————————————|
 16.  | ————————————————————————————————————————————————————————— |
 19.  | ————————————————————————————————————————————————————————— |
 54.  | ————————————————————————————————————————————————————————— |
 95.  | ————————————————————————————————————————————————————————— |
103.  | ————————————————————————————————————————————————————————— |
104.  | ————————————————————————————————————————————————————————— |
110.  | ————————————————————————————————————————————————————————— |
 11.  | ————————————————————————————————————————————————————————— |
158.  | ————————————————————————————————————————————————————————— |
---------------------------------------------------------------------
```

Here we determine the faunal similarity between 22 seamounts using distributional records of 163 genera of bottom invertebrates from depths of 162–1900 m. Fourteen of the 177 genera listed in Table 3 (Coelenterata, Tanaidacea and Amphipoda) were not included because we do not have enough distributional data. The similarity was evaluated using the Hacker and Dice index (D) (cited after Mirkin *et al.*, 1989, p. 79):

$$D(X,Y) = a/[a + \min(b,c)]$$

where *a* is the number of genera common for seamounts *X* and *Y*, *b* and *c*, the number of genera recorded on only one seamount. The Hacker and Dice index equals the Szymkiewicz–Simpson index when $b \geq c$ (Pesenko, 1982). This index was chosen because of its insensitivity to negative combinations of characters. It permits the most correct evaluation of faunistic similarity between unequally studied seamounts.

The dendrogram thus obtained (Figure 9) shows a clear separation of the area north of 23°S and east of 83°W (seamounts Nos 18–22). This corroborates the results obtained previously with the method of latitudinal transect. The faunistically separated northeastern region includes seamounts with both shallow (Nos 18–20) and deep (21–22) summits. The catches of bottom invertebrates were obtained on the seamounts Ekliptika, Professor Mesyatzev and Zvezda mostly at the depths of 200–800 m but on the Nachalnaya and Soldatov seamounts at 800–1058 m. Faunistic differences between these different vertical zones were found to be important but less than differences between the areas west and east of 83°W.

This analysis permits us to state with confidence that in the area studied two different faunistic complexes are represented, divided by a boundary at about 82–83°W. Mironov and Detinova (1990) have separated the Sala-y-Gómezian faunistic complex (including the seamounts of the Nazca/Sala y Gómez transitional area) from the Sala y Gómez Ridge westward of 83°W up to (but not including) the Sala y Gómez Is. The area of our investigation was limited by the exclusive economic zone (EEZ) of Chile, so we cannot judge the position of this boundary relative to Easter and Sala y Gómez Is to the west and San Félix, San Ambrosio and Juan Fernández Is to the south. The bathyal fauna of all these islands is poorly known and the data available are mainly for the intertidal and subtidal zones.

A sudden change near 83°W was recorded both for the taxonomic composition of the fauna and for the trophic structure of communities. This coincidence indicates once more that the position of biogeographic boundaries is determined by a complex of many recent factors and supports the idea of "ecological universality" of a biogeographic boundary (Mironov, 1990).

All species of cephalopods, except *Stoloteuthis* n.sp. aff. *leucoptera* (Verrill), were found on the western part of the Sala y Gómez Ridge, 97–101°W, three

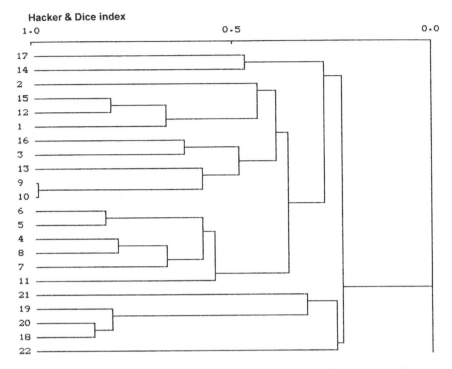

Figure 9.　A dendrogram of faunal similarity between 22 seamounts of the Nazca and Sala y Gómez ridges, based on distributional records of 155 genera of bottom invertebrates recorded from depths of 162–1900 m. The Hacker and Dice similarity index was calculated (cited after Mirkin *et al.* (1989)): $D(X,Y) = a/[a + \min (b,c)]$, where a is the number of common characters (= species) in the objects X and Y (= seamounts) and b and c are the number of characters (species) found in only one of these objects on only one of compared seamounts). Numbers in left column are seamount numbers named in the caption to Figure 1. The dendrogram shows a clear separation of north-eastern seamounts (Figure 1. Nos 18–22) located eastward 83°W and northward of 23°S from all others. This corroborates the results obtained previously by other methods. The faunistically separated north-eastern region includes both shallow (No 18–20) and deep (No 21–22) seamounts. The catches of bottom invertebrates were obtained on the Ekliptika, Professor Mesyatzev and Zvezda seamounts predominantly at depths of 200–800 m and on Nachalnaya and Soldatov seamounts at 800–1058 m. Faunistic differences between different vertical zones were found to be less important than those between areas westward and eastward of 83°W.

(*Heteroteuthis* n.sp. aff. *dispar, Nototodarus hawaiiensis* and *Scaeurgus unicirrhus patagiatus*) were also on the eastern part of this ridge or in the transitional Nazca/Sala y Gómez area at 84–90°W. *Stoloteuthis* sp. was found only on Zvezda Seamount on the Nazca Ridge at 81°W. All the species with Indo-West Pacific affinities (Hawaii or New Zealand), of the genera *Iridoteuthis, Heteroteuthis,*

Sepioloidea, *Nototodarus*, *Scaeurgus* and *Danoctopus* were not found on the Nazca Ridge and the only species present on this ridge (*Stoloteuthis* sp.) has no close relatives either in the Indo-West Pacific or in the Eastern Pacific fauna (see p. 220). The boundary near 83–84°W between the Nazca Ridge and Nazca/Sala y Gómez transitional area is the eastern boundary of distribution of *Nototodarus hawaiiensis* and *Scaeurgus unicirrhus patagiatus* and probably a western one for *Stoloteuthis* sp.; it is clearly an important zoogeographic boundary, as for the bottom invertebrates.

6.2. Benthopelagic and Near-bottom Fishes

In his analysis of the composition of the fish fauna of the Nazca and Sala y Gómez ridges, Parin (1990a, 1991a) drew the conclusion that the whole area under consideration should be included in the Indo-West Pacific Region by shifting its south-eastern boundary as a tongue-like protuberance from 106°W eastward to 80°W. This conclusion was based on the predominance of species of Indo-West Pacific origin along the entire chain of seamounts studied. A high degree of endemism indicates that this area should be considered as a separate zoogeographic province tentatively divided at approximately 83–84°W into two subareas: the Sala y Gómez Ridge and the Nazca Ridge (Parin, 1991a). The latter is characterized by an impoverished ichthyofauna and the presence of a few species in common with the South American neritic zone and slope (Parin, 1990a). However, most of these fishes are either widely distributed elsewhere (*Hexanchus*, *Echinorhinus*, *Nesiarchus*, etc.) or live in deep water (*Talismania*, *Nezumia*, *Dicrolene* and *Aphanopus*) or migrate sometimes from neritic Chilean waters (*Sardinops*, *Trachurus*, *Scomber* and maybe *Macruronus*). So only one relatively shallow-water (350–490 m) species, *Caelorinchus immaculatus*, is restricted in its distribution to the Nazca Ridge and Chilean slope (Sazonov and Iwamoto, 1992) (see below).

Distribution of benthopelagic and benthic fishes along the Nazca and Sala y Gómez ridges was analysed using the Hacker and Dice index (as above), and the results are shown in Figure 10. Two seamounts differ from the others in their fish fauna: the Stolbovaya (Pillar, No. 4) and Ikhtiandr (No. 17). The first has a very rough bottom relief, resembling a pillar, and the only Sigsbee trawl lowered here was badly torn and brought in only two fish species: *Bembradium roseum* and *Samariscus* sp. At the summit of the Ikhtiandr Seamount there are no good trawling grounds either; of the nine fish species known to occur there, two (*Aldrovandia affinis* and *A. phalacra*) were caught at great depths, two others (*Zenopsis oblongus* and *Glossanodon nazca*) were found in pelagic trawls, and five (*Synaphobranchus affinis*, *Bassanago nielseni*, *Physiculus hexacytus*, *Gaidropsarus parini* and *Caelorinchus immaculatus*) were taken with baited traps. Thus, the faunistic peculiarities of both these seamounts are probably artificial.

With the exclusion of these two seamounts and the Zapadnaya Seamount (No.

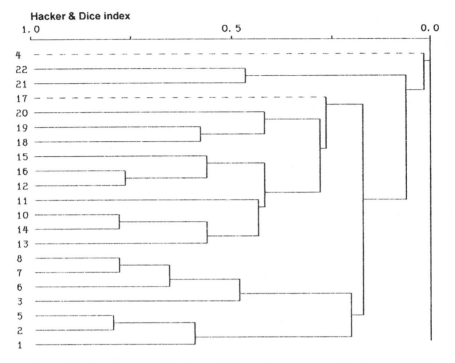

Figure 10 Dendrogram of faunistic similarity of seamounts of the Nazca and Sala y Gómez ridges based on the fish fauna. The similarity index was calculated after Hacker and Dice (see Figure 9) and 170 fish species are included. Numbers in left column are seamount numbers named in the caption to Figure 1. Dotted lines (seamounts Nos 4 and 17) represent seamounts with inadequately collected fishes.

9), from which we have no fish collections, all the underwater rises fall into five groups (Figure 10):

1 two deep seamounts of the Nazca Ridge (Nos 21 and 22);
2 four shallower seamounts of the same ridge, Ikhtiandr, Ekliptika, Professor Mesyatzev and Zvezda (Nos 17–20);
3 seven seamounts in the transitional Nazca/Sala y Gómez area (Nos 10–16), forming two slightly differentiated clusters, one with three seamounts (Bolshaya, Dorofeyev and Albert, Nos 12,15,16), another with four (Baral, Dlinnaya, Kommunar and Novaya, Nos 10,11,13,14);
4 three eastern seamounts of the Sala y Gómez Ridge located to the west of 88°W (Pervomayskaya, Zhemchuzhnaya and Yantarnaya, Nos 6–8) plus one of the westernmost section (Ichthyologists, No. 3);
5 three other western seamounts of the Sala y Gómez Ridge (Igolnaya, Utes and Kupol, Nos 1,2,5).

The second and third groups are separated by the important boundary near 83–84°W. For the fish fauna, this is a distinct boundary, between the Nazca Ridge (seamounts Nos 17–22) and the Nazca/Sala y Gómez transitional area (seamounts Nos 10–16) but, unlike the distribution of bottom invertebrates, it is less marked than the boundary between the transitional area and the Sala y Gómez Ridge proper (westward of 88°W) or the difference between the deep- and shallow-summit seamounts of the Nazca Ridge (respectively Nos 21–22 and 17–20).

However, the dendrogram under discussion (Figure 10) probably reflects both historical–zoogeographical heterogeneities and ecological (biotopic) differences, depending on differences in depths, bottom relief and bottom deposits (see Section 5.2.1) as well as in the productivity of the waters around the seamounts. Thus, the boundary at 88°W, dividing predominantly hard-bottom seamounts of the transitional area and soft-bottom eastern Sala y Gómez seamounts, may be considered as having little biogeographic importance. The faunistic similarity found between the Ichthyologists Seamount and three eastern Sala y Gómez seamounts is obviously explained by the presence of soft-ground areas on all their summits.

The problem of the zoogeographical distinctness of the Nazca Ridge proper (seamounts Nos 17–22) will be treated in more detail because Mironov and Detinova (1990) believe that it belongs to the Eastern Pacific Region. Among fish recorded on that part of the ridge system, 26 out of 57 species, or 45.6%, are not found on the Sala y Gómez Ridge or the transitional Nazca/Sala y Gómez area (Table 11). This distinctness (Figure 10) is determined primarily by the relative abundance of deep-water (lower bathyal) species, comprising 30% of the species list, of which 23% are species not found on other parts of the ridges. However, there were more deep trawl hauls on this part of the ridges than elsewhere (only two deep hauls were made on other seamounts, both in the transitional Nazca/Sala y Gómez area) because all seamounts with deep trawlable summits are located on the Nazca Ridge. Comparable depths of most other seamounts coincide with steep rough slopes, where bottom trawling is impossible. The distinctiveness of the Nazca Ridge is much less (28%) in the sublittoral–upper bathyal vertical zone (Table 12). Here, endemics of the Nazca and Sala y Gómez ridges compose half the species list but only five species (19% of all) were recorded on the Nazca Ridge only: *Mollisquama parini* (single record), *Muraenichthys profundorum* (belonging to a speciose Indo-West Pacific genus), *Facciolella castlei* (of a circumglobal bathyal genus, represented in the Pacific Ocean only off the American coast), *Gaidropsarus parini* (of a genus known in the Pacific only in the New Zealand waters), and *Plagiogeneion unispina* (also belonging to a southern–subtropical, but Indo-West Pacific genus). The other 13 endemics are known on both the Nazca and Sala y Gómez ridges and most of these have clear Indo-West Pacific roots. Only one (!) species, a rattail *Caelorinchus immaculatus*, is in common with the South American slope but this abundant species on the Nazca Ridge is probably rare off South America where it has been recorded only once (Sazonov and Iwamoto, 1992).

Table 11 Geographical and depth composition of the ichthyofauna of the Nazca Ridge.

Bathymetric and zoogeographic grouping	Number of species	
	Total	Not found on the Sala y Gómez Ridge
Sublittoral–upper bathyal including:	**36**	**10**
endemics	18	5
circumglobal	14	4
south subtropical	3	—
Eastern Pacific	1	1
Lower bathyal including:	**17**	**13**
circumglobal	14	10
Eastern Pacific	3	3
Neritic South American	**4**	**3**
Total	**57**	**26**

In addition, four other fish species recorded only on the Nazca Ridge (*Macruronus magellanicus, Sardinops sagax, Trachurus symmetricus* and *Scomber japonicus*) are not permanent inhabitants, but temporary migrants from Chilean neritic waters.

The conclusion is obvious. The ichthyofauna of the Nazca Ridge proper is an impoverished fauna of the Sala y Gómez Ridge. There are absolutely no reasons to consider it as a part of the Eastern Pacific Region. Moreover, it cannot be even treated as a "transitional zone" between the Indo-West Pacific and Eastern Pacific regions.

7. ENDEMISM

Endemic species represent the main part of the bottom and near-bottom invertebrates and fishes on the seamounts of the Nazca and Sala y Gómez ridges. Among 137 species of invertebrates from nine taxonomic groups the proportion of endemics varied from 22 to 96% (Table 3) being on average 52% (71 species). The proportion of endemics does not change (51%) if counted on the Sala y Gómez Ridge only (west of 83°W) while there are no endemics of the area eastward of 83°W. All 71 endemic species recorded in the area studied were found in the Sala-y-Gómezian Province but only four species from there are known both west and east of 83°W.

Table 12 Distribution in the subcontinental regions of the western and eastern Pacific of the species of bottom invertebrates known from the Nazca and Sala y Gómez ridges (from Mironov and Detinova, 1990, with additions).

Systematic groups	Total no. of species	Subcontinental regions			Endemics of Nazca and Sala y Gómez ridges	Other regions
		Western and Eastern Pacific	Western Pacific	Eastern Pacific		
Hyalospongia	6	—	2 (33%)	1 (17%)	3 (50%)	—
Gastropoda Turridae	24	1 (4%)	—	—	23 (96%)	—
Cirripedia	14	1 (7%)	1 (7%)	1 (7%)	11 (79%)	—
Tanaidacea	9	1 (11%)	—	1 (11%)	2 (22%)	5 (66%)
Macrura	30	—	14 (48%)	2 (7%)	9 (30%)	5 (15%)
Brachyura and Anomura	24	—	8 (33%)	3 (12%)	9 (37%)	4 (18%)
Bivalvia Septibranchia	7	—	1 (14%)	—	5 (72%)	1 (14%)
Brachiopoda	4	1 (25%)	—	—	1 (25%)	2 (25%)
Echinoidea	19	—	2 (11%)	1 (5%)	8 (42%)	8 (42%)
Total	137	4 (3%)	28 (20%)	9 (7%)	71 (52%)	25 (18%)

The degree of endemism on the generic level is low. Among 101 genera from the nine cited taxonomic groups (Table 3) only three (3%) were not found outside the Nazca and Sala y Gómez ridges, including the hexactinellid sponge *Pseudoplectella*, tanaidacean *Pseudobathytanais* and bivalve mollusc *Cribrosoconha*. One more genus, the echinoid *Scrippsechinus* has been recorded on the Sala y Gómez Ridge and off San Félix Is, which probably also belongs to the Sala-y-Gómezian Province (Allison *et al.*, 1967; Kudinova-Pasternak, 1990; Tabachnick, 1990; Krylova, 1995).

Among cephalopods all three sepiolids were new undescribed species, while *Sepioloidea* and *Danoctopus* were each represented in our collection by one juvenile female and it was uncertain whether they were both new species or belonged to known ones (*S. pacifica* and *D. hoylei*). Thus the degree of endemism among cephalopods is at least 42.9% but may be 71.4%.

Similarly, the degree of endemism on the species level is well established among fishes (Parin, 1990a, 1991a). There are 76 endemic species; that is, 44.2% of the species list for the area (Table 4) belonging to 31 families and 51 genera. Distribution of endemic species along ridges varies significantly (Figure 11). Forty endemic species are known at present from the westernmost area, seamounts Nos 1–5 (14 of these species are not found elsewhere); 23 from the central part of the Sala y Gómez Ridge, seamounts Nos 6–8 (six not found elsewhere); 43 (14) from the transitional Nazca/Sala y Gómez area (Nos 9–16) and 22 (five) from the Nazca Ridge proper (Nos 17–21). Almost half the endemic species are known from two or three groups of seamounts. It should be noted that eight endemic species are also known from near the islands lying to the east of the area studied, mainly from Juan Fernández Is, and one (*Pentaceros quinquespinis*) from Easter Is.

Most of the endemic species belong to the families Macrouridae (incl. Bathygadidae) (10), and Moridae (seven). Several families are represented only by endemic species (Torpedinidae, Nettastomatidae, Congridae, Argentinidae, Sternoptychidae, Polymixiidae and Bothidae) or a great majority of their species are endemics (Moridae, Serranidae, Epigonidae, Emmelichthyidae, Percophidae and Draconettidae). In contrast to the bottom invertebrates, there are endemic fish species (five) known at present only to the east of 83°W (*Mollisquama parini*, *Facciolella castlei*, *Gaidropsarus parini*, *Caelorinchus immaculatus* and *Plagiogeneion geminatus*).

There are three supposedly endemic fish genera, all monotypic: a shark *Mollisquama* (Squalidae) and two teleosts, *Anatolanthias* (Serranidae) and *Dactylopsaron* (Percophidae). The generic endemism is thus 2.3% (three out of 130 genera).

The endemism of the fauna of the Sala y Gómez and Nazca ridges seems to be unprecedently high, even in comparison with the very high endemism of the nearshore fauna of the Hawaiian Is (20% endemic molluscs and 29% shorefish species) or Easter Is (42% among nearshore molluscs, 14–50% in some other groups of invertebrates and 27% among fishes) (Schilder, 1965; Briggs, 1974;

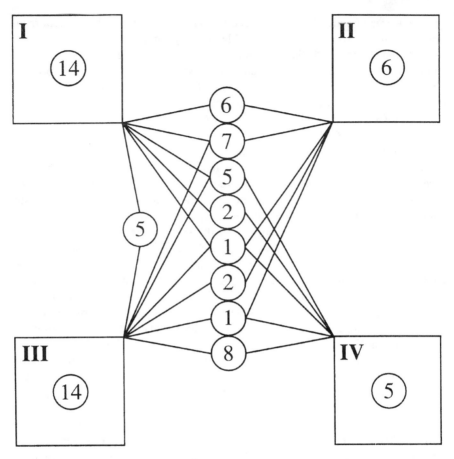

Figure 11 Number of endemic fish species on different seamounts and seamount groups of the Nazca and Sala y Gómez ridges. I, western Sala y Gómez ridge, seamounts Nos 1–5; II, eastern Sala y Gómez Ridge, seamounts Nos 6–8; III, transitional Nazca/Sala y Gómez area, seamounts Nos 9–16; IV, Nazca Ridge, seamounts Nos 17-22. Figures in circles inside squares show number of endemic species, recorded only in a given area; figures in circles on lines connecting squares show number of endemics common for two or more groups of seamounts.

Rehder, 1980; Springer, 1982; Newman and Foster, 1983; Osorio, 1992). Such a high level of endemism might be attributed to the isolation of the ridges from the main source of the fauna, the Indo-West Pacific, because the greater the isolation of an area, the higher is its endemism (MacArthur and Wilson, 1967; Barton, 1990; Cox and Moore, 1993), but the problem is not so simple.

Mironov (1985b, 1989, 1994) observed that open-oceanic areas with great faunal endemism are peripheral to the main biogeographic regions. For example,

Pacific open-ocean regions with a high level of endemism (Sala y Gómez Ridge, Hawaiian, Galapágos, Easter Is, New Zealand) are arranged along the peripheries of either the Indo-West Pacific or Eastern Pacific regions. Mironov termed this regularity the "edge effect". These rather small boundary areas are well-separated biogeographically and termed "exotic open-ocean biogeographical provinces" (Mironov, 1994; Figure 3). Other biogeographic provinces existing in the open ocean, he called "subordinate"; these differ from the exotic provinces by much larger size and a smaller degree of endemism.

An analogous but somewhat different phenomenon was described by Semenov (1977, 1982) on the basis of the distribution of bottom invertebrates on South American continental shelf, termed the "boundary effect". According to Semenov the biogeographic boundaries, the narrow zones of crowding of species range boundaries, have their own endemics with spatially small "pointwise" ranges.

8. FAUNISTIC RELATIONSHIPS

Genetic relationships of bottom invertebrates inhabiting the seamounts of the Nazca and Sala y Gómez ridges have been investigated in sea urchins (Mironov, 1989), decapod crustaceans (Burukovsky, 1990; Zarenkov, 1990) and endemic fauna in general (Mironov and Detinova, 1990). Mironov (1989, 1994) has proposed the following distributional types of genera and species: oceanic type, when a genus or species is found in open–oceanic regions only; panthalassic, the taxa are found in both subcontinental and open–oceanic regions; panthalassic–West Pacific, the taxa are found in open–oceanic and western subcontinental regions, but not in eastern subcontinental regions; panthalassic–East Pacific, the taxa are found only in the oceanic and eastern but not the western subcontinental regions; panthalassic–transoceanic, found both in the western and eastern subcontinental and open-oceanic regions (Figure 12; Table 12). The distributional types were determined from the distribution patterns of given taxa in the Pacific Ocean only. Echinoid genera are mainly of panthalassic–West Pacific distribution (10 out of 17 genera), and echinoid species are mainly of oceanic distribution (13 out of 19 species, including eight endemics of the Nazca and Sala y Gómez ridges). Only one species, *Tromikosoma hispidum*, has panthalassic–East Pacific distribution. Mironov (1989) came to the conclusion that the Sala y Gómez Ridge is the easternmost limit of the distribution of the Indo-West Pacific fauna.

If we consider the distribution of taxa outside the Pacific Ocean we find that most echinoid genera are widely distributed (Table 13). The number of genera in common with the Atlantic Ocean is the same (15) as with the subcontinental regions of the West Pacific. The similarity of the ridge fauna to the sublittoral fauna is more marked than to the lower bathyal fauna, both on the generic and species levels (Table 13).

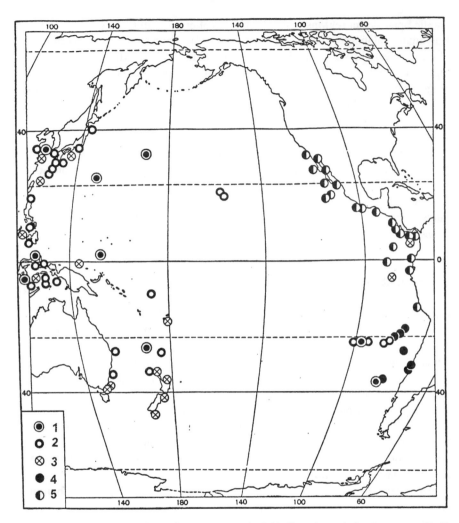

Figure 12   Panthalassic western asymmetrical (1–3) and panthalassic eastern (4–5) distributional types of sublittoral and bathyal animals in the Pacific Ocean. 1, Brittle stars of the genus *Ophiotholia* (after Litvinova, 1992); 2, sea urchins of the genus *Stereocidaris*; 3, sea urchins of the genus *Araeosoma*; 4, the spiny lobster *Projasus bahamondei* (after George, 1976; Rudjakov *et al.*, 1990); 5, sea urchins of the genus *Encope* (based on Mironov, 1994).

Burukovsky (1990) grouped 29 recorded shrimp species by their distributional patterns into: Pacific insular (12 species including endemics of the Nazca and Sala y Gómez ridges), West Pacific (two), Indo-West Pacific (seven), East Pacific (one), and cosmopolitan (seven). The number of shrimp species on the Sala y Gómez and

Table 13   Distribution in the Pacific Ocean of the Echinoidea known from Nazca and Sala y Gómez ridges. For descriptions of the distributional types see text.

Biogeographic characters	Number of taxa	
	Genera (17)	Species (19)
Endemics of Pacific open oceanic regions: Endemics of Sala y Gómez and		
Nazca ridges	—	8 (42%)
Common with other regions	1 (6%)	5 (27%)
Common to Pacific subcontinental regions only:		
West Pacific only	—	—
East Pacific only	—	1 (5%)
Indo-West Pacific only	1 (6%)	1 (5%)
Cosmopolitian:		
Panthalassic–West Pacific	10 (59%)	1 (5%)
Panthalassic–East Pacific	—	—
Panthalassic–Trans-Pacific	4 (23%)	—
Oceanic–Pacific	1 (6%)	—
Unknown distribution	—	3 (16%)

Nazca ridges decreases from west to east. According to Burukovsky (1990), this fact as well as the patterns of species ranges suggest that the Sala y Gómez Ridge lies at the eastern periphery of the Indo-West Pacific Region, particularly of its Pacific Insular Province, while the Nazca Ridge is an "impoverishment–transitional" zone between the Pacific Insular Province and the Eastern Pacific Tropical Region. With increasing depth, from 150 to 700 m, the number of shrimp species occurring only in the Pacific diminished more than three times. Pacific species predominate in depths of less than 300 m while widely distributed species are most common at greater depths.

Among tanaidaceans, two endemics, one Pacific species and six widely distributed species have been recorded (Kudinova-Pasternak, 1990). Six species of nine (67%) are in common with the North Atlantic. The predominance of widely distributed species was related to the depth of capture: seven species were caught in depths down 750 m. Widely distributed species also predominated among brachiopods (three out of four), they were caught, like tanaidaceans, in relatively large depths (Zezina, 1990). Among turrid gastropods and cirripedes, en-demic species dominated and genetic connections are weak (Sysoev, 1990; Zevina, 1990).

Zarenkov (1990) found that the ridge Brachyura and Anomura, both at species and genus levels, is closer to the West Pacific than to the East Pacific fauna. His list of hermit crab species was corrected and extended by D.G. Zhadan (personal communication) who moved three species referred to as *Parapagurus* into *Sympagurus*, re-identified *P. sculptochela* as *S. boletifer* and *P. dimorphus* as *S. africanus* subsp.n., and added four more hermit crab species: *Sympagurus affinis*, *S. ruticheles*, *S. wallisi* and *Bivalvopagurus sinensis*. Counting these additions, the fauna of Brachyura and Anomura includes 24 species in 18 genera. There are nine endemics, five Western Pacific, two Western Pacific–open oceanic, three Indo-West Pacific, three Eastern Pacific, and two are known outside the Indo-Pacific but unknown in the Eastern Pacific. Thus the number of species in common with the West Pacific is more than three times that in common with the East Pacific.

Mironov and Detinova (1990) had found that the number of genera and species with panthalassic–West Pacific distribution was approximately 25% of all taxa recorded in the area and greatly exceeded the number of taxa with panthalassic–East Pacific distribution. With the inclusion of new data the proportion of taxa with panthalassic–West Pacific distribution decreases to 20% (Table 12). Panthalassic–East Pacific distribution is characteristic for only seven species (sponge *Amphorme horrida*, cirripedian *Metaverruca tarasovi*, shrimp *Pasiphaea americana*, spiny lobster *Projasus bahamondei*, crabs *Chaceon chilensis* and *Mursia gaudichaudii*, and echinoid *Tromikosoma hispidum*).

Projasus presents a problem. The genus contains two species, *P. bahamondei*, found on the Nazca Ridge, San Félix, San Ambrosio, Juan Fernández Is and off the coast of central Chile between Huasco (Atakama) and Constitución (Maule), and *P. parkeri*, found on Valdivia Bank off Namibia, off the coast of south-eastern Africa (Cape Province north of East London and Natal), on some seamounts in the south-western Indian Ocean, off St Paul Is, south-eastern Australia and northern New Zealand (Webber and Booth, 1988; Holthuis, 1991; Parin *et al.*, 1993; Griffin and Stoddart, 1995). Thus it is a southern subtropical genus with an interrupted distribution, and absent from two huge areas: the south-western Atlantic and the southern Pacific between New Zealand and Nazca Ridge. Bearing in mind that *Projasus* has a long-lived planktonic larva (Webber and Booth, 1988) and that larval dispersion from New Zealand to the Nazca Ridge (with the current) is much easier than from South Africa to South America (against current), it is probable that *P. parkeri* is the ancestral species for *P. bahamondei* and thus *P. bahamondei*, although a typical representative of the Eastern Pacific fauna, is of western, not eastern genesis.

Mironov and Detinova (1990) noted that among endemics the most numerous group is formed of the species closely related to those known in the Indo-West Pacific. Examples are: the sponge *Pheronema naskaniensis* which is close to *P. placoideum* and *P. semiglobulosum* (Tabachnik, 1990); the gastropod *Sprendrillia basilirata*, close to *S. runcinata* (Sysoev, 1990), the bivalve *Rhinoclama similis*, close to *R. brevirostris* and *R. raoulensis* (Krylova, 1994), the cirripede *Paralepas naskai*, close to *P. xenophore* (Zevina, 1990), the shrimps *Stylodactylus pubescens*, close to *S. multidentatus* and *S. libratus*, *Pontophilus nikiforovi*, close to *P. japonica* and *P. prionolepis* (Burukovsky, 1990), the crabs *Cancer balssi*, close to *C. margaritarius* (Zarenkov, 1990), *Parthenope mironovi*, close to *P. stellata* (Garth, 1992), echinoid *Stereocidaris nascaensis*, close to *S. creptiferoides* and *S. japonica* (Allison *et al.*, 1967) and the brachiopod *Septicollarina oceanica*, close to *S. hemiechinata* (Zezina, 1990).

We know only one indication of a ridge endemic close to an Eastern Pacific species: this is a gastropod, *Mitromorpha maculata*, which is related to the Californian *M. aspera* (Sysoev, 1990). The endemics of the Sala y Gómez Ridge are closer to the Atlantic Ocean fauna than to those of the Eastern Pacific.

The fish fauna of the Nazca and Sala y Gómez ridges was divided into four groups by Parin (1990a, 1991a); that is, endemic, widely distributed (Atlanto-Indo-Pacific), Indo-West Pacific and southern subtropical ones comprising, correspondingly, 51.5, 22.8, 14.4 and 3.7% of the species list (without considering the deep-water forms). Parin postulated predominantly Indo-West Pacific and, to a lesser degree, southern subtropical (Australian–New Zealand) relationships of the fish fauna.

For this chapter an analysis of geographical distribution in the Pacific Ocean was undertaken for 100 non-endemic genera of fishes represented in sublittoral and upper bathyal zones of the ridges down to 800 m depth (Table 14). It was found that the greatest numbers of genera in common with the fish fauna of the Nazca and Sala y Gómez ridges are represented (in decreasing succession) in the waters of southern Japan, eastern and south-eastern Australia, Hawaii, New Zealand, central tropical region of the Western Pacific (Philippines to northern Australia), Kyushu–Palau Ridge, Emperor–Hawaiian ridges, and New Caledonia (Figure 13). Near the Polynesian islands (excluding Hawaii) only a quarter of Nazca/Sala-y-Gómezian genera are known, and along American coasts only one-fifth of these genera are recorded.

Among 20 Indo-West Pacific species inhabiting the south-eastern Pacific ridges, there is a predominance of species in common with Hawaii (16, four of them not known from any other place) and southern Japan (11) (Table 14). Some of the Nazca–Sala y Gómez endemics are closely related to western Pacific species, for example *Glossanodon nazca* related to Hawaiian *G. struhsakeri*, *G. danieli* related to Japanese *G. lineatus* (Parin and Shcherbachev, 1982), *Argyripnus electronus* related to an unnamed *Argyripnus* sp. from the Kyushu–Palau Ridge (Parin, 1992), *Polyipnus inermis* to *P. parini* from off Japan (Borodulina, 1981), *Hime macrops* to Japanese *H. damasi* (Parin and Kotlyar, 1989), *Kuronezumia pallida* to Japanese *K. dara* (Sazonov and Iwamoto, 1992), *Emmelichthys elongatus* to *E. stuhsakeri* from Hawaii and the Kyushu–Palau Ridge (Kotlyar, 1982b). Thus, there are quite obvious relationships between the Nazca–Sala y Gómez fish fauna, on the one hand, and the Hawaiian and Japanese (including Kyushu-Palau Ridge) faunas, on the other hand.

In addition, a small but well-defined group of the Nazca–Sala y Gómez fishes

Table 14   Other regions of the World Ocean where the Indo-West Pacific fish species that are recorded on the Sala y Gómez Ridge are also recorded.

Synodus doaki	Southern Japan, Australia, New Zealand, Hawaii
Laemonema rhodochir	Kyushu–Palau Ridge, Hawaii
Coryphaenoides paradoxus	Philippines, Hawaii
Hymenocephalus striatulus	Hawaii
Mataeocephalus acipenserinus	Hawaii
Nezumia propinqua	Kyushu–Palau Ridge, Hawaii
Trachonurus villosus	Southern Japan, Philippines, Australia, Hawaii
Ventrifossa johnboborum	New Guinea
Lophiodes mutilus	Taiwan, Philippines, Australia
Cyttomimus stelgis	Southern Japan, Australia, New Caledonia
Stethopristes eos	Hawaii
Bembradium roseum	Southern Japan, New Caledonia (?), Hawaii
Hoplichthys citrinus	New Caledonia, Hawaii
Grammatonotus laysanus	Hawaii
Symphysanodon maunaloae	Kyushu–Palau Ridge, Hawaii
Epigonus atherinoides	Philippines, Kyushu–Palau Ridge, Hawaii
Erythrocles scintillans	Hawaii, Tahiti, Easter Is
Chrionema chryseros	Southern Japan, Hawaii
Centrodraco nakaboi	Southern Japan
Schindleria praematurus	Southern Japan, Australia, New Guinea, Samoa, Tahiti, Hawaii

show a relationship, at both the generic and species level, to the peculiar warm-temperate–subtropical fauna of the southern hemisphere typical of New Zealand, southern and south-eastern Australia, South Africa and the oceanic islands and seamounts between, roughly, 30 and 40°S. Such southern subtropical genera are *Bassanago, Tripterophyces, Notopogon, Plagiogeneion, Acantholatris* and *Seriolella*, five of them with one species each in the area (*B. nielseni, T. svetovidovi, N. fernandinus, A. gayi* and *S. labyrinthica*) and one with two species (*P. geminatum* and *P. unispina*). Other members at this group are *Eurypleuron cinereum, Pterygotrigla picta, Caprodon longimanus, Rexea antefurcata, R. brevilineata* (a sister species of *R. solandri* from southern Australia and New Zealand), *Aphanophus capricornis,* and *Benthodesmus elongatus* (Table 14). The blue-mouth perch *Helicolenus lengerichi* seems to belong to the same group; its range is rather like that of the lobster *Projasus bahamondei* (see above) but unlike the lobster, the blue-mouth occurs not only off Juan Fernández Is (its type locality) and on the Nazca Ridge but also on the seamounts of the transitional Nazca/Sala y Gómez area, westward up to the Baral Seamount. This species has been reported from the continental coastal zone of Peru and Chile but, according to W.N. Eschmeyer (personal communication), the Peru-Chilean "population" belongs to a different, undescribed species. Relatively closely related species of *Helicolenus* are known from waters off south-eastern Australia and New Zealand (*H.*

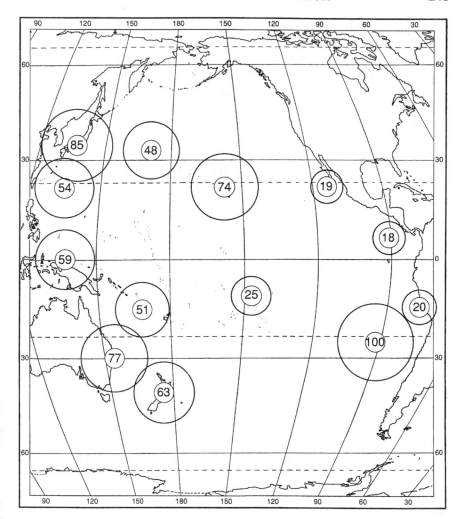

Figure 13   Number (per cent of total species list) of fish genera in the fish fauna of the Nazca and Sala y Gómez ridges, that occur also in the waters of the tropical western Pacific, southern Japan, New Caledonia, eastern Australia, New Zealand, Kyushu–Palau, Emperor and Hawaiian ridges, Hawaii, Polynesia, North, Central and South America. The area of the circles is proportional to the number of species.

percoides), South Africa and Uruguay (two subspecies of *H. dactylopterus*), and from islands and seamounts located between Tristan da Cunha Island, South Atlantic, and the St Paul and Amsterdam Is, southern Indian Ocean (*H. mouchezi*) (Paxton *et al.*, 1989; Andrew *et al.*, 1995). Thus, both *P. bahamondei* and *H. lengerichi* seem to be derived not from the South American fauna but from the circumglobal southern subtropical fauna. It should be noted that, unlike *P. bahamondei*, the blue-mouth has no long-lived pelagic larval stage and its larvae have been found immediately over the seamounts inhabited by the adult fish (Kotlyar, 1988a; Belyanina, 1989, 1990).

Relationships of the fish fauna with that of mainland South America are weak. As stated above (Section 6.2), the only bathyal species, *Caelorynchus immaculatus*, is common to both Nazca and Sala y Gómez ridges and Chile (Sazonov and Iwamoto, 1992) and the neritic species *Trachurus symmetricus murphyi*, *Scomber japonicus*, *Sardinops sagax* and probably *Macruronus magellanicus*, temporary migrants to the eastern part of the area studied, are considered to have no zoogeographical importance. However, in the deepest bottom trawlings at least three endemic American fish species were collected (*Talismania bussingi*, *Nezumia convergens* and *Dicrolene nigra*), indicating, probably, closer connections of the lower bathyal fauna with that of the mainland deep slope.

Among cephalopods the genera *Iridoteuthis* and *Nototodarus* are Indo-West Pacific, *Heteroteuthis*, *Scaeurgus* and *Danoctopus* are Atlanto-Indo-West Pacific in distribution. For *N. hawaiiensis*, *S. unicirrhus patagiatus* and *D. hoylei* the nearest records to the south-east Pacific ridges are those off the Hawaiian Is. Species of *Iridoteuthis* and *Heteroteuthis* are also present off the Hawaiian Is but those found on the ridges studied are closer to *I. maoria* from New Zealand and adjacent areas and *H. dispar* from the Atlantic Ocean and Mediterranean Sea, but having close relatives in the south-western Indian Ocean (Nesis, 1990, 1993a,b, 1994). *S. pacifica* is distributed around New Zealand and the genus *Stoloteuthis* includes one species in the North and south-east Atlantic and Mediterranean Sea and another, undescribed, in the subantarctic Indian Ocean (Nesis, 1990, 1993b). Thus, six of the seven species belong to genera distributed, either exclusively or also in the Indo-West Pacific. Three species (of the genera *Nototodarus*, *Scaeurgus* and *Danoctopus*) are identical or supposedly identical to those with the closest occurrence off Hawaii. Two species (of the genera *Iridoteuthis* and *Sepioloidea*) are identical or possibly identical to those with the nearest occurrence off New Zealand and the two last (*Heteroteuthis, Stoloteuthis*) have no close relatives in the western Pacific but have such relatives in the Atlantic and southern Indian Ocean.

Thus we may conclude that the bottom and near-bottom fauna, both invertebrates and fishes, of the Sala y Gómez and Nazca ridges is more closely related to the western Pacific than to the eastern Pacific fauna. It has been shown earlier, taking examples from many groups of animals (e.g. Ekman, 1953; Briggs, 1974; Springer, 1982) that the western Pacific fauna is much more widely distributed in the open Pacific Ocean than is the eastern Pacific fauna. The area around Easter Is has been considered to be the easternmost limit of the distribution of the western Pacific fauna (Rehder, 1980; Osorio, 1992). Investigations on the Sala y Gómez and Nazca ridges show that the meridional asymmetry in the distribution of the western and eastern Pacific faunas is even more pronounced and that the boundary between them lies much closer to South America. The figures for the relatedness of the Sala y Gómez Ridge fauna and the fauna of neighbouring bathyal areas of the open ocean are probably underestimated because of poor knowledge of the slope fauna of the San Félix, San Ambrosio, Juan Fernández, Easter and Sala y Gómez Is.

9. BIOGEOGRAPHICAL POSITION

The evident predominance of both invertebrates and fishes of Indo-West Pacific origin allows the investigated areas to be considered as an eastern outpost of the Indo-West Pacific Region (Figure 14). Moreover, the existence of very numerous endemic species in the area makes it possible to establish a separate biogeographic province (Parin, 1991a). However, the presence of a strongly marked biogeographic boundary at 83–84°W dividing the area studied into two parts permits more than one interpretation and the authors of this chapter have different opinions about it.

A.N. Mironov firmly adheres to his former opinion expressed first in his joint paper with N.N Detinova (Mironov and Detinova, 1990). They have separated "the Sala-y-Gómezian faunistic complex" including the Sala y Gómez Ridge from 83°W to about 101°W. This complex belongs to the Indo-West Pacific Region and does not include Sala y Gómez Is to the west and the Nazca Ridge to the east. The Nazca Ridge belongs to the subcontinental Peru–Chilean unit of the eastern Pacific region, and the boundary about 83°W is a part of a biogeographic boundary lying all along the Pacific continental slope. The existence of this boundary was suggested by Mironov (1989, 1994), who divided the Pacific Ocean into two circumcontinental biogeographic zones, subcontinental and oceanic. Sala y Gómez Is was excluded from the Sala-y-Gómezian faunistic complex on the basis of the differences in the echinoid and molluscan faunas. Among eight genera and eight species of echinoids known from Easter and Sala y Gómez Is (Fell, 1974; Mironov and Sagaidachny, 1984) only two genera and no species are common to the Nazca and Sala y Gómez ridges. Among four genera and five species of turrid gastropods known from the Sala y Gómez and Easter Is (Rehder, 1980; Sysoev, 1990) there are none in common with the seamounts of the Sala y Gómez Ridge. This may be the result of comparing animals that inhabit different vertical zones, some metres depth around the Easter and Sala y Gómez Is and some 100 m on the seamounts.

The classification of the Nazca Ridge as part of a subcontinental biogeographic unit (Mironov and Detinova, 1990) suggests a gradual faunistic transformation eastward from 83°W, without a sharp break in the bottom fauna near the mainland slope. However, there is insufficient information to confirm the presence or absence of a zone of crowding of species range boundaries between the mainland slope and the northeastern part of the Nazca Ridge.

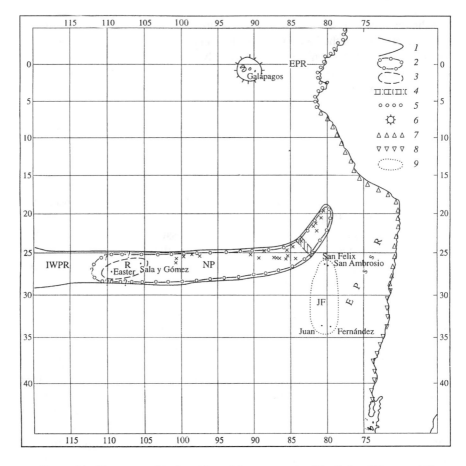

Figure 14 Zoogeographical position of the seamounts of the Sala y Gómez and Nazca ridges. 1, Eastern boundary of the Indo-West Pacific tropical Region (IWPR) in the shallow-water and lower sublittoral–upper bathyal vertical zones; the studied seamounts of the Nazca and Sala y Gómez ridges are indicated by "x" (after Parin, 1990a, 1991a); 2–3, Indo-West-Pacific tropical Region: 2, the Nazcaplatensis Province (NP) (lower sublittoral–upper bathyal vertical zone), 3, the Rapanuiian Province (R) (shallow-water); 4, boundary between the Nazca Ridge proper fauna and the fauna of the Sala y Gómez Ridge and geomorphologically transitional Nazca/Sala y Gómez area (the Nazca Ridge fauna proper belong to Nazcaplatensis Province from the fishes but to the subcontinental Peru–Chilean unit of the Eastern Pacific Region from bottom and near-bottom invertebrates); 5–6, Eastern Pacific tropical Region (EPR): 5, Panamanian Province; 6, Galapagos Province; 7–9, Eastern Pacific southern subtropical Region (EPssR); 7, Peruvian Province; 8, Central Chilean (Araucanian) Province; 9, Juan Fernández Province (JF) (shallow-water); in the lower sublittoral–upper bathyal vertical zone it may (or may not) belong to the NP. The zoogeographical divisions and boundaries are based on Nesis (1985) and Parin and Nesis (1986).

According to Burukovsky (1990), the Sala y Gómez Ridge is the easternmost periphery of the "Pacific Insular Province" of the Indo-West Pacific Region, while the Nazca Ridge is an "impoverishment-transitional zone" between the Pacific Insular Province and the Eastern Pacific Tropical Region.

Two other authors of the present chapter (N.V. Parin and K.N. Nesis) share the opinion that both Sala y Gómez and Nazca ridges, as studied here, should be placed in a single biogeographical unit ranked as a province of the Indo-West Pacific Region. This province might be named Sala-y- Gómezensis, as the biogeographic complex separated by Mironov and Detinova (1990), if it were not for the danger of confusion with the province including the shallow-water fauna of Sala y Gómez Is, and it belongs to a different province, the Rapanuiana or Rapanuiian Province (see below). Thus, they propose the name Nazcaplatensis Province, after the lithospheric Nazca Plate. Moreover, the area northeast of 83–84°W (the Nazca Ridge proper) may be considered, contrary to Mironov's proposal, as a faunistically impoverished part of this province because its peculiarity consists mainly of the absence of many species characteristic of both the transitional Sala y Gómez/Nazca area and the Sala y Gómez Ridge proper (including most of the endemic species) and in the presence of very few species common with mainland South America (Figure 14).

The other boundaries of the province under discussion are also debatable. It seems probable that the lower sublittoral and upper bathyal zones of San Félix, San Ambrosio and even Juan Fernández Is, to the east and south-east of the area, as well as that of Sala y Gómez and Easter Is, located to the west, may belong to the province under consideration. Also, there are many seamounts in the Chilean EEZ adjacent to the listed islands, which have not been investigated, and nothing can be said of their zoogeography. The Juan Fernández and San Félix Is have three genera and three species of echinoids in common with Sala y Gómez Ridge, among four genera and species recorded (Codoceo, 1976) (*Scrippsechinus fisheri*, *Centrostephanus sylviae* and *Clypeaster isolatus*). In the echinoid fauna of these islands the western faunistic element is also well marked: *Centrostephanus sylviae* is closest to the western Pacific *C. rodgersii* (Fell, 1975) and *Clypeaster isolatus* is closer to the western Pacific *C. australasiae* than to many eastern Pacific species of that genus (Serafy, 1971). The third species, *Scrippsechinus fisheri*, has only been recorded outside these islands on the Sala y Gómez Ridge.

The decapod fauna of these islands includes 26 species of 24 genera (Balss, 1924; Garth, 1957; George, 1976; Chirino-Galves and Manning, 1989). Zarenkov (1990) has indicated the presence of both western and eastern elements in the decapod fauna of the Juan Fernandez Is. The genera *Chaceon*, *Paramola*, *Projasus*, *Porcellanopagurus* and species *Chaceon chilensis* and *Projasus bahamondei* are shared with the Sala y Gómez Ridge.

The fishes recorded on the Nazca and Sala y Gómez ridges include the following species which are known from either San Félix and San Ambrosio or from Juan Fernández Is: *Squalus mitsukurii, Notopogon fernandezianus**,

Monocentris reedi*, Scorpaena uncinata*, Pterygotrigla picta, Caprodon lon-gimanus, Plectranthias exsul*, Emmelichthys cyaneus, Acantholatris gayi*, Parapercis dockinsi*, and *Seriolella labyrinthica*. Six of these, marked with asterisks, are known only on the ridges and off the islands included in this province. Two more "mountain" species are represented by "sister species" on the ridges and off Juan Fernández Is: *Muraenichthys profundorum* and *M. chilensis* (McCosker and Parin, 1995) and *Callanthias parini* and *C. platei* (Anderson and Johnson, 1984).

However, the coastal shallow-water fauna of the Juan Fernández Is and probably those of San Félix and San Ambrosio Is belong to the Eastern Pacific. According to the comprehensive book by Briggs (1974), endemic and Eastern Pacific species predominate in the marine fauna of Juan Fernández. Briggs considered these islands to be a distinct province of the Peru–Chilean Region. Existing knowledge refers mainly to the littoral and upper sublittoral fauna and is deficient for solving the question of biogeographic separation of the bathyal fauna of this area from that of the Nazca and Sala y Gómez ridges. If the lower sublittoral and bathyal fauna of these islands does in fact belong to the Nazcaplatensis Province of the Indo-West Pacific then adjoining vertical zones in the island region belong not only to different provinces but to different first order biogeographic regions. Such stratification has not been recognized before in tropical and subtropical areas of the World Ocean although it is rather usual in subpolar areas, for example along the boundaries of the Arctic and the Atlantic and Pacific Boreal regions where in many places the shelf or upper sublittoral fauna is of arctic character while the upper bathyal or also lower sublittoral zones are populated by boreal fauna (Nesis, 1985).

The position of the western boundary of the province remains uncertain because the lower sublittoral and bathyal fauna of Sala y Gómez and Easter Is and the nearby seamounts are almost unstudied. However, at least one fish species endemic to the Nazca and Sala y Gómez ridges, *Pentaceros decacanthus*, has been recorded near Easter Is (Parin and Kotlyar, 1988).

10. BIOGRAPHIC HISTORY

The most pronounced features of the Sala y Gómez and Nazca ridges fauna, including invertebrates and fishes, are the extremely high endemism and the very evident close relationships with the Indo-West Pacific fauna. The high degree of overall endemism requires the existence of biogeographical barriers surrounding the ridges. On the western side, such a barrier is provided by vast deep-water areas between the groups of islands and seamounts. The eastern barrier is formed by the subduction zone of the Peru–Chile Trench and the cold Peru Current running parallel to the South American coastline. The key problem in the biogeographical history of the Sala y Gómez and Nazca ridges is a question of the process of formation of the recent local fauna. There are two possibilities: it might be the result of either the crossing of the existing barriers by ancestral forms or their isolation after the formation of the barriers. These two choices correspond to postulates of either vicariance (Croizat, 1958; Foin, 1976; Rosen, 1976, 1978; Nelson and Platnick, 1981; Kohn, 1983; Nelson, 1984) or dispersal (Darlington, 1957; MacArthur and Wilson, 1967; Briggs, 1982, 1984, 1995) paradigms. In the first case the western Pacific relationships are explained as a more successful penetration of the western barrier than the eastern one. In the second case these relationships could be interpreted as resulting from a later development of the western barrier.

Hubbs (1959) suggested "stepping-stone dispersal" to account for the presence of Indo-West Pacific animals in the fauna of Shoal Guyot (Bolshaya Seamount) with numerous islands and underwater rises in the southern Pacific supporting the route of dispersal. Scheltema (1986, 1988) has shown that the planktonic larvae of many shallow-water benthic invertebrates are successfully transported between the islands of the central Pacific. The eastward direction of successful dispersal across the east Pacific Barrier has been explained by Briggs (1974) as follows: "successful (colonizing) migrations across the zoogeographic boundaries that delimit the Indo-West Pacific can apparently take place in one direction only, outward into areas where the fauna is poorer and the competition is less" (p. 114). Mironov (1985b, 1989), in accordance with Briggs, has suggested the following regularity: the higher is the species richness of a subcontinental fauna the further in the open ocean does the area of its distribution extend. Mironov and Detinova (1990) saw the corroboration of this rule in the affinity between the faunas of Sala y Gómez–Nazca ridges and the Indo-West Pacific. Springer (1982) attempted to link the distributions and relationships of the Pacific islands' shore fauna with the geotectonics of the Pacific Plate and explained the distinctions and endemism of southern Polynesia, including Easter Is, by vicariant events which divided ancestral populations and allowed these to diverge. This scenario was adopted by Parin (1991a) to explain the composition of the fish fauna of the Sala y Gómez and Nazca ridges. Newman and Foster (1983) suggest that the Easter and Sala y Gómez Is inherited their endemics from ancient islands, later submerged and now being part of the seamounts of the Nazca and Sala y Gómez ridges. However, Sysoev (1990) wrote "even if the propounded hypothesis is correct, in the modern fauna of the guyot summits not a vestige remained of the relict mollusc fauna of island shores, even on the generic level" (p. 260).

In our opinion the western Pacific affinities of the Sala y Gómez–Nazca ridge fauna are the result, in the first place, of more successful penetration of the western barrier. The dispersion of western fauna into the area of the ridges proceeded mostly from the north-west and west towards the east, from areas with high biotic pressure into those where pressure is low (Briggs, 1974, 1984). The stepping stones (*sensu* Hubbs, 1959) on the way were probably chains of islands and seamounts, including Cook Is, Austral (Tubuai) Is, Pitcairn Is, Henderson Is, Ducie Is, Easter Is and Sala y Gómez Is. The longest "mountainless" section of this chain is located between Ducie Is and Easter Is, on either side of the East Pacific Rise. The length of this section is approximately 840 nautical miles, or 1550 km, but it is in the part of the ocean with the world maximum spreading rate. Assuming a constant spreading rate of 180 mm year^{-1} (Krasnov, 1995) we find that in the Upper Pliocene, approximately 2.0–2.5 million years BP when such islands as Pitcairn, Henderson, Easter, Sala y Gómez, San Félix, San Ambrosio and others arose (some are younger, of Pleistocene age) (Newman and Foster, 1983; Foster and Newman, 1987; Pandolfi, 1995), this section was shorter than now by some 360–450 km, or by almost a quarter or a third of its modern extent. Even the smaller spreading rate cited by Geistdoerfer *et al.* (1995) (146 mm year^{-1}) is high enough to indicate that this section was shorter by 290–365 km in the Upper Pliocene than today.

The small number of species of eastern origin on seamounts of the Nazca Ridge may be explained by the comparatively low species richness (vs. the Indo-West Pacific) in the areas off the south American coast adjacent to the Nazca Ridge.

The rather substantial number of species of southern warm-temperate and subtropical origin in common with the fauna of southern and south-eastern Australia, New Zealand and the southernmost Polynesian islands may be explained by palaeo-oceanological features. At present the Sala y Gómez Ridge is located north of and not far from the northern periphery of the eastward-going South Pacific Current (Stepanov, 1974). During the cool epochs of the Pleistocene both the northern and southern subtropical anticyclonic circulations were narrower and shifted towards the equator, therefore the Sala y Gómez Ridge would have been in the middle of the South Pacific Current. This current may have carried the southern species' dispersal stages eastwards as it was shown for the West Wind Drift (Fell, 1962). The islands and seamounts lying on the southern periphery of Polynesia were possible stepping stones on the way from Australia–New Zealand to the Sala y Gómez Ridge, also located in the South Pacific Current (Galerkin *et al.*, 1982). Thus, the dispersal of southern warm-temperate and subtropical fishes and invertebrates would have been much easier then than now.

Dispersion of tropical elements of the fauna, at first glance, would not have followed the aforesaid scenario. However, the known distributions demonstrate that most of the Indo-West Pacific species of fishes common to Sala y Gómez–Nazca ridges and other localities in the Pacific Ocean are not really widespread but are found only in peripheral tropical areas, such as Hawaii, southern Japan, New Caledonia and south-eastern Australia (see Table 14) or they are represented there by the most closely related forms. It seems possible to hypothesize that these common species (or their ancestors) originated in the Indo-Malayan faunistic centre and spread from there first north (to Japan) and south (to south-eastern Australia) and then eastward (to Hawaii and Sala y Gómez Ridge) having been replaced by more successful competitors in some of the areas where they had become extinct. Their dispersal may, therefore, have followed almost the same route as the subtropical and warm-temperate animals but at the time of the extreme narrowing of the anticyclonic gyres.

Thus, the taxonomic composition of the Sala y Gómez– Nazca ridge fauna has been determined mostly by two processes; that is, dispersal of western Pacific invertebrates and fishes and active speciation *in situ*, as described by the classical process of speciation (Mayr, 1963), resulting in evolution of many endemic species. This hypothesis seems to be the most probable and allows for the faunal divergence on both sides of biogeographic barriers and the dispersal of the fauna across barriers.

11. CONCLUSIONS

The fauna of the submarine rises in the eastern South Pacific has remained almost unstudied until quite recently and Russian expeditions of 1973–1987 aboard R/V "Ikhtiandr", "Professor Mesyatzev", "Professor Shtokman", etc. discovered a practically unknown assemblage of both invertebrates and fishes at 22 seamounts investigated on the Nazca and Sala y Gómez ridges. As a result of the expeditions 177 genera and 192 species of benthopelagic and benthic invertebrates (including animals identified only to the genus) and 128 genera and 171 species of fishes were identified; seven genera and 150 species have been described as new to science: four and 74 among invertebrates, three and 76 among fishes.

Bottom invertebrate communities of the seamount summits are dominated by a few species. At depths less than 400 m the spiny lobster *Projasus bahamondei* and sea urchins predominate, eastward and westward respectively of approximately 83°W, whereas in greater depths sponges, gorgonarians, starfishes or shrimps dominate in various combinations. In benthopelagic fish communities the horse mackerel *Trachurus symmetricus murphyi* (a temporary visitor from Chilean waters), usually predominates above the seamounts, eastward of 85°W, other abundant species being *Emmelichthys cyanescens*, *E. elongatus*, *Decapterus muroadsi*, *Zenopsis oblongus*, *Epigonus elegans* and *Pentaceros quinquespinis* (all of them sometimes provide commercial catches). Among the bottom fishes *Caelorinchus immaculatus* (Nazca Ridge) and *Pterygotrigla picta* are predominant. The highest diversity of fishes is observed at depths shallower than 500–600 m, where the communities of soft-bottom and rocky biotopes are significantly different.

The fauna of benthic and benthopelagic invertebrates and fishes of the area is much more closely related to the Indo-West Pacific than to the eastern Pacific fauna and is characterized by very high degree of endemism at species level (51% among identified bottom invertebrates, 44% among fishes). The authors share the opinion that the area of Sala y Gómez Ridge and its transitional zone with the Nazca Ridge westward of 83–84°W should be considered as a separate province of the Indo-West Pacific Region. However, they have different opinions concerning the Nazca Ridge proper: N.V. Parin and K.N. Nesis include it in the same province (as its faunistically impoverished portion), A.N. Mironov adheres to his earlier opinion (Mironov and Detinova, 1990) considering it as a part of the Eastern Pacific subcontinental Region.

The composition of the fauna of the Sala y Gómez–Nazca ridges may be explained by two main processes: eastward dispersal of the western Pacific fauna across a biogeographic barrier (the mountainless area) and active speciation *in situ*.

ACKNOWLEDGEMENTS

The authors are thankful to N.V. Kucheruk and P.V. Rybnikov for great help in statistical calculations and preparation of figures.

REFERENCES

Allison, E.C., Durham, J.W. and Mintz, L.W. (1967). New southeast Pacific echinoids. *Occasional Papers of the California Academy of Sciences* **62**, 1–23.

Amaoka, K. and Parin, N.V. (1990). A new flounder, *Chascanopsetta megagnatha*, from the Sala y Gómez submarine ridge, eastern Pacific Ocean (Teleostei: Pleuronectiformes: Bothidae). *Copeia* **3**, 717–722.

Amaoka, K., Hoshino, K. and Parin, N.V. (1997). Description of a juvenile specimen of a rarely-caught, deep-sea species of *Samariscus* (Pleuronectiformes Samaridae) from Sala y Gómez Submarine Ridge, Eastern Pacific Ocean. *Ichthyological Research* **44**(1), 92–96.

Amarov, G.L. and Korostin, N.N. (1981). Some peculiarities of the water dynamics in the area of the underwater Nazca Ridge. TsNIITEIRKh (Moscow). *Express-Information, Series: Fishery Exploitation of the World Ocean Resources* **3**, 14–18. (In Russian).

Amarov, G.L. and Kuznetsov, A.N. (1986). Oceanological variability and its influence on the formation of fish concentrations in the Southeastern Pacific Ocean. TsNIITEIRKh (Moscow). *Express-Information, Series: Fishery Exploitation of the World Ocean Resources* **2**, 1–76. (In Russian).

Anderson, W.D. and Johnson, G.D. (1984). A new species of *Callanthias* (Pisces: Perciformes: Percoidei: Callanthiidae) from the southeastern Pacific Ocean. *Proceedings of the Biological Society of Washington* **97**(4), 942–950.

Anderson, W.D. and Randall, J.E. (1991). A new species of the anthiinae genus *Plectranthias* (Pisces: Serranidae) from the Sala y Gómez Ridge in the eastern South Pacific, with comments on *P. exsul. Proceedings of the Biological Society of Washington* **104**(2), 335–343.

Anderson, W.D., Parin, N.V. and Randall, J.E. (1990). A new genus and species of anthiine serranid (Pisces: Serranidae) from the eastern South Pacific with comments on anthiine relationships. *Proceedings of the Biological Society of Washington* **103**(4) 922–930.

Andrew, T.G.. Hecht, T., Heemstra, P.C. and Lutjeharms, J.R.E. (1995). Fishes of the Tristan da Cunha Group and Gough Island, South Atlantic Ocean. *Ichthyological Bulletin of the J.L.B. Smith Institute of Ichthyology*, Grahamstown, S.A., No. 63, 1–43.

Andriashev, A.P. (1979). On some problems of the vertical zonality of marine bottom fauna. *In* "Biological Resources of the World Ocean" (S.A. Studenetsky, ed.), pp 117–138. Nauka, Moscow. (In Russian).

Balss, H. (1924). Decapoden von Juan Fernandez. The natural history of Juan Fernandez and Easter Islands. *Fieldiana, Zoology* **3**(38), 1–71.

Barton, N.H. (1990). Speciation. *In* "Analytical Biogeography" (A.A. Myers and P.S. Giller, eds), pp. 185– 218. Chapman & Hall, London.

Becker, V.E. (1992). Benthopelagic fishes of the genera *Idiolychnus* and *Diaphus* (Myctophidae) from southeastern part of the Pacific Ocean with descriptions of two new species. *Voprosy Ikhtiologii* **32**(6), 3–10. (In Russian).

Belyanina, T.N. (1989). Ichthyoplankton of the regions of the Nazca and Sala y Gómez submarine ridges. *Voprosy Ikhtiologiii* **29**(5), 777–782. (In Russian; English translation *Journal of Ichthyology* **29**(7), 84–90).

Belyanina, T.N. (1990). Larvae and juveniles of poorly known bottom-dwelling and epibenthic fishes from the Nazca and Sala y Gómez Ridges. *Voprosy Ikhtiologii* **30**(6), 974–982. (In Russian; English translation *Journal of Ichthyology* **30**(8), 1–11).

Berman, I.S. (1969) Hydrology of the Peru Current. *Mirovoye Rybolovstvo* **11**, 22–36. (In Russian).

Berrisford, C.D. (1969). Biology and zoogeography of the Vema Seamount: a report of the first biological collection made on the seamount. *Transactions of the Royal Society of South Africa* **38**(4), 387–398.

Boehlert, G.W. (1998). Current–topography interactions at mid-ocean seamounts and the impact on pelagic ecosystems. *GeoJournal* **16**(1), 45–52.

Boehlert, G.W. and Genin, A. (1987). A review of the effects of seamounts on biological processes. *In* "Seamounts, Islands and Atolls" (B. Keating, P. Fryer, R. Batiza and G. Boehlert, eds), pp. 319–334. *Geophysical Monograph* **43**. American Geophysical Union, Washington.

Boldyrev, V.Z. (1986). Patterns of distribution of ichthyofauna of the seamounts in the Southeastern Pacific Ocean. *In* "Biological Resources of the Pacific Ocean" (M.E. Vinogradov, N.V. Parin and V.P. Shuntov, eds), pp. 520–527. Nauka, Moscow. (In Russian).

Boldyrev, V.Z., Darnitsky, V.B. and Kulikov, M.Yu. (1987). Forming up of the biological productivity in the areas of rises on the oceanic bottom. *In* "Biological Resources of the Open Ocean" (P.A. Moiseev, N.V. Parin and A.A. Elizarov, eds), pp. 31–64. Nauka, Moscow. (In Russian).

Borets, L.A. (1986). Ichthyofauna of the north-western and Hawaiian submarine ridges. *Voprosy Ikhtiologii* **26**(2), 208–220. (In Russian, English translation *Journal of Ichthyology* **26**(3), 1–13).

Borets, L.A. and Kulikov, M.Yu. (1986). Thalassobathyal. *In* "Biological Resources of the Pacific Ocean" (M.E. Vinogradov, N.V. Parin and V.P. Shuntov, eds), pp. 505– 520. Nauka, Moscow. (In Russian).

Borodulina, O.D. (1981). New species of the genus *Polyipnus – P. inermis* Borodulina (Sternoptychidae) from the southeastern Pacific. *Voprosy Ikhtiologii* **21**(3), 556–559. (In Russian).

Briggs, J.C. (1974). "Marine Zoogeography". McGraw-Hill, New York.

Briggs, J.C. (1982). Center of origin or vicariance? *In* "Fauna and Hydrobiology of the Shelf Zones of the Pacific Ocean" (O.G. Kussakin and A.I. Kafanov, eds). *Proceedings of the XIV Pacific Science Congress* (Khabarovsk, August 1979), Section "Marine Biology", N4, 17–28. Far Eastern Scientific Center, Acad. Sci. USSR, Vladivostok.

Briggs, J.C. (1984). "Centres of Origin in Biogeography". *Biogeographical Monographs* **1**. Biography Study Group, Leeds.

Briggs, J.C. (1995). "Global Biogeography". Elsevier, Amsterdam.

Burkov, V.A. (1972). "General Circulation of the Pacific Ocean". Nauka, Moscow. (In Russian).

Burkov, V.A. (1980). "General Circulation of the World Ocean". Hydrometeoizdat, Leningrad. (In Russian; English translation (1993) Amerind Publishing, New Delhi).

Burukovsky, R.N. (1990). Shrimps from the underwater Sala y Gómez and Nazca Ridges. *Trudy Instituta Okeanologii AN USSR* **124**, 187–217. (In Russian, English summary).

Chirino-Galvez, L.A. and Manning, R.B. (1989). A new deep-sea crab of the genus *Chaceon* from Chile (Crustacea, Decapoda, Geryonidae). *Proceedings of the Biological Society of Washington* **102**(2), 401–404.

Codoceo, R. (1976). Asteroidea, Echinoidea, and Holothurioidea of the Desventuradas and Juan Fernández Islands off Chile with new records for the last archipelago. *Thalassia Jugoslavica* **12**(1), 87–98. (Published 1978).

Cox, C.B. and Moore, P.D. (1993) "Biogeography". Blackwell Scientific Publications, Oxford.

Croizat, L. (1958). "Panbiogeography, or an Introductory Synthesis of Zoogeography, Phytogeography, and Geology; with Notes on Evolution, Systematics, Ecology, Anthropology etc." 3 vol. Caracas, L. Croizat.

Darlington, P.J. (1957). "Zoogeography: The Geographical Distribution of Animals". Wiley, New York.

Darnitsky, V.B. (1979). On baroclinic perturbations of synoptic scale in the areas of seamounts of the Southern Ocean and Tasman Sea. *Issledovaniya po Biologii Ryb i Promyslovoi Okeanografii* **10**, 14–25. TINRO, Vladivostok. (In Russian).

Darnitsky, V.B., Boldyrev, V.Z. and Pavlychev, V.P. (1986). Some patterns of forming up of the fishery productive zones and fish distribution in the areas of seamounts of the Pacific Ocean. *In* "Biological Resources of the Pacific Ocean" (M.E. Vinogradov, N.V. Parin and V.P. Shuntov, eds), 49–61. Nauka, Moscow. (In Russian).

Dolganov, V.N. (1984). A new shark of the family Squalidae, caught on the Nazca underwater ridge. *Zoologicheskii Zhurnal* **63**(10), 1589–1591. (In Russian, English summary).

Ekman, S. (1953). "Zoogeography of the Sea". Sidgwick & Jackson, London.

Evseenko, S.A. and Gorbunova, N.N. (1984). Ichthyoplankton of the area of the Nazca Ridge and of eastern part of the Subantarctic Frontal Zone. *In* "Frontal Zones of the Southeastern Pacific Ocean (Biology, Physics, Chemistry)" (M.E. Vinogradov and K.N. Fedorov, eds), pp. 303–314. Nauka, Moscow. (In Russian).

Fedorov, V.V. (1985). Morphosculpture of guyots of the Nazca Ridge. *Geomorphologiya* **3**, 62–69. (In Russian, English summary).

Fedorov, V.V. (1987). Landscapes of seamounts. *In* "III Congress of Soviet Oceanologists. Abstracts of Communications. Section: Biology of the Ocean" **3**, 131–132. Hydrometeoizdat, Leningrad. (In Russian).

Fedorov, V.V. (1990). Spiny lobsters distribution on the South Pacific Ocean seamounts. *In* All-USSR Conference "Reserve Food Biological Resources of the Open Ocean and USSR seas", Kaliningrad. *Abstracts of communications*, 163–164. (In Russian).

Fedorov, V.V. and Chistikov, S.D. (1985). Landscapes of seamounts as indicators of biological productivity of surrounding waters. *In* "Biological Bases of the Developing Fisheries of the World Ocean" (M.E. Vinogradov and M.V. Flint, eds), pp. 220–230. Nauka, Moscow (In Russian).

Fedorov, V.V. and Ivanov, V.E. (1981). New data on the morphosculpture of underwater Nazca Ridge. TsNIITEIRKh (Moscow). *Express-Information, Series: Fishery Exploitation of the World Ocean Resources* **11**, 18–21. (In Russian).

Fedorov, V.V. and Ivanov, V.E. (1984). Geomorphology of the seamounts of the Nazca Ridge. *In* "Fisheries-oceanographic Studies of the Productive Areas in the Seas and Oceans (D.E. Gershanovich, ed.), pp. 178–189. VNIRO, Moscow. (In Russian).

Fedorov, V.V. (1982) *Methodical Recommendations to Conduct Marine Landscape Investigations for Fishery Purposes*. VNIRO, Moscow. (In Russian).

Fell, H.B. (1962). West-wind-drift dispersal of echinoderms in the southern hemisphere. *Nature* **193** (4817), 759–761.

Fell, F.J. (1974). The echinoids of Easter Island (Rapa Nui). *Pacific Science* **28**(2), 147–158.

Fell, F.J. (1975). The echinoid genus *Centrostephanus* in the South Pacific Ocean with a description of a new species. *Journal of the Royal Society of New Zealand* **5**(2), 179–193.

Foin, T.C. (1976). Plate tectonics and the biogeography of the Cypraeidae (Mollusca: Gastropoda). *Journal of Biogeography* **3**(1) 19–34.

Foster, B.A. and Newman, W.A. (1987). Chthamalid barnacles of Easter Island, peripheral Pacific isolation of Notochthamalidae new subfamily and *hembeli*-group of Euraphiinae (Cirripedia: Chthamaloidea). *Bulletin of Marine Science* **41**(2), 322–336.

Fricke, R. (1992). Revision of the family Draconettidae (Teleostei), with description of two new species and a new subspecies. *Journal of Natural History* **26**, 165–195.

Galerkin, L.I., Barash, M.S., Sapozhnikov, V.V. and Pasternak, F.A. (1982). "Pacific Ocean". Mysl', Moscow. (In Russian).

Galkin, S.V. (1993). Starfishes distribution on the Nazca and Sala y Gómez Ridges. *In* "Feeding of Marine Invertebrates in Different Vertical Zones and Latitudes" (A.P. Kuznetsov and M.N. Sokolova, eds), pp. 101-111. P.P. Shirshov Institute of Oceanology, Russian Academy of Sciences. (In Russian, English summary).

Gar'kusha, A.V. and Kotelnikov, B.B. (1981). "Hydrological Regime of the Peruvian Fishery Subarea of the SEPO". Zapryba, Kaliningrad. (In Russian).

Garth, J.S. (1957). The Crustacea Decapoda Brachyura of Chile. Reports of the Lund University Chile Expedition 1948–49, 29. *Lunds Universitets Aarsskrift, N.F., Avd. 2* **53**(7), 1–130.

Garth, J.S. (1992). Some deep-water Parthenopidae (Crustacea, Brachyura) from French Polynesia and nearby Eastern Pacific ridges and seamounts. *Bulletin du Muséum National d'Histoire Naturelle* **14**A(3–4), 781–795.

Geistdoerfer, P., Auzende, J.-M., Batiza, R., Bideau, D., Cormier, M.-H., Fouquet, Y., Lagabrielle, Y., Sinton, J. and Spadea, P. (1995). Hydrothermalisme et communnautes animales associées sur la dorsale du Pacifique oriental entre 17°S et 19°S (campagne Naudur, décembre 1993). *Comptes Rendus de l'Académie des Sciences, Paris*, ser. IIa **320**(1), 47–54.

Genin, A., Dayton, P.K., Lonsdale, P.F. and Speiss, F.N. (1986). Corals on seamounts peaks provide evidence of current acceleration over deep-sea topography. *Nature* **322**, 59–61.

Genin, A., Haury, L. and Greenblatt, P. (1988). Interaction of migrating zooplankton with shallow topography: predation by rockfishes and identification of patchiness. *Deep-Sea Research* **35**(2), 151–175.

George, R.W. (1976). A new species of spiny lobster, *Projasus bahamondei* (Palinuridae "silentes"), from the South East Pacific region. *Crustaceana* **30**(1), 27–32.

Gevorkyan, V.H., Golovan', G.A., Kudryavtsev, A.M., Martynenko, S.V., Korostin, N.N., Parin, N.V. and Kolodnitsky, B.V. (1986). "Oceanographic Conditions and Biological

Productivity of Southeastern Pacific (the Nazca Ridge)". Institute of Geological Sciences of the Ukrainian Academy of Sciences, Preprint 86–13, 35 pp. (In Russian).

Gevorkyan, V.H., Golovan', G.A., Ivanov, V.E., Lomakin, I.E. and Vakaruyk, V.T. (1987). "Geological Structure and Conditions of Sedimentation in the Area of the Nazca Ridge (Pacific Ocean)". Institute of Geological Sciences of the Ukrainian Academy of Sciences, Preprint 87–16, 39 pp. (In Russian).

Golovan', G.A. and Pakhorukov, N.P. (1986). On new records of rare species of cartilaginous fishes. *Voprosy Ikhtiologii* **26**(1), 153–155. (In Russian; English translation, *Journal of Ichthology* **26**(2) 117–120).

Golovan', G.A. and Pakhorukov, N.P. (1987). Distribution and behaviour of fishes on the Nazca and Sala y Gómez underwater ridges. *Voprosy Ikhtiologii* **27**(3), 69–76. (In Russian).

Golovan', G.A., Pakhorukov, N.P. and Levin, A.B. (1982). Some peculiarities of the biogeocoenose on one seamount in the Southeastern Pacific. *In* "II All-USSR Congress of Oceanographers. Abstracts of Communications. Biology of the Ocean", 6, 46–47. (In Russian).

Grasshoff, M. (1972). Die Gorgonaria des östlichen Nordatlantik und des Mittelmeeres. I. Die Familie Ellisellidae (Cnidaria: Anthozoa). *"Meteor" Forschungsergebnisse* **D10**, 73–87.

Griffin, D.J.G. and Stoddart, H.E. (1995). Deep-water decapod Crustacea from Eastern Australia: lobsters of the families Nephropidae, Palinuridae, Polychelidae and Scyllaridae. *Records of the Australian Museum* **47**(3), 231–263.

Grigg, R.W. (1984). Resource management of precious corals: a review and application to shallow water reef building corals. *Marine Ecology* **5**(1), 57–74.

Grigg, R.W., Malahoff, A., Chave, E.H. and Landahl, J. (1987). Seamount benthic ecology and potential environmental impact from manganese crust mining in Hawaii. *In* "Seamounts, Islands and Atolls" (B. Keating, P. Fryer, R. Batiza and G. Boehlert, eds), 379–390. *Geophysical Monograph* **43**. Americlan Geophysical Union, Washington.

Grossman, N.S. (1978). Data about plankton in the areas of underwater rises of the Southern Pacific Ocean. *Issledovaniya po Biologii Ryb i Promyslovoi Okeanografii* **9**, 41–48. TINRO, Vladivostok. (In Russian).

Heemstra, P.C. and Anderson, W.D. (1983). A new species of the serranid fish genus *Plectranthias* (Pisces: Perciformes) from the southeastern Pacific Ocean, with comments on the genus *Ellerkeldia*. *Proceedings of the Biological Society of Washington* **96**(4), 632–637.

Heinrich, A.K. (1990). Calanoid copepods of the plankton collections over the submarime rises of the Nazca Ridge. *Trudy Instituta Okeanologii AN SSSR* **124**, 15–26. (In Russian, English summary).

Heinrich, A.K., Parin, N.V., Rudjakov, Yu.A. and Sazhin, A.F. (1993). Population of the near-bottom layer of the ocean. *Trudy Instituta Okeanologii RAN* **128**, 6–25. (In Russian, English summary).

Hensley, D.A. and Suzumoto, A.Y. (1990). Bothids of Easter Island, with the description of a new species of *Engyprosopon*. *Copeia* **1**, 130–137.

Holthuis, L.B. (1991). Marine lobsters of the World. *FAO Fisheries Synopsis* **13**(125) 1–292.

Hubbs, C.L. (1959). Initial discoveries of fish fauna on seamounts and offshore banks in the Eastern Pacific. *Pacific Science* **13**, 311–316.

Isaacs, J.D. and Schwartzlose, R.A. (1965). Migrant sound scatterers: Interaction with the sea floor. *Science* **150**(3705), 1810–1813.

Ivanov, V.E. (1988). "Geological Structure, Relief and Conditions of Sedimentation in the Area of the Nazca Ridge". Autoreferat Diss. Cand. Geol.-Mineralog. Sciences, Institute of Geological Sciences, Academy of Sciences of the Ukrainian SSR, Kiev. (In Russian).

Karmovskaya, E.S. (1990). New species of the conger eels (Congridae) from the submarine ridges of Southeastern Pacific. *Voprosy Ikhtiologii* **30**(5), 764–772. (In Russian).

Kashkin, N.I. (1984). Mesopelagic micronekton as a factor of fish productivity on the oceanic banks. *In* "Frontal Zones of the Southeastern Pacific Ocean (Biology, Physics, Chemistry)" (M.E. Vinogradov and K.N. Fedorov, eds), 285–291. Nauka, Moscow (In Russian).

Kashkin, N.I. and Kashkina, A.A. (1984). Macroplankton biomass in the Southeastern Pacific Ocean. *In* "Frontal Zones of the Southeastern Pacific Ocean (Biology, Physics, Chemistry)" (M.E. Vinogradov and K.N. Fedorov, eds), 277–285. Nauka, Moscow. (In Russian).

Kaufmann, R.S., Wakefield, W.W. and Genin, A. (1989). Distribution of epibenthic megafauna and lebensspuren on two central North Pacific seamounts. *Deep-Sea Research* **36**(12), 1863–1896.

Keating, B.H., Fryer, P., Batiza, R. and Boehlert, G.W. (eds) (1987). "Seamounts, Islands and Atolls". *Geophysical Monograph* **43**. American Geophysical Union, Washington.

Kohn, A.J. (1983). Marine biogeography and evolution in the tropical Pacific: zoological perspectives. *Bulletin of Marine Science* **33**(3), 528–535.

Kolodnitsky, B.V. and Kudryavtsev, A.M. (1982). Some peculiarities of the high bioproductivity zones over the guyots of the Nazca Ridge. *In* "II All-USSR Congress of Oceanographers. Abstracts of Communications. Biology of the Ocean", **6**, 41–42. (In Russian).

Konchina, Yu.V. (1990). Feeding ecology of pseudo-neritic fishes over the Nazca Ridge. *Voprosy Ikhtiologii* **30**(6), 983–993. (In Russian).

Konchina, Yu.V. (1992). Trophic status of Peruvian pseudo-neritic fishes in the oceanic epipelagic zone. *Voprosy Ikhtiologii* **32**(3), 67–82. (In Russian).

Kotlyar, A.N. (1982a). *Polymixia yuri*, sp.n. (Beryciforms, Polymixiidae) from the southeastern Pacific Ocean. *Zoologicheskii Zhurnal* **61**(9), 1380–1384 (In Russian, English summary).

Kotlyar, A.N. (1982b). A new species of *Emmelichthys* (Emmelichthyidae, Perciformes) from the southeastern Pacific Ocean. *Byulleten Moskovskogo Obshchestva Ispytatelei Prirody, Otdel Biologii* **87**(1), 48–52. (In Russian).

Kotlyar, A.N. (1988a). Materials on morphometry and patterns of colouration of *Helicolenus lengerichi* Norman (Scorpaenidae). *Voprosy Ikhtiologii* **28**(2), 325–329. (In Russian).

Kotlyar, A.N. (1988b). Materials on systematics and biology of *Monocentris reedi* Schultz (Monocentridae) and *Polymixia yuri* Kotlyar (Polymixiidae). *Voprosy Ikhtiologii* **28**(5), 853–856. (In Russian).

Kotlyar, A.N. (1990). Dogfish sharks of the genus *Etmopterus* Rafinesque from the Nazca and Sala y Gómez Ridges. *Trudy Instituta Okeanologii AN USSR* **125**, 127–147. (In Russian, English summary).

Kotlyar, A.N. (1991). A new species of the genus *Polymixia* from the Sala y Gómez underwater ridge. *Zoologicheskii Zhurnal* **70**(7), 83–86. (In Russian, English summary).

Kotlyar, A.N. and Parin, N.V. (1986). Two new species of *Chlorophthalmus* (Osteichthyes, Myctophiformes, Chlorophthalmidae) from submarine mountain ridges in the southeastern part of the Pacific Ocean. *Zoologicheskii Zhurnal* **65**(3), 369–377. (In Russian, English summary).

Kotlyar, A.N. and Parin, N.N. (1990). Otoliths and age of bottom and near-bottom fishes of the Nazca and Sala y Gómez submarine ridges. *Trudy Instituta Okeanologii AN USSR*

125, 97–126. (In Russian, English summary).

Krasnov, S.G. (1995). Huge sulphide deposits in the Ocean. *Priroda* **No. 2**(954), 3–14. (In Russian).

Krylova, H.M. (1991). A new genus and two new species of bivalve molluscs of the family Cetoconchidae (Bivalvia, Septibranchia, Poromyoidea). *Zoologicheskii Zhurnal* **70**(7), 132–136. (In Russian, English summary).

Krylova, H.M. (1994). Bivalve molluscs of the genus *Rhinoclama* (Septibranchia, Cuspidariidae). *Trudy Instituta Okeanologii RAN* **129**, 55–64. (In Russian, English summary).

Krylova, H.M. (1995). Clams of the family Protocuspidariidae (Septibranchia, Cuspidarioidea): taxonomy and distribution. *Zoologicheskii Zhurnal* **74**(9), 20–38.

Kudinova-Pasternak, R.K. (1990). Tanaidacea (Crustacea, Malacostraca) from the underwater Nazca Ridge. *Zoologicheskii Zhurnal* **69**(12), 135–140. (In Russian, English summary).

Kudryavtsev, A.M. (1981). Water masses and water structure in the area of the Nazca and Sala y Gómez Ridges, Pacific Ocean. *In* "Biological Resources of Large Depths and Pelagic Realm of the Open areas of the World Ocean. Abstracts of communications of the scientific-practical conference", Murmansk, 46–47. (In Russian).

Kuznetsov, A.P. (ed.) (1985). "Bottom fauna of the Open Oceanic Elevations (Northern Atlantic)". *Trudy Instituta Okeanologii AN USSR* **120**. (In Russian, English summaries).

Kuznetsov, A.P. and Mironov, A.N. (eds) (1981). "Benthos of the Submarine Mountains Marcus-Necker and adjacent Pacific regions". P.P. Shirshov Institute of Oceanology, USSR Academy of Sciences, Moscow. (In Russian, English summaries).

Kuznetsov, A.P. and Mironov, A.N. (eds) (1994). "Bottom Fauna of Seamounts". *Trudy Instituta Okeanologii RAN* **129**. (In Russian, English summaries).

Le Danois, Y. (1984). Description d'une nouvelle espece de Chaunacidae, *Chaunax latipunctatus*, des îles Galapagos. *Cybium* **8**(2), 95–101.

Levin, L.A., Huggett, C.L., Wishner, K. (1991). Control of deep-sea benthic community structure by oxygen and organic matter gradients in the Eastern Pacific Ocean. *Journal of Marine Research* **49**(4), 763– 800.

Littler, M.M., Littler D.S, Blair, S.M. and Norris, J.N. (1985). Deepest known plant life discovered on an uncharted seamount. *Science* **227**, 57–59.

Litvinov, F.F. (1990). Ecological characteristics of dogfish, *Squalus mitsukurii*, from the Nazca and Sala y Gómez underwater ridges. *Voprosy Ikhtiologii* **30**(4), 577–586. (In Russian).

Litvinova, N.M. (1992). Revision of the genus *Ophiotholia* (Echinodermata, Ophiuroidea). *Zoologicheskii Zhurnal* **71**(3), 47–57. (In Russian, English summary).

Lutjeharms, J.R.E. and Heydorn, A.E.F. (191). The rock-lobster *Jasus tristani* on Vema Seamount: drifting buoys suggest a possible recruiting mechanism. *Deep-Sea Research* **28A**(6), 631–636.

MacArthur, R.H and Wilson, E.O. (1967). "The Theory of Island Biogeography". Princeton, University Press, Princeton, NJ.

Mammerickx, J., Anderson, R.N., Menard, H.W. and Smith, S.M. (1975). Morphology and tectonic evolution of the East-Central Pacific. *Bulletin of the Geological Society of America* **86**(1), 111–117, map.

Mandritsa, S.A. (1992). New species and new records of the fishes from the genera *Phenacoscorpius* and *Plectrogenium* (Scorpaenidae) in the Pacific, Atlantic and Indian oceans. *Voprosy Ikhtiologii* **32**(4), 10–17. (In Russian).

Markle, D.F. and Olney, J.E. (1990). Systematics of the pearlfishes (Pisces: Carapidae). *Bulletin of Marine Science* **47**(2), 269–410.

Markov, A.V. (1988). Sea urchins of the family Saleniidae: composition and distribution. *Zoologicheskii Zhurnal* **67**(4), 567–575. (In Russian, English summary).

Markov, A.V. (1989). Species composition of the genus *Trigonocidaris* (Echinoidea). *Zoologicheskii Zhurnal* **68**(8), 75–84. (In Russian, English summary).

Mayr, E. (1963). "Animal Species and Evolution". Harvard University Press, Cambridge, MA.

McCosker, J.E. and Parin, N.V. (1995). A new species of deepwater worm-eel, *Muraenichthys profundorum* (Anguilliformes: Ophichthyidae), from the Nazca Ridge. *Japanese Journal of Ichthyology* **42**(3/4), 231–235.

Menard, H.W. (1964). "Marine Geology of the Pacific". McGraw-Hill, NY. (Russian transl. from the English, Mir Publishers, Moscow, 1966).

Mirkin, B.M., Rozenberg, G.S. and Naumova, L.G. (1989). "A Dictionary of Notions and Terms of Modern Phytocoenology". Nauka, Moscow. (In Russian).

Mironov, A.N. (1981). Sea urchins (Echinoidea). *In* "Benthos of the Submarine Mountains Marcus-Necker and Adjacent Pacific Regions" (A.P. Kuznetsov and A.N. Mironov, eds), pp. 131–140. P.P. Shirshov Institute of Oceanology, USSR Academy of Sciences, Moscow. (In Russian, English summary).

Mironov, A.N. (1985a). Sea urchins as potential indicators of the benthopelagic fish concentrations over the summits of oceanic underwater rises. *In* "Biological Bases of the Developing Fisheries of the World Ocean" (M.E. Vinogradov and M.V. Flint, eds), pp. 231–236. Nauka, Moscow. (In Russian).

Mironov, A.N. (1985b). Circumcontinental zonation in distribution of Atlantic echinoids. *Trudy Instituta Okeanologii AN USSR* **120**, 70–95. (In Russian).

Mironov, A.N. (1989). Meridional asymmetry and margial effect in the distribution of sea urchins in open ocean regions. *Okeanologiya* **29**(5), 845–854. (In Russian, English summary).

Mironov, A.N. (1990). Faunistic approach to the study of recent ecosystems. *Okeanologiya* **30**(6), 1006–1012. (In Russian, English summary).

Mironov, A.N. (1994). Bottom faunistic complexes from oceanic islands and seamounts. *Trudy Instituta Okeanologii RAN* **129**, 7–16. (In Russian, English summary).

Mironov, A.N. and Detinova, N.N. (1990). Bottom fauna of the Nazca and Sala y Gómez Ridges. *Trudy Instituta Okeanologii AN USSR* **124**, 269–278. (In Russian, English summary).

Mironov, A.N. and Pasternak F.A. (1981). Composition and distribution of the bottom fauna. *In* "Benthos of the Submarine Mountains Marcus-Necker and Adjacent Pacific Regions" (A.P. Kuznetsov and A.N. Mironov, eds), 10–28. P.P. Shirshov Institute of Oceanology Acad. Sci. USSR, Moscow. (In Russian, English summary).

Mironov, A.N. and Rudjakov, Yu.A. (eds) (1990). "Plankton and Benthos of the Nazca and Sala y Gómez Submarine Ridges". *Trudy Instituta Okeanologii AN USSR* **124**. (In Russian, English summaries).

Mironov, A.N. and Sagaidachny, A.Yu. (1984). Morphology and distribution of the recent echinoids of the genus *Echinocyamus* (Echinoidea: Fibulariidae). *Trudy Instituta Okeanologii AN USSR* **119**, 179–204. (In Russian, English summary).

Mironov, A.N. and Sagalevich, A.V. (1987). Life on the Seamounts. *Priroda* **No. 6**, 34–42. (In Russian).

Mironov, A.N., Sagaidachny, A.Yu. and Detinova N.N. (1990). Distribution patterns of commercial decapod crustaceans on the underwater Nazca Ridge and Walters Shoals. *In* "All-USSR Conference "Reserve food biological resources of the open ocean and USSR seas", Kaliningrad. Abstracts of communications", pp. 125–127. Moscow (In Russian).

Moskalev, L.I. and Galkin, S.V. (1986). Investigations of the fauna of submarine rises during 9th cruise of the R/V "Akademic Mstislav Keldysh". *Zoologicheskii Zhurnal* **65**(11), 1716–1720. (In Russian, English summary).

Nelson, G. and Platnick, N.I. (1981). "Systematics and Biogeography: Cladistics and Vicariance". Columbia University Press, NY.

Nelson, J.S. (1984). "Fishes of the World", 2nd edn. Wiley, NY.

Nesis, K.N. (1985). "Oceanic cephalopods: Distribution, Life Forms, Evolution". Nauka, Moscow. (In Russian).

Nesis, K.N. (1990). Bottom and near-bottom cephalopods of the submarine Nazca and Sala y Gómez Ridges (Pacific Ocean). *Trudy Instituta Okeanologii AN USSR* **125**, 178–191. (In Russian, English summary).

Nesis, K.N. (1993a). Cephalopods of the Saya de Malha Bank, Indian Ocean. *Trudy Instituta Okeanologii RAN* **128**, 26–39. (In Russian, English summary).

Nesis, K.N. (1993b). Cephalopods of seamounts and submarine ridges. *In* "Recent Advances in Cephalopod Fisheries Biology" (T. Okutani, R.K. O'Dor and T. Kubodera, eds), pp. 365–373. Tokai University Press, Tokyo.

Nesis, K.N. (1994). Teuthofauna of the Walters Shoals, a seamount in the Southwestern Indian Ocean. *Ruthenica* **4**(1), 67–77.

Newman, W.A. (1986). Origin the Hawaiian marine fauna: Dispersal and vicariance as indicated by barnacles and other organisms. *In* "Crustacean Biogeography" (R.H. Gore and K.L. Heck, eds), pp. 21–49. A.A. Balkema, Rotterdam, The Netherlands.

Newman, W.A. and Foster, B.A. (1983). The Rapanuian faunal district (Easter and Sala y Gómez): in search of ancient archipelagos. *Bulletin of Marine Science* **33**(3), 633– 644.

Newman, W.A. and Foster, B.A. (1987). Southern hemisphere endemism among the barnacles: Explained in part by extinction of northern members of amphitropical taxa? *Bulletin of Marine Science* **41**(2), 361–377.

Okutani, T. and Kuroiwa, M. (1985). The first occurrence of *Nototodarus* (Cephalopoda: Ommastrephidae) from off Chile, Southeast Pacific (Preliminary report). *Venus, Japanese Journal of Malacology* **44**(2), 95– 102.

Osorio R.C. (1992). Endemism and mollusks in Easter Island. *In* Abstracts of the 11th International Malacological Congress, Siena, 1992 (F. Giusti and G. Manganelli, eds), pp. 472–475.

Pakhorukov, N.P. (1976). Preliminary list of bathyal fishes of the Rio Grande underwater rise. *Trudy Instituta Okeanologii AN USSR* **104**, 318–331 (In Russian, English summary).

Pakhorukov, N.P. (1981). Deep-water near-bottom fishes of the Walvis Ridge and adjacent areas. *In* "Fishes of the Open Ocean" (N.V. Parin, ed.), pp. 19–31. Institut of Oceanology, Academy of Sciences USSR, Moscow. (In Russian, English summary).

Pakhorukov, N.P. and Golovan', G.A. (1987). Morphoecological differences of the blue-mouths on the seamounts of the Nazca Ridge. *In* "III Congress of Soviet Oceanologists. Abstracts of Communications. Section: Biology of the Ocean", **3**, 22–23. Hydrometeoizdat, Leningrad. (In Russian).

Pandolfi, J.M. (1995). Geomorphology of the uplifted Pleistocene atoll at Henderson Island, Pitcairn Group. *Biological Journal of the Linnean Society of London* **56**, 63–77.

Parin, N.V. (1982a). Additions to the list of fishes of the underwater Nazca Ridge and adjacent area. *In* "Insufficiently Studied Fishes of the Open Ocean" (N.V. Parin, ed.), pp. 72–78. Institute of Oceanology, Academy of Sciences USSR, Moscow. (In Russian, English summary).

Parin, N.V. (1982b). New species of the fish genus *Draconetta* and a key for the family Draconettidae (Osteichthyes). *Zoologicheskii Zhurnal* **61**(4), 554–563. (In Russian, English summary).

Parin, N.V. (1983). Two new species of bothid flatfishes (Bothidae, Pleuronectiformes) from the Nazca submarine ridge. *Byulleten Moskovskogo Obshchestva Ispytatelei Prirody, Otdel Biologii* **88**(4), 90–96. (In Russian).

Parin, N.V. (1984). Three new species of the genus *Physiculus* and other morid fishes (Moridae, Gadiformes) from submarine ridges of the southeastern Pacific Ocean. *Voprosy Ikhtiologii* **24**(4), 531–544. (In Russian).

Parin, N.V, (1985). A new hemerocoetine fish, *Osopsaron karlik* (Percophidae, Trachinoidei) from the Nazca submarine ridge. *Japanese Journal of Ichthyology* **31**(4), 358– 361.

Parin, N.V. (1987). On species identification of the dogfish, genus *Squalus*, inhabiting the seamounts of the Southeastern Pacific Ocean. *Voprosy Ikhtiologii* **27**(4), 531–538. (In Russian).

Parin, N.V. (1988). "Fishes of the Open Ocean". Nauka, Moscow. (In Russian).

Parin, N.V. (1989). A review of the genus *Rexea* (Gempylidae) with descriptions of three new species. *Voprosy Ikhtiologii* **29**(1), 3–23. (In Russian).

Parin, N.V. (1990a). Preliminary review of fish fauna of the Nazca and Sala y Gómez submarine ridges (Southestern Pacific Ocean). *Trudy Instituta Okeanologii AN USSR* **125**, 6–36. (In Russian, English summary).

Parin, N.V. (1990b). Percophidae from the Sala y Gómez submarine ridge. *Voprosy Ikhtiologii* **30**(1), 3–12. (In Russian).

Parin, N.V. (1991a). Fish fauna of the Nazca and Sala y Gómez submarine ridges, the easternmost outpost of the Indo-West Pacific zoogeographic Region. *Bulletin of Marine Science* **49**(3), 671–683.

Parin, N.V. (1991b). There new species of the benthopelagic fish genus *Plagiogeneion* from southern Pacific and Indian Oceans (Teleostei: Emmelichthyidae). *Proceedings of the Biological Society of Washington* **104**(3), 459–467.

Parin, N.V. (1992). *Argyripnus electronus*, a new sternoptychid fish from the Sala y Gómez submarine ridge. *Japanese Journal of Ichthyology* **32**(2), 135–138.

Parin, N.V. (ed.) (1993). "Biology of the Oceanic Fishes and Squids". *Trudy Instituta Okeanologii RAN* **128**. (In Russian, English summaries).

Parin, N.V. (1994). Three new species and new records of the scabbard-fishes, genus *Aphanopus* (Trichiuridae). *Voprosy Ikhtiologii* **34**(6), 740–746 (In Russian).

Parin, N.V. and Abramov, A.A. (1986). Materials for a revision of the genus *Epigonus* Rafinesque (Perciformes, Epigoniidae): species from submarine ridges of the southern East Pacific and preliminary review of the "*Epigonus robustus* species group". *Trudy Instituta Okeanologii AN USSR* **121** 173–194. (In Russian, English summary).

Parin, N.V. and Becker, V.E. (eds) (1990). "Seamount Fishes". *Trudy Instituta Okeanologii AN USSR* **125**. (In Russian, English summaries).

Parin, N.V. and Borodulina, O.D. (1986). Preliminary review of the benthopelagic fish genus *Antigonia* Lowe (Zeiformes, Caproidei). *Trudy Instituta Okeanologii AN USSR* **121**, 141–172. (In Russian, English summary).

Parin, N.V. and Borodulina, O.D. (1990). Review of the genus *Polymetme* with the descriptions of two new species. *Voprosy Ikhtiologii* **30**(3), 733–743. (In Russian).

Parin, N.V. and Golovan', G.A. (1982). Benthic and benthopelagic fishes from the seamounts of the underwater Nazca Ridge. *In* "II All-USSR Congress of Oceano-graphers. Abstracts of Communications. Biology of the Ocean", **6**, 49–50. (In Russian).

Parin, N.V. and Karmovskaya, E.S. (1985). Two new species of nettastomatid eels (Nettastomatidae, Anguilliformes) from submarine mountains of the southeastern Pacific. *Zoologicheskii Zhurnal* **64**(10), 1524–1530. (In Russian, English summary).

Parin, N.V. and Kobyliansky, S.G. (1993). Review of the genus *Maurolicus* (Sternoptychidae, Stomiiformes), with re-establishing validity of five species considered junior synonyms of *M. muelleri* and descriptions of nine new species. *Trudy Instituta Okeanologii RAN* **128**, 69–107. (In Russian, English summary).

Parin, N.V. and Kotlyar, A.N. (1985). Electric rays of the genus *Torpedo* in the open waters of the southeastern Pacific. *Voprosy Ikhtiologii* **25**(5), 707–718. (In Russian).

Parin, N.V. and Kotlyar, A.N. (1988). A new boarfish, *Pentaceros quinquespinis* (Pentacerotidae), from the Southeast Pacific. *Voprosy Ikhtiologii* **28**(3), 355–360. (In Russian).

Parin, N.V. and Kotlyar, A.N. (1989). A new aulopid species, *Hime macrops* from the eastern South Pacific, with comments on geographic variation of *H. japonica*. *Japanese Journal of Ichthyology* **35**(4), 407–413.

Parin, N.V. and Mironov, A.N. (1987). Formation of the fauna of underwater rises in the open ocean. *In* "III Congress of Soviet Oceanologists. Abstracts of Communications. Section: Biology of the ocean", **3**, 15–16. Hydrometeoizdat, Leningrad. (In Russian).

Parin, N.V. and Nesis, K.N. (1986). Biogeography of the Pacific Ocean. *In* "Biological Resources of the Pacific Ocean (M.E. Vinogradov, N.V. Parin and V.P. Shuntov, eds), 61–75. Nauka, Moscow. (In Russian).

Parin, N.V. and Sazonov, Y.I. (1990). A new species of the genus *Laemonema* (Moridae, Gadiformes) from the tropical southeastern Pacific. *Japanese Journal of Ichthyology* **37**(1), 6–9.

Parin, N.V. and Shcherbachev, Y.N. (1982). Two new argentinine fishes of the genus *Glossanodon* from the eastern South Pacific. *Japanese Journal of Ichthyology* **28**(4), 381–384.

Parin, N.V., Golovan', G.A., Pakhorukov, N.P., Sazonov, Y.I. and Shcherbachev, Y.N. (1980). Fishes from the underwater Nazca and Sala y Gómez Ridges collected in the cruise of the R/V "Ichthyandr". *In* "Fishes of Open Ocean" (N.V. Parin, ed.), the 5–18. Institute of Oceanology, Academy of Sciences USSR, Moscow. (In Russian, English summary).

Parin, N.V., Neiman, V.G. and Rudjakov, Yu.A. (1985). On the problem of biological productivity of waters in the areas of underwater rises in the open ocean. *In* "Biological Bases of the Developing Fisheries of the World Ocean" (M.E. Vinogradov and M.V. Flint, eds), pp. 192–203. Nauka, Moscow. (In Russian).

Parin, N.V., Nesis, K.N. and Rudjakov, Yu.A. (1988a). Investigation of the fauna of seamounts of the Nazca and Sala y Gómez Ridges, Southeastern Pacific Ocean (18th cruise of the R/V "Professor Shtockman", March 6–June 17, 1987). *Okeanologiya* **28**(6), 885–888. (In Russian).

Parin, N.V., Pavlov, Yu.P. and Andrianov, D.P. (1988b). Ecological characteristic of the mirror dory, *Zenopsis nebulosus*, from the Nazca underwater ridge. *Voprosy Ikhtiologii* **28**(5), 707–716 (In Russian).

Parin, N.V., Borodulina, O.D., Konovalenko, I.I. and Kotlyar, A.N. (1990a). Oceanic pelagic fishes of the Southeastern Pacific (composition of fauna and geographic distribution). *Trudy Instituta Okeanologii AN USSR* **125**, 192–222. (In Russian, English summary).

Parin, N.V., Gorelova, T.A. and Borodulina, O.D. (1990b). Feeding and trophic relationships of fishes inhabiting the Nazca and Sala y Gómez submarine ridges. *Trudy Instituta Okeanologii AN USSR* **125**, 37–57. (In Russian, English summary).

Parin, N.V., Konovalenko, I.I. and Nesterov, A.A. (1990c). Independent populations of neritic carangid fishes above the seamounts of the Nazca underwater ridge. *Biologiya morya* **2**, 16–20. (In Russian, English summary).

Parin, N.V., Nesis, K.N., Sagaidachny, A.Yu. and Shcherbachev, Yu.N. (1993). Fauna of

Walters Shoals, a seamount in the southwestern Indian Ocean. *Trudy Instituta Okeanologii RAN* **128**, 199–216. (In Russian, English summary).

Parker, T. and Tunnicliffe, V. (1994). Dispersal strategies of the biota on an oceanic seamount: Implications for ecology and biogeography. *Biological Bulletin* **187**(3), 336–345.

Paulin, C.D. (1991). Two new species of the genus *Physiculus* (Moridae) from seamounts of the Southeast Pacific. *Voprosy Ikhtiologii* **31**(3), 162–165. (In Russian).

Paxton, J.R., Hoese, D.F., Allen, G.R. and Hanley, J.E. (1989). Pisces. Petromyzonidae to Carangidae. *Zoological Catalogue of Australia* **7**, 1–665.

Pesenko, Yu.A. (1982). "Principles and Methods of Quantitative Analysis in Faunistic Investigations". Nauka, Moscow. (In Russian).

Phillips, B.F. (1982). The circulation of the Southeastern Indian Ocean and the planktonic life of the western rock lobster. *In* "Research Report 1970–1981. CSIRO Mar.Lab., Cronulla", pp. 43–52.

Phillips, B.F. and McMillan, G. (1986). The pelagic phase of spiny lobster development. *Canadian Journal of Fisheries and Aquatic Sciences* **43**(11), 2153–2163.

Prosvirov, S.E. (1990). On the biology of the spiny lobster *Projasus bahamondei* on underwater rises of the Nazca Ridge. *In* "Vth All-USSR Conference on Commercial Invertebrates, Minsk–Naroch. Abstracts of communications", pp. 42–43. Moscow. (In Russian).

Pshenichny, B.P., Kotlyar, A.N. and Glukhov, A.A. (1986). Fishery resources of the thalassobathyal of the Atlantic Ocean. *In* "Biological Resources of the Atlantic Ocean" (D.E. Gershanovich, V.I. Sauskan and O.A. Scarlato, eds), pp. 230–252. Nauka, Moscow. (In Russian).

Rehder, H.A. (1980). The marine mollusks of Easter Island (Isla de Pascua) and Sala y Gómez. *Smithsonian Contributions to Zoology* **289**, 1–167.

Robison, B.H. and Bailey, T.G. (1981). Sinking rates and dissolution of midwater fish fecal matter. *Marine Biology* **65**(2), 135–142.

Rogers, A.D. (1994). The biology of seamounts. *Advances in Marine Biology* **30**, 305–349.

Rosen, D.E. (1976). A vicariance model of Caribbean biogeography. *Systematic Zoology* **24**(4), 431–464.

Rosen, D.E. (1978). Vicariant patterns and historical explanation in biogeography. *Systematic Zoology* **27**(2), 159–188.

Rotondo, G.M., Springer, V.G.. Scott, G.A.J. and Schlanger, S.O. (1981). Plate movement and island integration – a possible mechanism in the formation of endemic biotas, with special reference to the Hawaiian Islands. *Systematic Zoology* **30**(1), 12–21.

Rudjakov, J.A. and Tseitlin, V.B. (1985). An imitation model of the independent fish population inhabiting an underwater rise. *Voprosy Ikhtiologii* **25**(6), 1031–1034. (In Russian).

Rudjakov, J.A. and Zaikin, A.N. (1990). 18th cruise of the R/V "Professor Shtockman", a next step of hydrobiological investigations of the near-bottom layer of the ocean. *Trudy Instituta Okeanologii AN USSR* **124**, 7–15. (In Russian, English summary).

Rudjakov, J.A., Kucheruk, N.V. and Chistikov, S.D. (1990). Population structure of *Projasus bahamondei* George (Crustacea, Decapoda, Palinuridae) from the underwater Nazca Ridge. *Trudy Instituta Okeanologii AN USSR* **124**, 156–160. (In Russian, English summary).

Sazonov, Y.I. (1989). A new species of the genus *Talismania* Goode et Bean (Alepocephalidae) from the Southeast Pacific. *Voprosy Ikhtiologii* **29**(3), 355– 360. (In Russian).

Sazonov, Y.I. and Iwamoto, T. (1992). Grenadiers (Pisces, Gadiformes) of the Nazca and

Sala y Gómez Ridges, southeastern Pacific. *Proceedings of the California Academy of Sciences* **48**(2), 27–95.

Sazonov, Y.I. and Shcherbachev, Y.N. (1982). A preliminary review of grenadiers related to the genus *Cetonurus* Günther (Gadiformes, Macrouridae). 1. Description of new taxa related to genera *Cetonurus* Günther and *Kumba* Marshall. *Voprosy Ikhtiologii* **22**(5), 707–721. (In Russian).

Scheltema, R.S. (1986). Long-distance dispersal by planktonic larvae of shoal-water benthic invertebrates among central Pacific islands. *Bulletin of Marine Science* **39**(2), 241–256.

Scheltema, R.S. (1988). Initial evidence for the transport of teleplanic larvae of benthic invertebrates across the East Pacific barrier. *Biological Bulletin* **174**, 145–152.

Schilder, F.A. (1965). The geographical distribution of cowries (Mollusca: Gastropoda). *The Veliger* **7**(3), 171–183.

Semenov, V.N. (1977). The boundary effect in biogeography, its role in the evolution of faunas and possible commercial importance. *In* "Soviet–American Symposium of Water Ecosystems Responses to Introduction of Alien Species", Tallinn, October, 1977. Abstracts of Communications, pp. 117– 121. Moscow. (In Russian and English).

Semenov, V.N. (1982). Biogeographic regionalization of the South American continental shelf on the base of classification of species ranges of the bottom invertebrates. *In* "Marine Biogeography: Subject, Methods, Principles of Regionalization" (O.G. Kussakin, ed.), pp. 184–269. Nauka, Moscow. (In Russian).

Serafy, D.K. (1971). A new species of *Clypeaster* (Ecinodermata, Echinoidea) from San Felix Island, with a key to the recent species of the Eastern Pacific Ocean. *Pacific Science* **25**(2), 165–170.

Shcherbachev, Y.N., Kotlyar, A.N. and Abramov, A.A. (1989). Ichthyofauna and fish resources of underwater rises of the Indian Ocean. *In* "Biological Resources of the Indian Ocean" (N.V. Parin and N.P. Novikov, eds), pp. 159–185. Nauka, Moscow. (In Russian).

Sheremet, I.I. and Zalesinsky, Yu.M. (1981). "Hydrological regime of the Chilean fishery Subarea of the SEPO". Zapryba, Kaliningrad. (In Russian).

Simpson, E.S.W. and Heydorn, A.E.F. (1965). Vema Seamount. *Nature* **207**, 249–251.

Sokolov, V.A., Kotenev, B.N., Kamenskaya, E.A. and Neiman, A.A. (eds) (1986). "Fisheries Resources of the Rises in the Open Part of the Indian Ocean". VNIRO, Moscow. (In Russian).

Sokolova, M.N. (1987). On the feeding of sea urchins *Stereocidaris nascaensis* Durham, Mintz in the southern part of the Nazca Ridge. *In* "Feeding of Marine Invertebrates and its Role in the Formation of the Communities" (A.P. Kuznetsov and M.N. Sokolova, eds), pp. 38–44. P.P. Shirshov Institute of Oceanology, Acad. Sci. USSR, Moscow. (In Russian, English summary).

Sokolova, M.N. (1990). Feeding of different species of carnivorous sea urchins in the areas of the Nazca Ridge and Sala y Gómez Rise. *In* Feeding and Bioenergetics of Bottom Invertebrates" (A.P. Kuznetsov, ed.), pp. 28–46. P.P. Shirshov Institute of Oceanology, Acad. Sci. USSR Moscow. (In Russian, English summary).

Sokolova, M.N. (1994). Feeding peculiarities of macrobenthos from underwater rises as seen on sea urchins. *Trudy Instituta Okeanologii RAN* **129**, 31–43. (In Russian, English summary).

Springer, V.G. (1982). Pacific Plate biogeography, with special reference to shorefishes. *Smithsonian Contributions to Zoology* **367**, 1–182.

Stepanov, V.N. (1974). "World Ocean". Znaniye, Moscow. (In Russian).

Svetovidov, A.N. (1986). A review of the three-bearded marine rocklings of the genus *Gaidropsarus* Rafinesque (Gadidae), with a description of a new species. *Voprosy*

Ikhtiologii **26**(1), 3–23. (In Russian).

Sysoev, A.V. (1990). Gastropods of the fam. Turridae (Gastropoda: Toxoglossa) from the underwater Sala y Gómez Ridge. *Trudy Instituta Okeanologii AN USSR* **124**, 245– 260. (In Russian, English summary).

Sysoev, A.V. and Ivanov, D.L. (1985). New taxa of Turridae (Gastropoda, Toxoglossa) from the Nazca Ridge (Southeastern Pacific). *Zoologicheskii Zhurnal* **64**(2), 194–205. (In Russian, English summary).

Tabachnick, K.R. (1990). Hexactinellid sponges from the Nazca and Sala y Gómez Ridges. *Trudy Instituta Okeanologii AN USSR* **124**, 161–173. (In Russian, English summary).

Tabachnick, K.R. (1991). Adaptation of the hexactinellid sponges to the deep-sea life. *In* "Fossil and Recent Sponges" (J. Reitner und H. Keupp, eds), pp. 378–386. Springer, Berlin.

Thiel, H. (1970). Bericht über die Benthosuntersuchungen wahrend der "Atlantischen Kuppenfahrten 1967" von F.S. "Meteor". *Meteor-Forschungsergebnisse* **D7**, 23–42.

Trofimov, M.N., Pomazanova, N.P., Tyuleva, L.S. and Rozhkov, Ye.G. (1989a). Southwestern part of the Indian Ocean. *In* "Biological Resources of the Indian Ocean" (N.V. Parin and N.P. Novikov, eds), pp. 385–415. Nauka, Moscow. (In Russian).

Trofimov, M.N., Pomazanova, N.P., Tyuleva, L.S. and Rozhkov, Ye.G. (1989b). Southeastern part of the Indian Ocean. *In* "Biological Resources of the Indian Ocean" (N.V. Parin and N.P. Novikov, eds), pp. 415–425. Nauka, Moscow. (In Russian).

Tseitlin, V.B. (1985). The energetics of the fish population inhabiting the underwater mountain. *Okeanologiya* **25**(2), 308–311. (In Russian, English summary).

Uchida, R.N., Hayasi, S. and Boehlert, G.W. (eds) (1986) "Environment and Resources of Seamounts in the North Pacific". *NOAA Technical Report NMFS* **43**.

Udintsev, G.B. (1972). "Geomorphology and Tectonics of the Bottom of the Pacific Ocean". Nauka, Moscow (In Russian).

Vereshchaka, A.L. (1990a). Distribution of euphausiids over the underwater rises of the Nazca and Sala y Gómez Ridges. *Trudy Instituta Okeanologii AN USSR* **124**, 112– 117. (In Russian, English summary).

Vereshchaka, A.L. (1990b). Mysids over the underwater rises of the Nazca and Sala y Gómez Ridges. *Trudy Instituta Okeanologii AN USSR* **124**, 118–128. (In Russian, English summary).

Vereshchaka, A.L. (1990c). Pelagic decapods over the underwater rises of the Nazca and Sala y Gómez Ridges. *Trudy Instituta Okeanologii AN USSR* **124**, 129–155. (In Russian, English summary).

Vereshchaka, A.L. (1995). Macroplankton in the near-bottom layer of continental slopes and seamounts. *Deep-Sea Research* **42**(9), 1639–1668.

Vinogradov, G.M. (1990). Amphipods (Amphipoda, Crustacea) in the pelagic realm of the Southeastern Pacific Ocean. *Trudy Instituta Okeanologii AN USSR* **124**, 27–104. (In Russian, English summary).

Vinogradov, M. E. and Vinogradov, G.M. (1991). Amphipods caught by a bottom trap on the underwater Nazca Ridge. *Zoologicheskii Zhurnal* **70**(6), 32–38. (In Russian, English summary).

Vinogradov, M.E., Shushkina, E.A. and Musaeva, E.I. (1984). Influence of a frontal division near the Nazca Ridge on the distribution of microplankton. *In* "Frontal Zones of the Southeastern Pacific Ocean (Biology, Physics, Chemistry)" (M.E. Vinogradov and K.N. Fedorov, eds), pp. 239–243. Nauka, Moscow. (In Russian).

Vozzhinskaya, V.B. (1990). Deep-water calcareous macrophyte algae. *Trudy Instituta Okeanologii AN USSR* **124**, 261–263. (In Russian, English summary).

Webber, W.R. and Booth, J.D. (1988). *Projasus parkeri* (Stebbing, 1902) (Crustacea, Decapoda, Palinuridae) in New Zealand and description of a *Projasus puerulus* from

Australia. *Records of the National Museum of New Zealand* **3**(8), 81–92.

Williams, J.T. and Machida, Y. (1992). *Echiodon anchipterus*: a valid western Pacific species of the pearlfish family Carapidae with comments on *Eurypleuron. Japanese Journal of Ichthyology* **38**(4), 367–373.

Wilson, R.R., Jr. and Kaufmann, R.S. (1987). Seamount biota and biogeography. *In* "Seamounts, Islands and Atolls" (B. Keating, P. Fryer, R. Batiza and G. Boehlert, eds), 355–377. *Geophysical Monograph* **43**. American Geophysical Union, Washington.

Wilson, R.R., Jr., Smith, K.L., Jr and Rosenblatt, R.H. (1985). Megafauna associated with bathyal seamounts in the central North Pacific Ocean. *Deep-Sea Research* **32**(10), 1243–1254.

Wishner, K., Levin, L., Gowing, M. and Mullineaux, L. (1990). Multiple roles of the oxygen minimum in benthic zonation on a deep seamount. *Nature* **346**, 57–59.

Wishner, K.N., Ashjian, C.J., Gelfman, C., Gowing, M.M., Kann, L., Levin, L.A., Mullineaux, L.S. and Saltzman, J. (1995). Pelagic and benthic ecology of the lower interface of the Eastern Tropical Pacific oxygen minimum zone. *Deep-Sea Research* **42**(1), 93–115.

Wyrtki, K. (1966). Oceanography of the eastern equatorial Pacific Ocean. *Oceanography and Marine Biology, Annual Review* **4**, 33–68.

Zakharov, L.A., Rudenko, M.V. and Soldatov, A.V. (1984). Morphometric characteristics of the Walvis and Nazca block-volcanic ridges. *Okeanologiya* **24**(6), 929–935. (In Russian, English summary).

Zarenkov, N.A. (1990). Decapods (Stenopodidea, Brachyura, Anomura) of the underwater Nazca and Sala y Gómez Ridges. *Trudy Instituta Okeanologii AN USSR* **124**, 218–244 (In Russian, English sunnnary).

Zevina, G.B. (1983). The Cirripedia from peaks of the Nazca Ridge mountains (Pacific Ocean). *Zoologicheskii Zhurnal* **62**, 1635–1642.

Zevina, G.B. (1990). Cirripedia Thoracica from the guyot summits of the Nazca and Sala y Gómez Ridges. *Trudy Instituta Okeanologii AN USSR* **124** 174–218. (In Russian, English summary).

Zevina, G.B., Nikitina, E.N., Goryachev, V.N. and Marakuyev, V.I. (1979). Quantitative investigation of the benthos of the banks in the area of the Canary Islands. *In* "Complex Investigations of the Nature of the Ocean" **6**, 196–218. Moscow State University, Moscow. (In Russian, English summary).

Zezina, O.N. (1990). Composition and distribution of articulate brachiopods on the underwater rises of the Eastern Pacific. *Trudy Instituta Okeanologii AN USSR* **124**, 264–268. (In Russian, English summary).

Zullo, V.A. and Newman, W.A. (1964). Thoracic Cirripedia from a Southeast Pacific guyot *Pacific Science* **18**(4), 355–372.

Gonatid Squids in the Subarctic North Pacific: Ecology, Biogeography, Niche Diversity and Role in the Ecosystem

K.N. Nesis

P.P. Shirshov Institute of Oceanology, Russian Academy of Sciences,
Nakhimovsky Prospekt 36, Moscow 117851, Russia

ADVANCES IN MARINE BIOLOGY VOL. 32
ISBN 0-12-026132-4

ABSTRACT

This review is based on the author's own data and all available published data
(mostly Russian and Japanese) on the ecology, biogeography and role in the
ecosystem of gonatid squids in the northern North Pacific. For the best studied
species, *Berryteuthis magister*, information is given on size, horizontal and
vertical distribution, diel and ontogenetic vertical migrations, maturation, mating,
spawning, fecundity, population structure, age, growth, life cycle, horizontal
migrations, underwater behaviour, food and feeding, and predators. The assessed
biomass and its interannual dynamics and the fisheries importance are also
covered.

For other, less studied, species of the genera *Berryteuthis (B. anonychus)*,
Gonatopsis (three species) and *Gonatus* (seven species) all available ecological
and biogeographical data are included. All species are compared according to their
size, horizontal and vertical distribution, spawning habitats, diel vertical
migrations and gelatinous degeneration associated with maturation. The "ecological individuality" of each species is evaluated. It is shown that each occupies
its own ecological niche but these niches overlap to different degrees. The history
of niche divergence in North Pacific gonatids during the Neogene–Pleistocene
period is briefly reviewed.

Common features are described of the horizontal and vertical distribution,
relative abundance and biomass of North Pacific gonatids in general. Their roles
in the ecosystem, as predators, prey, competitors and hosts of parasites is
evaluated.

The total biomass of gonatid squids in the whole subarctic North Pacific and
the Russian Far Eastern seas is estimated as approximately 15–20 million t. They
contribute some 10–15% of the total production of mesopelagic cephalopods in
the World Ocean. Their yearly food consumption is assessed at 100–200 million
t. The life cycle of gonatids is shorter and their P/B-coefficient much higher than
that of subarctic mesopelagic fishes. As a result, though the squid biomass
(calculations for the Okhotsk Sea) is less than 10% of the total mesopelagic fish
and squid population, they form 58–67% of annual total fish and squid production.
Thus gonatid squids have an important place in the ecosystem of the northern
North Pacific and the Far Eastern seas of Russia.

1. INTRODUCTION

The oegopsid squid family Gonatidae includes three genera, *Berryteuthis*, *Gonatopsis* and *Gonatus*, the two latter with two subgenera each (*Gonatopsis* s.str. and *Boreoteuthis*; *Gonatus* s.str. and *Eogonatus*), and 17 species (Nesis, 1985, 1987; Okutani *et al.*, 1988; Okutani and Clarke, 1992). It has a bipolar distribution: the whole North Polar Basin, boreal North Atlantic, boreal (subarctic) North Pacific, and notalian (subantarctic) belt of the Southern Ocean (Nesis, 1985). The southern boundary of its range in the North Atlantic is between Cape Cod and the Bay of Biscay. In the North Pacific the southern boundary is in the Japan Sea and between the Pacific coast of eastern Honshu and the area south-west of the southern tip of Baja California; while in the southern hemisphere the gonatids reach the subequatorial area off northern Peru, under the influence of the cold Peru Current. Only two species of *Gonatus* (*Gonatus*) are found in the Arctic (one species) and North Atlantic and one more, of the same genus and subgenus, in the southern hemisphere; all other genera and species inhabit the subarctic North Pacific (Figure 1). This is evidently the homerange of the family Gonatidae (Nesis, 1973a, 1985).

Ten species of gonatids inhabit the subarctic Northwestern Pacific and Russia's Far Eastern seas, the Bering, Okhotsk and Japan Seas. They all inhabit the upper water layers as larvae and juveniles (Kubodera and Okutani, 1981b; Kubodera and Jefferts, 1984) and most species descend in adulthood to great depths or as far as the seabed, where they mate and spawn (Nesis, 1985; Nesis and Nikitina, 1996; Roper and Young, 1975).

The gonatid squids are very common animals both in juvenile and adult life, and play an important role as food of many fish (including commercial species), seabirds, whales (primarily toothed whales) and seals (Akimushkin, 1963; Clarke, 1966, 1977; Zuev and Nesis, 1971; Nesis, 1985; Okutani *et al.*, 1988; Radchenko, 1992; Shuntov *et al.*, 1993c; Ilyinsky and Gorbatenko, 1994; Lapko, 1995; Sobolevsky, 1996; Sobolevsky and Senchenko, 1996).

One of the gonatid species, the Commander squid *Berryteuthis magister*, has been fished commercially in the Northwestern Pacific and adjacent areas since the 1960s by the Russian, Japanese and other fleets, predominantly as a by-catch of various demersal fishes.

The Russian fisheries of this species began in the 1960s off the Commander Is as a by-catch to Alaska pollack. Catches in the 1960–1970s were 2.7–2.8 thousand t (Fedorets, 1987). Later, a specialized fishery began along the Pacific side of the Central Kurile Is (from 1977), then the northern Kurile Is (from 1986) and in the late 1970s and early 1980s the catches reached some 25 thousand t (Fedorets, 1987; Zuev and Nesis, 1971). In the mid-1980s good fishing grounds were discovered in the western Bering Sea in Olyutorsky Bay and between Cape Olyutorsky and Cape Navarin. Here squids are fished by the Russian, Japanese and

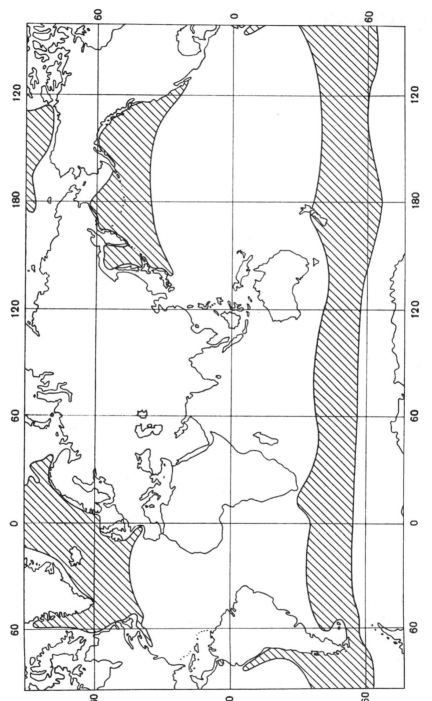

Figure 1   The range of the genus *Gonatus* (modified from Nesis, 1985).

other fleets and catches are approximately 9 thousand t year^{-1} (Fedorets and Kozlova, 1987; Nesis, 1995).

Some good catches have been obtained in the Japan Sea, in its northern part and on the seamounts of the Yamato Bank complex, where Russian and Japanese ships have worked from time to time, beginning in 1987 (Naito et al., 1977a; Shevtsov, 1988a,b, 1991; Okiyama, 1993b). Russian ships have worked on the slopes of the Kita–Yamato Bank, and more rarely in the Tatar Strait off the coasts of western Sakhalin and Primorye (Maritime Province) (Skalkin, 1977; Shevtsov, 1988a,b).

The total Russian catch of B. magister reached about 65.7 thousand t in 1985, but the next year it dropped to 13.5 thousand t. In 1990 it reached a maximum value of 69.1 thousand t but now it does not exceed 22–27 thousand t (Bizikov, 1987; Bizikov and Arkhipkin, 1996).

The maximum Japanese catch was 9 thousand t in 1978 (Osako and Murata, 1983). The squid catch in the eastern Bering Sea, mostly Japanese and consisting mainly of B. magister, was 4.0–6.9 thousand t in 1977–1980 (peak in 1978), then it decreased from 3.8 thousand t in 1981 to 1.6 thousand t in 1985 (Bakkala and Balsinger, 1987). The catch of squids in the Gulf of Alaska was 14.3 thousand t in 1978, including 13.8 thousand t taken by Japanese, 0.3 thousand t by South Korean and 0.2 thousand t by USSR vessels (Wall et al., 1981). It may be supposed that this catch consisted mostly of B. magister. If so, the total Japanese bycatch of B. magister in 1978 may be estimated at approximately 30 thousand t, but this figure may include some other species too.

This species is fished by bottom trawls in all areas, although catches by pelagic trawls in the Okhotsk Sea may reach up to 0.5 t h^{-1} (Ilyinsky, 1991). The ground rope of the bottom trawl must pass close to the bottom to avoid the squid escaping under the net (Railko, 1987) but this risks tearing or even losing the net on rough ground (Bizikov, 1987; Shevtsov, 1988b, 1991). Fishing on the steep slopes of islands and underwater rises requires a highly skilful shipmaster.

The possible yearly catch of B. magister in the western Bering Sea is calculated at 30 thousand t at least; oceanward of the Kurile Is it may be 40–60 thousand t and in the Russian EEZ of the Japan Sea (Kita–Yamato Bank) it may be 6–8 thousand t (Shevtsov et al., 1988; Shevtsov, 1991; Bizikov and Arkhipkin, 1996).

Other North Pacific gonatids are not fished commercially although Gonatopsis (Boreoteuthis) borealis is caught by jigging and is a substantial by-catch in gillnets set for salmonids and the neon squid Ommastrephes bartramii (Naito et al., 1977a,b; Kubodera et al., 1983; Osako and Murata, 1983).

The gonatids are not fished outside the north-western Pacific but Gonatus (Gonatus) fabricii (Lichtenstein, 1818) in the north-western Atlantic and G. (G.) antarcticus Lönnberg, 1898 in the Southern Ocean could form important resources for commercial fisheries (Wiborg, 1979; Kristensen, 1983).

2. DATA SOURCES

Most of the Russian collections of postlarval (juvenile and adult) gonatids from the sub-arctic North Pacific and Far Eastern seas were made either during standard pelagic surveys or during commercial or scientific bottom operations. The pelagic surveys were made with large pelagic trawls having an opening 45–50 m (vertical) by 50–80 m (horizontal), the mesh size diminishing from 11–12 m in the wings, near the mouth, to 60 mm at the rear end, which is covered by fine-meshed netting with minimal size 10 mm (all meshes measured knot to knot). Trawl hauls were made day and night in the subsurface layer (horizontal from 0 to 45–50 m); and by oblique hauls in the 200–0 m layer, the 500–200 m and 1000–500 m layers. Sometimes the 1500–1000 m layer was investigated. Bottom surveys were made in the near-bottom layer or on the bottom by commercial bottom trawls with groundrope length 27–43 m or more, at depths from 50–100 to 500–800 m and, in some surveys, to 2000 m. Larvae and early juveniles were collected using the Isaacs–Kidd Midwater Trawl (IKMT) and Bongo net.

Japanese, USA and Canadian collections were made mostly by *Maruchi-A* larva net, multiple plankton samplers and IKMT (larvae, postlarvae and early juveniles), gillnets and commercial bottom trawls (late juveniles and adults).

The routine methods used for field study of squids during the Russian expeditions were those recommended by Filippova (1972, 1983) and Shevtsov (1972), but in recent years the methods recommended by Zuev *et al.* (1985) have been widely used, supplemented by studies of gladii and statoliths (Arkhipkin and Bizikov, 1991; Arkhipkin *et al.*, 1996; Bizikov, 1991, 1996; Natsukari *et al.*, 1993a).

3. SYSTEMATICS

The family Gonatidae is represented in the subarctic North Pacific by the following species (Okutani, 1968; Nesis, 1973a,b, 1985, 1987; Jeffcrts, 1988; Okutani *et al.*, 1988):

Berryteuthis Naef, 1921.
   B. magister (Berry, 1913), *B. anonychus* (Pearcy and Voss, 1963).
Gonatopsis Sasaki, 1920.
   Subg. *Boreoteuthis* Nesis, 1971. *G.(B.) borealis* Sasaki, 1923.
   Subg. *Gonatopsis* s.str. *G.(G.) japonicus* Okiyama, 1969; *G.(G.) octopedatus* Sasaki, 1920.
Gonatus Gray, 1849
   Subg. *Eogonatus* Nesis, 1972. *G.(E.) tinro* Nesis, 1972

Subg. *Gonatus* s.str. *G.(G.) onyx* Young, 1972; *G.(G.) kamtschaticus* (Middendorff, 1849) = *G. middendorffi* Kubodera and Okutani, 1981; *G.(G.) californiensis.* Young, 1972; *G.(G.) berryi* Naef, 1923; *G.(G.) pyros* Young, 1972; *G.(G.) madokai* Kubodera and Okutani, 1977 = ?*Gonatopsis okutanii* Nesis, 1972; ?*G.(G.) ursabrunae* Jefferts, 1985; ?*G.(G.) oregonensis* Jefferts, 1985.

The subspecies *B. magister nipponensis* Okutani and Kubodera, 1987 is not justified as yet as a valid taxon. *Gonatopsis (G.) makko* Okutani and Nemoto, 1964 is a doubtful species: many specimens identified as *G. makko* probably belong to *G. japonicus*, but some others may be *Gonatus kamtschaticus* or *G. madokai.* *Gonatus oregonensis* and *G. ursabrunae* were described from larval and early postlarval specimens, of dorsal mantle length (DML) 24–46 and 12–24 mm respectively (Jefferts, 1985) and have not been recognized since the first description. The first is very close to *G. californiensis* (Jefferts, 1985; Okutani and Clarke, 1992) and may be the northern subspecies or a form of the latter. The second may be either a different species, as yet not recognized in adult state, or a transient juvenile stage of a known species, maybe *G. kamtschaticus*, judging by data presented by Jefferts (1985). Both *G. oregonensis* and *G. ursabrunae* are probably absent from the north-western Pacific and will be not treated further.

4. SPECIES ACCOUNTS

4.1. *Berryteuthis magister (Berry, 1913)*

The Commander squid (in Russian literature) or schoolmaster gonate squid (in English publications) is the only gonatid squid fished in commercial quantities and thus has been studied intensively. The amount of literature devoted to this squid, mostly in Russian and Japanese, is rather large, therefore this section is much longer than those describing other, less thoroughly studied, species.

4.1.1. *Size*

B. magister is a large species, having a maximum recorded DML of 42–43 cm and weighing up to 2.2–2.6 kg. In the western Bering Sea and off the Commander Is the maximum size of males is 29–30 cm, of females 37–42 cm; in the Okhotsk Sea they are respectively 31 and 38 cm (Bizikov, 1987; Fedorets, 1987; Fedorets and Kozlova, 1987; Nesis, 1989a, 1995; Arkhipkin *et al.*, 1996; Bizikov and Arkhipkin, 1996). In the Japan Sea the squids are smaller, the maximum size of males is 19–25 cm, of females 29–31 cm (Yuuki and Kitazawa, 1986; Shevtsov, 1988a; Okiyama, 1993b).

4.1.2. *Horizontal distribution*

The Commander squid has two different habitats. The larvae, juveniles and immature squid live in midwater, mostly in the mesopelagic layer, while maturing and the mature squids inhabit the bottom and near-bottom layer from the lower shelf to lower slope, predominantly in the upper bathyal zone (Kasahara *et al.*, 1978; Nesis, 1985, 1989a; Fedorets, 1986; Okutani *et al.*, 1988). This is reflected in the pattern of its range (Figure 2a). The range of adult squids is characteristic for other bottom animals living on the slopes while the total range of all specimens, including juveniles, covers the whole northern North Pacific, Bering, Okhotsk and Japan seas, but the greatest concentration of squids is over the slopes and they are scarce over the shelves and areas far from slopes in the open ocean (Kubodera *et al.*, 1983; Nesis, 1985; Okutani *et al.*, 1988). In the pelagic realm of the Okhotsk Sea *B. magister* is most common over depths of 400–600 m (Ilyinsky, 1991). In the western Bering Sea it is also most common over the slope and the Shirshov Ridge and rarer in the open part of the sea (Radchenko, 1992).

In general *B. magister* is distributed from the northern Bering Sea to Korea (Tsushima) Strait, eastern Honshu (southern part of the Tohoku area) and central California. The northern limit of distribution of adults is off Cape Navarin, while for juveniles it is east of the Gulf of Anadyr (Kondakov, 1941; Akimushkin, 1963; Okutani *et al.*, 1976, 1987, 1988; Naito *et al.*, 1977a; Bernard, 1980; Arkhipkin *et al.*, 1996). It is very common in the following areas: the slopes of the western (particularly from Cape Navarin to Olyutorsky Bay) and eastern Bering Sea, areas off Commander Is and oceanward from the northern and central Kurile Is, the western, southern and eastern parts of the Okhotsk Sea slope (particularly along western Kamchatka), the northern and north-western Japan Sea slope, the Yamato Bank complex, areas westward from the La Perouse (Soya) Strait (Rebun Bank) and off the Oki Is, the area seaward of eastern Hokkaido (Cape Erimo), the Gulf of Alaska and British Columbia (Skalkin, 1977; Bernard, 1980; Fedorets, 1986, 1991; Yuuki and Kitazawa, 1986; Bizikov, 1987; Shevtsov, 1988a,b, 1990; Nesis, 1989a, 1995; Ilyinsky, 1991; Kubodera, 1992, 1996; Radchenko, 1992; Okiyama, 1993a,b; Arkhipkin *et al.*, 1996). The range of this species is panboreal.

Figure 2 Ranges of some species of Gonatidae (schematized). (a) *Berryteuthis magister* (dots) and *Gonatopsis borealis* (vertical lines); (b) *Berryteuthis anonychus*; (c) *Gonatus onyx*; (d) *Gonatus madokai* (modified from Nesis, 1985).

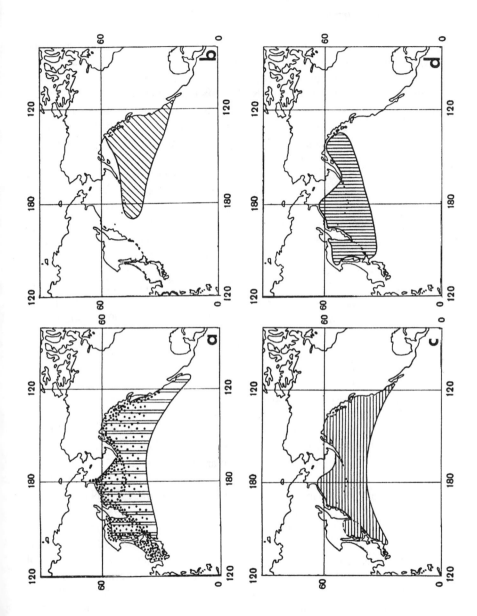

4.1.3. *Vertical Distribution and Diel Vertical Migrations*

Larval, postlarval and juvenile Commander squids inhabit the water column from the surface to the bathypelagic layer. In the Okhotsk Sea these squids have been caught down to 1500 m but the average abundance in the 300–500 m layer was twice that in the 500–1000 m layer (Nesis, 1989a). In the epipelagic (0–200 m) layer the total biomass of this species was 11 times less than in the mesopelagic (200–500 m) layer (Didenko, 1991a). In the western Bering Sea the total biomass of *B. magister* in the mesopelagic layer (200–500 m) was 3.5 times that in the epipelagic (0–200 m) and 6 times that in the bathypelagic layer (500–1000 m) (Radchenko, 1992); Didenko (1991b) indicated an even greater difference. Thus *B. magister* in the pelagic life stage is predominantly mesopelagic.

In the benthic life stage this squid is closely associated with the bottom in the upper bathyal zone, from 100–150 to 1200–1500 m. The shallowest recorded depth (in various areas) is approximately 50 m but this is clearly unusual. For example, in the Gulf of Anadyr juvenile *B. magister* were recorded at depths of 50–60 m but this is only excursion from the spawning zone to the south, near Cape Navarin (Radchenko, 1992). In the western Bering Sea the usual depth zone is from 100–200 to 800 m and the greatest catches usually were obtained between 200–250 and 400–600 m (Fedorets, 1977, 1979, 1983; Nesis, 1995; Arkhipkin *et al.*, 1996). In the eastern Bering Sea maximum catches were recorded at 200–700 m (Bakkala *et al.*, 1985). Off the Commander Is *B. magister* was found at 100–800 m, mainly at 100–300 m; off the Kurile Is it was at 50–1500 m, mostly at 150–600 m (Shevtsov, 1974; Naito *et al.*, 1977a; Railko, 1983; Malyshev and Railko, 1986; Alexeyev and Bizikov, 1987; Fedorets, 1987, 1991).

In the Okhotsk Sea it was recorded at the bottom in depths of 145–1500 m, mainly at 200–1000 m, with maximum catches at 400–900 m (Nesis, 1989a); a similar depth distribution was recorded at Kitamiyamato Bank in the south-western part of this sea: 500–950 m (Natsukari *et al.*, 1993a).

On the shelves and slopes of the Japan Sea it is distributed between 50 and 700 m, mainly at 100–400 m (Skalkin, 1977; Kasahara *et al.*, 1978; Nazumi *et al.*, 1979; Railko, 1979; Yuuki and Kitazawa, 1986; Shevtsov, 1988b) but on the offshore banks (Yamato complex, Rebun Bank) its depth range is wider, 225–1225 m, mostly 300–500 m (Ogata *et al.*, 1973; Shevtsov, 1988a; Natsukari *et al.*, 1993; Okiyama, 1993a,b). Off British Columbia it was recorded at 80–400 m (Bernard, 1980). The bottom temperatures in the Bering and Okhotsk seas and the north-western Pacific off the Commander and Kurile Is where *B. magister* occurred were −1 to +4.5°C, predominantly +1 to 3.6°C; in the Japan Sea they were predominantly 0 to +1.5°C (Fedorets, 1977, 1979, 1983; Railko, 1979). Pelagic juveniles were recorded in the open north-western Pacific at surface temperatures of approximately 3–10°C (Kubodera *et al.*, 1983).

Thus *B. magister* is widely distributed on continental and island slopes and the depths of maximal catches varied in different areas. This predominance in catches is probably determined more by the distribution of trawlable bottom than by the actual distribution of squids but in all areas *B. magister* tends to prefer the upper part of its depth range, being predominantly a mesobenthic species.

Larvae and postlarvae identified as *B. magister*, of dorsal mantle length (DML) 7–30 mm, are common and widely distributed in the near-surface layer of the Bering and Okhotsk seas and across the northern North Pacific from off the southern Kurile Is to central California (Kubodera and Okutani, 1981b; Kubodera and Jefferts, 1984).

Diel vertical migrations with night ascent are known but not well studied. All the life stages inhabiting the pelagic region migrate, even mature males (Shevtsov, 1990). Juveniles of DML 3–6 cm may occur in the upper water layer both by day and night though the larger ones (from 5 to 6 cm) occur at night only. Vertical migrations of juveniles probably do not exceed the limits of the epipelagic zone (Shuntov *et al.*, 1993a,b; Bizikov and Arkhipkin, 1996; Nesis and Nikitina, 1996).

During their benthic life stage squids usually concentrate at the bottom during the day and disperse in the near-bottom layer at night (Railko, 1983; Alexeyev *et al.*, 1989). This is correlated with day/night differences in the foraging activity, active feeding occurring at night (Fedorets, 1986). High catches at the bottom may be obtained in any time of day, more usually in the morning; concentrations of squid are observed during the morning and day at lesser depths than during the evening and night and the differences may be as much as 100–150 m (Fedorets *et al.*, 1985; Fedorets, 1987; Fedorets and Didenko, 1991; Fedorets and Luchin, 1991).

4.1.4. *Maturation and Ontogenetic Descent*

The sex ratio in *B. magister* populations is variable. In the western Bering Sea during the spawning season females greatly outnumbered males, but this ratio significantly changed with time, being usually 1:1.2 to 1:2.2. Before the main spawning time the male:female ratio was close to one but during the period of high activity it decreased to approximately 10 to one and there were some catches without any males (Fedorets, 1983; Fedorets and Kozlova, 1987, 1988; Nesis, 1995; Arkhipkin *et al.*, 1996). Off northern Hokkaido on both sides of the La Perouse Strait females were also much more abundant than males though the ratio changed with time and depth (Kubodera, 1992). In the deep layers of the Okhotsk Sea and oceanward of the Kurile Is this ratio was almost equal (Alexeyev and Bizikov, 1987; Nesis, 1989a). The cause of the change in ratio is that during spawning the males exhaust their reserves of sexual products and die earlier than females.

The sex is identifiable at DML 5 cm in males and 6 cm in females but may be observed by the naked eye at DML 10–11 and 12–13 cm respectively (Fedorets and Kozlova, 1987; Nesis, 1995). The beginning of maturation (the transition from I to II maturity stage) was observed in the western Bering Sea at DML 16–18 cm in males and 17–19 cm in females. Mass maturation occurs in males at DML 19–20 cm and in females at 23–24 cm (Fedorets and Kozlova, 1987; Nesis, 1995). In the Okhotsk Sea maturation begins at DML 14–18 cm in both sexes, and mass maturation occurs at 20–22 cm in males and 26–28 cm in females (Nesis, 1989a). In the north-western Pacific off Hokkaido all males larger than 20 cm and females larger than 27 cm are mature (Naito et al., 1977b). In the Japan Sea squids mature at a much smaller size: males at DML 13–15 cm, females at 17–18 cm (Railko, 1979; Yuuki and Kitazawa, 1986; Shevtsov, 1988a), but on the Rebun Bank in the northeastern Japan Sea all squids smaller than 20 cm were immature (Kubodera, 1992).

The process of maturation is rather protracted and squids of both sexes exceeding the size of mass maturation, but not yet mature, were observed in all areas, for example males of 23–24 cm and females of 28–30 cm in the western Bering and Japan seas or males larger than 25 and females larger than 30 cm in the Okhotsk Sea (Yuuki and Kitazawa, 1986; Nesis, 1989a; Kubodera, 1992). In every seasonal group early- and late-maturing females were observed (Arkhipkin et al., 1996), resulting in the wide size-range of mature females (to a lesser degree, males), for example from 17 to 41 cm in the Olyutoro–Navarin region (Fedorets and Kozlova, 1987; Nesis, 1995).

The general morphology of the reproductive system and the oo- and spermatogenesis of B. magister were described by Nazumi et al. (1979) and Reznik (1981a,b, 1982, 1983). Oogonia and stage I oocytes are present in the ovaries of juvenile females; at maturity stage I females contain oocytes of stage II in the transition from protoplasmic to trophoplasmic growth phase. In the next stages trophoplasmatic growth continues, then yolk accumulates and in the mature females the mature oocytes are ovulated and fall into the oviducts. The modal size of oocytes at I, II, III and IV stages of maturity are approximately 0.3; 0.4–0.5; 1 and 2 mm respectively (Reznik, 1981a; Bizikov and Arkhipkin, 1996).

The nidamental glands are large, and their length may reach 50–60% DML (Shevtsov, 1974; Nesis, 1995). They grow with positive allometry; growth rate increases after maturation (24–25 cm in the Okhotsk Sea squids) and then becomes linear. But the correlation between nidamental gland length, maturity stage and DML is weak, thus the nidamental gland index, widely used in some studies (Kubodera et al., 1983; Kubodera, 1992), is uninformative (Nesis, 1989a, 1995).

Mature eggs in the oviducts are all approximately of the same size, oval, light-yellow and transparent. In mature females the mass of mature eggs fills 20–60% of the volume of the oviducts. Some oocytes (c. 10–15%) in maturing and mature females cease development at the protoplasmatic growth stages and are resorbed without maturation (Reznik, 1981a, 1983).

The size of mature eggs is 3.0–4.2, averaging 3.5 mm in squids from the Bering and Okhotsk seas (Arkhipkin *et al.*, 1996'; Voight, 1996; Fedorets and Kozlova, 1987; Nesis, 1989a). In females from the Japan Sea they are larger: 4.2–5.9 mm (Nazumi *et al.*, 1979; Yuuki and Kitazawa, 1986). This size difference may be due to a difference in the measurement conditions. The average long axis length of freshly removed mature non-fertilized oviducal eggs measured in a small volume of sea water is 3.5 mm. Fresh non-fertilized eggs do not swell but soon after removal, or in defrosted eggs, the envelope deteriorates and when measured out of water the egg size increases.

Spermatocytes at stages I and II were observed in males at maturity stages I and II. Transition from maturity stage II to III and IV is rapid, spermatids appear followed by the first spermatophores (stage V). Spermatophores are 15–21 mm in length, averaging 18.6–18.8 mm in males of DML 19–29 cm. Variation in the length of spermatophores in one male may be as much as 3–8 mm (Shevtsov, 1974; Nazumi *et al.*, 1979; Fedorets and Kozlova, 1987).

Squids begin to descend from the pelagic region to the bottom at DML approximately 6–8 cm, but the main period of descent coincides with the beginning of mass maturation. In the western Bering and Okhotsk seas most squids are settled on the bottom at DML 14–16 cm and in pelagic tows from the Okhotsk Sea only 5% of squids were larger than 20 cm, the size at which the largest juveniles and first mature males were found (Nesis, 1989a, 1995; Fedorets, 1991). In pelagic tows most squids were juvenile or at maturity stage I, rarely II, whereas on the bottom maturing and mature squids predominated (Nesis, 1989a). Mature males may be found in pelagic tows (Fedorets and Kozlova, 1988; Shevtsov, 1990) but mature females are not, although there has been a report of one spawned female in a pelagic tow (Shevtsov, 1990). Thus the pelagic phase of the life cycle is over when squids reach approximately half their definitive size (Nesis, 1995). The feeding rate diminishes greatly after their descent to the bottom and during the second half of their life the squids depend mostly on accumulated lipid reserves in the digestive gland.

4.1.5. *Mating, Spawning and Fecundity*

Reproduction in *B. magister* is protracted and can occur during most or all the year, but nevertheless definite spawning seasons exist, usually two, but in the western Bering Sea probably three, and during each season one seasonal spawning group predominates (Fedorets and Kozlova, 1987; Nesis, 1989a, 1995; Arkhipkin *et al.*, 1996). In the western Bering Sea the main spawning seasons are: summer (June–August), autumn (September or October–November) and winter (December–April) or summer and autumn–winter; off the Commander Is they are November–January and April; in the Okhotsk and Japan Seas and in the Pacific oceanward of the Kurile Is they are summer (May or June–October) and winter–spring (December–April, with peak in January–March), the summer is the main spawning season (Panina, 1971; Shevstov, 1974, 1988a,b, 1990; Naito *et al.*,

1977a,b; Skalkin, 1977; Kasahara et al., 1978; Fedorets, 1979, 1983, 1987, 1991; Nazumi et al., 1979; Railko, 1979, 1983; Yuuki and Kitazawa, 1986; Bizikov, 1987; Fedorets and Kozlova, 1987; Nesis, 1989a, 1995; Kubodera, 1992; Okiyama, 1993b; Fedorets et al., 1994; Arkhipkin et al., 1996; Bizikov and Arkhipkin, 1996). The larvae of B. magister were recorded in March–April off British Columbia (Okutani, 1988) but in the northern North Pacific most were found in June–September (Kubodera and Jefferts, 1984). Back-calculation indicates two weakly separated periods of hatching in the squids on the Kitamiyamato Bank, south-western Okhotsk Sea (April–September and November) and Rebun Bank, north-eastern Japan Sea (May–June and December) (Natsukari et al., 1993a). In the western Bering Sea each month squids of 5–12, usually 7–8 monthly classes were encountered, with three predominating seasonal groups, winter-, summer- and autumn-hatched squids (Arkhipkin et al., 1996).

In areas of commercial fisheries the seasonal dynamics of squid concentrations of B. magister reflect the dynamics of spawning. For example, in the Olyutoro–Navarin region the density of squid shoals was low from late April to June, then the abundance steadily increased during July–September and reached the annual maximum in October, when the spawning activity of the most abundant autumn–spawned group approached maximal intensity and the catches sometimes exceeded $1 \, t \, h^{-1}$. By mid-November or late in November the spawning activity abated and the catches dropped suddenly to very low figures, sometimes less than $1 \, kg \, h^{-1}$. This picture was repeated from year to year with minor differences (Fedorets, 1983; Fedorets and Kozlova, 1987; Nesis, 1995; Arkhipkin et al., 1996).

Mating in B. magister occurs immediately before spawning. Only mature (stage V) females mate. Only one mated female was seen at stage IV (Nesis, 1995). During mating a male places a bunch of spermatophores, presumably with the long distensible penis, into the mantle cavity of a female and glues it to the inner mantle wall in the vicinity of the openings of the oviducal glands (Nazumi et al., 1979; Yuuki and Kitazawa, 1986; Fedorets and Kozlova, 1987; Nesis, 1989a, 1995; Okiyama, 1993b). The bunches are placed on one (either right or left) or both sides and there may be more than one (up to four) on one side, but rarely more than one bunch on both sides. Each bunch represents one mating event, thus a female rarely mates more than twice. There is no hectocotylus but about 10 pairs of sucker pedicels are enlarged in both dorsal rows in the central part of the left or right ventral arm in mature males (Voight, 1996). This may be an adaptation to control the position of the penis during copulation.

The average number of spermatangia in one mated female is 76–190 and is maximal in the smallest (22–23 cm) and largest (37–38 cm) females and during the peak of spawning: July and November in the western Bering Sea, November–January in the eastern Bering Sea, November–January and April off the Commander Is (Fedorets and Kozlova, 1987).

The number of spermatophores in males of DML 20–27 cm from the western

Bering Sea varies from 78 to 478 independently of male size but has a seasonal maximum in September–October (Fedorets and Kozlova, 1987). In the males from the Japan Sea and off the Commander Is there are, independently of male size, 100–240 spermatophores (Shevtsov, 1974, 1988a; Railko, 1979; Yuuki and Kitazawa, 1986). Penis length in mature males of DML 19–29 cm, with the Needham sac filled with spermatophores, is shorter (2–3 cm) than in those in the spent or close to spent stage (3.5 cm) (Fedorets and Kozlova, 1987).

The number of mature eggs in the oviducts varies from 3400 to 27 000 (Shevtsov, 1974; Fedorets, 1979; Fedorets and Kozlova, 1987). It increases with female size up to a maximum at DML of about 32–35 cm (in the western Bering Sea) and then sharply decreases. Average fecundity (sum of egg number in both oviducts and of mature eggs in ovary) increases from 5000 in females DML 23–24 cm to 29 000 at DML 34–35 cm (Fedorets and Kozlova, 1987). In females from the Japan Sea the number of mature eggs does not exceed 6000 (1200–5800) and the number of mature eggs plus oocytes larger than 0.5 cm is 4000–123 000, usually 6000–10 000 (Nazumi et al., 1979; Yuuki and Kitazawa, 1986; Shevtsov, 1988a).

Males are capable of many copulations and females produce not less than 6–8 and sometimes as many as 15 egg masses, 1000–1200 eggs in each, in the rather short spawning time (Reznik, 1981a,b, 1982; Fedorets and Kozlova, 1987). Spawning occurs at depths of predominantly 400–600 m in the Bering and Okhotsk seas and off the Commander and Kurile Is; in the Japan Sea spawning takes place from about 500 to more than 1000 m (Bizikov, 1987; Shevtsov, 1988a; Okiyama, 1993a,b; Bizikov and Arkhipkin, 1996).

The female cannot spawn without mating. Usually it mates and begins to spawn at the substage V_1 of full maturity. But if the mating is delayed for some reason, the female continues to accumulate mature eggs in the oviducts, going from substage V_1 to V_2 and finally to V_3 of maturation stage V, when its oviducts are overfilled with eggs. After it mates it begins to spawn eggs and its oviducts quickly empty. At the beginning of the spawning period not all mature females have mated and the proportion of mated females may be lower among the stage V_3 and in some cases among V_2 than among V_1 females. Only during the spawning rush (in mid-November in the Olyutoro–Navarin region during the autumn season) were 100% mature females mated (Nesis, 1995). The spawning proceeds with decreasing intensity and the first egg masses are far the largest. Spent females are not larger than the bulk of mature females and it is probable that the average egg number in an egg mass is independent of or weakly dependent on the female's size; thus smaller females may finish spawning and die earlier than large ones (Nesis, 1995).

The total number of eggs spawned by the entire population of B. magister in the western Bering Sea during the time of its maximal abundance (October, 1976) was estimated at 5.3–5.9×10^{12} (Fedorets and Kozlova, 1987).

There is no gelatinous degeneration in B. magister and mature individuals of both sexes have tough, palatable mantle tissues and large digestive glands. Spent squids are very different. In fully mature and spent males the testis is thin, semi-transparent and in a state of resorption (Reznik, 1981b). In some spent females only the oviducts are emptied and some yolky oocytes remain in the ovary, in others only a few small, whitish, yolkless oocytes remain. Nidamental glands in spent females are flabby, the mantle muscles are thin and flabby, and the body wet weight is half what it was before spawning. Some spent females have a rather large and strong digestive gland but in some others it is very thin. Such females often lose tentacles which are either autotomized as in mature females of *Gonatus* spp., or are torn off at random, frequently only one tentacle being lost (Okiyama, 1993b; our observations). The latter is a result of mechanical damage to the weakened tentacular stems. In some spent females the entire stock of spermatangia is exhausted and only the pedicels of the bunches of spermatangia remain (Nesis, 1995).

The smallest known larva of *B. magister* had DML 4.9 mm (Okutani, 1988). Taking into consideration the large egg size this larva was probably just hatched.

4.1.6.   *Population Structure*

The population structure of *B. magister* is very complex and can be analysed at three levels: spatial, temporal and dimensional (small-sized/large-sized groupings).

The spatial structure was investigated by genetic–biochemical methods using gel-electophoretic allozyme analysis (Glazachev and Fedorets, 1981; Katugin, 1988, 1990, 1993b, 1995a,b) and substrate-specific properties of cholinesterases (Kovalev and Rosengart, 1987; Kovalev, 1990).

Detailed investigations by Katugin (1993b, 1995a), based on a study of 2100 specimens from 18 samples, showed four loci with variant allele frequencies out of 24 presumptive loci of 14 enzymes and an unidentified ganglion protein found to be useful for population studies. No significant deviations from Hardy–Weinberg equilibrium were found at any locus and no evidence of genetic difference between sexes was found. Genetic variability is high compared with other squid species; the mean observed heterozygosity per locus was 0.13, the mean effective number of alleles 1.33. Significant differences were found between squids from three major geographic regions: (i) the Japan Sea; (ii) the Kurile, Commander Is and the western Bering Sea; and (iii) the Gulf of Alaska. Minor genetic differences exist within the second area but these were not spatial; they were differences between successive generations of squids in the same macrogeographic locality. The differences between the three geographically distinct regions were caused by genetic isolation; differences inside the second area may be caused by natural selection on early life stages or by some degree of isolation between successive breeding stocks/cohorts. Glazachev and Fedorets (1981) showed significant genetic differences not only between squids from the Gulf of Alaska and the western Bering Sea (Olyutoro–Navarin region) but also between the western and eastern Bering Sea (Navarin–Alaska region); Katugin had no samples from the latter area.

Kovalev (1990) and Kovalev and Rosengart (1987) found that the substrate-specific properties of cholinesterase from squid optic ganglia showed that the *B. magister* populations from the western (Olyutoro–Navarin), the eastern (Navarin–Alaska) Bering Sea and the Northwestern Pacific off the central Kurile Is. are all different, revealing intraspecific polymorphism.

Temporal population structure is represented by different seasonal spawning groups. As was stated above, in most areas of the *B. magister* range there are two or three main spawning periods: winter–spring and summer in the Japan and Okhotsk seas, autumn, winter and summer (or autumn–winter and summer) in the western Bering Sea (Fedorets and Kozlova, 1987; Nesis, 1989a, 1995; Fedorets, 1991; Natsukari *et al.*, 1993a; Arkhipkin *et al.*, 1996). Two or more spawning periods are known in very many squids, primarily Loliginidae and Ommastrephidae (Nesis, 1985), for example, in *Todarodes pacificus* there are three or four such groups (Shevtsov, 1978; Okutani, 1983). Probably this is a common feature for large, abundant, active and muscular cephalopods (Nesis, 1995).

In summer 1992, when the abundance of *B. magister* in the Okhotsk Sea was unusually low, six specimens of this species were caught off eastern Sakhalin at advanced maturity stages but of very small size: two males at stages II and III, DML 80–85 mm, and four females at stage II, DML 62–99 mm (Nesis and Nezlin, 1993). The transition from maturity stage I to II usually occurs at DML 140–180 mm (Nesis, 1989a); squids of this usual size range were also caught in the Okhotsk Sea though in very small quantity. It was supposed that the remarkable small but maturing specimens belonged to a previously unknown dwarf early-maturing group, analogous to an early-maturing group without dorsal photophore in the tropical Indo-Pacific ommastrephid squid *Sthenoteuthis oualaniensis* (Nesis, 1993b), but this group appears only in periods of very low abundance of the common late-maturing group (Nesis and Nezlin, 1993).

The stocks of *B. magister* in the Japan and Okhotsk Seas are isolated by the shallows of the La Perouse (Soya) Strait. They are morphologically and biologically different: the Japan Sea squids are smaller than those from the Okhotsk and Bering Seas and the areas off the Kurile and Commander Is, they mature earlier and have larger eggs (see Sections 4.1.1 and 4.1.4) (Nesis, 1989a; Kubodera, 1992). Although the maximum depth of the La Perouse Strait (140 m) is deeper than the upper limit of habitation of *B. magister* in the bottom layer (50–100 m), the larvae of this species inhabit the upper water layer and gene exchange between the Okhotsk and Japan Sea stocks is possible through the deep Kurile straits, the area off eastern Hokkaido and another deep strait, the Tsugaru Strait. The Okhotsk and Japan Sea stocks are probably effectively isolated because the shallows (30–60 m) of the La Perouse Strait are impassable to these squids. Squids from the Japan Sea and the vast region including the Bering and Okhotsk Seas and the Northwestern Pacific off the Commander, Kurile Is and eastern Kamchatka should be treated as belonging to two geographic subspecies, and genetic differences between them as suprapopulational ones. Whether this is applicable to the stock of the Gulf of Alaska is unknown but probable according to isozyme data (Katugin, 1995a).

There is no unequivocal answer to the question: are the seasonal spawning groups genetically isolated stock units or they are only parts of a non-subdivided population with protracted spawning? Among loliginid and ommastrephid squids there are examples of both situations (Nesis, 1985). Different seasonal groups are not clearly separated, squid hatching dates in a given area may cover the whole or almost the whole year (Natsukari *et al.*, 1993a; Arkhipkin *et al.*, 1996), and early and late-maturing individuals are found in some groups (Arkhipkin *et al.*, 1996). The regular sudden and strong decrease of squid concentration density in the western Bering Sea during late October–November, between the spawning times of the autumn- and winter-spawning groups (Fedorets, 1983; Fedorets and Kozlova, 1987; Nesis, 1995; Arkhipkin *et al.*, 1996), is evidence of some degree of isolation. This is consistent with biochemical–genetic data (Katugin, 1993b, 1995a).

Genetic differences between successive generations in the summer-, autumn- and winter-spawning groups of squids are less than between the stocks from the Japan and, say, the Okhotsk Sea and may be termed interpopulational. It is not possible to speculate about the differences between usual (late-maturing) and supposed early-maturing Okhotsk Sea grouping of *B. magister* (Nesis and Nezlin, 1993).

4.1.7. *Age, Growth and Life Cycle*

The life span of *B. magister* was first assessed using the dynamics of modal size and form of size distributions. Two different hypotheses were formulated: 1-year cycle (Naito *et al.*, 1977b; Fedorets, 1991) or 2-year cycle (Fedorets, 1979; Railko, 1979, 1983; Yuuki and Kitazawa, 1986; Okutani, 1988; Nesis, 1989a). The definitely three-modal size structure of squid population from the Okhotsk Sea during the summer favours the hypothesis of a 2-year life cycle; the three peaks representing 0–group, 1- and 2-year squids (Nesis, 1989a).

The first ageing data using statoliths and gladii produced inconsistent results: Bizikov (1991, 1996) showed that the age of squids at DML 197–233 mm, was 246–265 days by statoliths and 200–268 days by gladii, while Natsukari *et al.* (1993a) calculated the age of squids of nearly the same size from the Okhotsk and Japan seas as 4 or more years. The reason is that statoliths of *B. magister* have three kinds of growth increments, first, second and third order, with each second-order increment including 3–6 (usually 4–5) increments of the first order. Natsukari *et al.* (1993a) considered the narrowest (first-order) increments to be daily, while

Bizikov (1991) and Arkhipkin *et al.* (1996) took the second-order increments to be daily and treated the first-order ones as subdaily (discussion: Natsukari *et al.*, 1993b; Jackson, 1994). The analysis of data on the microstructure of statoliths and gladii by Arkhipkin *et al.* (1996) shows that the fall-hatched group of the western Bering Sea *B. magister* has a 2-year life cycle, including 6 months of embryonic development and 18 months of post-embryonic growth.

The growth of *B. magister*, assessed by the shifting of modal size, is very different according to different authors: 20 mm month^{-1} for the area off the southern Kurile Is (Petrov, 1988); 7.0 mm month^{-1} in males and 8.3 mm month^{-1} in females during summer off the Commander Is according to Shevtsov (1974) but only 2.5 and 5.8 mm month^{-1} for males and females respectively according to Fedorets (1987). Rates of 7.0–12.5 and 8.4–12.6 mm month^{-1} for males and females during summer and 1–3 mm month^{-1} during winter in Japan Sea squids were calculated by Railko (1979); but lower rates of summer to autumn growth of 2.4 mm month^{-1} in males and 4.6 mm month^{-1} in females on the Kita–Yamato Bank were found by Shevtsov (1988a).

Results of studies by Kubodera (1992) and Natsukari *et al.* (1993a) treated *B. magister* as a very slow-growing and long-lived species. The squids from the Japan Sea have DML 14–15 cm at the age of 3 years and 21–23 cm at 4 years, those from the south-western Okhotsk Sea have DML 17–18.5 cm at 3 years and 25–30 cm at 4 years. The growth was approximated by a double logarithmic model. Growth rate between the 2nd and 3rd year was evaluated at 55 mm year^{-1} in males and 60 mm year^{-1} in females in the Japan Sea and 64 and 82 mm year^{-1} respectively in the south-western Okhotsk Sea (Natsukari *et al.*, 1993a).

Data presented by Arkhipkin *et al.* (1996) confirmed that the Commander squid is a slow-growing and slow-maturing species but not as slow as it was supposed by Kubodera (1992) and Natsukari *et al.* (1993a). The growth was fitted to a logarithmic curve. Daily growth rate in DML in the western Bering Sea increased from 0.27–0.35 mm day^{-1} in juveniles to 0.9 mm day^{-1} in 6–7-month-old males and 1.05–1.08 mm day^{-1} in 7–9-month-old females (this corresponds to 8.1–10.5, 27, and 31.5–32.4 mm month^{-1}) and then decreased to the juvenile rate in females and even less in males of 1-year or older. Daily growth rates in body weight were greatest in males aged 9–10 and females 10–11 months.

Males began maturing earlier than females. All fall-hatched males were at stage I in 7 months, half reached stage II in 8 months and most were mature in 10 months. The first spent fall-hatched males were 330–340 days old and most transit into the spent condition probably much later than 12.5 months. In fall-hatched females maturation began at 8–9 months and the transit from stage I to II occurred at approximately 10 months At 11 months there were some immature females but most were maturing and a few females were spent. The proportion of females at stage II decreased and those at stage IV increased between 11 and 12 months. Most fall-hatched females matured and mated in approximately 1 year (11–13 months) (Arkhipkin *et al.*, 1996; Bizikov and Arkhipkin, 1996).

Squids aged 1 year (after hatching) have DML of about 23 cm (males) and 26 cm (females). The bulk of the spent autumn-hatched squids are 13–14 months, summer-hatched are 12–14 months and the oldest specimens are about 16 months old. Growth rates may differ between seasonal groups because the summer-hatched females mature being 2 cm smaller than the autumn-hatched ones (Arkhipkin *et al.*, 1996). A 2-year life cycle is reported also for *Gonatus fabricii* (Kristensen, 1983; see discussion: Jackson, 1994) and *G. madokai* (see Section 4.12).

4.1.8. *Horizontal Migrations*

The study of active horizontal migrations in *B. magister* has been impeded by low survival of squids in bottom trawls and the absence of marking experiments. All conclusions about migrations are based on circumstantial evidence.

There are indications that migrations up and down the slope occur in *B. magister*. For example, in the central part of the Japan Sea squids are concentrated during the summer in depths of 100–250 m and during winter in 200–350 m (Railko, 1979). On the Kitamiyamato Bank in the south-western Okhotsk Sea juveniles smaller than 20 cm tend to concentrate at depths less than 500 m while maturing and mature females (larger than 27 cm) are deeper than 500 m (Kubodera, 1992). The same was observed in the Pacific off Simushir Is (central Kurile Is): immature squids are concentrated at 200–300 m while mature ones are at 400–500 m (Alexeyev and Bizikov, 1987). In the Olyutoro–Navarin region in the western Bering Sea during early autumn the immature foraging squids are most common on the outer shelf at 100–200 m, prespawning mature females are at 300–400 m and mature males are at 400–600 m; later both sexes mix on the spawning grounds. In general, squids are most common during summer at 400–500 m and during autumn at 300–400 m (Fedorets, 1977, 1979, 1983; Bizikov and Arkhipkin, 1996).

The spawning and foraging areas of the Commander squid may coincide, as in the Okhotsk Sea or in the central part of the Japan Sea, but they may be spatially separated. The Olyutoro–Navarin region is primarily a spawning area while the areas off northern and central Kurile Is are primarily foraging ones and those off the Commander Is serve sometimes as spawning and at other times as foraging areas. Seasonal dynamics of catches show that squid shoals move in waves to and from the foraging and spawning areas. When these areas coincide spatially, squids may use them successively for feeding and spawning and when they are different the squids migrate actively from one to another (Arkhipkin *et al.*, 1996). It is believed that in the Olyutoro–Navarin region the squids of the autumn-spawned group migrate to the spawning sites from the east, from St Matthew Is area (Arkhipkin *et al.*, 1996; Bizikov and Arkhipkin, 1996).

It is possible that there are active migrations of squids between the areas off the Kurile/Commander Is and the western Bering Sea (Fedorets, 1987; Fedorets *et al.*, 1994). Seasonal dynamics of abundance and genetic–biochemical data (Katugin, 1993b, 1995a) do not contradict this proposal, but the direction of the sea currents would hinder squid migration between the Kurile and Commander Is. The drift of squid larvae and postlarvae from the Commander Is to the area oceanward of the Kurile Is would be easy as it coincides with the Kamchatka and Kurile currents, the sources of the cold Oyashio Current (Dodimead *et al.*, 1963; Favorite *et al.*, 1976), but the return of squids to the Commander Is would hardly be possible because they would have to swim against the current and cross very deep straits (Krusenstern, Fourth Kurile and Kamchatka straits).

Exchange would be possible between the Commander Is and the Olyutoro–Navarin region using two main counterclockwise circulations, westward of the Shirshov Ridge and between the Shirshov and Bowers ridges (Shuntov *et al.*, 1993c). Such a migration of squids from the feeding grounds in the Olyutoro–Navarin region to the Commander Is for spawning is hypothesized by Arkhipkin *et al.* (1996). Assuming that the migratory swimming speed of *B. magister* is the same as that of *Illex argentinus*, 0.46 total length per second, they calculate that females and males might reach the spawning areas off the Commander Is in 66–73 and 80–92 days, respectively. The larvae hatched here might travel with the current to the Olyutoro–Navarin region and thus complete their life cycle. In fact, *B. magister* cannot swim as swiftly as *I. illecebrosus* because its mantle is less muscular and more watery but, nevertheless, the temporal dynamics of squid concentrations (Fedorets, 1987) support the supposition about active migrations of *B. magister* in the western Bering Sea (Arkhipkin *et al.*, 1996).

Exchange between squid stocks of the Okhotsk Sea and the north-western Pacific off the Kurile Is is possible since there are strong ingoing and outgoing currents in the Kurile Straits. Their strength and direction as well as the overall circulation pattern in the Okhotsk and Bering seas show interannual and longer fluctuations (Shuntov *et al.*, 1993c; Radchenko, 1994) and this may lead to redistribution of the larvae and juveniles and to large-scale mixing of squid populations (Fedorets, 1991). This may be the reason for the absence of large genetic differences between squids from the Okhotsk, western Bering Sea and the north-western Pacific off the Commander Is, eastern Kamchatka and the Kurile Is (Katugin, 1993b, 1995a). However, squids from the Olyutoro–Navarin region cannot reach even the westernmost of the Aleutian Is. Thus they are as isolated from the Aleutian–Gulf of Alaska stocks as the Okhotsk Sea squids are isolated from those in the Japan Sea (see Section 4.1.6).

4.1.9. *Underwater Behaviour and Behavioural Responses to Meteorological and Oceanographic Factors*

The underwater behaviour of *B. magister* was studied from the manned submersible "Sever 2" in the north-western Pacific off the northern and central

Kurile Is at depths of 100–900 m, mostly 200–600 m, and in the Japan Sea off Cape Gamov and on the Kita–Yamato Bank at 140–700 m, mostly 400–540 m (Alexeyev *et al.*, 1986, 1989). Squids (DML > 13–14 cm) were observed during the day to be 5–10 m off the bottom; they were rarely as much as 20–30 m off the bottom or very close to the bottom; at night they were far above the bottom. They were seen singly or in swarms of 2–3 to 30 and more, usually 5–7 squids in a swarm. They were oriented in the same direction and moved together. Average size of a swarm was $2 \times 3 \times 6$ m and distances between neighbouring squids were 0.3–3 m. Squids swam either by jetting (average speed of a squid DML 25–30 cm was up to 1 m s^{-1}, usually to $0.6–0.8 \text{ m s}^{-1}$) or by fin wave (average speed $0.2–0.3 \text{ ms}^{-1}$). The latter mode was more usual than jetting. Squids were concentrated predominantly in areas with complex bottom relief.

Six types of response were found to the constant light of the submersible's searchlight: (i) squids attack the submersible; (ii) they approach the submersible and detour; (iii) they cross the light zone without approaching the apparatus; (iv) they approach from behind and touch the lighting appliance; (v) they lie on the bottom; and (vi) they lose orientation. Squids attacking the submersible usually bumped into it, inked and vomited food. During this accident they commonly experienced a light shock and stopped motionless in a characteristic posture with extended arms and tentacles for a time, from 5 to 20–30 s, then escaped. The squids responded to the flashlight either with a sudden change of swimming direction, increased speed and inking, or with light shock with a typical latency time of 1–2 s. A colour change from dark-brown to purplish-white was usual; it was strongest in the state of light shock, but never observed in squids lying on the bottom (Alexeyev and Bizikov, 1987; Alexeyev *et al.*, 1986, 1989).

In all areas of its range, except the Japan Sea, *B. magister* occurs mostly in the warm intermediate layer and mostly (in particular during foraging) closer to its upper boundary, where the feeding conditions are best (Fedorets, 1986; Fedorets and Kun, 1988). The fishery conditions for the Commander squid off the northern and central Kurile Is are very closely correlated with tides, moon phases, winds and other factors of sea state. The catches increase 2–3 h before or after high tide. They fall significantly during both full and new moon. During the lunar month (28 days) the catches of squids increased in the days of lower (neap) tides. Wind changes may alter the time of squid concentration relative to tides by as much as 3 days earlier or later. Long periods of strong north and north-west winds resulted in the outward flow of cold water from the Okhotsk Sea into the Pacific and a decrease in squid catches. Catches increased after long periods of moderate south and south-east winds but this connection depends on the season (Fedorets *et al.*, 1985; Fedorets, 1991; Fedorets and Didenko, 1991). Such non-seasonal change in fishery conditions is not characteristic in other fishing areas. The cause is that small areas of commercial fishery off the northern and central Kuriles are situated near the straits connecting the Okhotsk Sea with the Pacific Ocean with strong deep currents. Small sporadic cyclonic eddies arise there, associated with enhanced productivity and concentrations of plankton (Malyshev and Railko, 1986). In most other areas the fishery conditions are more stable.

4.1.10. Food and Feeding

The food of the Commander squid in the mesopelagic layer of the Okhotsk Sea during June–August consists of planktonic (76% of food by mass) and nektonic (24%) organisms. Among the former the most common food items are euphausiids (52.0%, including *Thysanoessa longipes*, 48.3%) and hyperiids (17.2%, including *Themisto japonica* 12.3% and *T. pacifica* 3.6%). Other planktonic organisms in the food are copepods (3.7%), chaetognaths (1.3%) and coelenterates (1.1%). Among nektonic animals there are small mesopelagic fish (the commonest fish in that sea) *Leuroglossus schmidti* (14.4%), and squids (*B. magister* 4.9% and *G. borealis* 4.8%). The daily food ration was assessed at 1.98% of fresh body weight (Ilyinsky and Gorbatenko, 1994).

During autumn–winter, in the western Bering and Okhotsk Seas juvenile squids consumed euphausiids (71–89% by weight in the Bering and 44–47% in the Okhotsk Sea, mostly *Thysanoessa raschii* and *T. inermis* in the former and *T. longipes* in the latter), hyperiids (*Themisto japonica*, *T. libellula* and *Primno macropa*, 47% in the epipelagic layer of the Okhotsk Sea), smaller quantities of copepods (*Calanus cristatus*, *C. plumchrus*, *Pareuchaeta elongata* and *Metridia pacifica*), chaetognaths (*Parasagitta elegans*), crab larvae, juvenile fishes (*Leuroglossus* and others), squids and coelenterates. Fifty to a hundred per cent of stomachs contained food and the weight of the stomach contents reached 0.7–5.4% body weight. The size range of the food organisms was approximately 2–100 mm. Daily food ration was assessed at 0.8% (Gorbatenko *et al.*, 1995).

The Commander squid feeds less actively in its benthic life stage, in most areas. The proportion of empty stomachs in bottom-caught squids in spawning areas is usually more than 70% (Naito *et al.*, 1977b; Kuznetsova and Fedorets, 1987; Nesis, 1989a, 1995; Okiyama, 1993b). Feeding activity clearly decreases in the course of maturation. In the Okhotsk Sea (pelagic and bottom tows together) the proportion of empty stomachs was 52% in males at maturity stages I–IV and 56% in females at stages I–III, but 83% in females at stage IV, 92% in mature males and 100% in mature females. In the western Bering Sea it varied from 69.5% in females at stages I–IV and 79% in males at stages I–IV, to 90.5–90.7% in mature males and females, also in juveniles (Nesis, 1989a, 1995). In foraging areas, particularly off the central Kuriles and Commander Is, feeding is active even on the seabed (Shevtsov, 1974; Alexeyev and Bizikov, 1987; Kuznetsova and Fedorets, 1987; Fedorets and Kun, 1988; Fedorets, 1991). The daily ration of maturing and mature squids during autumn–winter time was evaluated at 0.9–1.4% in the western Bering and Okhotsk seas (Gorbatenko *et al.*, 1995).

In most areas crustaceans (mostly euphausiids and hyperiids), fish and squids predominated in the food of benthic adults (Shevtsov, 1974; Fedorets, 1977; Naito *et al.*, 1977b; Skalkin, 1977; Railko, 1983; Okutani *et al.*, 1988; Nesis, 1989a, 1995; Gorbatenko *et al.*, 1995). In the Bering and Okhotsk Seas and off the Commander and Kurile Is euphausiids included *Thysanoessa longipes*, *T. raschii*, *T. inermis*, *T. inspinata* and *Euphausia pacifica*; hyperiids included *Themisto japonica*, *T. pacifica*, *Hyperia galba*, *Primno macropa* and *Cyphocaris* sp., among fish – juvenile Alaska pollack (*Theragra chalcogramma*), *Lipolagus ochotensis*, *Leuroglossus schmidti*, Myctophidae (incl. *Stenobrachius leucopsarus*), Sternoptychidae (all these fishes are pelagic), benthic cottoid fishes and fish fry. Squids were mostly conspecifics. There were also shrimps *Pandalus goniurus*, crabs, pelagic gastropods *Clione limacina*, and small quantities of copepods (*Pareuchaeta elongata*, *Calanus cristatus*, *Eucalanus bungii* and *Metridia pacifica*), ostracods *Conchoecia*, chaetognaths (*Parasagitta elegans*), coelenterates and others. The size range of food organisms was wide, 3 to 100–150 mm (Shevtsov, 1974; Malyshev and Railko, 1986; Alexeyev and Bizikov, 1987; Kuznetsova and Fedorets, 1987; Fedorets and Kun, 1988; Nesis, 1989a, 1995; Gorbatenko *et al.*, 1995;).

In the Japan Sea the food of *B. magister* consists of small fishes, *Maurolicus japonicus* and *Bothrocara hollandi*, squids *B. magister* and enoploteuthids *Watasenia scintillans* and *Enoploteuthis chuni*, and various crustaceans: euphausiids (*Thysanoessa longipes* and *Euphausia japonica*), hyperiids (*Themisto japonica* and *Primno macropa*), mysidaceans (*Meterythrops microphthalma*), shrimps (?*Argis* sp.) and copepods (*Pareuchaeta japonica*) (Kasahara *et al.*, 1978; Yuuki and Kitazawa, 1986; Okiyama, 1993b). Thus the food of *B. magister* near the seabed consists mainly of pelagic and benthopelagic organisms while benthic organisms are a minor component.

As in most squids, the juveniles feed primarily on crustaceans and as they grow the proportion of squid and fish eaten increases, particularly in squids larger than about 16 cm. Thus, crustaceans are usually the main food in the pelagic stage, fish and squids in the benthic stage (Nesis, 1989a; Bizikov and Arkhipkin, 1996). However, on the north-western and north-eastern slopes of the Bering Sea squids of all sizes consume large quantities of squids (sometimes up to 100% of the food in juveniles and to 85% in adults). The proportion of euphausiids is largest on the upper slope (80–200 m); at greater depths (500–800 m) the proportion of decapod crustaceans and squids increases (Fedorets and Kun, 1988; Kuznetsova and Fedorets, 1987; Gorbatenko *et al.*, 1995). In the Pacific off the Kurile Is the intensity of feeding is fairly constant during January–May; average stomach contents weight index is 0.1%, but increases to 2.25% in June, and when *Lipolagus* is the main food: 55.4% (Fedorets and Kun, 1988). In the western Bering Sea feeding is most active in March, June–July and November. The proportion of crustaceans is greatest during April–May, the proportion of squids is greatest during January–February, August and October–November; fish are eaten intensively all the year round. When spawning intensifies, feeding activity decreases dramatically.

Squids feed more actively at night than in the day: maximum feeding activity in pelagic juveniles was recorded (during the autumn-winter season) near midnight in the western Bering Sea and before sunrise in the Okhotsk Sea (Fedorets, 1986; Gorbatenko et al., 1995). However, at the bottom off the Kurile Is the squids feed more actively during the day or in the evening (Railko, 1983; Alexeyev and Bizikov, 1987). At the bottom during the day squids eat mostly juvenile gadoid fish and conspecifics, while at night, being off the bottom, they feed predominantly upon crustaceans and pelagic fish.

Foraging by B. magister occurs mostly in the pelagic life stage (Shevtsov, 1990). Juvenile and maturing squids consume mostly crustaceans (euphausiids and hyperiids) and accumulate reserves in their voluminous digestive gland (liver), whose weight may reach up to 25–40% of wet body weight (Kubodera, 1992), sufficient for use after descent to the seabed, during final maturation and spawning. Feeding in the benthic stage is less active in foraging areas (where fish and squids usually predominate) and very low during the spawning period. Mature squids do not feed (Nesis, 1989a, 1995).

4.1.11. Predators

The Commander squid is preyed on by a very large number of fishes, squids, marine mammals, seabirds and other predators. Among them are: Alaska pollack (Theragra chalcogramma), Pacific cod (Gadus macrocephalus), chinook salmon (Oncorhynchus tschawytscha), chum salmon (O. keta), coho salmon (O. kisutsch), pink salmon (O. gorbuscha), red salmon (O. nerka), Pacific pomfret (Brama japonica), Greenland halibut (Reinhardtius hippoglossoides matsuurae), various grenadiers (Macrouridae), a liparid (Polypera simushirae), (baird's beaked whale (Berardius bairdi), short-finned pilot whale (Globicephala macrorhynchus), Dall's porpoise (Phocoenoides dalli), northern fur seal (Callorhynus ursinus), sperm whale (Physeter catodon), thick-billed murre (Uria lomvia) and B. magister itself (Panina, 1966, 1971; Nishimura, 1970; Pearcy and Ambler, 1974; Okutani et al., 1976, 1988; Okutani and Satake, 1978; Ogi, 1980; Fiscus, 1982; Makhnyr et al., 1983; Dolganova, 1988; Sanger and Ainsley, 1988; Yang and Livingston, 1988; Kubodera and Miyazaki, 1993; Kuramochi et al., 1993; Chuchukalo et al., 1994a,b; Ilyinsky and Gorbatenko, 1994; Sinclair et al., 1994; Sokolovsky and Sokolovskaya, 1995; Orlov and Pitruk, 1996; Savinykh and Chuchukalo, 1996; Sobolevsky and Senchenko, 1996). This list is far from complete.

4.1.12. Assessed Biomass and its Interannual Population Dynamics

The values for biomass of B. magister in the bottom layer, from various surveys, are as follows (thousand metric tonnes): Olyutoro–Navarin region of the western Bering Sea, 1976, 350–390; Commander Is, 1983–1984, 8–11; Kurile Is off

Simushir Is, 1985–87, 109–233; eastern Bering Sea, US zone, 1979, 15–30, average 22.6; Aleutian Is area, 1980, 18.1; Japan Sea, Kita–Yamato Bank 12.7 (Bakkala *et al.*, 1985; Ronholt *et al.*, 1986; Bizikov, 1987; Fedorets, 1987; Shevtsov *et al.*, 1988; Shevtsov, 1988b, 1991; Orlov, 1990). These figures are not strictly comparable because, first, they were obtained using different catchability coefficients and second, the biomass of *B. magister* changes from year to year. For example, the assessed total biomass of this species in the mesopelagic layer of the Okhotsk Sea varied from 219 thousand t during February–April, 1990 to 1627 thousand t during August, 1989 and the squid concentration varies from 481 to 1818 kg km^{-2} (Lapko, 1995). The assessed biomass in the bottom layer off the Commander Is varied from 47.5 thousand t in 1978 to 11.0 thousand t in 1979; in the bottom layer of the western Bering Sea it varied from 350–390 thousand t in 1976 to 66–75 thousand t in 1977 (Fedorets, 1987). In summer 1992 the abundance of *B. magister* in the epipelagic layer of the Okhotsk Sea and the north-western Pacific off the Kurile Is was unusually low (Nesis and Nikitina, 1996), although in the previous summer (1991) juvenile *B. magister* were the commonest squids oceanward of the Kuriles (Shuntov *et al.*, 1993b). The biomass of this species in the upper epipelagic layer on the oceanic side of the Kurile Is fell from 109 thousand t to 35 thousand t in 1993 (Savinykh and Chuchukalo, 1996). Substantial interannual changes were reported in the quantity of this species in the food of northern fur seals in the Japan Sea (Panina, 1966). The causes of these rapid changes are unknown.

4.2. *Berryteuthis anonychus (Pearcy and Voss, 1963)*

The smallfin gonate squid is a small, rather muscular species, DML to 15 cm (Roper *et al.*, 1984; the 20 cm in Kuramochi *et al.* (1993) is doubtful). Its distribution is of low-boreal distant-neritic type, not truly oceanic (Figure 2b). It is found from the south-eastern Bering Sea and the Gulf of Alaska to southern Baja California but becomes rare south of Oregon. It extends westward to 160–180°W between approximately 40 and 53°N and is most abundant in the Gulf of Alaska and the open subarctic domain to the south and south-west of this gulf. It occurs in the open ocean over great depths but is commonest over the slope and seamounts: Eikelberg Ridge, seamounts Warwick, Miller and Morton (Pearcy and Voss, 1963; Nesis, 1985; Okutani *et al.*, 1988; Didenko, 1990). The juveniles are widely dispersed over the north-eastern Pacific and the south-eastern Bering Sea (Kubodera and Jefferts, 1984; Kubodera, 1986;). This is a gregarious species (Kubodera and Shimazaki, 1989).

The spawning habitat is unknown. The squids have been recorded from depths as great as 1000–1500 m (Anderson, 1978; Roper *et al.*, 1984; Okutani *et al.*, 1988) but adult squids are always caught in the epipelagic layer, 0–200 m, and are concentrated mostly over the seamounts (Didenko, 1990) and slope (Pearcy and

Voss, 1963). The distant-neritic range of this species is different from the trans-oceanic ranges of many other gonatids. It may spawn either in the epipelagic realm or near the seabed on or over the continental slope and seamounts. If it spawns in the epipelagic layer, it does not undergo ontogenetic descent but if it spawns near the seabed, it must undergo such a descent, but only shortly before spawning. Maturation begins at DML 6–7 cm (Roper *et al.*, 1984). There is no gelatinous degeneration.

Spawning, judging from the occurrence of larvae, is year-round with two peaks: February–April and late June–September (Kubodera and Jefferts, 1984). Assuming the growth rate is 10 mm month^{-1} (Kubodera and Shimazaki, 1989), the life cycle may be assessed at approximately 1 year.

This species undergoes diel vertical migrations in the epipelagic zone, living during the day mostly in the 50–200-m layer and at night beween 0 and 150 m, including the surface layer (Pearcy and Voss, 1963; Roper and Young, 1975; Didenko, 1990). Thus, it is a first-order diel vertical migrator according to the classification by Roper and Young (1975) or belongs to the first type of diel vertical migrators according to Nesis (1985).

Concentrations of *B. anonychus* were observed in 1984 and 1986–1989 between 48 and 53°N at 170–176 and 130–145°W. The abundance, size and maturity of squids during spring and summer increased towards the north. Over the seamounts listed above there were concentrations of mostly mature squids, DML 85–128 mm (Didenko, 1990).

Squids fed actively in rings and vortices where there was a high concentration (to 600 mg m^{-3}) of intermediate and larger zooplankton; 58–72% of stomachs were full or semi-full. In 56 specimens studied in early July, 1986 (DML 75–127, average 100.4 mm, weight 12–64, average 30.6 g) caught in daytime at 100–150 m, the weight of the stomach contents was 0.043–4.21% of body weight (Lapshina, 1988; Didenko, 1990). Sixteen taxa were identified: copepods, amphipods, euphausiids, ostracods, pteropods, siphonophores, and chaetognaths. Copepods (*Calanus cristatus*, *C. plumchrus*, *C. pacificus* and *Metridia lucens*) were found in 56% of stomachs; hyperiid amphipods (*Themisto pacifica*, *Hyperia galba*, *Oxycephalus* sp. and others) were found in 52%; pteropods (*Limacina helicina* and *Euclio sulcata*) in 40%; and euphausiids (*Euphausia pacifica* and others) in 28%. *Berryteuthis anonychus* is a macroplanktophagous squid (Lapshina, 1988).

During zoological investigations the squids were fished with large pelagic trawls with a mesh too large for such small animals; in fact they were caught only by the fine-meshed (10 mm) cover. Nevertheless, the catches reached 1–3 t per haul and 5–7 t day^{-1}. The total stock in the area studied was assessed at 200–250 thousand t (Didenko, 1990).

B. anonychus of all sizes, juveniles to adults, are preyed on by pomfrets *Brama japonica*, Pacific salmon (*Oncorhynchus nerka*, *O. keta*, *O. gorbuscha*, *O. tschawytscha* and *O. kisutsch*), steelhead trout *Salmo gairdneri*, thick-billed murre

Uria lomvia and short-tailed shearwaters *Puffinus tenuirostris*. They were also eaten by neon squid *Ommastrephes bartramii*, albacore *Thunnus alalunga*, lancetfish *Alepisaurus ferox*, Dall's porpoise *Phocoenoides dalli*, fur seal *Callorhinus ursinus*, fin whale *Balaenoptera physalis* and others (Ogi, 1980; Ogi *et al.*, 1980; Kawamura, 1982; Okutani *et al.*, 1988; Pearcy *et al.*, 1988; Sanger and Ainsley, 1988; Kubodera and Shimazaki, 1989; Kuramochi *et al.*, 1993; Savinykh, 1993, 1995).

4.3. *Gonatopsis (Boreoteuthis) borealis* Sasaki, 1923

The boreopacific, or northern gonatopsis is a robust, muscular, nektonic, gregarious squid of intermediate size (DML to 27 cm in males and 33 cm in females). *G. borealis* is one of the commonest and most widely distributed gonatids. It is a panboreal species distributed (Figure 2a) from the northern Bering Sea slope and the Gulf of Alaska to north-eastern Honshu (40°N) and the area southwestward of the southern end of Baja California (20°N). It is very eurybathic and can be found from the surface to bathypelagic and abyssal depths. One maturing male (DML 16 cm) was taken in a bottom tow at the greatest depth of the Okhotsk Sea, 3400 m (Nesis and Shevtsov, 1977), but it may have been caught during the hauling of the trawl.

  G. borealis is represented in the north-western Pacific and Okhotsk Sea by two partly sympatric groups, small-sized early-maturing and large-sized late-maturing. Squids of the former group are usually less than 18 cm DML, those of the latter, larger than 19–20 cm. They are separated spatially in open areas of the northern North Pacific although there is a wide area of mixing: the former group are distributed predominantly to the north, the latter to the south of 45–47°N (Naito *et al.*, 1977a,b; Kubodera *et al.*, 1983; Nesis, 1989a; Nesis and Nezlin, 1993). In the Okhotsk Sea, where the former group completely dominates juvenile squids much smaller than the size at which maturation begins (DML 10–12 cm) were the most common and mature males were absent. In the North Pacific oceanward of the Kurile Is and southward of the Aleutians, juveniles and also maturing and mature males and maturing females of both groups were common in the upper epipelagic layer (Kubodera *et al.*, 1983; Nesis and Nezlin, 1993).

  In our samples taken during July–August 1992, *G. borealis* was the most abundant squid both in the Okhotsk Sea (68.2%) and in the north-western Pacific off the Kurile Is (28.8%); its biomass shared first and second place with *Todarodes pacificus* (35.4% in the former and 35.9% in the latter area). The occurrence, abundance, biomass and average size in the epipelagic tows were much higher in the north-western Pacific than in the Okhotsk Sea, where the small-sized squid group predominated (Nesis and Nikitina, 1996).

  In the deep layers of the Okhotsk Sea in September–October, 1984 *G. borealis* was recorded in midwater at 300–1500 m and at the bottom at 400–1375 m, but it was much more common in pelagic tows, where the average occurrence was 95% against 19% in bottom tows. The average abundance was greatest at 300–500 m while only single specimens were caught deeper than 1000 m and on the bottom. The average size was the same, about 80 mm, both in midwater and bottom tows and did not change significantly with depth (Nesis, 1989a).

In other years, *G. borealis* held either first or second (after *Berryteuthis magister*) place as biomass in the Okhotsk Sea. Its distribution in this sea mirrors that of all squids: mostly over the slopes of eastern Sakhalin, the Kurile Is and western Kamchatka. The densest concentrations are recorded usually along the western Kamchatka slope over depths of 400–800 m (Ilyinsky, 1991). Its total biomass during various surveys undertaken in 1989–1991 was estimated at 267–285 thousand t; its density was assessed at 159–296 t km^{-2} in the mesopelagic and 332 t km^{-2} in the bathypelagic layer (Didenko, 1991a; Ilyinsky and Gorbatenko, 1994; Lapko, 1995). In the north-western Pacific off the Kurile Is its biomass in the epipelagic layer was asessed at 100 thousand t (Shuntov *et al.*, 1993b).

G. borealis is also very abundant in the western Bering Sea. It is distributed from the epipelagic to the bathypelagic layer, but mainly in the mesopelagic layer. High concentrations were recorded in the mesopelagic layer over and near the slope off the Koryak Shelf and the Shirshov Ridge but it was also found over the whole deep-water part of the sea. Its biomass in two surveys during 1990 was assessed at 142–210 thousand t (Didenko, 1991b; Radchenko, 1992; Shuntov *et al.*, 1993a).

G. borealis is common and widely distributed in the upper water layers off eastern Hokkaido and in the northern North Pacific between 43° and 52°N at 155°E–175°W, mostly north of about 44–45°N in the western North Pacific and north of 50°N in the Gulf of Alaska. Here it is caught by jigs and surface gillnets with *Onychoteuthis borealijaponica* and *Ommastrephes bartramii*, but in much smaller quantities. It is caught in the upper layers only to the north of the subarctic front. The small-sized early-maturing group inhabits the Bering and Okhotsk seas and the North Pacific north of 43°N; the large-sized late-maturing group only occurs in the North Pacific from off north-eastern Honshu, Hokkaido and the southern Kurile Is and eastward across the ocean, mainly south of 48°N (Murakami, 1976; Murata *et al.*, 1976; Naito *et al.*, 1977a,b; Okutani, 1977; Kubodera *et al.*, 1983; Osako and Murata, 1983; Kobayashi *et al.*, 1986; Kubodera, 1986; Okutani *et al.*, 1987; Yatsu *et al.*, 1993).

In the near-surface layer of the northern North Pacific south of the Aleutians and east of the Kurile Is *G. borealis* is widely distributed between the surface isotherms 2 and 9°C in May, 2 and 11°C in June, 5 and 13°C in July and 8 and 14°C in August. These squids are caught year-round, but particularly between June and August (Fiscus and Mercer, 1982; Kubodera *et al.*, 1983; Yatsu *et al.*, 1993). From fall to spring they inhabit the water masses of the Subarctic and Transitional Domains, during early summer they leave the southern part of Transitional Domain and in late summer are distributed near the surface in the Subarctic Domain only. This may not be a horizontal migration but the result of the descent of squids from the warm upper layer to colder, deeper layers (Kubodera *et al.*, 1983; Kubodera, 1986). The general coincidence of the distributional areas of larvae, juveniles and subadults is evidence against the existence of long horizontal migrations.

The larvae, postlarvae and juveniles of *G. borealis*, DML up to 60 mm, are very abundant in the uppermost water layer throughout the northern North Pacific. The smallest larvae (< 10 mm) are specially abundant in the Bering Sea and south of the western Aleutians. They have been found year-round, mostly in summer (June–September) and late winter, while in the Bering Sea they are common during June–August. The hatching period in the North Pacific is very extended, reproduction occurs supposedly in late summer (Kubodera and Okutani, 1981b; Kubodera *et al.*, 1983; Kubodera and Jefferts, 1984).

Postlarval *G. borealis* smaller than 3–4 cm remain in the upper layer of the Okhotsk Sea while the larger squid undertake diel vertical migrations (Lapko, 1995).

In our epipelagic tows in the Okhotsk Sea and the north-western Pacific off the Kurile Is the occurrence, abundance, biomass and average size of *G. borealis* were significantly higher at night than in the daytime. The biomass (average catch, kg h^{-1}) at night was two orders of magnitude more than during the day in the Okhotsk Sea and three orders of magnitude greater in the north-western Pacific. In mesopelagic tows in the north-western Pacific the abundance and biomass were much higher during the day than at night. The average size was nearly the same (51–56 mm) day and night in the Okhotsk Sea epipelagic layer, but 15% larger at night (96 mm) than during the day (84 mm) in the north-western Pacific mesopelagic layer and more than twice as large at night (107 mm) than in daytime (44 mm) in the north-western Pacific epipelagic layer (Nesis and Nikitina, 1996).

Off Oregon juvenile *G. borealis* were caught during the day at 200–500 m and at night at 0–300 m; off California juveniles (DML 16–47 mm) were deeper than 300 m during the day, mostly at 400–700 m, while at night they were at 0–500 m (Roper and Young, 1975; Pearcy *et al.*, 1977). This species is a first order diel vertical migrator (Roper and Young, 1975).

Thus, diel vertical migrations are very pronounced and span the epi- and mesopelagic layers. Even the small juveniles migrate every day through the whole epi- and mesopelagic zones.

The male-to-female ratio in the early-maturing group of *G. borealis* in the Okhotsk Sea was 1:1.35 in the deep layers and 1:2.9 in the epipelagic layers; for the late-maturing group it was 1:1.2; in the open waters of the North Pacific the ratio was 1:2–1:6 (Kubodera *et al.*, 1983). The sexes may be visually identified in both groups at DML 7–10 cm in males and at 9.5–10 cm in females. The largest juveniles have DML 10–12.5 cm. Only mature males can be divided into two groups but females can be distinguished at maturity stage II onward, by the nidamental gland length at a given DML (Nesis and Nezlin, 1993). The sizes of males in early and late-maturing groups are 11–18 and 19–26 cm respectively, the females 10–17 and 19–27 cm respectively (Naito *et al.*, 1977b; Kubodera *et al.*, 1983; Nesis and Nezlin, 1993).

In the males of the early-maturing group maturation begins at 10–11 cm, while in females it begins at approximately 12 cm and proceeds quickly. Mass maturation occurs in males at 12–14 cm and in females at 13–17 cm. In the larger late-maturing group the males mature at 20–26 cm and maturation of females begins at 22–27 cm.

The growth of the penis in males begins during the transition from maturity stage I to II and slows down at DML 19–20 cm when its tip reaches the anterior mantle margin (this is the size of maturation in late-maturing males). Penis growth in early-maturing males may be described by a linear function, in late-maturing ones and in all males of both groupings by a von Bertalanffy-like equation (Nesis and Nezlin, 1993).

Mature females of both groups were absent from our samples and from those of Kubodera *et al.* (1983). Mated females were very rare (I saw one, DML 164 mm). The spermatophores are transferred to the buccal membrane. Nidamental glands are rather small, up to a third of DML. Their growth may be approximated by a linear or power function (Nesis, 1989a; Nesis and Nezlin, 1993) and proceeds in parallel in females of both groupings. The nidamental gland length index in most mature females in our material (transitional stage III–IV) is on average 0.29–0.30 in both groups. In late-maturing females the nidamental glands have the same length at the same maturity stage as in early-maturing females, but at a much larger DML. Thus the late-maturing females have much less developed reproductive organs than the early-maturing ones of the same size (Nesis and Nezlin, 1993). The largest known female *G. borealis* was caught in the Okhotsk Sea. At DML 330 mm it was only at maturity stage III and had a nidamental gland length only 12% DML (Nesis, 1989a). The gonadosomatic index (gonad weight/total weight) in late-maturing males is somewhat higher than in early-maturing ones at the same maturity stage, but it is lower in late-maturing females than in early-maturing ones.

If it is true that the females of the two groups may be differentiated by the nidamental gland length at a given DML, then the differences between early- and late-maturing groups are distinguishable long before maturation, at least in females. They are probably determined genetically and these two groups may be considered different stock units of suprapopulational rank (Nesis and Nezlin, 1993). Because of the absence of mature females in our own and the Japanese collections, the size of mature eggs is unknown in *G. borealis*.

Although not only juveniles but also submature squids and even mature males of the late-maturing group may be caught in the uppermost layer, ontogenetic descent is pronounced and more clearly expressed in the early-maturing group. In the early-maturing group the squids leave the epipelagic layer before maturation begins, but in the late-maturing group the females descend during the late stages of maturation. The most advanced females (of both groups) caught in the upper epipelagic layer were at stage II of maturity in the Okhotsk Sea and at transitional stage III–IV in the Northwestern Pacific (Nesis, 1989a; Nesis and Nezlin, 1993; Nesis and Nikitina, 1996).

The feeding of *G. borealis* has been rather well studied. Its diet in the mesopelagic layer of the Okhotsk Sea during June–August consists predominantly of euphausiids (23.3% of food by the weight), hyperiids (20.6%), fish (18.3%) and squid (34.6%). The most common prey were *Thysanoessa longipes* (21.5%), *Themisto japonica* (16.8%) and *Leuroglossus schmidti* (13.2%). The daily food ration was assessed at 5.47% of wet weight.

During autumn–winter (October–November) *G. borealis* in the western Bering Sea consumed euphausiids *Thysanoessa longipes* (38–39%), hyperiids (*Themisto pacifica, Hyperia galba* and *Primno macropa*, from 0.7 to 42%), fish (mostly Myctophidae, 5–20%) and squids (13.5–35.5%), small quantities of copepods (*Calanus plumchrus, C. cristatus, Pareuchaeta elongata* and others), decapod crustaceans and coelenterates. In the Okhotsk Sea (November–January) it consumed euphausiids (*Thysanoessa longipes, T. raschii, T. inermis, T. inspinata* and *Euphausia pacifica*, 10–20%), hyperiids (*Themisto pacifica, T. japonica* and *Primno macropa*, 18–23%), fish (*Leuroglossus schmidti, Lipolagus ochotensis* and Myctophidae, 12–30%) and squids (29–32% in the open sea and 58–60% near the Kurile Is), small quantities of copepods (*C. plumchrus, C. cristatus, Metridia okhotensis, M. pacifica, Gaidius variabilis* and *Gaetanus intermedius*) and chaetognaths (*Parasagitta elegans*). The proportion of empty stomachs in the open waters of the western Bering and Okhotsk Seas was not more than 3–4%. The daily food ration was assessed at 4.8–5.3% of wet weight in the open areas of the western Bering and Okhotsk Seas and 11.3–11.9% near the Kurile Is (Okhotsk Sea side), where *G. borealis* with modal DML 7–12 cm consumed mostly squids (Gorbatenko *et al.*, 1995).

In our material from the upper layers of the Okhotsk Sea and the north-western Pacific off the Kurile Is the proportion of empty stomachs was 21.3% in the former and 6.4% in the latter area; the average stomach fullness was 1.5 and 2.6 respectively. The most active feeding (average fullness almost 3) was observed in the upper water layer on the ocean side of the southern Kuriles. Males foraged somewhat more actively than females in the epipelagic layer and less actively in the mesopelagic layer. In the Okhotsk Sea euphausiids were recorded in 62.7% of stomachs, in the north-western Pacific fish predominated (31.2%), then squids (23.6%) and crustaceans (20.7%). The stomachs of some squids of the large-sized group were literally stuffed with food, mostly lanternfish; the food weight reached 80–100 g, or 12–16% of squid weight, so that the squids were barrel-shaped. The food was totally undigested and evidently swallowed in the trawl (our unpublished observations).

In the deeper layers of the Okhotsk Sea the proportion of empty stomachs was

17.2% and the average fullness was 1.6. Fish were recorded in 33.3% of stomachs, euphausiids in 29.2%, unidentified crustaceans in 8.3%, and squids in 4.2% (Nesis, 1989a). In the upper layers of the open ocean 92–100% of stomachs in squids (DML 13–31 cm) were empty and in the others only small fishes were found (Naito *et al.*, 1977b).

The feeding period of *G. borealis* in the western Bering Sea is mostly around sunset and sunrise (18–22 and 6–10 h), but in the Okhotsk Sea it feeds both at night (22–24 h) and during the day (12–14 h) (Gorbatenko *et al.*, 1995).

The annual consumption of zooplankton by *G. borealis* in the Okhotsk Sea was estimated at 4.4 million t, including 2.2 million t of euphausiids and 1.9 million t of hyperiids (Ilyinsky and Gorbatenko, 1994; Ilyinsky, 1995). Evidently, *G. borealis* do not stop feeding during maturation.

The life span of the early-maturing group of *G. borealis*, judging by the unimodal and rather narrow size distribution, is 1 year. The late-maturing group was at first thought to have a 2-year life span, based on the clearly bimodal size distribution (Kubodera *et al.*, 1983; Nesis, 1989a). However, these squids inhabit a much warmer environment than the early-maturing individuals and they are very voracious, so we cannot exclude the possibility of a 1-year life cycle in this group too, as proposed by Naito *et al.* (1977b). They live at about the same surface temperature and mature at nearly the same size as *Todarodes pacificus* which has a one-year life cycle (Okutani, 1983).

G. borealis is preyed on by many fish, squids, seabirds and marine mammals, including *Berryteuthis magister*, *G. borealis*, the pink, chum, coho, sockeye and chinook salmon, steelhead trout, Alaska pollack, pomfret, albacore, whiptails, grenadiers, murres, northern fur seal, California sea lion, northern elephant seal, Dall's porpoise, Pacific white-sided dolphin, pilot whale, sperm whale and many others (Pearcy and Ambler, 1974; Kajimura *et al.*, 1980; Ogi, 1980; Antonelis *et al.*, 1987; Okutani *et al.*, 1988; Pearcy *et al.*, 1988; Sanger and Ainsley, 1988; Fiscus *et al.*, 1989; Kubodera and Shimazaki, 1989; Lowry *et al.*, 1991; Shuntov, 1993; Kubodera and Miyazaki, 1993; Kuramochi *et al.*, 1993; Walker and Jones, 1993); Chuchukalo *et al.*, 1994a,b; Ilyinsky and Gorbatenko, 1994; Sinclair *et al.*, 1994; Savinykh, 1995).

4.4. *Gonatopsis (Gonatopsis) japonicus* Okiyama, 1969 (=*Gonatopsis makko* Okutani and Nemoto, 1964, *partim*)

The taxonomy of this species is confused. *G. japonicus* was described (Okiyama, 1969) based on two immature females of which the holotype was jigged and in very fine condition. For rather a long time all specimens attributed to *G. japonicus* (DML to 17 cm) were immature. Larger specimens were usually mentioned under the name *G. makko* and were either caught in trawls or found in sperm whale stomachs; thus they were damaged or semi-digested. *G. makko* was described as

G. borealis makko based on three very mutilated specimens from sperm whale stomachs off the Western Aleutian Is (Okutani and Nemoto, 1964). DML of the holotype of *G. japonicus* was 15 cm, whereas the holotype of *G. makko* had DML 35 cm. I had supposed (Nesis, 1989b, 1990) that *G. japonicus* was a juvenile and *G. makko* an adult stage of the same species, but this is true only for squids mentioned as *G. makko* from the Japan Sea and probably from the Okhotsk Sea, caught with trawls. The specimens from sperm whale stomachs, including the holotype of *G. makko*, may be part-digested females of *Gonatus madokai* (see Section 4.12). Thus I am using the name *G. japonicus* instead of the earlier name *G. makko* (Nesis, 1989b).

G. japonicus is the largest of all gonatid squids. My largest specimen has DML 62 cm but Akimushkin (1963) mentioned a specimen supposedly of this species, from a sperm whale stomach, with a reconstructed total length approximately 1.5 m, which corresponds to DML about 75 cm.

The Japanese gonatopsis is recorded from the Bering Sea to the Japan Sea and north-western Pacific off eastern Honshu; it probably inhabits the Gulf of Alaska too (Nesis, 1985, 1989b,c, 1990; Okutani *et al.*, 1988). Thus it is a panboreal species in the western and upper-boreal in the eastern part of the northern North Pacific.

The larval and early juvenile stages have not been described (Okutani and Clarke, 1992). Immature specimens with DML 15–17 cm have been jigged or gillnetted in the Japan and southern part of the Okhotsk Sea (Okiyama, 1969, 1993a; Murata and Okutani, 1975). In our collections from the upper epipelagic layer of the Okhotsk Sea and the north-western Pacific off the Kurile Is the species was uncommon and was caught mostly over the southern deep-water basin of the Okhotsk Sea. In the north-western Pacific single specimens were caught in the nearshore area off Urup and Simushir Is. The size range was very wide, DML 2–30 cm, with two peaks, at 3–6 and 11–17 cm. The sex in almost all squids smaller than 20 cm was indeterminable; only two males (DML 163 and 183 mm) and one female (301 mm) were at maturity stage I (Nesis and Nikitina, 1996).

In the epipelagic zone of the Okhotsk Sea the abundance of this species at night was three times and the biomass five times higher than during the day, but in the Northwestern Pacific it was caught in the epipelagic zone only during the day. The largest female was also caught in the daytime while other large specimens were caught at night. Diel vertical migration is well developed although subadult squids may remain in the near-surface layer by day and night.

In deep trawl hauls in the Okhotsk Sea *G. japonicus* was caught in midwater at 300–1000 m and at the bottom at 400–2000 m (Nesis, 1989b, 1990). Its biomass in the mesopelagic layer was an order of magnitude higher than in the epipelagic layer (Didenko, 1991a). The average abundance in midwater was much greater than at the bottom and increased with depth in both realms. This species was caught in both the deepest bottom trawl hauls (1500–2000 m). Average and modal sizes were 14 and 5–10 cm in midwater, and 25.3 and 15–20 cm in bottom trawls and increased with depth in both regions, as did the average abundance. Specimens smaller than 27.5 cm were either juveniles or immature females at maturity stage I; the larger specimens were males at stages IV and V (DML 278–478 mm) and females at stages II and III (DML 398–582 mm); the largest specimen (620 mm) was not opened. Nidamental gland length in females at stages II and III was 13–14% DML. All juveniles were muscular; large specimens were soft and semi-gelatinous (Nesis, 1989b, 1990).

In the Japan Sea most large specimens were caught near the bottom at the Yamato Bank complex and north of the Oki Is at depths of 665–1100 and 900–1125 m, respectively. The lower limit is approximately 1200 m. Here *G. japonicus* occurred deeper than the peak of abundance of *Berryteuthis magister* (300–600 m) and its abundance increased with depth particularly below 900 m. All specimens (DML 212–405+ mm) were immature (Ogata *et al.*, 1973; Okiyama, 1993a). The largest animal described as *Gonatopsis* sp., which belonged most probably to *G. japonicus*, was caught near the surface between the North Yamato Bank and the Primorye (Maritime Province) of Russia. It was not opened or measured but eaten; its DML was approximately (in a photograph) 45 cm and the flesh "was very soft and not so delicious" (Okiyama, 1969). Maybe it was a spent female?

We may suppose that *G. japonicus* is pelagic when juvenile and bottom-associated when adult. In the juvenile stage it inhabits the epi-, meso- and bathypelagic layers and undergoes diel vertical migrations although some specimens may be found near the surface even during the day. The ontogenetic descent is probably extended and begins at a DML between 20 and 30 cm, simultaneously with the onset of maturation, probably between stages I and II of maturity. This coincides with the start of gelatinous degeneration. Maturing animals live either in the deep pelagic (mostly bathypelagic) layers or at the bottom. The spawning habitat is likely to be at or near the bottom in lower bathyal. Spent females probably rise to the surface.

Two large specimens, probably of *G. japonicus* (total length of one, 100 cm; mantle length of the other, 60 cm) were observed by divers nearshore off Rausu, northern Hokkaido, Okhotsk Sea, in 1991 and 1995, floating near the surface and moribund. The large one trailed a gelatinous egg mass (Okutani *et al.*, 1995). These exhausted females appear to lay their first and largest egg-mass on the bottom, and egg extrusion may have disrupted their buoyancy.

The feeding activity of juveniles in the Okhotsk Sea seems rather low; the proportion of empty stomachs among 30 opened specimens was 57%, average fullness was 1.3. Crustaceans were found in 61.5% of stomachs and fish in 23.1% (our data).

G. japonicus were observed underwater in the Japan Sea (on the Kita–Yamato Bank and off Cape Gamov) from the manned submersible "Sever-2" (Alexeyev *et al.*, 1989). They were identified by the elongated mantle, arrow-shaped fin and absence of tentacles. Juveniles of DML 7–15 cm and adults of DML 20–30 cm were observed, the former in midwater at 440–600 m, the latter both in midwater and at the bottom at 700–1150 m (the same depths as off the Japanese coasts). They were single or in groups of three to five squids. Their behaviour and reactions to the submersible are like those of *Berryteuthis magister*. Squids actively hunted shrimps or fish that appeared in the illuminated zone. Once (at 810 m) a squid was observed eating a rockfish lying on the bottom; it abandoned its prey only when the submersible approached as close as 1 m (Alexeyev *et al.*, 1989).

"*G. makko*" is preyed on by sperm whales in the Bering Sea, the Gulf of Alaska, off the Aleutian Is, off Honshu and probably throughout its range (Akimushkin, 1963; Okutani and Nemoto, 1964; Okutani *et al.*, 1976, 1988; Okutani and Satake, 1978).

4.5. *Gonatopsis (Gonatopsis) octopedatus* Sasaki, 1920

The short-tailed gonatopsis is a large species, of DML up to 39 cm (Nesis, 1993a) with semi-gelatinous tissues in juveniles and almost comb-jelly-like in degenerated adults. It is distributed mainly in the north-western Pacific (southward to northeastern Honshu), in the Bering, Okhotsk and Japan Seas. In the north-eastern Pacific it is unknown south of the Gulf of Alaska (Nesis, 1985). Thus it is a panboreal species in the west and an upper-boreal one in the eastern part of the northern North Pacific.

It is very common in the western part of the North Pacific and in adjacent seas, particularly in the Okhotsk Sea, where it was found in almost 97% of trawls made in the meso- and bathypelagic layers over great depths (Nesis, 1989b). In the Okhotsk and Japan Seas and the north-western Pacific it was recorded in the epi-, meso- and bathypelagic zones and near the bottom on the slope and on seamounts (Akimushkin, 1963; Okiyama, 1970, 1993a; Okutani *et al.*, 1976; Nesis, 1985, 1989b; Didenko, 1991a,b; Lapko and Ivanov, 1993; Nesis and Nikitina, 1996). It is much commoner in the meso- and bathypelagic layers than in the epipelagic layer. In the bottom layer of the Okhotsk Sea it has been caught at depths from approximately 400 to 2000 m but its occurrence at the bottom was only about 50% of that in the deep pelagic layers and its abundance was very much lower than in pelagic tows. Both in pelagic and bottom tows the occurrence and abundance did not change with depth but the average size increased slightly with depth and was a third greater in bottom than in pelagic tows. In the Japan Sea this species is recorded in the near-bottom layer down to 1100 m (Okiyama, 1993a). In the Bering Sea and in the north-western Pacific oceanward of the Kurile Is it is much rarer than in the Okhotsk Sea (Didenko, 1991a,b; Nesis and Nikitina, 1996).

In the upper epipelagic layer usually only juveniles are caught, of DML 3–6, mostly 4–5 cm. Both in epipelagic tows in the Okhotsk Sea and in mesopelagic tows in the north-western Pacific off the Kurile Is these juveniles were more common at night than during the day and in the latter area they were caught in epipelagic tows only at night. Thus the juveniles are performing diel vertical migrations. Juveniles of modal size 5–7 cm predominate at all depths down to 1500 m so it may be supposed that only a part of the population ascends to the upper layers at night (Nesis and Nikitina, 1996). Their feeding was not studied but their digestive glands were usually filled with a liquid orange fat characteristic of macroplanktophagous animals.

The size of maturing and mature males in our materials from the deep pelagic and bottom tows in the Okhotsk Sea was DML 9–12 cm, maturing females (stages III and IV) were DML 11–16 cm. Four mated spent females were caught near the surface of the Okhotsk Sea, three of them over the shelf (depth 92–126 m) and one over great depths (> 3000 m). They were much larger, DML 24–39 cm, with empty or almost empty oviducts (not more than three to five mature eggs) and with only a limited number of small oocytes in the ovaries (Nesis, 1993a).

Three more mated spent females were caught in the western Japan Sea, on the surface, over depths of more than 3000 m. They were like those from the Okhotsk Sea, but much smaller, DML 10.5–13.5 cm (Nesis, 1993a; Okiyama, 1993a) and only a little larger than the largest adult male recorded in that sea: 99+ mm (Okiyama, 1970). The largest specimen of *G. octopedatus* recorded in the Japan Sea had DML 24 cm and the largest in the north-western Pacific, off northeastern Honshu, had DML about 20 cm (Okutani *et al.*, 1976).

All mated spent females, particularly the large ones, were so weak that they literally came to pieces in the hand, with the ends of all the arms and almost all the arm armament lost. The males had not degenerated and most of their arm hooks and suckers were present (Okiyama, 1970; Nesis, 1993a). During mating the spermatophores are transferred to the buccal membrane of the female. The size of mature eggs is 4.0–4.3 by 2.5–3.0 mm. Nidamental gland length in maturing females is on average 15–25% of DML and is weakly correlated with the female's size. In spent females the gland is 11–18% of DML. The male-to-female ratio in our Okhotsk Sea samples was 1:8. The stomachs of all opened maturing and spent specimens ($n = 28$) were empty (Nesis, 1989b, 1993a).

The ontogenetic vertical descent in *G. octopedatus* is thought to start early, at the onset of maturation. Gelatinous degeneration begins rather early in females, at maturity stage II or III, but at very different sizes. Feeding probably stops when they begin their descent. The spawning habitat is probably in the deep midwater (meso- and bathypelagic) layers. All the eggs are spawned in a short time. After spawning the females come passively to the surface and die. The juveniles hatch in the deep layers (Nesis, 1989b, 1993a; Nesis and Nikitina, 1996).

G. octopedatus has been recorded in the stomachs of the Alaska pollack (Ilyinsky and Gorbatenko, 1994) and sperm whale (Okutani *et al.*, 1976).

4.6. *Gonatus (Eogonatus) tinro* Nesis, 1972

The hookless gonatid is a rather small (maximal recorded DML 14 cm), not muscular, species distributed from off British Columbia to the area south-eastward of the southern Kurile Is, including the Gulf of Alaska, Bering and Okhotsk Seas (Fields and Gauley, 1971; Nesis, 1972, 1973a, 1985; Nesis and Nikitina, 1996). It is an upper-boreal, primarily meso-bathypelagic, species. It is uncommon; in the Okhotsk Sea it occurred in approximately half of the meso- and bathypelagic tows and almost exclusively in the pelagic realm (Nesis, 1989b). The occurrence and abundance of juveniles in the meso- and bathypelagic layers are greater than in the epipelagic; their abundance and modal size usually increase slightly with depth. The calculated biomass in the mesopelagic layer of the Okhotsk Sea is two orders of magnitude larger than in the epipelagic layer, while in the western Bering Sea this difference is three orders of magnitude. In the western Bering Sea *G. tinro* was second in abundance, while in the Okhotsk Sea it was third in abundance (epi- plus mesopelagic tows) among all squids (Didenko, 1991a,b).

Juveniles of DML 2–5 cm predominated. In the Okhotsk Sea juveniles in epipelagic tows were more common at night than in the daytime, in the north-western Pacific off the Kurile Is they were caught in epipelagic tows at night only. In the mesopelagic tows taken in the latter area the abundance and biomass were two to three times higher by day than at night. A mature male (DML 111 mm) and one larger immature specimen (female?) with DML 127 mm were caught in a mesopelagic tow (Nesis, 1972, 1989b; Nesis and Nikitina, 1996). It appears that there are diel vertical migrations, but ontogenetic descent occurs at an early stage. The spawning habitat is probably deep in the pelagic realm. Penis length in a mature male was very long, 40.5% of DML. Mature females are unknown. The main reproductive period is probably in the summer (Nesis, 1972). Seventy per cent of opened stomachs were empty, planktonic crustaceans predominating in the others (our data).

This species is reported in the stomachs of the Alaska pollack and northern fur seal (Ilyinsky and Gorbatenko, 1994; Sinclair *et al.*, 1994).

4.7. *Gonatus (Gonatus) onyx* Young, 1972

The one-hooked gonatid is one of the commonest gonatids in the North Pacific. It is a muscular panboreal species of intermediate size (DML to 26 cm) distributed (Fig. 2c) from the northern Bering Sea slope to southern Honshu (Suruga Bay) and northern Baja California. The southern boundary of distribution of larvae lies off Izu Is and off Southern Baja California (Okutani, 1969; Young, 1972; Nesis, 1973a,b, 1985; Jefferts, 1988; Okutani *et al.*, 1988). *G. onyx* is absent from the Japan Sea.

In the Okhotsk Sea this species is rather uncommon and occurs mostly along the Kurile Is; it is more common in the north-western Pacific off the Kurile Is, particularly off the middle Kuriles and off the ocean side of the southern Kuriles (Nesis, 1989b; Nesis and Nikitina, 1996). In both areas it occurs from the epipelagic to the bathypelagic zone but it is rare in the former (Nesis, 1989b;

Didenko, 1991a; Lapko and Ivanov, 1993; Nesis and Nikitina, 1996). In the Bering Sea it is much more abundant than in the Okhotsk Sea but also predominantly in the mesopelagic zone (Didenko, 1991b).

In the epipelagic layers of the Okhotsk Sea and the north-western Pacific off the Kurile Is the juveniles predominate, of modal size 5–6 cm. They are also caught in the meso- and bathypelagic layers. Squids larger than 10 cm are caught in the mesopelagic zone at night (Nesis and Nikitina, 1996).

The occurrence, abundance and biomass of *G. onyx* in the epipelagic zone are much higher at night than by day and the average size at night is smaller. In the mesopelagic zone of the north-western Pacific the occurrence and abundance at night are lower, the biomass and average size are higher than during the day. All figures were higher in mesopelagic than in epipelagic tows, the average size in mesopelagic tows was much larger than in epipelagic ones, both by day and night (Nesis and Nikitina, 1996).

The larvae and postlarvae of *G. onyx* are very widely distributed through the northern North Pacific from the Okhotsk Sea to the coast of North America and from the Bering Sea to off Hokkaido and California (Kubodera and Okutani, 1981b; Kubodera and Jefferts, 1984). Larvae less than 10 mm long are particularly common in the Okhotsk Sea south-west of southern Kamchatka and off Washington and Oregon. They and the postlarvae occur year-round but are more abundant in the Bering and Okhotsk Seas from June to August and in the eastern North Pacific from April to September. The largest specimens caught in the upper layer were only 26 mm DML (Kubodera and Jefferts, 1984).

Off California this species is the most abundant gonatid. Juveniles larger than 20 mm live during the day mostly at 400–800 m and at night mostly at 300–500 m but some have been caught in the upper 100 m (Roper and Young, 1975; Anderson, 1978). Off Oregon *G. onyx* were fished by day at 50–500 m and at night at 0–400 m (Pearcy *et al.*, 1977). Thus *G. onyx* may be classified as close to the second-order diel vertical migrators (Roper and Young, 1975).

Diel vertical migrations and ontogenetic descent are pronounced in *G. onyx*. The juvenile squids live in the epi- and mesopelagic layers, the maturing and mature ones in the meso- and bathypelagic layers. Ontogenetic descent begins rather early.

The beginning of maturation in males is at about DML 10 cm and in females it is at 20 cm. Males have a long penis; its length in a mature male (DML 242 mm) being 28.9% of DML (our data). The spermatophores are transferred to the female's buccal membrane. Mature females undergo degeneration (Young, 1973). Spawning probably takes place in the pelagic habitat.

The proportion of empty stomachs in specimens caught in the Okhotsk Sea and north-western Pacific off the Kurile Is is about 50%, crustaceans being recorded in 60% of full stomachs.

This species was recorded in the food of the Alaska pollack *Theragra chalcogramma*, lancetfish *Alepisaurus ferox*, pomfret *Brama japonica*, a zoarcid

Lycodapus mandibularis and Dall's porpoise *Phocoenoides dalli* (Okutani and Kubota, 1976; Anderson, 1981; Kubodera and Shimazaki, 1989; Kuramochi *et al.*, 1993 our data). The lancetfish and pomfret consume mostly postlarval and juvenile squids with DML less than 40 mm.

4.8. *Gonatus (Gonatus) kamtschaticus* (Middendorff, 1849) (=*Gonatus middendorffi* Kubodera and Okutani, 1981a)

The Kamchatka gonatus is a large (DML to 46 cm) muscular and gregarious squid of nektonic type. Its geographic range is of nerito-oceanic type or intermediate between truly oceanic and nerito-oceanic because it is pelagic in habit but common in the "nearshore oceanic" zone not far from continental and island slopes and rare in the open ocean (Nesis, 1985, 1989b,c). It is an upper-boreal species distributed from the Bering Sea to the Gulf of Alaska and off Tsugaru Strait, mainly along the Aleutian Is, eastern Kamchatka and the Kurile Is (Nesis, 1973a,b, 1985; Kubodera and Okutani, 1981a; Roper *et al.*, 1984; Okutani *et al.*, 1988). Jefferts (1988) has listed *G. kamtschaticus* (as *G. middendorffi*) among those species distributed as far as southern California (if this were right, the species would be characterized as panboreal, but this is contradicted by the data of Okutani and Clarke (1992) and Okutani *et al.* (1988)).

Up to the late 1980s this species was considered to be rather rare but in the 1990s it became very abundant in the western Bering Sea, the north-western Pacific off Kamchatka and along the Kurile Is on both the Okhotsk Sea and oceanic sides (a population explosion?).

In summer, 1991 *G. kamtschaticus* was common in the upper epipelagic layer (0–50 m) from the western Bering Sea to the area oceanward of the southern Kurile Is, but not in the Okhotsk Sea. Squids were found near the surface both by day and night; juveniles with DML 3–8 (mostly 4–6) cm predominated but adult squids, DML 20–30 and to 37.5 cm were fished in the upper layer. The juveniles were most common oceanward of the northern Kuriles (Shuntov *et al.*, 1993a,b).

In summer 1992, in the epipelagic zone, *G. kamtschaticus* accounted for 9% of all squids in the Okhotsk Sea and 8% in the Northwestern Pacific off the Kurile Is. The maximum catch was obtained on the ocean side of the middle Kuriles: 2600 specimens and 5.35 kg h^{-1} trawling. Juveniles predominated so the average individual weight in epipelagic tows was only 2 g. Probably most of the squids belonged to one cohort, being the result of simultaneous spawning and high survival rate. In mesopelagic tows in the north-western Pacific off the Kurile Is *G. kamtschaticus* was much less abundant than in epipelagic ones. Both in the Okhotsk Sea epipelagic tows and in the north-western Pacific epi- and mesopelagic tows the squids were fewer at night than during the day, though the occurrence was somewhat higher at night in mesopelagic tows. The average size in the epipelagic tows was the same (3–5 cm) by day and night in both areas; whereas in mesopelagic tows it was larger, 6–7 cm (Nesis and Nikitina, 1996).

In the deeper layers of the Okhotsk Sea in 1984 this species was uncommon, but it was commoner in the mesopelagic (300–500 m) than in the bathypelagic (500–1000 m) layer; the average size was the same in both layers, 93–94 mm (Nesis, 1989b). In autumn 1990 and winter 1990/91 it was still rather rare in the Okhotsk Sea and its biomass was the same in the epi- and mesopelagic zones while in the western Bering Sea it was very common (third place after *Gonatopsis borealis* and *Gonatus tinro* in the meso- and second place in the epipelagic zone) and its biomass in the mesopelagic zone was more than 10 times higher than in the epipelagic (Didenko, 1991a,b).

Postlarval and juvenile squids, DML 6–60 mm, were common in the upper layer, particularly in the Bering Sea and the open ocean southward of the Aleutian Is. They were recorded only during June–August and were most common in July–August, thus the spawning time is short and the larvae hatch primarily during May and June (Kubodera and Okutani, 1981a,b; Kubodera and Jefferts, 1984). Juveniles with the modal size 4–6 cm were common in the food of the pomfret in September–October (Kubodera and Shimazaki, 1989).

In our samples from the Okhotsk Sea and the north-western Pacific off the Kuriles the squids smaller than 12–13 cm were juveniles of indeterminate sex. The largest female (DML 322 mm) was at maturity stage II and was caught in a mesopelagic tow. Another female at the same stage, caught in the western Bering Sea at the depth 400–480 m, was even larger, 380 mm (our data). Mature males have DML 183 and 245 mm (Kubodera and Okutani, 1981a). Squids in these samples were not feeding actively.

G. kamtschaticus clearly do not undergo diel vertical migrations. Our data show that their swarms are more dense during the day than at night. Probably this is an antipredatory adaptation. The postlarvae, juveniles and submature squids live in the epi- and mesopelagic layers. Even the submature squids may be fished in the upper epipelagic layer by day and night. Ontogenetic descent is well pronounced but adult squids inhabit the mesopelagic rather than the meso- and bathypelagic layers as in most other species of *Gonatus*. The descent begins at a rather large size (Nesis and Nikitina, 1996).

G. kamtschaticus was found in the stomachs of sperm whales, fur seals, Dall's porpoises and their juveniles, Pacific salmon (pink *Oncorhynchus gorbuscha*, sockeye *O. nerka*, chinook *O. tschawytscha* and coho *O. kisutsch*), steelhead trout *Salmo gairdneri*, Alaska pollack *Theragra chalcogramma* and pomfret *Brama japonica*; they were the major squid in the stomach of pomfret. Squids of almost all sizes, DML from 2 to 22 cm were found in the stomachs of Dall's porpoises *Phocoenoides dalli* but juveniles 6–8 cm predominated (Kubodera and Okutani, 1981a; Dolganova, 1988; Okutani *et al.*, 1988; Pearcy *et al.*, 1988; Kubodera and Shimazaki, 1989; Kuramochi *et al.*, 1993; Chuchukalo *et al.*, 1994b; Sinclair *et al.*, 1994; Savinykh, 1995).

4.9. *Gonatus (Gonatus) californiensis* Young, 1972

This is a small (DML to 12 cm) species distributed off western North America from Vancouver Is to Baja California and (perhaps) in the Gulf of Panama; it is rather uncommon (Young, 1972; Roper and Young, 1975; Jefferts, 1988). It is a rather muscular, low-boreal, pseudo-oceanic (distant-neritic) meso-bathypelagic species.

The larvae have not been recorded in the near-surface layer (Kubodera and Jefferts, 1984). The juveniles live deeper than 400 m during the day, mostly at 700–800 m, at night they are below 100 m, mainly at 300–500 m. Thus this species is a diel vertical shifter (Roper and Young, 1975), beginning its ontogenetic vertical descent rather early. The larvae and postlarvae were recorded in March, May and September (Kubodera and Jefferts, 1984).

G. californiensis has been recorded in the food of sperm whales and beaked whales (Fiscus *et al.*, 1989).

4.10. *Gonatus (Gonatus) berryi* Naef, 1923

The large club gonatus is a robust muscular squid of intermediate size (largest known DML 23.5 cm). It is a common panboreal meso-bathypelagic species, distributed from the Bering Sea to eastern Hokkaido and northern Baja California but not recorded with certainty in the Okhotsk Sea (Young, 1972; Nesis, 1973b, 1985, 1989b,c; Jefferts, 1988; Okutani *et al.*, 1988; Kubodera, 1996; Nesis and Nikitina, 1996). Judging by the distribution of larvae and postlarvae, this species is more common in the eastern than in the western part of the northern North Pacific (Kubodera and Jefferts, 1984).

During our work in the Okhotsk Sea and the north-western Pacific off the Kurile Is this species was recorded only in the latter area and only in mesopelagic tows. Juveniles predominated, modal DML 7–8 cm. The biomass was higher at night than by day but occurrence, abundance and average size were nearly the same at night and in the daytime (Nesis and Nikitina, 1996).

Off California *G. berryi* was not found in the 0–100 m layer. The larvae (DML 6–15 mm) were concentrated between 300 and 700-m while the juveniles and subadults were at 500–800 m in the daytime and 400–800 m at night; many squids of all sizes were recorded day and night deeper than 900 m and one was photographed from a manned submersible at 915 m (Roper and Young, 1975). According to Roper and Young (1975) this species does not migrate vertically and this conclusion is corroborated by our data (Nesis and Nikitina, 1996). However, the larvae and juveniles have been recorded, although in small quantities, in the upper layers of the north-western, north-central and north-eastern Pacific and in the Bering Sea (Kubodera and Okutani, 1981b; Kubodera and Jefferts, 1984; Didenko, 1991b; Lapko and Ivanov, 1993) it seems that ontogenetic descent does occur in this species but it happens at a very early stage, earlier than in other gonatids. The development of arm hooks and club armature (central hook) finishes much sooner than in other gonatids (Young, 1972; Nesis, 1979; Okutani *et al.*, 1988; Okutani and Clarke, 1992).

The largest reported female (DML 185 mm) was fully mature and degenerated, the mature egg size being 3.5 mm (Young, 1973). The only large specimen (140 mm) in our samples from the north-western Pacific off the Kurile Is was a juvenile.

The larvae of *G. berryi* were recorded throughout most of the year, with peaks in June and August; the later stages were common in September–November (Kubodera and Jefferts, 1984; Kubodera and Shimazaki, 1989).

Feeding of *G. berryi* in the mesopelagic layer oceanward of the Kurile Is was rather active; the proportion of empty stomachs was 41.7%, average fullness was 1.83. Crustaceans were found in 43% of stomachs, fish in 14% (our data).

The juveniles of this species are very common in the food of the pomfret (*Brama japonica*) to the south of the Aleutians (Kubodera and Shimazaki, 1989) and of the Dall's porpoise (*Phocoenoides dalli*) in the Bering Sea and the north-western Pacific eastward of the Kurile Is and northern Japan (Kuramochi *et al.*, 1993). They have been recorded also in the food of sperm whales, northern fur seals and northern elephant seals (Okutani and Satake, 1978; Antonelis *et al.*, 1987; Fiscus *et al.*, 1989; Sinclair *et al.*, 1994).

4.11. *Gonatus (Gonatus) pyros* Young, 1972

This species, the only gonatid having an eye photophore (Young, 1972) and probably also larval photophores (Loffler and Vecchione, 1993), is of intermediate size (maximal reported DML is 28 cm but the largest squid I studied had DML 158 mm) and rather muscular before adulthood. It is distributed from the Bering Sea to northern Baja California, being most common in the open ocean south of the Gulf of Alaska and west of Oregon; it is absent in the Okhotsk Sea (records from this sea are probably mis-identifications) and Japan Sea and in the cold Oyashio Current along the Kurile Is (Young, 1972; Nesis, 1973a,b, 1985; Kubodera and Jefferts, 1984; Jefferts, 1988; Okutani *et al.*, 1988). The record off north-eastern Japan (Kubodera, 1996) is not confirmed as yet. Thus, its panboreal range is skewed eastward.

G. pyros is predominantly a meso-bathypelagic species. The larvae and juveniles are rarely found in epipelagic tows (Kubodera and Okutani, 1981b; Kubodera and Jefferts, 1984). The juveniles are found by day mostly deeper than 300–400 m, predominantly at 400–700 m, and at night between 0 and 700 m, mostly at 100–500 m (Roper and Young, 1975; Pearcy *et al.*, 1977). Thus, this species is a diel vertical migrator.

The larvae were found in February, June, July and September, the postlarvae and juveniles throughout the year, with peaks from late spring to late summer, and juveniles till November (Kubodera and Jefferts, 1984). The ontogenetic descent probably begins rather early. DML of mature females is 13.0–13.5 cm. They undergo gelatinous degeneration. The spermatophores are transferred to the buccal membrane of the female. The size of mature eggs is 3.0×1.7 mm (Young, 1973).

The juveniles of this species were found in large quantities in the food of the pomfret (*Brama japonica*) south of the Aleutian Is (Kubodera and Shimazaki, 1989). It was recorded also in the food of the sockeye salmon *Oncorhynchus nerka*, northern fur seal, Dall's porpoise and sperm whale (Fiscus *et al.*, 1989; Kuramochi *et al.*, 1993; Sinclair *et al.*, 1994; Sobolevsky and Senchenko, 1996).

4.12. *Gonatus (Gonatus) madokai* Kubodera and Okutani, 1977 (=? *Gonatopsis okutanii* Nesis, 1972)

The long-armed gonatus is a large (DML to 44 cm), not gregarious but very common planktonic or semi-planktonic squid with thin, soft and flabby mantle (specially in larvae and juveniles). It is an upper-boreal species distributed (Figure 2d) from the Bering Sea and the Gulf of Alaska to the areas eastward of Hokkaido and westward of Oregon; south of the Gulf of Alaska only postlarvae are known (Nesis, 1973a,b, 1985; Kubodera and Jefferts, 1984; Roper *et al.*, 1984; Jefferts, 1988; Okutani *et al.*, 1988; Okutani and Clarke, 1992). It is one of the commonest squids in the Okhotsk Sea (Kubodera and Okutani, 1977, 1981b; Kubodera and Jefferts, 1984; Nesis, 1989b; Nesis and Nikitina, 1996).

In epipelagic tows in the Okhotsk Sea and in epi- and mesopelagic tows in the north-western Pacific off the Kurile Is *G. madokai* was represented by specimens of all sizes from DML 15 to 340 mm. The juveniles predominated, modal DML 2–3 cm, but two fully mature males were caught in the upper layer (Nesis and Nikitina, 1996). One mature male (DML 33 cm) was caught near the surface in the northern North Pacific (Fiscus and Mercer, 1982).

In our epipelagic tows in the north-western Pacific these squids were caught more rarely, than in the Okhotsk Sea but in higher numbers and greater average size. The occurrence, abundance, biomass and average size were notably higher at night than during the day. In mesopelagic tows in the north-western Pacific they were caught only during the day; but in epipelagic tows at the same stations the occurrence, abundance, biomass and average size were again higher at night (biomass – 230 times) than during the day (Nesis and Nikitina, 1996).

In the deeper layers of the Okhotsk Sea *G. madokai* was very abundant and distributed throughout the whole of the deeper part of the sea. Squids with DML 20–410 mm were caught in midwater at 300–1500 m and on the bottom at 400–3400 m. The size distribution was extended, with many peaks, but juveniles of 7–9 cm predominated. The occurrence in pelagic trawls was 100% in all depths from 300 to 1500 m, the average abundance and size were almost the same at all depths. In bottom tows the occurrence increased from 25% at depths of 400–500 m to 100% at 1500–2000 m (two trawl hauls); the average abundance was almost 10 times less than in pelagic tows but the average size was larger, DML 168.0 mm in all bottom tows and 96.4 mm in deep pelagic tows (Nesis and Shevtsov, 1977; Nesis, 1989b).

The abundance of *G. madokai* in the western Bering Sea was much less than in the Okhotsk Sea, particularly in the epipelagic layer where it accounted for 0.1% of all squids (by biomass) in the former and 9.1% in the latter; in both the biomass was much higher in the meso- than in the epipelagic layer (Didenko, 1991a,b).

The larvae, very characteristic in outer appearance, postlarvae and juveniles with DML up to 7 cm were very common in the epipelagic layer including the uppermost 1 m. The morphological changes during growth were well described by Kubodera and Okutani (Kubodera and Okutani, 1977; Okutani, 1987; Okutani *et al.*, 1988). Early stages of *G. madokai* were distributed through the whole North Pacific from Hokkaido to Oregon and were most common in the eastern part of the Okhotsk Sea (Kubodera and Okutani, 1977; Kubodera and Jefferts, 1984; Kubodera, 1986). Larvae and postlarvae were recorded from March to September but were common only in June–August in the Okhotsk and Bering seas and early June to September in the Northeastern Pacific. Thus the reproductive time is short and hatching occurs in spring: March to May with the peak in April in the Okhotsk Sea, February to early June in the North Pacific. The early growth of *G. madokai* in the Okhotsk Sea has been estimated from the shift of modal size in postlarvae (DML < 30 mm) during June to September and described by an exponential equation (Kubodera and Jefferts, 1984). Adopting the growth rate calculated by Kubodera and Jefferts (about 9–10 mm month^{-1}), our specimens of modal size (70–90 mm) caught in this sea in September–October, 1984, should have hatched in January–February, at a time without mass spawning (Nesis, 1989b). Evidently, growth rate in juveniles is higher than calculated by Kubodera and Jefferts, perhaps 12–18 mm month^{-1}.

G. madokai is a typical meso-bathypelagic squid but its larvae and juveniles are very common in the epipelagic layer and even mature males may be found there. Both diel vertical migration and ontogenetic descent are clearly evident.

The largest immature squids in our samples have DML 109–143 mm but one immature specimen from the western Bering Sea had DML 240 mm. Males and females at stage I have DML 126–138 and 171–200 mm respectively. The male-to-female ratio is approximately 1:2 in the Okhotsk Sea. The size distribution of maturing squids of both sexes and of mature males (there were no fully mature females in our collections) is very wide. Females at maturity stages II, III and IV have DML 19–41 cm; a male at stage IV has 28.6 cm and males at stage V (full maturity) have 13.5–34 cm. Mature males of DML 305–340 mm have a penis length index of 10–22% DML and a gonadosomatic index of 0.8–0.9%. The spermatophores are transferred mainly onto the female's buccal membrane, but some spermatangia were also found at the arm bases (Nesis, 1972, 1989b; Nesis and Shevtsov, 1977). The nidamental gland is rather short; its length in the largest (> 30 cm) females was 13–14% DML and up to 25% in smaller females (DML 26 cm). The size of maturing (not mature) oviducal eggs is 2.1–2.5 by 1.5–1.8 mm, the largest being found in a female at stage III, DML 36 cm (Nesis, 1972, 1989b).

The gelatinous degeneration in *G. madokai* is obvious, but only in females. Our one female from a mesopelagic tow oceanward of the Kurile Is, had DML 281 mm, was at maturity stage III, but already fully degenerated, without tentacles or most arm armature. Its tissues were so watery that its weight (937 g) was 2.5 times more than in much longer (305–340 mm) mature males (average weight 375 g).

The characteristic feature of this species is that when the tentacles autotomize during maturation, they usually (but not always) break off, not at the base (as happens in other species of *Gonatus*), but a little farther out, leaving the short remnants of the tentacular stems hidden between the bases of the third and fourth arms (Nesis, 1989b). This is a diagnostic feature of the form originally described from a sperm whale stomach off the Aleutian Is as *Mastigoteuthis* sp.? (Okutani and Nemoto, 1964), then as *Gonatopsis* sp. (Okutani, 1967) and then, based on new specimens from the Kurile–Kamchatka region and the western Bering Sea, as *Gonatopsis okutanii* (Nesis, 1972). The type series of *G. okutanii* consisted of soft, degenerated females with DML 12+ to 23+ cm, tentacleless and without arm suckers but with intact arm hooks. Our data (Nesis, 1989b) show that among the tentacleless degenerated females from the deep layers of the Okhotsk Sea there were specimens with remnants of both tentacles (*okutanii*-like), the remnant of only one tentacle and those without such rudiments (*makko*-like, see Section 4.8). Thus, it is possible that *G. okutanii* and some *G. makko* are in fact degenerated maturing or mature females of *Gonatus madokai* (Nesis, 1989b). Because the features useful for identification of species of *Gonatus* at the stage of gelatinous degeneration are not stated as yet, I prefer to retain the name *Gonatus madokai* and not to rename it *Gonatus okutanii*.

Two of the females from the type series of *G. okutanii*, the smallest and the largest, DML 120+ and 230+ mm, were mated, although only the largest was nearly mature (Nesis, 1972). In our collections from the Okhotsk Sea there were some maturing but mated females, DML 19–24 cm; two small ones were at maturity stage IV (Nesis and Shevtsov, 1977; Nesis, 1989b;). So mating in *G. madokai* may occur long before maturation. The sizes of nearly mature females and fully mature males are very different, DML 23–41 and 13.5–34 cm, respectively; thus, maturation may occur at very different sizes. The spawning habitat is probably in deep midwater.

The life span is assessed as at least 2 years (Kubodera and Jefferts, 1984).

The specimens of *G. madokai* in our tows were not feeding actively; the proportion of empty stomachs in epipelagic tows was 33% in the Okhotsk Sea and

70% in the Northwestern Pacific, 89% in deep tows. One juvenile specimen caught in a mesopelagic tow had the stomach full (fullness 5) of salps, others contained crustaceans, fish and squid (Nesis, 1989b; unpublished data). In another juvenile (DML 72 mm) the stomach was distended to half the volume of the mantle cavity and filled mostly with chaetognaths, *Sagitta* sp. (Kubodera and Okutani, 1977). The abundance of gelatinous zooplankton in the food, although based on only two examples, differentiates this species from the other Gonatidae.

G. *madokai* has been recorded in the food of Alaska pollack, pink and coho salmon, pomfret, northern fur seal and sperm whale (Okutani *et al.*, 1988; Pearcy *et al.*, 1988; Kubodera and Shimazaki, 1989; Ilyinsky and Gorbatenko, 1994; Sinclair *et al.*, 1994).

The gills of almost all G. *madokai* from deep tows in the Okhotsk Sea were literally covered (Nesis, 1989b) by many hundreds of small brightly yellow encysted parasites of unknown taxonomic position, probably *Hochbergia moroteuthensis* Shinn and McLean, 1989 (Hochberg, 1990). These were rare or absent in other gonatids in our collections.

5. ECOLOGICAL DIVERGENCE OF GONATID SQUIDS

Ten species of gonatid squids inhabit the subarctic north-western Pacific and Far Eastern seas of Russia and up to six species may be caught in the same trawl. Such a large number of co-occurring species is usual for benthic and nektobenthic cephalopods (Sepiidae, Loliginidae, Octopodidae) in tropical and subtropical areas, but is rare among pelagic oceanic squids. In the arctic and boreal regions the Gonatidae is the only speciose squid family and its members are overwhelmingly abundant. According to general features of their distribution, reproduction and feeding, they belong to the same guild of swarming pelagic animals, being planktonic in the young and nektonic in the adult stage, feeding on macroplankton and micronekton. Thus, it may be expected that the species are more or less different ecologically and occupy separate ecological niches. Investigations made mainly in very recent years show that this is right and that almost every species has its own "ecological face".

The morphological, biogeographic and ecological characters that differ in various species of gonatids are listed below. These differences may represent the basic niche divergences among co-occurring species.

5.1. Size

Size differences in co-occurring animals belonging to the same guild are important because it is well known that this character may well indicate divergence of ecological niches. The North Pacific gonatids differ a great deal in size (Table 1).

Table 1 Maximum known size (dorsal mantle length (DML)) of North Pacific gonatids. Sources of data are listed in Nesis and Nikitina (1996).

Species	Maximum known DML, cm
Berryteuthis magister	43
B. anonychus	15
Gonatopsis borealis	33
G. japonicus	62+
G. octopedatus	39
Gonatus tinro	14
G. onyx	26
G. kamtschaticus	46
G. californiensis	12
G. berryi	23.5
G. pyros	28 (16?)
G. madokai	44

Small, intermediate, large and very large species may be distinguished. Three species (*Berryteuthis anonychus*, *Gonatus tinro* and *G. californiensis*) are small, DML 10–15 cm, four species (*Gonatopsis borealis*, *Gonatus onyx*, *G. pyros* and *G. berryi*) are intermediate, DML 20–35 cm, four others are large (*Berryteuthis magister*, *Gonatopsis octopedatus*, *Gonatus madokai* and *G. kamtschaticus*) DML from 35–40 to 50 cm, and one (*Gonatopsis japonicus*) very large, DML at least to 62 cm. It should be noted that fully mature specimens of *G. tinro* and *G. californiensis* have not been reported; thus these species may in fact fall into the intermediate group, but clearly they are not large. In general the species are fairly equally divided into the three categories. The category of intermediate species also contains two of the three non-North Pacific species of *Gonatus* (*G. fabricii* and *G. antarcticus*) with DML up to 30–35 cm, while the third, *G. steensrupi*, is small, DML to 13 (?19) cm (Nesis, 1987).

5.2. Horizontal Distribution

Among the North Pacific gonatids, panboreal, upper-boreal and low-boreal species are distinguished (terminology after Nesis (1985)). All panboreal and upper-boreal species are distributed in the western and at least the northern part of the eastern North Pacific but the low-boreal species are distributed in the eastern North Pacific only (Figure 2). Thus, the latitudinal distribution of gonatids in the western and eastern halves of the North Pacific is asymmetrical. The differences in distribution of North Pacific gonatids and particularly in their southern and (for the low-boreal species) northern boundaries are clearly connected with differences in their thermal ranges.

Data about the ranges of North Pacific gonatids are presented in Jefferts (1988), Nesis (1973b, 1985, 1989c), Okutani and Clarke (1992), Okutani *et al*. (1988) and Young (1972).

The panboreal species are *Berryteuthis magister, Gonatopsis borealis, Gonatus onyx, G. berryi* and *G. pyros*. They are distributed from the northern slope of the Bering Sea to eastern Honshu and California. Only *B. magister* inhabits the Japan Sea. The southern boundaries of these species off North America are: *B. magister* – central California, *G. onyx, G. pyros* and *G. berryi* – northern Baja California, *G. borealis* – southward of the southern tip of Baja California, approximately 20°N (Young, 1972; Nesis, 1985). The larvae of *G. onyx* have been found as far south as the Izu Is (Japan) and southern Baja California (Okutani, 1969; Okutani *et al*., 1988). All panboreal species have oceanic ranges except *B. magister*, the adult animals of which are associated with the bottom. *G. pyros* differs from other species of this group in that its range is skewed eastward, it is absent in the area of the cold Oyashio Current but widely distributed from the Bering Sea to California.

The upper-boreal species are *Gonatopsis japonicus, G. octopedatus, Gonatus tinro, G. kamtschaticus* and *G. madokai*. The first two species are present in the Japan Sea and reach the area off eastern Honshu and may be termed panboreal in the western and upper-boreal in the eastern North Pacific. The southern boundary of distribution of the three species of *Gonatus* is off eastern Hokkaido. In the north-eastern Pacific they are not often found south of the Gulf of Alaska; *G. tinro* reaches British Columbia and *G. madokai* reaches Oregon (Nesis, 1985; Jefferts, 1988; Okutani *et al*., 1988). Here also most species have oceanic ranges except *G. kamtschaticus*. The range of this species is of nerito-oceanic type or intermediate between truly oceanic and nerito-oceanic because it is rare in the open ocean far from continental and insular slopes (Nesis, 1985, 1989c).

The low-boreal species are *Berryteutis anonychus* and *Gonatus californiensis* + *G. oregonensis*. Their ranges are not truly oceanic. The former species is distributed from the south-eastern Bering Sea and the Gulf of Alaska to southern Baja California and westward to 160–180°W between 40 and 53°N. The latter pair are pseudo-oceanic (= distant-neritic) species distributed along the coasts of North America from off Vancouver Is to Baja California and (with a question mark) the Gulf of Panama (Young, 1972; Nesis, 1985; Jefferts, 1988; Okutani *et al*., 1988).

It is impossible to characterize the range of *G. ursabrunae* because of its doubtful taxonomic position.

Of 12 species distributed in the northern and western parts of the northern North Pacific all species are known from the Bering Sea although *Berryteuthis anonychus* and *G. ursabrunae* are only in its south-eastern part. Eight species are known from the Okhotsk Sea, where *Gonatus berryi* and *G. pyros* are absent (*G. berryi* is, however, found oceanward of the Kurile Is). Only three species, *B. magister, Gonatopsis octopedatus* and *G. japonicus*, are found in the Japan Sea (Nesis, 1985, 1993a; Kubodera, 1986; Okutani *et al*., 1988; Okiyama, 1993a).

These patterns are clearly connected with the thermal ranges of different species and differences in the depths of adult habitats.

These are not the only factors. Why do only three species of gonatids inhabit the Japan Sea? This sea is characterized by a very strong contrast between the warm upper and the very cold deep layers (Leonov, 1960). The gonatids can enter this sea only through the Tsugaru (the best route) and La Perouse (Soya) straits because the Nevelskoi Strait is too shallow and brackish for them and the Korea (Tsushima) Strait is too warm (there are no gonatids southwestward of the Korea (Tsushima) Strait). Immediately to the east of Tsugaru and La Perouse straits there are 10 species of Gonatidae (Nesis, 1985, 1989a–c; Okutani *et al.*, 1988). They all are eurybathic and sufficiently eurythermic to live in the cold depths of the Japan Sea and their larvae (in most species the juveniles also) inhabit the upper water layers and thus could be carried by the currents into the Japan Sea. Nevertheless, all the species of *Gonatus* and *Gonatopsis borealis* are absent there (Nesis, 1985; Okiyama, 1993a). The juvenile stages of *Berryteuthis magister* and *G. octopedatus* are rarely found in the upper layers but both live in the Japan Sea. In contrast, the juveniles of, for example, *G. borealis* and *G. onyx* are tolerant enough of the high surface temperature and common enough in the upper layers (Kubodera and Jefferts, 1984; Nesis, 1985) to live in that sea but they are not found there.

The notable absence of many common species of gonatids in the Japan Sea may be caused by the relatively low productivity of this sea in comparison with the Okhotsk Sea and the north-western Pacific off the Kurile Is and Hokkaido (Shuntov, 1985; Shuntov *et al.*, 1993c). Perhaps the relative scarcity of squid food, particularly macroplankton and mesopelagic fishes, in the deep layers of the Japan Sea, in contrast to their abundance in the Okhotsk Sea and the north-western Pacific, is the cause of the scarcity of gonatid fauna in the Japan Sea (Nesis, 1993a).

5.3. Vertical Distribution

5.3.1. *Spawning Habitats*

The spawning habitats of gonatids are unknown. Mature animals of most species are caught in deep midwater; thus the spawning occurs presumably in midwater in the mesopelagic and/or bathypelagic layers. *Berryteuthis magister* spawns at or near the bottom and at the beginning of maturation descends to the bottom layer (Roper and Young, 1975; Naito *et al.*, 1977a; Nesis, 1985, 1989a, 1995; Fedorets and Kozlova, 1987; Okutani *et al.*, 1988; Kubodera, 1992).

The largest specimens of *Gonatopsis japonicus* were also caught in bottom trawls, their occurrence increased with depth both in midwater and at the bottom, and their average size in bottom trawl hauls is very much larger than in pelagic ones (Nesis, 1989b). Thus, it is probable that *G. japonicus* also reproduces at the bottom or in the near-bottom layer (Nesis, 1989b, 1990; Nesis and Nikitina, 1996).

Berryteuthis anonychus may spawn either in the epipelagic layer or near the bottom on or over continental slopes and seamounts. Judging by their range, the *Gonatus californiensis* + *G. oregonensis* may be benthic-spawners too but I have no data on the reproductive biology of this species group.

5.3.2. Diel Vertical Migration

The diel vertical migrations have been investigated in four species of *Gonatus* and in *Gonatopsis borealis*, off California with non-closing Isaacs–Kidd Midwater Trawls (IKMTs) (Roper and Young, 1975). I have investigated these migrations in the southern Okhotsk Sea and north-western Pacific off the Kurile Is by comparing day and night catches of large non-closing pelagic trawls operating in the upper epipelagic layers and also in mesopelagic tows in the ocean off the Kurile Is (Nesis and Nikitina, 1996).

According to Roper and Young (1975) and Nesis and Nikitina (1996), all species exhibit notable diel vertical migrations at least in the juvenile stage of the life cycle. In the north-western Pacific and the Okhotsk Sea vertical migrations are particularly pronounced in *Gonatopsis borealis* where its biomass in night catches was two orders of magnitude higher than in day catches in the Okhotsk Sea and three orders higher in the north-western Pacific. There were also clear differences in other species, for example, in *Gonatus onyx* the abundance and biomass were one or two orders of magnitude higher at night than by day (Nesis and Nikitina, 1996).

Roper and Young (1975) showed that *G. onyx*, *G. pyros* and *G. californiensis* undergo a diel vertical migration, living by day predominantly at 400–800 m and at night at 300–500 m, and *Gonatopsis borealis* is a first-order diel vertical migrator: living at night at 0–400 m and by day at 400–700 m. In contrast to the California data, my catches from the Okhotsk Sea and north-western Pacific showed that all species except one were caught during the day in the upper epipelagic layer. Their habitat spans the epi-, meso- and often also bathypelagic layers. *Berryteuthis anonychus* lives mostly, if not exclusively, in the epipelagic realm but also undergoes diel migrations; during the day it is mostly in the layer between 50 and 200 m, while at night it is between 0 and 150 m (Didenko, 1990).

There are two notable exceptions. *Gonatus kamtschaticus* does not undergo distinct vertical migrations. The juveniles of this epi-mesopelagic species were found in upper water layer in smaller quantities by day than at night and the average DML was the same during day and night. The increase in abundance in daytime was probably because swarms of juveniles were denser during the day than at night. On some occasions even submature squids with DML 10–30 cm and more were caught in the upper layer (Nesis and Nikitina, 1996).

Gonatus berryi was not caught in the epipelagic layer in the north-western Pacific off Kurile Is, but in the mesopelagic layer its occurrence, abundance and average size were almost the same by day and night (Nesis and Nikitina, 1996). It does not migrate vertically off California and is absent from the upper 0–100 m layer there (Roper and Young, 1975). But early juvenile *G. berryi* have been caught, though rarely, right at the surface (Kubodera and Jefferts, 1984). Thus, it is possible that only postlarval *G. berryi* are non-migratory.

5.3.3. *Ontogenetic Vertical Descent*

Probably all gonatids spawn in the deep layers and thus all exhibit ontogenetic vertical descent (Roper and Young, 1975; Nesis, 1985). This descent begins at different ages and sizes in different species, and the time and size at the beginning of their descent into greater depths are characteristic for each species.

Gonatus berryi is rarely found in the epipelagic layer even as a juvenile. Probably it descends from the epipelagic to mesopelagic zone as a young postlarva, earlier than other gonatids.

The descent into great depths begins rather early in *Gonatus onyx*, *G. tinro* and *Gonatopsis octopedatus*. In these species only juveniles with modal DML of 3–6 cm inhabit the upper water layers. According to Kubodera and Jefferts (1984) and Roper and Young (1975), this may also be true for *Gonatus pyros* and *G. californiensis*.

In *Gonatus madokai*, *G. kamtschaticus* and *Gonatopsis japonicus* juveniles also predominate in the upper layers but some much larger specimens with DML up to 20–30 and even 35 cm also occur here (Okiyama, 1969; Murata and Okutani, 1975; Didenko, 1991a,b; Shuntov *et al.*, 1993a,b; Nesis and Nikitina, 1996). All *Gonatopsis japonicus* caught in the upper layers were immature but among *Gonatus madokai* and *G. kamtschaticus* some mature males (but never mature females) were recorded in the upper water layers (Kubodera and Okutani, 1981a; Nesis and Nikitina, 1996).

In *Berryteuthis magister* most individuals descend from the mesopelagic layer to the seabed at a DML no larger than 15–20 cm, at the beginning of maturation (Nesis, 1989a).

Gonatopsis borealis is a very eurybathic species and juveniles with average DML approximately 6–8 cm predominate at all depths down to more than 1000 m (Nesis, 1989a). Mature males and maturing females of the large-sized late-maturing group are common in the upper epipelagic layer but mature females have not been not found here. In our samples the most advanced females caught in the upper layer were at stage II of maturity in the Okhotsk Sea and at transitional stage III–IV in the north-western Pacific (Nesis, 1989a; Nesis and Nezlin, 1993; Nesis and Nikitina, 1996). Thus *G. borealis* is a species with very late ontogenetic descent.

Berryteuthis anonychus finishes this continuum: all stages, including mature females, have been caught in the upper 200 m (Didenko, 1990). If this species spawns in the 0–200-m layer we could say that it is the only gonatid performing diel but not ontogenetic vertical migration. However, the exact spawning habitat of *B. anonychus* is still unknown.

Thus, we may distinguish six categories of ontogenetic descent: very early descent, at postlarval stage (*Gonatus berryi*); early, at juvenile stage (*Gonatus onyx*, *G. tinro*, *Gonatopsis octopedatus*, probably *Gonatus pyros* and *G. californiensis*); at intermediate size, in juvenile and subadult stages (*Gonatus madokai*, *G. kamtschaticus*, *Gonatopsis japonicus*); at the beginning of maturation (*Berryteuthis magister*); in the course or at the end of maturation (*Gonatopsis borealis*); and no descent or descent immediately before the spawning (*Berryteuthis anonychus*).

5.4. Gelatinous Degeneration

The family Gonatidae includes two groups of taxa. The genus *Berryteuthis* and subgenus *Gonatopsis* (*Boreoteuthis*) have a seven-membered radula, which is usual for squids and evolutionarily primitive, while the genus *Gonatus* and subgenus *Gonatopsis* s.str. have an aberrant and evolutionarily advanced radula with only five rows of teeth (Nesis, 1973a).

Gelatinous degeneration with loss of tentacles (except for *Gonatopsis* which loses the tentacles at the end of the larval stage) and most of the arm armature (hooks and suckers) is characteristic of all the species with a five-membered radula (Nesis, 1973a, 1985, 1989b; Young, 1973; Nesis and Nikitina, 1996). In these squid degeneration begins long before final maturation (Nesis, 1973a, 1989b, 1993a).

All species of this group are soft and their meat is unpalatable because of its watery consistence. The softness of different species prior to gelatinous degeneration is not the same. The softest and most gelatinous tissues are found in *Gonatus madokai*. The most muscular and nektonic in outer appearance is *G. kamtschaticus*, although its tissues are far softer than, for example, in Loliginidae and Ommastrephidae. In second place are juvenile specimens of *Gonatopsis japonicus* (Okiyama, 1969; Kubodera and Okutani, 1977, 1981b; Nesis, 1985, 1989b, 1990). Other species of *Gonatus* and *Gonatopsis* s.str. are intermediate in this character.

Among species with the usual seven-membered radula, *Berryteuthis* and *Gonatopsis* (*Boreoteuthis*), mature squids of both sexes are not degenerated even at the beginning of spawning. Squids of this group are rather muscular, though not as muscular as Loliginidae and Ommastrephidae, their meat is delicious and mature squids are edible. In *B. magister* degeneration is evident only at the end of the spawning time (Nesis, 1989a, 1995). It should be noted that the early-maturing and late-maturing intraspecies groupings are known only in this group (Nesis and Nezlin, 1993). In species with a five-membered radula the size distributions of squids at the beginning and end of maturation are usually extended but do not separate into different groups.

The spawned and spent females of species undergoing gelatinous degeneration rise to the surface and die (Young, 1973; Nesis, 1993a). Spent females of *Berryteuthis magister* remain at the bottom after death. The fate of the spent females of the other two species of the group with seven-membered radula (*B. anonychus* and *Gonatopsis borealis*) is unknown but they have never been caught at the surface and probably fall to the bottom after death.

5.5. "Ecological Individuality" of Gonatid Species

Most species of the North Pacific Gonatidae differ from the others by at least one character and generally by two, three or more (Table 2).

The two ecologically closest species are *Gonatus onyx* and *G. pyros*. They differ in that *G. pyros* is probably a more stenothermal–warm-water species than *G. onyx*, being absent from the Okhotsk Sea and the area of the Oyashio Current while *G. onyx* is very eurythermal. Also, *G. onyx* is much commoner than *G. pyros* in the upper water layers. *G. pyros* is the only luminescent gonatid (Young, 1972). However, the ecology of *G. pyros* is not well known and there could be more differences between the two species.

Two other ecologically rather similar pairs of species are *Gonatus madokai* and *G. kamtschaticus*; and *G. tinro* and *G. californiensis*. The former pair are largely sympatric and both are very common. They differ in that *G. kamtschaticus* is a more muscular, more nektonic-like squid with an intermediate type of range, between truly oceanic and nerito-oceanic, while *G. madokai* is a semi-gelatinous planktonic squid with an oceanic range (Kubodera and Okutani, 1977, 1981a; Nesis, 1985; Okutani *et al.*, 1988). Species of the second pair are almost allopatric, their ranges coinciding only in a small area off British Columbia.

Thus almost every species of the North Pacific Gonatidae has its own "ecological individuality".

5.6. The History of Niche Divergence in North Pacific Gonatids

Palaeontological information on gonatid squids is scanty and shows only that the statoliths of *Berryteuthis* sp. (definitely not *B. magister*) are present in the Late Pliocene of California (Clarke *et al.*, 1980; Clarke and Maddock, 1988a,b). Thus, when discussing the history of ecological evolution of the family I shall base my

Table 2 Ecological characteristics of the North Pacific gonatid squids.

Species	Size	Horizontal distribution	Spawning habitat	Diel vertical migration	Ontogenetic descent	Gelatinous degeneration
Berryteuthis magister	la	pbba	bt	+/emb	lt	–
B. anonychus	sm	lb	pel/bt?	+/e	no?	–
Gonatopsis borealis	int	pb	pel	+/emb	vlt	+
G. japonicus	vl	pbub	bt?	+/emb	int	+
G. octopedatus	la	pbub	pel	+/emb	ea	+
Gonatus tinro	sm	ub	pel	+/emb	ea	+
G. onyx	int	pb	pel	+/emb	ea	+
G. kamtschaticus	la	ubno	pel	–/em	int	+
G. californiensis	sm	lbpo	pel	+/(e)mb	ea?	+
G. berryi	int	pb	pel	–/mb	ve	+
G. pyros	int	pbse	pel	+/(e)mb	ea?	+
G. madokai	la	ub	pel	+/emb	int	+

Size: sm, small; int, intermediate; la, large; vl, very large.

Horizontal distribution: pb, panboreal; pbba, panboreal bottom-associated; pbub, panboreal in the western and upper-boreal in the eastern North Pacific; pbse, panboreal with the range skewed eastward; ub, upper-boreal; ubno, upper-boreal with the range of nerito-oceanic type; lb, low-boreal; lbpo, low-boreal pseudo-oceanic.

Spawning habitat: bt, bottom; pel, pelagic (midwater).

Diel vertical migration: +, present; –, absent; main vertical habitation zones: e, epipelagic; m, mesopelagic; b, bathypelagic.

Ontogenetic descent: ve, very early; ea, early; int, at intermediate size; lt, lately; vlt, very lately; no, without descent.

Gelatinous degeneration: +, before full maturation; –, after full maturation.

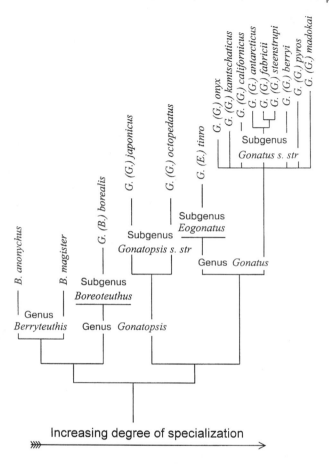

Figure 3   Scheme of phylogenetic relationships of the genera, subgenera and species in the family Gonatidae (modified from Nesis, 1985).

considerations on morphological (Nesis, 1973a) and general paleoclimatological and biogeographic data (Kafanov, 1982; Nesis, 1985).

Because the family Gonatidae is a cold-water one (Figure 1) it could not have been established before the Late Oligocene when cooling began in the North Pacific. The recent diversity of the family, including the diversification of low-boreal, panboreal and upper-boreal species, may have evolved only after the present oceanographic structure of the subarctic North Pacific was established in general terms, in the Late Miocene (Kafanov, 1982). On the other hand, the common ancestor of *Gonatus fabricii* and *G. steenstrupi* had migrated through the Bering Strait into the Arctic Ocean and then into the North Atlantic in the early Late Pliocene, during the Beringian Transgression, approximately 3.5–3.2 million years BP (Nesis, 1985). Somewhat later, in the Pleistocene, the ancestor of the notalian *G. antarcticus* had migrated from the north-eastern Pacific into the Southern Ocean along the western coasts of the Americas (Nesis, 1985).

It is very important that, according to purely morphological data (Figure 3), all three species distributed outside the North Pacific (*G. fabricii*, *G. steenstrupi* and *G. antarcticus*) are very closely related to each other and to *G. californiensis* (Young, 1972; Nesis, 1973a, 1985; Kristensen, 1981). The latter species has the warmest habitat of the whole family and was undoubtedly the ancestor of *G. antarcticus* (Young, 1972; Nesis, 1973a, 1985). The ancestor of *G. fabricii* was also a relative of *G. californiensis* and *G. steenstrupi* is a descendant of *G. fabricii* (Nesis, 1985). *G. fabricii* has the coldest habitat of the family but this does not imply that the common ancestor of this species and of *G. steenstrupi* was also a cold-water one. Probably, this ancestor was pre-adapted to be a cold-water species but its tolerance of conditions in the Arctic Ocean evolved after its migration through this ocean (being ice free during the time of the Beringian Transgression), some time in the Pleistocene.

Thus, most of the branches leading to the recent species of Gonatidae or to their immediate progenitors should have been in existence in approximately mid-Pliocene.

We may conclude that the radiation of the Gonatidae as a whole took place between the Late Miocene and mid-Pliocene. *Berryteuthis* is probably the oldest taxon (Figure 3). The formation of the upper-boreal species probably took place during the Pleistocene cooling. So the evolution of gonatids resulted in the appearance of species which were rather well differentiated morphologically and ecologically; evolution was rapid and extended over only a few million years.

Katugin (1993a, 1995b) has investigated relationships between eight species representing all genera and subgenera of Gonatidae, using starch gel electrophoresis. It was shown (Katugin, 1993a) that *Gonatus* (*Eogonatus*) *tinro* is very close to *G.* (*Gonatus*) *pyros*, and *G.* (*G.*) *berryi* is close to *G.* (*G.*) *kamtschaticus*; *Gonatopsis* (*Gonatopsis*) *octopedatus* is close to *Gonatus* spp. while *Gonatopsis* (*Boreoteuthis*) *borealis* is distant from *Gonatopsis* s.str. and *Gonatus*. *Berryteuthis magister* and *B. anonychus* are most distantly related to each other and to other gonatids (Katugin, 1993a, 1995b) and *B. magister* "is evidently the most unique ecologically, morphologically, and genetically among the gonate species studied", diverging from the main family stem before the outbranching of the ancestors of *B. anonychus* and *G. borealis* (Katugin, 1995b). This is in accordance with the phylogenetic scheme based on morphological characters (Nesis, 1985). A large distance between *Berryteuthis* and other gonatids has also been shown using the structure of statoliths and beaks (Clarke, 1988; Clarke and Maddock, 1988a,b).

The pattern of distribution of both species of *Berryteuthis* shows that this genus is associated with the seabed at least in the adult stage of the life cycle while most (though not all) other gonatids are pelagic. The species of *Berryteuthis* are nektonic animals while among other gonatids there are both nektonic (as *G. kamtschaticus* or juvenile *G. japonicus*) and more or less planktonic forms (as *G. madokai*). This is in accordance with a general scheme of ecological evolution of nekton from the shallow water benthic through neritic benthopelagic and nerito-oceanic (living over the slopes) nektonic, to oceanic nektonic forms and then to midwater and deep-water planktonic forms (Nesis, 1978, 1985). The evolution of Gonatidae represents a rather common evolutionary pattern among oceanic animals.

6. HORIZONTAL AND VERTICAL DISTRIBUTION OF RELATIVE ABUNDANCE AND BIOMASS

Quantitative assessment of gonatid biomass is not possible at present, first, because of differences between the catching devices used in various countries and institutions and, second, because no device is capable of catching squids of all sizes equally effectively. Macroplankton nets are good for quantitative assessment of larvae and postlarvae, while large midwater and bottom commercial trawls may well catch large adult squids but their effectiveness in catching gonatid squids is low. In the works of Pacific Research Institute of Fisheries and Oceanography (TINRO) (Vladivostok, Russia) in the north-western Pacific, Okhotsk and Japan seas the catching ability of large midwater trawls for gonatids was assessed to be 0.1, and this figure is used for all pelagic surveys (Shuntov *et al.*, 1993c).

I can only use relative assessments based on catches with the same devices. These assessments are reasonably good for the Okhotsk Sea, western and central Bering Sea and the north-western Pacific off the Kurile Is. The data for the northern North Pacific in general are scanty and scattered; that for the eastern Bering Sea, eastern North Pacific off Canada and USA, and the Japan Sea, are even less complete.

In the Okhotsk Sea the biomass of gonatid squids in the upper epipelagic (0–60 m) and whole epipelagic (0–200 m) layer is an order of magnitude less than in the mesopelagic (200–500 m) and bathypelagic (500–1000 or 500–1500 m) layers; these differences are smaller at night than in daytime because of the diel vertical migration of the large mass of juvenile squids (Nesis, 1989c; Ilyinsky, 1991; Lapko, 1995; Nesis and Nikitina, 1996). The squid concentration in the bathypelagic layer is somewhat higher than in the mesopelagic layer. For example, in September–October, 1984, the average catch of squids (99% of which, by biomass, consisted of Gonatidae) was 2.5 kg h^{-1} at 300–500 m, 3.1 kg h^{-1} at 500–1000 m and 3.0 kg h^{-1} at 1000–1500 m (Nesis, 1989c). In February–April, 1990; November, 1990–January, 1991; and August, 1989 the assessed concentration of all squids was respectively 962; 1004; 2330 kg km^{-2} in the 200–500 m layer, and 1224; 1129; 759 kg km^{-1} in the 500–1000 m layer (Lapko, 1995). During November, 1990–January, 1991 the biomass of squids in the epipelagic layer (0–200 m) was 13 times less than in the mesopelagic layer (200–500 m); gonatids accounted for 99.9% of biomass in the epi- and 96.8% in the mesopelagic layer. *Berryteuthis magister* and *Gonatopsis borealis* predominated, forming respectively 76.5 and 76.8% of biomass (Didenko, 1991a). The abundance and biomass of squids varies greatly between seasons and from year to year.

The areas of high concentrations of gonatid squids form a ring between the shelves of the Okhotsk Sea, where these squids are very uncommon, and the southern deep-water basin, over which the catches, particularly in mesopelagic layer, are usually rather low, as, for example, in September–October, 1984 (Nesis, 1989c). The highest catches in August, 1989 in the 200–500 m layer, were observed over the slope of western Kamchatka, particularly off the south-western part of this peninsula and over the TINRO Trough at 55–58°N (Ilyinsky, 1991), where some catches exceeded 0.5 t h^{-1}. High squid abundance off south-western Kamchatka corresponds with the extremely high abundance of gonatid larvae in this area (Kubodera and Jefferts, 1984). The lowest catches were observed over the deep areas between southern Kamchatka and Sakhalin (Ilyinsky, 1991).

Throughout the Okhotsk Sea the juveniles of *Berryteuthis magister* and *Gonatopsis borealis* strongly predominated and the catches of over 0.5 t h^{-1} consisted of *B. magister* only (Nesis, 1989a,c; Didenko, 1991a; Ilyinsky, 1991; Shuntov et al., 1993c; Lapko, 1995). In areas with low catches of squids, particularly over the deep-water southern basin, *B. magister* and *G. borealis* are uncommon and other gonatids predominate (Ilyinsky, 1991).

In the benthic layers of the Okhotsk Sea the largest catches were obtained between 250 and 1250 m. Maturing and mature *B. magister* and large near-bottom cirrate octopods *Opisthoteuthis* spp. predominated in the catches (Nesis, 1989b,c).

In the pelagic realm of the western Bering Sea during October–November, 1990, the biomass in the epipelagic layer was seven times less than in the mesopelagic layer; *G. borealis* predominated in the epipelagic layer (86.2%) but in the mesopelagic layer the species composition was more diverse: *Gonatopsis borealis*, *Gonatus tinro*, *G. kamtschaticus* together accounted for 71.3% of the biomass (Didenko, 1991b). In April–June, 1990, the differences in biomass between the epi-, meso-, and bathypelagic zones were less pronounced and the species composition was diverse in all zones. *B. magister* and juvenile gonatids predominated in the epi- and mesopelagic layers while juvenile gonatids and cranchiids predominated in the bathypelagic layer. Gonatids formed 99.9% of the biomass in the epipelagic, 91.5% in the meso- and 65.7% in the bathypelagic zone. They were commonest over the slope between the 100 and 500 m isobaths, but some were also found over great depths: over the Shirshov Ridge and the northwestern part of the Aleutian Basin (Radchenko, 1992). Here again large seasonal and yearly differences in squid abundance, biomass and species composition were noted.

In the benthic realm of the western Bering Sea *B. magister* is the only common species of gonatid. Here this species is distributed on the slope at depths from 150–200 (only rarely from 50 m) to approximately 800 m, being most common at 200–400 or 200–500 m (Fedorets, 1983; Nesis, 1995; Arkhipkin et al., 1996; Bizikov and Arkhipkin, 1996). In the eastern Bering Sea this species was also the predominant gonatid in bottom trawls and it was associated with the upper slope (Bakkala et al., 1985; Ronholt et al., 1986).

In the Japan Sea there are small concentrations of gonatids, primarily B. magister, on the upper slope in the Tatar Strait (Skalkin, 1977), on the Yamato Bank complex (Panina, 1971; Railko, 1979; Shevtsov, 1988a), on the Rebun Bank (Kubodera, 1992) and around the Oki Is (Kasahara et al., 1978; Nazumi et al., 1979; Yuuki and Kitazawa, 1986).

In the north-western Pacific off the Kurile Is, gonatids are much more abundant than in the Okhotsk Sea (Nesis and Nikitina, 1996; Shuntov et al., 1993b). They are most common 100–200 nautical miles from the Kurile Is, in the area of confluence of the cold Oyashio Current with the north-western branch of the warmer Subarctic Current, the continuation of the Kuroshio (Dodimead et al., 1963; Favorite et al., 1976; Shuntov et al., 1993c). The abundance of squids in the nearshore zone off the southern Kurile Is is rather low but oceanwards off the middle and northern Kuriles high concentrations of squids have been observed both offshore and inshore (Nesis and Nezlin, 1993; Shuntov et al., 1993b; Nesis and Nikitina, 1996). The abundance of squids in the epipelagic and upper mesopelagic layers is of approximately the same order of magnitude. The commonest species here was Gonatopsis borealis (Lapko and Ivanov, 1993; Nesis and Nezlin, 1993; Nesis and Nikitina, 1996).

The early life stages of gonatids are widely distibuted in the northern North Pacific particularly during summer and autumn. The areas of high abundance are the Okhotsk Sea off western Kamchatka, the deep-water part of the Bering Sea between shelf break and the Aleutian Is, particularly eastward of 180°, and the north-eastern Pacific along the Aleutians (Kubodera and Jefferts, 1984). Larvae and juveniles of Gonatidae are also very common off Oregon (Kubodera and Jefferts, 1984), where they are clearly concentrated over the upper slope (Pearcy, 1976). Concentrations of mature Berryteuthis anonychus have been recorded in the open part of the north-eastern Pacific, mostly over seamounts (Didenko, 1990).

Gonatid squids are generally abundant through the whole subarctic North Pacific, in the Bering and Okhotsk seas, but they are most common over the continental and insular slopes, near the Subarctic Front and on the boundary between the Oyashio and Subarctic Currents and in some places over the underwater ridges and seamounts. They are mostly very eurybathic and a high abundance of squids may be observed right from the surface to depths of 1000–1500 m (Pearcy, 1976; Kubodera et al., 1983; Kubodera and Jefferts, 1984; Nesis, 1985, 1989c; Didenko, 1990; Ilyinsky, 1991; Radchenko, 1992; Nesis and Nezlin, 1993; Nesis and Nikitina, 1996; Sobolevsky, 1996). This corresponds rather well with the distribution of the biomass of macroplankton and particularly of predatory plankton and mesopelagic fish (Nesis, 1989c; Ilyinsky, 1991, 1995; Radchenko, 1992; Shuntov et al., 1993c, 1994a,b).

7. THE ROLE OF GONATID SQUIDS IN THE ECOSYSTEM

7.1. Prey

The food and feeding are adequately known for *Berryteuthis magister*, they are reasonably well studied in *B. anonychus* and *Gonatopsis borealis* but only scanty evidence exists for many other species while some are almost unstudied (Okiyama, 1969; Fedorets, 1977, 1986; Naito et al., 1977b; Kuznetsova and Fedorets, 1987; Fedorets and Kun, 1988; Lapshina, 1988; Okutani et al., 1988; Nesis, 1989a,b; Radchenko, 1992; Ilyinsky and Gorbatenko, 1994; Gorbatenko et al., 1995; Ilyinsky, 1995; Lapko, 1995; our unpublished observations).

The main food of gonatids, as of most squids (Nixon, 1987), consists of crustaceans, squid and fish. Small quantities of pelagic gastropods (*Clione, Euclio* and *Limacina*), pelagic polychaetes (*Tomopteris*), chaetognaths (*Sagitta* and *Parasagitta*), coelenterates and other planktonic animals are eaten. Only *Gonatus madokai* consumes quantities of gelatinous plankton, chaetognaths (Kubodera and Okutani, 1977) and salps (our data), but this evidence is based on two full stomachs only. As in most other squids, the juveniles feed on meso- and macroplankton, during growth the proportion of fish and squids increases (Nixon, 1987; Vecchione, 1987).

Among crustaceans, copepods (*Calanus, Eucalanus, Pareuchaeta, Metridia*, rarely *Gaidius, Gaetanus* and others), euphausiids (*Thysanoessa* and *Euphausia*) and hyperiid amphipods (*Themisto, Hyperia, Primno, Cyphocaris* and others) predominate. The food of benthic adult *Berryteuthis magister* includes pandalid shrimps (*Pandalus goniurus*) and other decapods. In the pelagic realm of the Okhotsk Sea the commonest crustaceans in the food of gonatids are *Thysanoessa longipes* and *Themisto japonica*; in the western Bering Sea *Thysanoessa inermis*, *T. raschii* and *Themisto pacifica* predominate (Ilyinsky and Gorbatenko, 1994; Gorbatenko et al., 1995).

Fish eaten in the pelagic realm include small mesopelagic fishes such as Myctophidae (mostly *Stenobrachius* spp.) and Microstomatidae (*Leuroglossus schmidti, Lipolagus ochotensis* and others); in the benthic realm *B. magister* consumes juvenile Alaska pollack (*Theragra chalcogramma*) and cottoid fishes.

Cephalopods in the food of gonatids are mostly gonatids of the same and other species; in some places small enoploteuthid squids *Watasenia scintillans* and *Enoploteuthis chuni* are also eaten (Kasahara et al., 1978; Yuuki and Kitazawa, 1986). Squids are a usual component of the food and, although it is likely that many of them were devoured in the trawls (our observations), cannibalism in general is fairly important for many gonatids, particularly for *Berryteuthis magister* and *Gonatopsis borealis* (Radchenko, 1992; Ilyinsky and Gorbatenko, 1994; Gorbatenko *et al.*, 1995; Lapko, 1995). Squids compose 20% of the food of juvenile *B. magister* and from 20–30 to 50–60% of the food of *G. borealis* (Gorbatenko *et al.*, 1995; Lapko, 1995).

The juveniles of all the species studied consumed meso- and, mostly, macroplankton. Among adult gonatids three weakly differentiated groups can be distinguished: (i) macroplanktophages (*Berryteuthis anonychus*); (ii) macroplankto-ichthyophages (*B. magister*, most species of *Gonatus* and probably *Gonatopsis octopedatus*); and (iii) ichthyo-teuthophages with a large proportion of macroplankton in the food (*Gonatopsis borealis*, *Gonatus kamtschaticus* and probably *Gonatopsis japonicus*) (Naito *et al.*, 1977b; Kasahara *et al.*, 1978; Kuznetsova and Fedorets, 1987; Fedorets and Kun, 1988; Lapshina, 1988; Nesis, 1989a,b, 1995; Okutani *et al.*, 1988; Ilyinsky and Gorbatenko, 1994; Gorbatenko *et al.*, 1995).

Daily food rations were assessed for the two most common squid species of the Okhotsk and western Bering seas: approximately 2% in juvenile and 1% in mature *B. magister* and 4.8–5.5% in juvenile *G. borealis*, but near the Kurile Is the daily ration of the latter species increased to 11–12% (Ilyinsky and Gorbatenko, 1994; Gorbatenko *et al.*, 1995). Laevastu and Larkins (1981) presented the following figures for daily food rations of Bering Sea and the Gulf of Alaska squids (mostly gonatids): supporting ration 0.51%, growth ration 1.94%. These figures as well as those assessed for *B. magister* by Gorbatenko *et al.* (1995) and Ilyinsky and Gorbatenko (1994) are less than in temperate and tropical squids (O'Dor and Wells, 1987) but rates of digestion and metabolism are greatly dependent on temperature (Boucher-Rodoni *et al.*, 1987; O'Dor and Wells, 1987), thus they will be less in squids inhabiting cold waters. Nevertheless, the figures for *G. borealis* are of the same order or higher than in cephalopods of temperate and tropical areas (O'Dor and Wells, 1987).

It should be specially noted that active feeding is observed only during part of gonatid life. *Berryteuthis magister* feeds actively only during its pelagic life stage. After descending to the bottom, in the spawning areas, its feeding diminishes dramatically and the proportion of empty stomachs increases to 70–90% or even more (Naito *et al.*, 1977b; Kuznetsova and Fedorets, 1987; Nesis, 1989a, 1995). Gonatids with a five-membered radula, *Gonatus* and *Gonatopsis* s.str., stop or almost stop feeding after gelatinous degeneration because it is impossible to catch prey when the tentacles are autotomized and most arm armature is lost (Nesis, 1989b, 1993a). Thus, if we integrate over the whole life cycle the daily rations calculated for the period of active feeding, we obtain a much lower figure. Only two species with a seven-membered radula, *Berryteuthis anonychus*, which probably does not undergo ontogenetic descent, and *Gonatopsis borealis*, which descends very late in the life cycle, feed actively during most if not their whole life (Lapshina, 1988; Nesis, 1989a; Ilyinsky and Gorbatenko, 1994). Maturing females and mature males of the large-sized group of *G. borealis* are particularly voracious (our data).

7.2. Predators

Almost all predatory animals of the subarctic North Pacific, squids, fishes, seals, whales, dolphins and seabirds, consume gonatids of appropriate size, larvae, juveniles or adults. The list of predatory species is very long and the amount of literature is immense (reviews: Akimushkin, 1963; Ito, 1964; Clarke, 1966, 1977, 1983, 1987; Panina, 1966, 1971; Zuev and Nesis, 1971; Takeuchi, 1972; Klumov, 1978; Kawakami, 1980; Kawamura, 1980; Fiscus, 1982; Sobolevsky, 1983a,b, 1984, 1996; Markina, 1987; Okutani *et al.*, 1988; Pearcy *et al.*, 1988; Radchenko, 1992; Ilyinsky, 1995; Lapko, 1995, 1996). The most important consumers of Gonatidae are Alaska pollack (*Theragra chalcogramma*), Pacific cod (*Gadus macrocephalus*), macrourids, Pacific halibut (*Reinhardtius hippoglossoides mat-suurae*), pomfret (*Brama japonica*), some Zoarcidae, Liparidae, sperm whale, killer whale, beaked whales, Dall's porpoise, right whale and white-sided dolphins, northern fur seal, Steller sea lion, northern elephant seal, murres and shearwaters. The juveniles are eaten also by the Pacific salmon, mostly chinook (*Oncorhynchus tschawytscha*), coho (*O. kisutsch*), also pink (*O. gorbuscha*) and chum salmon (*O. keta*), by many common mesopelagic fishes (*Leuroglossus schmidti, Stenobrachius leucopsarus, Lipolagus ochotensis* and others) (Ito, 1964; Larkin and Ricker, 1964; Takeuchi, 1972; Klumov, 1978; Antonelis *et al.*, 1987; Okutani *et al.*, 1988; Pearcy *et al.*, 1988; Sanger and Ainsley, 1988; Yang and Livingston, 1988; Kubodera and Shimazaki, 1989; Kubodera and Miyazaki, 1993; Kuramochi *et al.*, 1993; Savinykh, 1993; Shuntov *et al.*, 1993c; Walker and Jones, 1993; Ilyinsky and Gorbatenko, 1994; Lapko, 1994, 1995, 1996; Sinclair *et al.*, 1994; Sokolovsky and Sokolovskaya, 1995; Orlov and Pitruk, 1996; Savinykh and Chuchukalo, 1996; Sobolevsky, 1996; Sobolevsky and Senchenko, 1996).

The paralarvae are preyed upon not only by fishes and seabirds but by macroplankton predators. Arkhipkin and Bizikov (1996) observed paralarvae of eight species of western Bering Sea gonatids in small aquaria aboard ship. In response to sudden disturbance these larvae took up a special body posture, by relaxing the mantle and pulling the head and arms inside the mantle cavity. By doing this they became similar in size, appearance and colour to the medusa *Aglantha digitale*, with its stinging tentacles. A similar defensive posture, imitating a dangerous animal, is shown by cranchiid squid paralarvae.

Traditionally, marine mammals were considered to be the most important consumers of squids in general and North Pacific gonatids in particular (Akimushkin, 1963; Clarke, 1966, 1977; Panina, 1966, 1971; Zuev and Nesis, 1971; Klumov, 1978). Sobolevsky (1983a,b, 1984; Nesis, 1985) has assessed the

Table 3 Approximate yearly consumption of squids (mostly gonatids) by various predators in the Okhotsk Sea, thousand t (data from Sobolevsky, 1983a; Shuntov *et al.*, 1993c; Lapko, 1995, 1996).

Alaska pollack (stock state for late 1980s)	4500
Pacific salmons (*Oncorhynchus* spp.)	25–30
Leuroglossus schmidti	2000
Lipolagus ochotensis	144
Coryphaenoides pectoralis	550–1000
C. acrolepis	160
Reinhardtius hippoglossoides matsuurae	490
Lycogramma brunnea	45
Lycodes soldatovi	25
Berryteuthis magister	1150
Gonatopsis borealis	2740
Toothed whales (in present state of stocks)	125–155
Seals (in present state of stocks)	123–151
Seabirds	30
Total	12 100–12 600

amount of all squids eaten by marine mammals (whales, dolphins and seals) in three Far Eastern seas of Russia for the stock state at the beginning of 20th century (the virgin stock) and in the early 1980s: in general, 9–15 and 6.5–13 thousand t respectively in the Japan Sea, 314–362 and 248–306 thousand t in the Okhotsk Sea, 670–930 and 520–740 thousand t in the Bering Sea. Squid consumption by toothed whales (mostly sperm and killer whales) in the Bering Sea was assessed at 480–700 thousand t before massive whaling after World War II and at 92–174 thousand t at present stock state (Sobolevsky and Mathisen, 1996).

More recent data show that, on the contrary, it is not marine mammals, but fish and squids that are the most important consumers of squids (Table 3). In the Okhotsk Sea two-thirds of total squid consumption (8.4 out of 12.5 million t) is eaten by various fishes, nearly one-third (3.9 million t) by the squids and only 2% by seabirds and marine mammals (Lapko, 1995, 1996). The common fishes (Alaska pollack and others) consume squids predominantly during winter because juvenile gonatids appear on the shelf mostly during autumn and winter. The role of squids in the food of common pelagic fish in the Okhotsk Sea rises from 0.9% during summer to 3.6% during winter, including from 1.1 to 3.6% in the food of the Alaska pollack (Lapko, 1994, 1996).

Radchenko (1992) assessed the quantity of squids, again mostly gonatids, eaten in the western and central Bering Sea by the Alaska pollack at 800 thousand t, by the Pacific salmon at 85 thousand t and by gonatids themselves at 2400 thousand

t. Shuntov *et al.* (1993c) assessed the squid consumption by the Alaska pollack in the whole Bering Sea in the1980s at 6.6 million t. The amount of gonatid squids, mostly *Berryteuthis magister*, eaten by fur seals breeding on the Commander, Kurile and Tyuleny Is, was estimated at 41.4 thousand t year^{-1} (Makhnyr *et al.*, 1984). Toothed whales, at the present state of stock, consumed in the Bering Sea 300–480 thousand t and seals consumed 220–260 thousand t (Sobolevsky, 1983b). Total yearly consumption by all predators was assessed at 4.2 million t in the western and central Bering Sea (Radchenko, 1992) and at 12.5 million t in the Okhotsk Sea in the late 1980s (Lapko, 1995). Later, squid consumption by the Alaska pollack decreased because of a stock decrease (2–3 times) in the western Bering and Okhotsk Seas (Lapko, 1996; Shuntov, 1996).

7.3. Competitors

The most important competitors for food with gonatid squids are the Alaska pollack, which feeds on almost the same food as juvenile *Berryteuthis magister* (Shuntov *et al.*, 1993c; Ilyinsky and Gorbatenko, 1994), and various mesopelagic fishes. Of the latter the most important (being the most common) in the Okhotsk Sea is *Leuroglossus schmidti*, the biomass of which is, in different years, from 8 to 20 million t (Ilyinsky, 1995) and which is also very close to gonatids in its diet (Ilyinsky and Gorbatenko, 1994). Other competitors are: in the Okhotsk Sea, *Lipolagus ochotensis*, *Stenobrachius nannochir*, *Bothrocarichthys microcephalus* and *Lycogrammoides schmidti*; in the Bering Sea, *Stenobrachius leucopsarus*, *Leuroglossus schmidti*, *Bathylagus pacificus* and others (Balanov, 1995). The main food of all these fishes and squids are *Thysanoessa longipes*, *Themisto japonica*, *Primno macropa* and *Metridia okhotensis*, but *Leuroglossus*, *Lipolagus* and *Bathylagus* consume many gelatinous animals (Coelenterata, Ctenophora, Larvacea and others) that form only a rather insignificant part in the diet of gonatids, with the probably exception of *Gonatus madokai* (Ilyinsky and Gorbatenko, 1994; Balanov, 1995; Ilyinsky, 1995). The very high biomass and production of zooplankton in the Bering and Okhotsk seas (Shuntov, 1985; Shuntov *et al.*, 1993c) makes it unlikely that there is severe competition for food between fish and squids.

7.4. Parasites

The parasitic fauna in gonatids has been studied mainly in the commercial species *Berryteuthis magister*. Ciliates, digenean trematodes *Derogenes varicus*, cestodes *Phyllobothrium* sp. (order Tetraphyllidea) and *Nybelinia surmenicola* (order Trypanorhynchidea), acanthocephalans *Echinorhynchus cotti*, anisakid nematodes *Anisakis simplex* and *Hysterothylacium reliquens* (= *Thynnascaris* sp.), and protistans *incertae sedis Hochbergia* sp. were recorded (Hochberg, 1990).

The larvae of *Phyllobothrium* and *Nybelinia* were found in large quantities in squids from the western Bering Sea and the north-western Pacific off the Kurile Is: *Nybelinia* occurred in 87% of squids from the former and in 92% from the latter area; *Phyllobothrium* in 95 and 91% respectively (Avdeyeva *et al.*, 1982). *N. surmenicola* was recorded also in the Japan Sea (Kasahara *et al.*, 1978). *Anisakis* were rarer (13.6%), *Hysterothylacium*, *Echinorhynchus* and *Derogenes* occurred as single specimens (Avdeyeva *et al.*, 1982). The abundance of larval cestodes increased with squid age. Larvae of nematodes were found only in squids larger than DML 17 cm and their abundance slowly increased with age (Avdeyeva *et al.*, 1982). The incidence of *Anisakis simplex* varied from 8% in the Pacific Ocean off the northern Kuriles to 23% in the north-western Bering Sea between Olyutorsky Bay and Cape Navarin (Bagrov, 1982).

Squids were infected by anisakid nematodes from euphausiids and hyperiids; the adult worms parasitize various marine mammals, including toothed whales, dolphins and seals, and seabirds. *Anisakis* are found typically in hosts feeding on planktonic and nektonic prey and *Thynnascaris* are found in those feeding at the bottom. The life cycle of *Derogenes varicus* includes the benthic gastropod family Naticidae (first intermediate host), planktonic copepods, chaetognaths, various fish and squids and then some piscivorous and teuthivorous fishes (incl. Gadidae) as final hosts. Definitive hosts of the trematodes *Phyllobothrium* and *Nybelinia* are sharks (Carcharinidae, Isuridae, Lamnidae and others). Squids receive them mostly from euphausiids, also from small fishes and squids, and are transport- or reservoir-hosts (Hochberg, 1990).

A ciliate, *Chromidina* sp. was found in *Gonatus* sp. from off Oregon and an undescribed ciliate was found in *Gonatopsis* sp. (Hochberg, 1990). *Hochbergia* sp. were found in various gonatids: *Berryteuthis magister*, *Gonatopsis borealis* and *Gonatus* sp. (McLean *et al.*, 1987; Hochberg, 1990) and were particularly common in *G. madokai* (Nesis, 1989b). *Nybelinia* sp. was recorded in *Gonatus* sp. off British Columbia (Hochberg, 1990).

In general, the parasitic fauna of gonatid squids is like that in other oceanic squids. Squids are the second intermediate or reservoir-hosts of most common parasites, sharing them with various pelagic and demersal fishes. Squid are thus the intermediate link in the chain leading from planktonic animals to large predatory (mostly elasmobranch) fish and marine mammals, the main predators of squids. It may be supposed that the parasitic fauna is more diverse in *B. magister* which inhabits both the pelagic and benthic realms, than in the gonatids which are not associated with the bottom at the adult stage.

7.5. Biomass, Production and Food Consumption of Gonatids

The biomass of squids was assessed for the Okhotsk and Bering seas and the Northwestern Pacific oceanward of the Kurile Is on the results of many standard epi-, meso- and sometimes also bathypelagic surveys using large pelagic trawls, assuming that their catching power for gonatid squids is 0.1 (Shuntov *et al.*, 1990, 1993a,b,c, 1994a,b; Didenko, 1991a,b; Ilyinsky, 1991; Radchenko, 1992, 1994; Lapko and Ivanov, 1993; Lapko, 1995, 1996). The data presented by these authors give a wide range of values. For example, the biomass of squids in the epipelagic zone of the western and central Bering Sea in seven surveys, 1986–1991, varied from 21.4 to 123.9 thousand t and of the Okhotsk Sea in six surveys, 1990–1993, from 24.3 to 122.0 thousand t (Radchenko, 1992; Lapko, 1995).

Production was assessed on the basis of the supposed production/biomass coefficients (B/P) for squids. The P/B-coefficients for squids with a 1-year life cycle are 3.5–7 and for those with a 2-year cycle they are 2–4, as in gastropods and bivalves with the same length of life cycle (Nesis, 1985). Some gonatids have a 2-year life cycle, for example *Berryteuthis magister*, others a 1-year cycle, as *Gonatopsis borealis* (see Section 4.1.7 and 4.3). Radchenko (1992) and Shuntov and Dulepova (1995) assumed the average P/B to be 3.0; Lapko (1995, 1996), supposing that the ratio of biomass of squids with 2- and 1-year cycles is 2:1, assumed the average P/B as 3.5–4.

The food consumption was assessed from the consumer biomass and the daily food ration, at approximately 2% for juvenile *B. magister* and 5.5% for *G. borealis* (Ilyinsky and Gorbatenko, 1994). Radchenko (1992) used lower figures, 1.1% and 4.2% respectively.

The results are as follows:

- For the Okhotsk Sea the total biomass of all squids is approximately 3 million t, including 2 million t of *B. magister* and 0.5 million t of *G. borealis*. Yearly production of all squids is 10.5–12 million t. Yearly food consumption by these two species alone is 19 million t of plankton (mostly euphausiids and hyperiids) and 10 million t of nekton (fish and squid). Yearly consumption of gonatids by various predators is approximately 12.5 million t including 3.9 million t by *B. magister* and *G. borealis* (Lapko, 1995, 1996). Ilyinsky and Gorbatenko (1994) have presented somewhat lower figures for the same sea: total biomass of squids 2.3 million t, including *B. magister* 1.73 and *G. borealis* 0.28 million t; yearly consumption of zooplankton by *B. magister* 9.4 million t, by *G. borealis* 2.7 million t.
- For the western and central Bering Sea the total biomass of gonatids is 3.1 million t. Yearly production of gonatids is 9.4 million t. Yearly consumption of gonatids by various predators is 6.6 million t. The consumption of zooplankton by gonatids, mostly *B. magister* and *G. borealis*, is 6.4 million t (Radchenko, 1994). The total squid biomass in the whole Bering Sea together with the waters off Aleutians is approximately 4 million t and total yearly production is 12 million t (Shuntov *et al.*, 1993c; Shuntov and Dulepova, 1995).

In both seas gonatid squids play a large role in the mesopelagic and bathypelagic layers and a much smaller role in the epipelagic layer.

There is no evaluation of squid biomass in the north-western Pacific off eastern Kamchatka and the Kurile Is, but it is known to exceed that in the Okhotsk Sea (Nesis and Nikitina, 1996). The concentration of mesopelagic fish in the upper mesopelagic layer (200–350 m) oceanward of the middle and southern Kuriles is approximately 1.5 times more than off Oregon (Lapko and Ivanov, 1993), and the area off Oregon is rich in gonatid squids (Pearcy, 1976). The total biomass of mesopelagic fish in the Northwestern Pacific off the Kurile Is is not less than in the Okhotsk Sea (Shuntov *et al.*, 1993c). The biomass of squids in the eastern part of the northern North Pacific is probably less than in the western part (Nesis, 1985) but the area is greater. The biomass of *Berryteuthis anonychus* alone was assessed at 200–250 thousand t in this area (Didenko, 1990).

We may roughly estimate ("guesstimate") the total biomass of gonatid squids in the whole subarctic North Pacific at approximately 15–20 million t. Yearly production may be 50–80 million t, some 10–15% of world total production of mesopelagic cephalopods, which has been assessed at 350–700 million t (Nesis, 1985). Based on calculations made by Ilyinsky and Gorbatenko (1994), the yearly food consumption by gonatids in the subarctic North Pacific may be assessed at 100–200 million t.

Compared with fish, squids are much faster growing and shorter-lived (Calow, 1987; O'Dor and Wells, 1987; Saville, 1987). The P/B-coefficients in squids are much higher than in fish. The biomass of squids in the Okhotsk Sea is less than 10% of fish biomass but their production is 58–67% of total fish production (Lapko, 1995). This emphasizes the great role of gonatid squids in the ecosystems of the northern North Pacific Ocean and the far eastern seas of Russia.

ACKNOWLEDGEMENTS

I am very thankful to Ch.M.Nigmatullin for many inspiring discussions and to N.P. Nezlin for his help in preparing illustrations. This work was supported in part by the Russian Foundation for Basic Research, grant No. 95-04-11072a and 97-05-64461.

REFERENCES

Akimushkin, I.I. (1963). "Cephalopods of the Seas of the USSR". Izdatel'stvo AN SSSR, Moscow. (In Russian; English translation by A. Mercado. IPST, Jerusalem, 1965).
Alexeyev, D.O. and Bizikov, V.A. (1987). Some aspects of biology and ecology of Commander squid *Berryteuthis magister* off Simushir Island in January, 1985. In

"Resources and Perspectives of the Use of Squids in the World Ocean" (B.G. Ivanov, ed.), 50–57. VNIRO, Moscow. (In Russian, English summary).

Alexeyev, D.O., Bizikov, V.A. and Khromov, D.N. (1986). Underwater observations on the Commander squid. *In* "IVth All-USSR Conference on commercial Invertebrates, Sevastopol. Abstracts of Communications", Moscow, **1**, pp. 126–127. (In Russian).

Alexeyev, D.O., Bizikov V.A., Khromov D.N. and Pomozov A.A. (1989). Underwater observations on behaviour and distribution of Commander squid and other cephalopod species in the north-western part of the Pacific Ocean. *In* "Underwater Observations in Bio-oceanologic and Fishery Research" (D.E. Gershanovich, ed.), pp. 66–77. VNIRO, Moscow. (In Russian, English summary).

Anderson, M.E. (1978). Notes on the cephalopods of Monterey Bay, California, with new records for the area. *Veliger* **21**(2), 255–262.

Anderson, M.E. (1981). Aspects of the natural history of the midwater fish, *Lycodapus mandibularis* (Zoarcidae) in Monterey Bay, California. *Pacific Science* **34**(2), 181–194.

Antonelis, G.A., Jr, Lowry, M.S., DeMaster, D.P. and Fiscus, C.H. (1987). Assessing northern elephant seal feeding habits by stomach lavage. *Marine Mammal Science* **3** (4), 308–322.

Arkhipkin, A.I. and Bizikov, V.A. (1991). Comparative analysis of age and growth rates estimation using statoliths and gladius in squids. *In* "Squid Age Determination using Statoliths" (P. Jereb, S. Ragonese and S.v. Boletzky, eds.), 19–33. N.T.R.-Istituto di Tecnologia della Pesca e del Pescato, *Special Publication* **1**. Mazzara del Vallo, Italy.

Arkhipkin, A. and Bizikov, V. (1996). Possible imitation of jellyfish by the squid paralarvae of the family Gonatidae (Cephalopoda: Oegopsida). *Polar Biology,* **16**(7), 531–534.

Arkhipkin, A.I., Bizikov, V.A., Krylov, V.V. and Nesis, K.N. (1996). Distribution, stock structure, and growth of the squid *Berryteuthis magister* (Berry, 1913) (Cephalopoda, Gonatidae) during summer and fall in the western Bering Sea. *Fishery Bulletin, U.S.* **94**(1), 1–30.

Avdeyeva, N.V., Vlasova, L.P. and Shevchenko, G.G. (1982). On the infection rate of the Commander squid *Berryteuthis magister* of the Northwestern Pacific Ocean. In "Problems of Rational Use of Commercial Invertebrates. Abstracts of Communications of III All-USSR Conference", Kaliningrad, pp. 214–215. (In Russian).

Bagrov, A.A. (1982). On the infection rate of squids in the North Pacific by the anisakid larvae (Nematoda, Anisakidae). *Parasitologiya* **16**(3), 200–203. (In Russian).

Bakkala, R.G. and Balsinger, J.W. (1987). Condition of groundfish resources of the eastern Bering Sea and Aleutian Islands region in 1986. *NOAA Technical Memorandum NMFS/NWC-117*, 187 pp.

Bakkala, R.G., Wakabayashi, K. and Sample T.M. (1985). Results of the demersal trawl surveys. *Bulletin of International North Pacific Fisheries Commission* **44**, 39–191.

Balanov, A.A. (1995). "Composition and structure of the nekton community in mesopelagic zone of the Bering Sea". Dissertation Abstract, Candidate in Biological Science, Insitute of Marine Biology, Russian Academy of Science, Vladivostok. (In Russian).

Bernard, F.R. (1980). Preliminary report on the potential commercial squid of British Columbia. *Canadian Technical Report of Fisheries and Aquatic Sciences* **942**, 51 pp.

Bizikov, V.A. (1987). "Recommendation for Development of the Commander Squid Fisheries in Northwestern Pacific and the State of its Stocks". VNIRO, Moscow. (In Russian).

Bizikov, V.A. (1991). "Squid gladius: its use for the study of growth, age, intraspecies structure and evolution (on the example of the family Ommastrephidae)". Dissertation

Abstract, Candidate in Biological Science, Institute of Oceanology, Moscow. (In Russian).

Bizikov, V.A. (1996). "Atlas of Morphology and Anatomy of the Gladius of Squids". VNIRO, Moscow. (In Russian, English summary).

Bizikov, V.A. and Arkhipkin, A.I. (1996). Distribution, stock structure and fishery prospects for the Commander squid. *Rybnoye Khozyaistvo* **1**, 42–45 (In Russian).

Boucher-Rodoni, R., Boucaud-Camou, E. and Mangold, K. (1987). Feeding and digestion. *In* "Cephalopod Life Cycles. Vol. 2. Comparative Reviews" (P.R. Boyle, ed.), pp. 85–108. Academic Press, London.

Calow, P. (1987). Fact and theory – an overview. *In* "Cephalopod Life Cycles. Vol. 2. Comparative Reviews" (P.R. Boyle, ed.), pp. 351–365. Academic Press, London.

Chuchukalo, V.I., Volkov, A.F., Efimkin, A.Ya. and Kuznetsova, N.A. (1994a). Feeding and daily rations of red salmon (*Oncorhynchus nerka*) in summer time. *Izvestiia TINRO* **116**, 122–127. (In Russian).

Chuchukalo, V.I., Volkov, A.F., Efimkin, A.Ya. and Blagoderov, A.I. (1994b). Distribution and feeding of chinook salmon (*Oncorhynchus tschawytscha*) in the north-western part of the Pacific Ocean. *Izvestiia TINRO* **116**, 137–141. (In Russian).

Clarke, M.R. (1966). A review of the systematics and ecology of oceanic squids. *Advances in Marine Biology* **4**, 91–300.

Clarke, M.R. (1977). Beaks, nets and numbers. *Symposia of The Zoological Society of London* **38**, 89–126.

Clarke, M.R. (1983). Cephalopod biomass – estimation from predation. *Memoirs of the National Museum of Victoria* **44**, 95–107.

Clarke, M.R. (1987). Cephalopod biomass – estimation from predation. In "Cephalopod Life Cycles. Vol. 2. Comparative Reviews" (P.R. Boyle), pp. 221–237. Academic Press, London.

Clarke, M.R. (1988). Evolution of recent cephalopods – A brief review. *In* "Paleontology and Neontology of Cephalopods" (M.R. Clarke and E.R. Trueman, eds), pp. 331–340. *The Mollusca* **12**. Academic Press, San Diego.

Clarke, M.R. and Maddock, L. (1988a). Statoliths of fossil coleoid cephalopods. *In* "Paleontology and Neontology of Cephalopods" (M.R. Clarke and E.R. Trueman, eds), pp. 153–168. *The Mollusca* **12**. Academic Press, San Diego.

Clarke, M.R. and Maddock, L. (1988b). Statoliths from living species of cephalopods and evolution. *In* "Paleontology and Neontology of Cephalopods" (M.R. Clarke and E.R. Trueman, eds), 169–184. *The Mollusca* **12**. Academic Press, San Diego.

Clarke, M.R., Fitch, J.E., Kristensen, T., Kubodera, T. and Maddock, L. (1980). Statoliths of one fossil and four living squids (Gonatidae: Cephalopoda). *Journal of the Marine Biological Association of the United Kingdom* **60**(2), 329–347.

Didenko, V.D. (1990). Prospects of fishery for the squid *Berryteuthis anonychus* in the north-eastern Pacific. *In* "Vth All-USSR Conference on Commercial Invertebrates, Minsk – Naroch. Abstracts of Communications", pp. 82–83. (In Russian).

Didenko, V.D. (1991a). Biological resources of cephalopods in the pelagic zone of the Okhotsk Sea during the winter season of 1990/91. *In* "Rational Use of Bioresources of the Pacific Ocean. Abstracts of Communications of All-USSR Conference", Vladivostok, pp. 88–90. (In Russian).

Didenko, V.D. (1991b). Biological resources of squids in the western Bering Sea during the autumn season of 1990. *In* "Rational Use of Bioresources of the Pacific Ocean. Abstracts of Communications of All-USSR Conference", Vladivostok, pp. 90–92. (In Russian).

Dodimead, A.J., Favorite, F. and Hirano T. (1963). Salmon of the North Pacific Ocean, Part II. Review of oceanography of the subarctic Pacific region. *Bulletin of International*

North Pacific Fisheries Commission **13**, 1–195.

Dolganova, N.T. (1988). Alaska pollack feeding in the autumnal time. *In* "Populational Structure, Abundance Dynamics and Ecology of the Alaska pollack", TINRO, Vladivostok, pp. 166–173. (In Russian, English summary).

Favorite, F., Dodimead, A.J. and Nasu, K. (1976). Oceanography of the subarctic Pacific region, 1960–71. *Bulletin of International North Pacific Fisheries Commission* **33**, 1–187.

Fedorets, Yu. A. (1977). Biological characteristics of *Berryteuthis magister* from the Bering Sea slope during autumn, 1976. *In* "All-USSR Scientific Conference on the Use of Commercial Invertebrates, Odessa," Moscow, pp. 97–98. (In Russian).

Fedorets, Yu. A. (1979). Some results of the studies of the Commander squid *Berryteuthis magister* in the Bering Sea. *In* "Molluscs. Main Results of their Investigations. Abstracts of Communications", **6**, 122–123. (In Russian).

Fedorets, Yu. A. (1983). Seasonal distribution of the squid (*Berryteuthis magister*) in the western Bering Sea. *In* "Taxonomy and Ecology of Cephalopods" (Ya.I. Starobogatov and K.N. Nesis, eds), p. 129. Zoological Institute AN SSSR, Leningrad. (In Russian).

Fedorets, Yu. A. (1986). Feeding rhythms in the Commander squid (*Berryteuthis magister*) in the western Bering Sea. *In* "IVth All-USSR Conference on Commercial Invertebrates, Sevastopol. Abstracts of Communications", Moscow, **1**, 158–159. (In Russian).

Fedorets, Yu. A. (1987). Biology and resources of the squid *Berryteuthis magister* off Commander Islands. *In* "Resources and Perspectives of the Use of Squids in the World Ocean" (B.G. Ivanov, ed.), pp. 57–66. VNIRO, Moscow. (In Russian, English summary).

Fedorets, Yu. A. (1991). Biology, distribution, resources and fishery of the Commander squid off Kurile Islands. *In* "Abstracts of Communications at a Report Session of TINRO and its Branches on the Results of Investigations made in 1990", pp. 36–41. TINRO, Vladivostok. (In Russian).

Fedorets, Yu. A. and Didenko, V.D. (1991). A practical application of the method of forecasting of the changes of Commander squid catches off the Central Kuriles. *In* "Problems of Fisheries Forecasting. Abstracts of Communications of All-USSR Scientific Conference", Kaliningrad, pp. 42–44 (In Russian).

Fedorets, Yu. A. and Kozlova, O.A. (1987). Spawning, fecundity and abundance of the squid *Berryteuthis magister* (Gonatidae) in the Bering Sea. *In* "Resources and Fishery Perspectives of Squids in the World Ocean" (B.G. Ivanov, ed.), pp. 66–80. VNIRO, Moscow. (In Russian, English summary).

Fedorets, Yu. A. and Kozlova, O.A. (1988). Distributional patterns of the Commander squid in the western Bering Sea and off the eastern Kamchatka. *In* "Fishery Resources and Biological Bases of Rational Use of Commercial Invertebrates. Abstracts of Communications of All-USSR Conference", 22–24 November 1988, Vladivostok, pp. 74– 76. (In Russian).

Fedorets, Yu. A. and Kun, M.S. (1988). Food composition and feeding of the Commander squid *Berryteuthis magister* off Kurile Islands. *In* "Fishery Resources and Biological Bases of Rational Use of Commercial Invertebrates. Abstracts of Communications of All-USSR Conference", Vladivostok, pp. 70–72 (In Russian).

Fedorets, Yu. A. and Luchin, V.A. (1991). Interaction between oceanological conditions and density of the Commander squid concentrations off the Kurile Islands. *In* "Problems of Fisheries Forecasting. Abstracts of Communications of All-USSR Scientific Conference", Kaliningrad, pp. 120–122. (In Russian).

Fedorets, Yu. A., Grenkin, V.V. and Railko, P.P. (1985). Results of the study of biology and stock size of the Commander squid off the Central Kurile Islands. *In* "Investigations and Rational Use of Bioresources of the Far Eastern and Northern Seas of the USSR.

Abstracts of Communications of All-USSR Conference", Vladivostok, pp. 102–103. (In Russian).

Fedorets, Yu. A., Didenko, V.D. and Railko, P.P. (1994). Dynamics of *Berryteuthis magister* spawning off Commander Islands (Pacific Ocean). *In* "The Behaviour and Natural History of Cephalopods. Programme and Abstracts", Vico Equense (Napoli), Italy, p. 9.

Fields, W.G. and Gauley, V.A. (1971). Preliminary description of an unusual gonatid squid (Cephalopoda: Oegopsida) from the North Pacific. *Journal of the Fisheries Research Board of Canada* **28**(11), 1796–1801.

Filippova, Yu. A. (1972). "Methods of the Study of Cephalopods". VNIRO, Moscow. (In Russian).

Filippova, Yu. A. (1983). "Recommendations for the Study of Cephalopods". VNIRO, Moscow. (In Russian).

Fiscus, C.H. (1982). Predation by marine mammals on squids of the eastern North Pacific Ocean and the Bering Sea. *Marine Fisheries Review* **44**(2), 1–10.

Fiscus, C.H. and Mercer, R.W. (1982). Squids taken in surface gillnets in the North Pacific Ocean by the Pacific Salmon Investigations Program, 1955–72. *NOAA Technical Memorandum* NMFS, F/NWC-28, 1–32.

Fiscus, C.H., Rice, D.W. and Wolman, A.A. (1989). Cephalopods from the stomachs of sperm whales taken off California. *NOAA Technical Report* NMFS **83**, 1–12.

Glazachev, V.M. and Fedorets, Yu. A. (1981). Intraspecies groupings of the Commander squid *Berryteuthis magister* in the Bering Sea and the Gulf of Alaska. *In* "Biological Resources of Large Depths and Pelagic Realm of the Open Areas of the World Ocean. Abstracts of Communications of the Scientific – Practical Conference", Murmansk, pp. 90–91 (In Russian).

Gorbatenko, K.M., Chuchukalo, V.I. and Shevtsov, G.A. (1995). Daily food ration in two abundant squid species in the Bering and Okhotsk seas during autumn-winter time. *In* "Complex Studies of the Bering Sea Ecosystem" (B.N. Kotenev and V.V. Sapozhnikov, eds), pp. 349–357. VNIRO, Moscow. (In Russian, English summary).

Hochberg, F.G. (1990). Diseases of Mollusca: Cephalopoda. 1.2. Diseases caused by protistans and metazoans. *In* "Diseases of Marine Animals" (O. Kinne, ed.), pp. 47–202. Biol. Anstalt Helgoland, Hamburg.

Ilyinsky, E.N. (1991). Summer distribution of squids in mesopelagic zone of the Okhotsk Sea. *Okeanologiya* **31**(1), 151–154. (In Russian, English summary).

Ilyinsky, E.N. (1995). "Composition and structure of nektonic community in the mesopelagic zone of the Okhotsk Sea". Abstr. Diss. Cand. Biol. Sci., Inst. Mar. Biol. Russ. Acad. Sci., Vladivostok. (In Russian).

Ilyinsky, E.N. and Gorbatenko, K.M. (1994). Main trophic connections of the mesopelagic nekton of the Okhotsk Sea. *Izvestiia TINRO* **116**, 91–104. (In Russian).

Ito, J. (1964). Food and feeding habit of Pacific salmon (genus *Oncorhynchus*) in their oceanic life. *Bulletin of Hokkaido Regional Fisheries Research Laboratory* **29**, 85–97.

Jackson, G.D. (1994). Application and future potential of statolith increment analysis in squids and sepioids. *Canadian Journal of Fisheries and Aquatic Sciences* **51**(11), 2612–2625.

Jefferts, K. (1985). *Gonatus ursabrunae* and *Gonatus oregonensis*, two new species of squids from the north-eastern Pacific Ocean (Cephalopoda: Oegopsida: Gonatidae). *Veliger* **28**(2), 159–174.

Jefferts, K. (1988). Zoogeography of cephalopods from the north-eastern Pacific Ocean. *Bulletin of the Ocean Research Institute, University of Tokyo* **26**(I), 123–157.

Kafanov, A.I. (1982). Cenozoic history of the malacofaunas of the North Pacific shelves.

In "Marine Biogeography" (O.G. Kusakin, ed.), pp. 134–176. Nauka, Moscow. (In Russian).

Kajimura, H., Fiscus, C.H. and Stroud, R.K. (1980). Food of the Pacific white-sided dolphin, *Lagenorhynchus obliquidens*, Dall's porpoise, *Phocoenoides dalli*, and northern fur seal, *Callorhinus ursinus*, off California and Washington, with appendices on size and food of Dall's porpoise from Alaskan waters. *NOAA Technical Memorandum* NMFS F/NWC-2, 1–30.

Kasahara, S., Nazumi, T., Shimizu, T. and Hamabe M. (1978). Contributions of biological information useful for development of inshore squid fishery in the Japan Sea. II. A note on reproduction and distribution of *Berryteuthis magister* (Berry) assumed from biological observations on trawl catches in the waters around the Oki Islands, Japan Sea. *Bulletin of Japan Sea Regional Fisheries Research Laboratory* **29**, 159–178. (In Japanese, English summary).

Katugin, O.N. (1988). Polymorphism of 6–phosphogluconatedehydrogenase in the squid *Berryteuthis magister*. *In* "Fishery Resources and Biological Bases of Rational Use of Commercial Invertebrates. Abstracts of Communications of All-USSR Conference", Vladivostok, pp. 80–81. (In Russian).

Katugin, O.N. (1990). Polymorphic fermentative systems in the Commander squid *Berryteuthis magister* (Berry). *In* "Ecology, Migration and Distributional Patterns of Marine Fisheries Objects. Functioning of Marine Systems and Anthropogenic Impact Upon Them. Abstracts of Communications of TINRO Young Scientists Conference, Vladivostok", pp. 25–27 (In Russian).

Katugin, O.N. (1993a). Study of relationships between different species of squid, family Gonatidae, by protein electrophoresis. *In* "Biology and Rational Use of Hydrobionts. Their role in Ecosystems. Abstracts of Communications of the TINRO Young Scientists Conference, Vladivostok", pp. 14–15 (In Russian).

Katugin, O.N. (1993b). Genetic variation in the squid *Berryteuthis magister* (Berry, 1913) (Oegopsida: Gonatidae). *In* "Recent Advances in Cephalopod Fisheries Biology" (T. Okutani, R.K. O'Dor and T. Kubodera, eds), pp. 201–213. Tokai University Press, Tokyo.

Katugin, O.N. (1995a). Genetic differentiation in *Berryteuthis magister* from the North Pacific. *ICES Marine Science Symposia* **199**, 459–467.

Katugin, O.N. (1995b). Morphology, genetic difference, and evolution of *Berryteuthis magister*, *Berryteuthis anonychus*, and *Gonatopsis borealis* (Cephalopoda: Oegopsida). *In* "Unitas Malacologica. 12th International Malacological Congress, Vigo, Spain. Abstracts", pp. 312–313.

Kawakami, T. (1980). A review of sperm whale food. *Scientific Report of the Whales Research Institute, Tokyo* **32**, 199–218.

Kawamura, A. (1980). A review of food of balaenopterid whales. *Scientific Report of the Whales Research Institute, Tokyo* **32**, 155–197.

Kawamura, A. (1982). Food habits and prey distributions of three rorqual species in the North Pacific Ocean. *Scientific Report of the Whales Research Institute, Tokyo* **34**, 59–91.

Klumov, S.K. (1978). Feeding of sperm whale in northern half of the Pacific Ocean. *In* "Marine Mammals. Results and Methods of Investigations" (V.E. Sokolov, ed.), pp. 175–213. Nauka, Moscow. (In Russian).

Kobayashi, Y., Masuda, K., Anma, G., Meguro, T., Yamaguchi, H. and Takagi, Sh. (1986). Distribution and abundance of three species of squids along 155°W longitude. *Bulletin of the Faculty of Fisheries, Hokkaido University* **37**, 181–189.

Kondakov, N.N. (1941). Cephalopods of the Far Eastern seas of the USSR. *Issledovaniya dalnevostochnykh morei* **1**, 216–255. (In Russian, English summary).

Kovalev, N.N. (1990). "Properties of cholinesterases in the Commander squid *Berryteuthis magister* from different parts of the species range". Abstr. Diss. Cand. Biol. Sci., Inst. Evolut. Physiol. Biochemistry Acad. Sci. USSR, Leningrad. (In Russian).

Kovalev, N.N. and Rosengart, E.V. (1987). Interaction of reversible inhibitors with cholinesterase of the squid *Berryteuthis magister* from different areas of the Bering Sea. *Zhurnal Evolutsionnoi Biochimii i Physiologii* **23**(4), 548–550. (In Russian, English summary).

Kristensen, T.K. (1981). The genus *Gonatus* Gray, 1849 (Mollusca: Cephalopoda) in the North Atlantic. A revision of the North Atlantic species and description of *Gonatus steenstrupi* n.sp. *Steenstrupia* **7**(4), 61–99.

Kristensen, T.K. (1983). *Gonatus fabricii*. In "Cephalopod Life Cycles. Vol. 1. Species Accounts" (P.R. Boyle, ed.), 159–173. Academic Press, London.

Kubodera, T. (1986). Relationships between abundance of epipelagic squids and oceanographical–biological environments in the surface waters of the subarctic Pacific in summer. *Bulletin of International North Pacific Fisheries Commission* **47**, 215–228.

Kubodera, T. (1992). Biological characteristics of the gonatid squid *Berryteuthis magister magister* (Cephalopoda: Oegopsida) off Northern Hokkaido, Japan. *Memoirs of the National Science Museum, Tokyo* **25**, 111–123.

Kubodera, T. (1996). Cephalopod fauna off Sanriku and Joban Districts, north-eastern Japan. *Memoirs of the National Science Museum, Tokyo*, **29**, 187–207.

Kubodera, T. and Jefferts, K. (1984). Distribution and abundance of the early life stages of squid, primarily Gonatidae (Cephalopoda, Oegopsida), in the northern North Pacific, **1**, 2. *Bulletin of the National Science Museum, Tokyo*, ser. A **10**(3), 91–106; **10**(4), 165–193.

Kubodera, T. and Miyazaki, N. (1993). Cephalopods eaten by short-finned pilot whales, *Globicephala macrorhynchus*, caught off Ayukawa, Ojika Peninsula, in Japan, in 1982 and 1983. *In* "Recent Advances in Cephalopod Fisheries Biology" (T. Okutani, R.K. O'Dor and T. Kubodera, eds), pp. 215–227. Tokai University Press, Tokyo.

Kubodera, T. and Okutani, T. (1977). Description of a new species of gonatid squid, *Gonatus madokai* n.sp. from the north-west Pacific, with notes on morphological changes with growth and distribution in immature stages (Cephalopoda: Oegopsida). *Venus* **36**(3), 123–151.

Kubodera, T. and Okutani, T. (1981a). *Gonatus middendorffi*, a new species of gonatid squid from the northern North Pacific, with notes on morphological changes with growth and distribution in immature stages (Cephalopoda, Oegopsida). *Bulletin of the National Science Museum, Tokyo*, ser. A **7**(1), 7–26.

Kubodera, T. and Okutani, T. (1981b). The systematics and identification of larval cephalopods from the northern North Pacific. *Research Institute of the North Pacific Fisheries, Faculty of Fisheries, Hokkaido University*, special volume, 131–159.

Kubodera, T. and Shimazaki, K. (1989). Cephalopods from the stomach contents of the pomfret (*Brama japonica* Hilgendorf) caught in surface gillnets in the northern North Pacific. *Journal of Cephalopod Biology* **1**(1), 71–83.

Kubodera, T., Pearcy, W.G., Murakami, K., Kobayashi, T., Nakata J. and Mishima S. (1983). Distribution and abundance of squids caught in surface gillnets in the subarctic Pacific, 1977–1981. *Memoirs of the Faculty of Fisheries, Hokkaido University* **30**(1–2), 1–49.

Kuramochi, T., Kubodera, T. and Miyazaki, N. (1993). Squids eaten by Dall's porpoises, *Phocoenoides dalli* in the north-western North Pacific and in the Bering Sea. In "Recent Advances in Cephalopod Fisheries Biology" (T. Okutani, R.K. O'Dor and T. Kubodera, eds), pp. 229–240. Tokai University Press, Tokyo.

Kuznetsova, N.A. and Fedorets, Yu. A. (1987). On the feeding of the Commander squid *Berryteuthis magister*. *Biologiya Morya* (1), 71–73 (In Russian, English summary).

Laevastu, T. and Larkins, H.A. (1981). "Marine Fisheries Ecosystem. Its Quantitative Evaluation and Management". Fishing News Books, Farnham.

Lapko, V.V. (1994). Trophic relationships in the epipelagic ichthyocene of the Okhotsk Sea. *Izvestiia TINRO* **116**, 168–177 (In Russian).

Lapko, V.V. (1995). Role of squids in the communities of the Okhotsk Sea. *Okeanologiya* **35** (5), 737–742. (In Russian, English summary).

Lapko, V.V. (1996). "Composition, structure, and dynamics of the epipelagic nekton in the Okhotsk Sea". Dissertation Abstract, Candidate in Biological Science, Institute of Marine Biology, Russian Academy of Sciences, Vladivostok. (In Russian).

Lapko, V.V. and Ivanov, O.A. (1993). Composition and distribution of the fauna in the sound-scattering layer in the Pacific Ocean near Kurile Islands. *Okeanologiya* **33**(4), 574–578. (In Russian, English summary).

Lapshina, V.I. (1988). On the feeding of the squid *Berryteuthis anonychus* Pearcy and Voss from the area of north-eastern Pacific. Deposited MS, Dep. VINITI 25.10.88, No. 7673–B 88, 1–5. (In Russian).

Larkin, P.A. and Ricker, W.E. (eds.) (1964). Canada's Pacific marine fisheries. Past performances and future prospects. *In* "Inventory of the Natural Resources of British Columbia". *Studies of the Fisheries Research Board of Canada* **1963**(2), No. 851 194–268.

Leonov, A.K. (1960). "Regional Oceanography. Pt 1. Bering, Okhotsk, Japan, Caspian and Black Seas". Hydrometeoizdat, Leningrad. (In Russian).

Loffler, D.L. and Vecchione, M. (1993). An unusual squid paralarva (Cephalopoda) with tentacular photophores. *Proceedings of the Biological Society of Washington* **106**(3), 602–605.

Lowry, M.S., Stewart, B.S., Heath, C.B., Yochem, P.K. and Francis, J.M. (1991). Seasonal and annual variability of the diet of California sea lions *Zalophus californianus* at San Nicolas Island, California, 1981–86. *Fishery Bulletin, U.S.* **89**(2), 331–336.

Makhnyr, A.I., Kuzin, A.E. and Perlov, A.S. (1983). Quantitative characteristic of the feeding of otariid seals in the north-western Pacific. *In* "Ecological–Faunistic Investigations of Some Mammals on the Sakhalin and Kurile Islands", pp. 83–89. Far Eastern Scientific Center of the USSR Academy of Sciences, Vladivostok. (In Russian).

Makhnyr, A.I., Kuzin, A.E. and Perlov, A.S. (1984). Seasonal variation in the prey biomass for otariid seals in the north-western Pacific. *In* "Marine Mammals of the Far East", pp. 3–13. TINRO, Vladivostok. (In Russian).

Malyshev, A.A. and Railko, P.P. (1986). Oceanological patterns of the formation of the Commander squid concentrations off the Simushir Island. *In* "IVth All-USSR Conference on Commercial Invertebrates, Sevastopol. Abstracts of Communications", Moscow, **1**, 153–154. (In Russian).

Markina, N.P. (1987). Alaska pollack role in the trophic structure of Far Eastern seas. *In* "Populational Structure, Abundance Dynamics and Ecology of the Alaska Pollack", pp. 144–157. TINRO, Vladivostok. (In Russian, English summary).

McLean, N., Hochberg, F.G. and Shinn, G.L. (1987). Giant protistan parasites on the gills of cephalopods (Mollusca). *Diseases of Aquatic Organisms* **3**(2), 119–125.

Murakami, K. (1976). Distribution in relation to environment of squid in the north-west Pacific and the Okhotsk Sea. *FAO Fisheries Report* **170** (Suppl. 1), 9–17.

Murata, M. and Okutani, T. (1975). Rare and interesting squid from Japan. IV. An occurrence of *Gonatopsis japonicus* Okiyama in the Sea of Okhotsk (Oegopsida: Gonatidae). *Venus* **33**(4), 210–211.

Murata, M., Ishii, M. and Araya, H. (1976). The distribution of the oceanic squids, *Ommastrephes bartrami* (Lesueur), *Onychoteuthis borealijaponicus* Okada, *Gonatopsis borealis* Sasaki and *Todarodes pacificus* Steenstrup in the Pacific Ocean off north-eastern Japan. *Bulletin of Hokkaido Regional Fisheries Research Laboratory* **41**, 1–29.

Naito, M., Murakami, K., Kobayashi, T., Nakayama, N. and Ogasawara, J. (1977a). Distribution and migration of oceanic squids (*Ommastrephes bartrami*, *Onychoteuthis borealijaponicus*, *Berryteuthis magister* and *Gonatopsis borealis*) in the western subarctic Pacific region. *Research Institute of the North Pacific Fisheries, Faculty of Fisheries, Hokkaido University*, special volume, pp. 321–337. (In Japanese with English summary).

Naito, M., Murakami, K. and Kobayashi, T. (1977b). Growth and food habits of oceanic squids (*Ommastrephes bartrami*, *Onychoteuthis borealijaponicus*, *Berryteuthis magister* and *Gonatopsis borealis*) in the western subarctic Pacific region. *Research Institute of the North Pacific Fisheries, Faculty of Fisheries, Hokkaido University*, special volume, pp. 339–351. (In Japanese with English summary).

Natsukari, Y., Mukai, H., Nakahama, S. and Kubodera, T. (1993a). Age and growth estimation of a gonatid squid, *Berryteuthis magister*, based on statolith microstructure (Cephalopoda: Gonatidae). *In* "Recent advances in Cephalopod Fisheries Biology" (T. Okutani, R.K. O'Dor and T. Kubodera, eds), pp. 351–364. Tokai University Press, Tokyo.

Natsukari, Y., Boletzky, S.v. and Clarke, M.R (1993b). Age session. *In* "Recent Advances in Cephalopod Fisheries Biology" (T. Okutani, R.K. O'Dor and T. Kubodera, eds), pp. 668–696. Tokai University Press, Tokyo.

Nazumi, T., Kasahara, S. and Hamabe, M. (1979). Contribution of biological information useful for development of inshore squid fishery in the Japan Sea. III. Supplements and amendments to the previous paper on reproduction and distribution of *Berryteuthis magister* (Berry). *Bulletin of the Japan Sea Regional Fisheries Research Laboratory* **30**, 1–14. (In Japanese with English abstract).

Nesis, K.N. (1972). Two new species of squids, family Gonatidae, from the North Pacific. *Zoologicheskii Zhurnal* **51**(9), 1300–1307. (In Russian, English summary).

Nesis, K.N. (1973a). Taxonomy, phylogeny and evolution of squids of the family Gonatidae (Cephalopoda). *Zoologicheskii Zhurnal* **52**(11), 1626–1638. (In Russian, English summary).

Nesis, K.N. (1973b). Types of distribution areas of cephalopod molluscs in the North Pacific. *Trudy Instituta Okeanologii AN SSSR* **91**, 213–239. (In Russian, English summary).

Nesis, K.N. (1978). Evolutionary history of nekton. *Zhurnal Obshchei Biologii* **39**(1), 53–65. (In Russian, English summary).

Nesis, K.N. (1979). The larvae of cephalopods. *Biologiya Morya* No. 4, 26–37 (In Russian, English summary; English translation in *Soviet Journal of Marine Biology* **5**(4), 267–275).

Nesis, K.N. (1985). "Oceanic Cephalopods: Distribution, Life Forms, Evolution". Nauka, Moscow. (In Russian).

Nesis, K.N. (1987). "Cephalopods of the World". T.F.H. Publications, Neptune City, NJ.

Nesis, K.N. (1989a). Teuthofauna of the Okhotsk Sea. Biology of squids *Berryteuthis magister* and *Gonatopsis borealis* (Gonatidae). *Zoologicheskii Zhurnal* **68**(9), 45–56. (In Russian, English summary).

Nesis, K.N. (1989b). Teuthofauna of the Okhotsk Sea. Distribution and biology of non-coastal species. *Zoologicheskii Zhurnal* **68**(12), 19–29. (In Russian, English summary).

Nesis, K.N. (1989c). Cephalopods of the open waters of the Okhotsk Sea: general distribution and zoogeography. *Okeanologiya* **29**(6), 999–1005. (In Russian, English summary).

Nesis, K.N. (1990). Japanese gonatopsis – the largest gonatid squid. *In* "All-USSR Conference "Reserve Food Biological Resources of the Open Ocean and USSR Seas", Kaliningrad, Abstracts of communications", Moscow, pp. 133–136. (In Russian).

Nesis, K.N. (1993a). Spent females of deep-water squid *Gonatopsis octopedatus* Sasaki, 1920 (Gonatidae) in the epipelagic layer of the Okhotsk and Japan seas. *Ruthenica* **3**(2), 153–158.

Nesis, K.N. (1993b). Population structure of oceanic ommastrephids, with particular reference to *Sthenoteuthis oualaniensis*: A review. *In* "Recent Advances in Cephalopod Fisheries Biology" (T. Okutani, R.K. O'Dor and T. Kubodera, eds), pp. 375–383. Tokai University Press, Tokyo.

Nesis, K.N. (1995). Population dynamics of the commander squid, *Berryteuthis magister* (Berry) in the western Bering Sea during the autumn spawning season. *Ruthenica* **5** (1), 55–69.

Nesis, K.N. and Nezlin, N.P. (1993). Intraspecific groupings in gonatid squids. *Russian Journal of Aquatic Ecology* **2**(2), 91–102.

Nesis, K.N. and Nikitina, I.V. (1996). Vertical distribution of squids in the southern Okhotsk Sea and north-western Pacific off Kurile Islands (summer 1992). *Russian Journal of Aquatic Ecology* **4**(1), 9–24 (1995).

Nesis, K.N. and Shevtsov, G.A. (1977). First data on abyssal cephalopods of the Okhotsk Sea. *Biologiya Morya* No. 5, 76–77. (In Russian, English summary).

Nishimura, S. (1970). Recent records of Baird's beaked whale in the Japan Sea. *Publications of the Seto Marine Biological Laboratory* **18**(1), 61–68.

Nixon, M. (1987). Cephalopod diets. *In* "Cephalopod Life Cycles. Vol. 2. Comparative Reviews" (P.R. Boyle, ed.), pp. 201–219. Academic Press, London.

O'Dor, R.K. and Wells, M.J. (1987). Energy and nutrient flow. *In* "Cephalopod Life Cycles. Vol. 2. Comparative Reviews" (P.R. Boyle, ed.), pp. 109–133. Academic Press, London.

Ogata, T., Okiyama, M. and Tanino, Y. (1973). Diagnoses of the animal populations in the depths of the Japan Sea, chiefly based on the trawling experiments by the R/V "Kaiyo-Maru". *Bulletin of the Japan Sea Regional Fisheries Research Laboratory* **24**, 21–51. (In Japanese with English summary).

Ogi, H. (1980). The pelagic feeding ecology of thick-billed murres in the North Pacific, March–June. *Bulletin of the Faculty of Fisheries, Hokkaido University* **31**(1), 50–72.

Ogi, H., Kubodera, T. and Nakamura, K. (1980). The pelagic feeding ecology of the short-tailed shearwater *Puffinus tenuirostris* in the subarctic Pacific region. *Journal of the Yamashita Institute of Ornithology* **12**(3), 50–72.

Okiyama, M. (1969). A new species of *Gonatopsis* from the Japan Sea, with the record of a specimen referable to *Gonatopsis* sp. Okutani, 1967 (Cephalopoda: Oegopsida, Gonatidae). *Publications of the Seto Marine Biological Laboratory* **17**(1), 19–32.

Okiyama, M. (1970). A record of the eight-armed squid, *Gonatopsis octopedatus* Sasaki, from the Japan Sea (Cephalopoda, Oegopsida, Gonatidae). *Bulletin of the Japan Sea Regional Fisheries Research Laboratory* **22**, 71–80.

Okiyama, M. (1993a). Kinds, abundance and distribution of the oceanic squids in the Sea of Japan. *In* "Recent Advances in Cephalopod Fisheries Biology" (T. Okutani, R.K. O'Dor and T. Kubodera, eds), pp. 403–415. Tokai University Press, Tokyo.

Okiyama M. (1993b). Why do gonatid squid *Berryteuthis magister* lose tentacles on maturation? *Nippon Suisan Gakkaishi* **59**(1), 61–65.

Okutani, T. (1967). Preliminary note on a hitherto unknown form of the eight-armed squid,

320   K.N. NESIS

genus *Gonatopsis*, from northern North Pacific. *Venus* **25**(2), 65–68.
Okutani, T. (1968). Review of Gonatidac (Cephalopoda) from the North Pacific. *Venus* **27** (1), 31–34 (In Japanese).
Okutani, T. (1969). Studies on early life history of decapodan Mollusca – IV. Squid larvae collected by oblique hauls of a larva net from the Pacific coast of eastern Honshu, during the winter seasons, 1965–1968. *Bulletin of Tokai Regional Fisheries Research Laboratory* **58**, 83–96.
Okutani, T. (1977). Stock assessment of cephalopod resources fished by Japan. *FAO Fisheries Technical Paper* **173**, 1–62.
Okutani, T. (1983). *Todarodes pacificus*. In "Cephalopod Life Cycles. Vol. 1. Species Accounts" (P.R. Boyle, ed.), pp. 201–214. Academic Press, London.
Okutani, T. (1987). Juvenile morphology. In "Cephalopod Life Cycles. Vol. 2. Comparative Reviews" (P.R. Boyle, ed.), pp. 33–44. Academic Press, London.
Okutani, T. (1988). Evidence of spawning of *Berryteuthis magister* in the north-eastern Pacific. *Bulletin of the Ocean Research Institute, University of Tokyo* **26**(I), 193–200.
Okutani, T. and Clarke, M.R. (1992). Gonatidae. In "Larval and Juvenile Cephalopods: A Manual of Their Investigation" (M.J. Sweeney, C.F.E. Roper, K.M. Mangold, M.R. Clarke and S.v. Boletzky, eds). pp. 139–156. *Smithsonian Contributions to Zoology* **513**.
Okutani, T. and Kubota, T. (1976). Cephalopods eaten by lancetfish, *Alepisaurus ferox* Lowe, in Suruga Bay, Japan. *Bulletin of Tokai Regional Fisheries Research Laboratory* **84**, 1–9.
Okutani, T. and Nemoto, T. (1964). Squids as the food of sperm whales in the Bering Sea and Alaskan Gulf. *Scientific Reports of the Whales Research Institute* **18**, 111–122.
Okutani, T. and Satake, Y. (1978). Squids in the diet of 38 sperm whales caught in the Pacific waters off north-eastern Honshu, Japan, February 1977. *Bulletin of Tokai Regional Fisheries Research Laboratory* **93**, 13–27.
Okutani, T., Satake Y., Ohsumi, S. and Kawakami, T. (1976). Squids eaten by sperm whales caught off Joban District, Japan, during January–February, 1976. *Bulletin of Tokai Regional Fisheries Research Laboratory* **87**, 67–113.
Okutani, T., Tagawa M. and Horikawa, H. (1987). "Cephalopods From Continental Shelf and Slope Around Japan". Japan Fish. Resource Conserv. Assoc., Tokyo.
Okutani, T., Kubodera, T. and Jefferts K. (1988). Diversity, distribution and ecology of gonatid squids in the subarctic Pacific: A review. *Bulletin of the Ocean Research Institute, University of Tokyo* **26**(I), 159–192.
Okutani, T., Nakamura, I. and Seki, K. (1995). An unusual egg-brooding behavior of an oceanic squid in the Okhotsk Sea. *Venus* **54**(3), 237–239.
Orlov, A.M. (1990). On the possibilities to expand the Commander squid fishery in the Far Eastern seas. In "Vth All-USSR Conference on Commercial Invertebrates, Minsk – Naroch. Abstracts of Communications", pp. 82–83. (In Russian).
Orlov, A.M. and Pitruk, D.L. (1996). Materials on the feeding of *Polypera simushirae* (Liparidae) and notes on its distribution in the area of Northern Kurile Islands. *Voprosy Ichthyologii* **36**(6), 821–826. (In Russian).
Osako, M. and Murata M. (1983). Stock assessment of cephalopod resources in the north-western Pacific. *FAO Fisheries Technical Paper* **231**, 55–144.
Panina, G.K. (1966). Feeding of fur seals in the western Pacific. *Izvestiia TINRO* **58**, 23–40. (In Russian with English and French summaries).
Panina, G.K. (1971). Feeding and food base of fur seals in the Japan Sea. *Trudy VNIRO* **82**, 65–76. (In Russian).
Pearcy, W.G. (1976). Seasonal and inshore–offshore variations in the standing stocks of micronekton and macroplankton off Oregon. *Fishery Bulletin, U.S.* **74**(1), 70–80.

Pearcy, W.G. and Ambler, J.W. (1974). Food habits of deep-sea macrourid fishes off the Oregon coast. *Deep-Sea Research* **21**(9), 745–759.

Pearcy, W.G. and Voss, G.L. (1963). A new species of gonatid squid from the north-eastern Pacific. *Proceedings of the Biological Society of Washington* **76**, 105–112.

Pearcy, W.G., Brodeur, R.D., Shenker, J.M., Smoker, W.W. and Endo, Y. (1988). Food habits of Pacific salmon and steelhead trout, midwater trawl catches and oceanographic conditions in the Gulf of Alaska, 1980–1985. *Bulletin of the Ocean Research Institute, University of Tokyo* **26**(II), 29–78.

Pearcy, W.G., Krygier, E.E., Mesecar, R. and Ramsey, F. (1977). Vertical distribution and migration of oceanic micronekton off Oregon. *Deep-Sea Research* **24**(3), 223–245.

Petrov, O.A. (1988). Interannual variability of distribution, abundance and size composition of the Commander squid off Southern Kurile Islands. *In* "Raw Resources and Biological Bases of Rational Use of Commercial Invertebrates. Abstracts of Communications of All-USSR Conference", Vladivostok, pp. 91–93. (In Russian).

Radchenko, V.I. (1992). The role of squids in the pelagic ecosystem of the Bering Sea. *Okeanologiya* **32**(6), 1093–1101. (In Russian, English summary). English translation in *Oceanology* **32**(6), 762–767.

Radchenko, V.I. (1994). "Composition, structure and dynamics of nektonic communities in the epipelagic zone of the Bering Sea". Dissertation Abstract, Candidate in Biological Science, Institute of Marine Biology, Russian Academy of Sciences, Vladivostok. (In Russian).

Railko, P.P. (1979). Distribution and some features of biology of the Commander squid *Berryteuthis magister* in the Japan Sea. *In* "Molluscs. Main Results of Investigations. Abstracts of Communications" **6**, 128–129. (In Russian).

Railko, P.P. (1983). Biology and distribution of the Commander squid *Berryteuthis magister* in the area off the Kurile Islands. *In* "Taxonomy and Ecology of Cephalopods" (Ya.I. Starobogatov and K.N. Nesis, eds), pp. 97–98. Zoological Institute AN SSSR, Leningrad. (In Russian).

Railko, P.P. (1987). Method of evaluation of trawl catchability during squid fishing. *In* "Resources and Perspectives of the Use of Squids in the World Ocean" (B.G. Ivanov, ed.), pp. 94–100. VNIRO, Moscow. (In Russian, English summary).

Reznik, Ya.I. (1981a). Oogenesis of the Commander squid *Berryteuthis magister* Berry, 1913, in the Olyutoro–Navarin region of the Bering Sea. *In* "Biological Resources of Large depths and Pelagic Realm of the Open Areas of the World Ocean. Abstracts of Communications of the Scientific–Practical Conference", Murmansk, pp. 85–86. (In Russian).

Reznik, Ya.I. (1981b). Some data about spermatogenesis in the Commander squid *Berryteuthis magister* (Berry, 1913). *In* "Biological Resources of the Shelf, their Rational Use and Protection. Abstracts of Communications of Regional Conference of Young Researchers and Specialists of the Far East", Vladivostok, pp. 123–125. (In Russian).

Reznik, Ya.I. (1982). Some results of histological investigation of the gonads of female Commander squid *Berryteuthis magister* in the Olyutoro–Navarin region of the Bering Sea. *Izvestiia TINRO* **106**, 62–69. (In Russian).

Reznik, Ya.I. (1983). General morphology of the reproductive system in the Commander squid *Berryteuthis magister*. *In* "Taxonomy and Ecology of Cephalopods" (Ya.I. Starobogatov and K.N. Nesis, eds), pp. 64–66. Zoological Institute AN SSSR, Leningrad. (In Russian).

Ronholt, L.L., Wakabayashi, K., Wilderbuer, T.K., Yamaguchi, H. and Okada, K. (1986). Groundfish resource of the Aleutian Islands waters based on the U.S.–Japan trawl survey, June–November 1980. *Bulletin of International North Pacific Fisheries Commission* **48**, 251 pp.

Roper, C.F.E., Sweeney, M.J. and Nauen, C.E. (1984). Cephalopods of the World. *FAO Fisheries Synopsis* **125**, Vol. 3, 277 pp.

Roper, C.F.E. and Young, R.E. (1975). Vertical distribution of pelagic cephalopods. *Smithsonian Contributions to Zoology* **209**, 51 pp.

Sanger, G.A. and Ainsley, D.G. (1988). Review of the distribution and feeding ecology of seabirds in the oceanic subarctic North Pacific Ocean. *Bulletin of the Ocean Research Institute, University of Tokyo* **26**(II), 161–186.

Saville, A. (1987). Comparisons between cephalopods and fish of those aspects of the biology related to stock management. *In* "Cephalopod Life Cycles. Vol. 2. Comparative Reviews" (P.R. Boyle, ed.), pp. 277–290. Academic Press, London.

Savinykh, V.F. (1993). Feeding of Japanese pomfret *Brama japonica. Voprosy Ichthyologii* **33**(5), 644–650. (In Russian).

Savinykh, V.F. (1995). "Ecology and Fisheries Prospects for the Japanese Sea Bream". Abstr. Diss. Cand. Biol. Sci., Inst. Mar. Biol. Russ. Acad. Sci., Vladivostok. (In Russian).

Savinykh, V.F. and Chuchukalo, V.I. (1996). Daily ration of the Pacific pomfret *Brama japonica* in the north-western Pacific Ocean during feeding period. *Biologiya morya,* **22**(6), 359–364. (In Russian, English summary).

Shevtsov, G.A. (1972). "Instructions for the Collection and Identification of Commercial Squid Species in the Pacific Ocean". TINRO, Vladivostok. (In Russian).

Shevtsov, G.A. (1974). Some features of the biology of the squid *Berryteuthis magister* from the area off Commander Islands. *In* "Hydrobiology and Biogeography of the Shelves of Cold and Temperate Waters of the World Ocean. Abstracts of Communications". Leningrad, pp. 68–69. (In Russian).

Shevtsov, G.A. (1978). "Pacific squid *Todarodes pacificus* Steenstrup, 1880 (Cephalopoda, Ommastrephidae) of the north-western Pacific Ocean (biology, distribution, state of stocks)". Dissertation Abstract, Candidate in Biological Sciences, Far Eastern Science Center, Academy of Science, USSR, Vladivostok. (In Russian).

Shevtsov, G.A. (1988a). Commander squid (*Berryteuthis magister*) of the Kita–Yamato Bank in the USSR economic zone of the Japan Sea. *In* "Fishery Resources and Biological Bases of Rational use of Commercial Invertebrates. Abstracts of Communications of All-USSR Conference", Vladivostok, pp. 78–79. (In Russian).

Shevtsov, G.A. (1988b). "Instructions for Fishery of Commander Squid in the Japan Sea on the Kita–Yamato Bank". TINRO, Vladivostok. (In Russian).

Shevtsov, G.A. (1990). Patterns of distribution of the Commander squid in the pelagic zone off 9 Kurile Islands. *In* "Vth All-USSR Conference on Commercial Invertebrates, Minsk – Naroch. Abstracts of Communications", 99–100. (In Russian).

Shevtsov, G.A. (1991). Cephalopod resources in the Far Eastern seas. *In* "Rational Use of Bioresources of the Pacific Ocean. Abstracts of Communications of All-USSR Conference", Vladivostok, pp. 16–18. (In Russian).

Shevtsov, G.A., Petrov, O.A., Fedorets, Yu.A., Slobodskoy E.V. and Mokrin, N.M. (1988). Prospects for squid fishery in the Pacific Ocean. *In* "Conference of Specialists of the USSR Ministry of Fisheries' Associations, Fishery Reconnaissance Services and Basin Institutes on Expanding of Fisheries for Valuable Fishes and Marine Products", Kerch, 62–67. (In Russian).

Shuntov, V.P. (1985). Summer population of seabirds in the Pacific waters off Kamchatka and Kurile Islands. *Zoologicheskii Zhurnal* **71**(11), 77–88. (In Russian, English summary).

Shuntov, V.P. (1993). Recent distribution of whales and dolphins in the Far Eastern seas and adjacent waters of the Pacific Ocean. *Zoologicheskii Zhurnal* **72**(7), 131–141. (In Russian, English summary).

Shuntov, V.P. (1996). State of pelagic nektonic communities of the Far Eastern seas. *Rybnoye Khozyaistvo* **1**, 35–37. (In Russian).

Shuntov, V.P. and Dulepova, E.P. (1995). Contemporary state, bio- and fish-productivity of the Bering Sea ecosystem. *In* "Complex Studies of the Bering Sea Ecosystem" (B.N. Kotenev and V.V. Sapozhnikov, eds), pp. 358–387. VNIRO, Moscow. (In Russian, English summary).

Shuntov, V.P., Volkov, A.F., Abakumov, A.I., Shvydky, G.V., Temnykh, O.S., Vdovin, A.N., Startsev, A.N. and Shebanova, M.A. (1990). Composition and present state of the fish communities in the epipelagic zone of the Okhotsk Sea. *Voprosy Ichtyologii* **30**(4), 587–597. (In Russian; English translation in *Journal of Ichthyology* **30**(4), 116–129).

Shuntov, V.P., Radchenko, V.I., Chuchukalo, V.I., Efimkin, A.Ya., Kuznetsova, N.A., Lapko, V.V., Poltev, Yu.N. and Senchenko, I.A. (1993a). Composition of planktonic and nektonic communities of upper epipelagic zone in the western Bering Sea and Pacific waters off Kamchatka in the period of anadromous migrations of salmons. *Biologiya Morya* No. 4, 19–31. (In Russian, English summary: English translation in *Russian Journal of Marine Biology,* No. 4, 231–239).

Shuntov, V.P., Radchenko, V.I., Chuchukalo, V.I., Efimkin, A.Ya., Kuznetsova, N.A., Lapko, V.V., Poltev, Yu.N. and Senchenko, I.A. (1993b). Composition of planktonic and nektonic communities of upper epipelagic zone in the Sakhalin–Kurile region in the period of anadromous migrations of salmons. *Biologiya Morya* No. 4, 32–43. (In Russian, English summary; English translation in *Russian Journal of Marine Biology* No. 4, 240–247).

Shuntov, V.P., Volkov, A.F., Temnykh, O.S. and Dulepova, Ye.P. (1993c). "Alaska Pollack in the Ecosystems of the Far Eastern seas". TINRO Press, Vladivostok. (In Russian).

Shuntov, V.P., Lapko, V.V., Nadtochiy, V.V. and Samko, Ye.V. (1994a). Interannual changes in the ichthyocenes of upper epipelagic zone in the western Bering Sea and Pacific waters off Kamchatka. *Voprosy Ichtyologii* **34**(5), 642–648. (In Russian).

Shuntov, V.P., Lapko, V.V., Nadtochiy, V.V. and Samko, Ye.V. (1994b). Interannual changes in the ichthyocenes of upper epipelagic zone in the Sakhalin–Kurile region. *Voprosy Ichtyologii* **34**(5), 649–656. (In Russian).

Sinclair, E., Loughlin, Th. and Pearcy, W.G. (1994). Prey selection by northern fur seals (*Callorhinus ursinus*) in the eastern Bering Sea. *Fishery Bulletin, U.S.* **92**(1), 144–156.

Skalkin, V.A. (1977). To the distribution and biology of Commander squid in the Tatar Strait. *In* "All-USSR Scientific Conference on the Use of Commercial Invertebrates, Odessa", Moscow, pp. 87–88. (In Russian).

Sobolevsky, Ye.I. (1983a). Role of marine mammals in trophic chains of the Bering Sea. *Izvestiia TINRO* **107**, 120–132. (In Russian).

Sobolevsky, Ye.I. (1983b). Marine mammals of the Okhotsk Sea, their distribution, abundance and importance as consumers of other animals. *Biologiya Morya* No. 5, 13–20 (In Russian, English summary).

Sobolevsky, Ye.I. (1984). Distribution of marine mammals, their abundance and importance as consumers of other animals in the Japan Sea. *In* "Marine Mammals of the Far East", pp. 39–53. TINRO, Vladivostok. (In Russian).

Sobolevsky, Ye.I. (1996). Species composition and distribution of squids in the western Bering Sea. *In* "Ecology of the Bering Sea: A review of Russian literature" (O.A. Mathisen and K.O. Coyle, eds). *Alaska Sea Grant Report* No. 96–01, 135–141. Fairbanks, Alaska.

Sobolevsky, Ye.I. and Mathisen, O.A. (1996). Distribution, abundance, and trophic relationships of Bering Sea cetaceans. *In* "Ecology of the Bering Sea: A review of Russian literature" (O.A. Mathisen and K.O. Coyle, eds), pp. 265–275. *Alaska Sea Grant Report* No. 96–01, Fairbanks, Alaska.

Sobolevsky, Ye.I. and Senchenko, I.A. (1996). Spatial structure and trophic connections of common species of pelagic fishes in the waters off Eastern Kamchatka in autumn-winter season. *Voprosy Ichtyologii* **36**(1), 34–43. (In Russian).

Sokolovsky, A.S. and Sokolovskaya, T.G. (1995). Some features of feeding of the Pacific cod *Gadus macrocephalus* in the area off Commander Islands. *Voprosy Ichtyologii* **35**(4), 543–545. (In Russian).

Takeuchi, I. (1972). Food animals collected from the stomachs of three salmonid fishes (*Oncorhynchus*) and their distribution in the natural environments in the northern North Pacific. *Bulletin of Hokkaido Regional Fisheries Research Laboratory* **38**, 1–119.

Vecchione, M. (1987). Juvenile ecology. *In* "Cephalopod Life Cycles. Vol. 2. Comparative Reviews" (P.R. Boyle ed.), pp. 61–84. Academic Press, London.

Voight, J.R. (1996). The hectocotylus and other reproductive structures of *Berryteuthis magister* (Teuthoidea: Gonatidae). *Veliger* **39**(2), 117–124.

Walker, W.A. and Jones, L.L. (1993). Food habits of northern right whale dolphin, Pacific white-sided dolphin and northern fur seal caught in the high seas driftnet fisheries of the North Pacific Ocean, 1990. *Bulletin of International North Pacific Fisheries Commission* **53**(II), 285–295.

Wall, J., French, R. and Nelson, R., Jr (1981). Foreign fisheries in the Gulf of Alaska, 1977–78. *Marine Fisheries Review* **43**(5), 20–35.

Wiborg, K.F. (1979). *Gonatus fabricii* (Lichtenstein), en mulig fiskeriressurs i Norskehavet. *Fisken og Havet* **1979**(1), 33–46.

Yang, M.S. and Livingston, P.A. (1988). Food habits and daily ration of Greenland halibut, *Reinhardtius hippoglossoides*, in the eastern Bering Sea. *Fishery Bulletin, U.S.* **86**(4), 675–690.

Yatsu, A., Shimada, H. and Murata, M. (1993). Distributions of epipelagic fishes, squids, marine mammals, seabirds and sea turtles in the central North Pacific. *Bulletin of International North Pacific Fisheries Commission* **53**(II), 111–146.

Young, R.E. (1972). "The Systematics and Areal Distribution of Pelagic Cephalopods from the Seas off Southern California". *Smithsonian Contributions to Zoology* **97**, 159 pp.

Young, R.E. (1973). Evidence for spawning by *Gonatus* sp. (Cephalopoda: Teuthoidea) in the High Arctic Ocean. *Nautilus* **87**(2), 53–58.

Yuuki, Y. and Kitazawa, H. (1986). *Berryteuthis magister* in the southwestern Japan Sea. *Bulletin of the Japan Sea Regional Fisheries Research Laboratory* **52**(4), 665–672. (In Japanese with English summary).

Zuev, G.V. and Nesis, K.N. (1971). "Squids (Biology and Fisheries)". Pishchevaya Promyshlennost, Moscow. (In Russian).

Zuev, G.V., Nigmatullin, Ch.M. and Nikolsky, V.N. 1985. "Nektonic Oceanic Squids (genus *Sthenoteuthis*)". Agropromizdat Press, Moscow. (In Russian, English summary).

NOTE

A very informative book was recently published, unfortunately it was published too late to be used in this contribution. Jelizarov, A.A. (ed.) (1997). *Commercial Aspects of Biology of Commander Squid* Berryteuthis magister *and of Fishes of Slope Communities in the western Part of the Bering Sea. Scientific Results of VNIRO Expedition in the Bering Sea According to Programme of Joint Russian–Japanese Research of Commander Squid* Berryteuthis magister *in the Bering Sea.* VNIRO Publishing, Moscow. (In Russian with English contents.)

Zoogeography of the Abyssal and Hadal Zones

N.G. Vinogradova

P.P. Shirshov Institute of Oceanology, Russian Academy of Sciences, Nakhimovsky Prospekt 36,Moscow 117851, Russia

ABSTRACT

Deep-sea bottom-living macroinvertebrates occurring below 3000 m depth include some species with a broad cosmopolitan geographical distribution and others with more limited, sometimes local, ranges. Taxa containing a large number of species with a wide vertical range (eurybathic) have a wider horizontal distribution than those dominated by species with narrow vertical ranges

ADVANCES IN MARINE BIOLOGY VOL. 32
ISBN 0-12-026132-4

(stenobathic abyssal forms). Examination of distribution records of more than 1000 species from different taxa, mainly from Russian collections, confirms that the extent of species' ranges is related to their degree of eurybathy. Groups with a high proportion of truly abyssal species show a high level of endemism.

Taxonomic links vary between the main ocean areas. A hierarchical scheme of regionation of the abyssal fauna, comprising regions, subregions and provinces, is provided, based on quantitative analysis of the fauna from different areas. This scheme is compared with other known schemes. The Arctic, the Caribbean and the Mediterranean are discussed in more detail. In analysing relationships between adjacent regions and factors controlling distribution, special attention is given to the near-continental (ring-like) species ranges, determined by nutritional conditions at the base of the continental slope. Bipolar and amphi-oceanic species ranges are analysed, together with the probable determining factors.

The hadal or ultra-abyssal fauna (deeper than 6000 m) demonstrates restriction of species to local areas within an ocean; 95% of hadal species occur only in a single trench or a group of adjacent trenches. This separation of the trench faunas gives them the status of independent zoogeographic provinces in a joint scheme of abyssal and hadal regionation.

1. INTRODUCTION

The abyssal and hadal zones of the ocean, occupying depths below 3000 m, comprise a vast biotope that extends over about two-thirds of the surface of the globe. The distribution of the fauna of the deep-sea bottom has thus attracted much attention from oceanographers and zoologists. The biotope is characterized by rather monotonous environmental conditions without sharp gradients. The conditions include low temperature (except at hydrothermal vents), absence of light, hydrostatic pressure increasing with depth, mainly soft muddy bottoms, and extremely limited food resources. An interesting fauna has become adapted to live under these conditions, including representatives of practically all the major classes of marine invertebrates. This fauna was inaccessible to investigators for a long time owing to the technical difficulties of deep-sea sampling.

The main aim of this review is to introduce the world community of deep-sea researchers to the results of Russian marine zoogeographers and it dwells primarily on analysis of lesser-known Russian contributions. It deals with the current distribution of deep-sea bottom fauna, specifically the macrobenthic metazoan fauna, but does not discuss the problems of origin and dispersal of deep-sea fauna or their history, despite the interesting data accumulated in this field. For a review of other aspects of deep-sea zoogeography, including a discussion of the role of plate tectonics, readers are referred to Gage and Tyler (1991).

2. ZOOGEOGRAPHIC DISTRIBUTION OF ABYSSAL BOTTOM FAUNA

2.1. Extent of Ranges of the Deep-sea Bottom Fauna

Early investigations of the deep-sea fauna off England and Norway showed that many species were common to both regions. It was suggested that the entire Atlantic deep-sea fauna was homogeneous and a theory was widely accepted by 19th-century zoologists that the deep-sea fauna had a cosmopolitan distribution. When proposing his well-known scheme of oceanic regions, Ortman (1896) took the abyssal zone fauna to be common to the whole World Ocean. At that time, the available information was insufficient for analysis of the worldwide distribution of deep-sea organisms, and the known homogeneity of abiotic abyssal factors was assumed to be responsible for the existence of a common abyssal faunal zone.

However, Murray and Hjort (1912), basing their ideas on evidence from the "Challenger" Expedition of 1874–1876, suggested that there was rather more heterogeneity of deep-sea fauna in different parts of the ocean. Subsequent deep-sea expeditions ("Valdivia", "Albatross") demonstrated some similarities in the distributions of various animal groups. Murray and Hjort had noted that, though the various groups of animals are widely distributed in ocean depths, particular species, genera and families of these groups may be limited in their distribution to specific areas. There are few cosmopolitan deep-sea species, though genera are usually cosmopolitan. Ekman (1935) also argued for non-uniformity of the deep-sea fauna, despite the seemingly homogeneous environmental conditions. According to Ekman, abyssal species are more widely distributed than archibenthal, and the size of their ranges increases at greater depths; he divided the World Ocean abyssal into three or, possibly, four zoogeographic regions – Atlantic, Pacific, Arctic and Antarctic. Although the theory of cosmopolitan distribution of deep-sea bottom fauna was rejected in the 1970s, this by no means denies the existence of cosmopolitan ranges in some species. Menzies *et al.* (1973) for example, believed that the diversity of opinions regarding the extension of abyssal species distribution resulted primarily from the differing use of the term "deep-sea fauna". In Menzies' opinion, the fauna of the continental slope is more widely distributed than that in any other vertical faunistic zone, and Bruun (1957) had this in mind when speaking about the cosmopolitanism of deep-sea fauna. Menzies *et al.* (1973) defined the bathymetric distribution of the archibenthal zone (roughly equivalent to some authors' bathyal zone, as noted by Madsen, 1961) as 12–360 m in the Arctic; 100–900 m in the Antarctic; 446–940 m in the north-west Atlantic and 1920–3300 m off Peru. The lower limits of these ranges mark the upper limit of the abyssal province in Menzies classification.

However, Briggs (1974) demonstrated for bottom fish that the bathyal fauna has limited ranges in different oceanic regions. In regard to invertebrates, Hansen (1967) noted that none of the species of bathyal holothurians is cosmopolitan, compared with abyssal species; Wolff (1962) concluded that 60% of bathyal species of Isopoda are limited to one section of the Atlantic slope. Similar distributional data have been reported for Amphipoda (Barnard, 1962), *Xylophaga* (Knudsen, 1961), other molluscs (Clarke, 1962a) and pogonophorans (Southward and Southward, 1967).

The following schools of thought exist, at least to some extent, in deep-sea marine zoogeography:

- Proponents of the idea that there is a very wide distribution of bottom fauna because of the absence of ecological barriers and the uniformity of the ocean floor.
- Supporters of the theory of extremely fractional zoogeographic division, who point to the the existence of very narrow ranges resulting from topographic division of the ocean floor into about 50 separate oceanic basins. Menzies *et al.* (1973) formed a hypothetical scheme of zoogeographic zonation of the World Ocean at depths exceeding 4000 m. This scheme recognized five large deep-water zones corresponding to the number of oceans and divided them into 13 provinces and 17 regions and subregions (see below).
- Finally, a third group of investigators holds a viewpoint close to that of Ekman, but modified to some extent by recent evidence.

In the 1950s, the idea of a widely distributed uniform deep-sea fauna was held by some well-known scholars, including Bruun (1957). Later, the idea of wide cosmopolitanism in deep-sea fauna was more cautiously expressed (Bruun and Wolff, 1961), and Wolff (1962) noted the absence of cosmopolitan forms in deep-sea isopods. These Danish studies, following the "Galathea" Expedition, included suggestions of a wide geographical distribution of abyssal Polychaeta (Kirkegaard, 1954, 1995). However, the data indicate that less than 5% of these species are truly cosmopolitan. Knudsen (1970) considered that there was wide distribution of abyssal Bivalvia. However, only three of 193 species of this group appear to be deep-sea cosmopolites. Madsen (1954) thought that deep-sea genera of echinoderms are typically cosmopolitan, as were a large percentage of species of the greatest depths. According to his calculations, about 3% of deep-sea starfish species, 8% of holothurians and 11% of ophiurans are cosmopolites. According to our data (Vinogradova, 1959a), the echinoderms of the World Ocean (excluding ophiurans) include 3.4% cosmopolitan species. Nevertheless, even with the existence of widely distributed species, Madsen was able to define individual faunas inhabiting specific deep-sea zoogeographic regions.

Having analysed the geographic distribution of deep-sea Pennatularia of the genus *Umbellula*, Pasternak (1964, 1975a) pointed out that there was no tendency to form narrow local ranges in this group. Based on this fact, it has been suggested that cosmopolitan distribution is characteristic of the whole abyssal fauna. Widely distributed abyssal species are common in Cnidaria (Vervoort, 1966; Keller *et al.*,

1975). Examples can be found in other animal groups including Polychaeta (Levenstein, 1975; Uschakov, 1975), Sipuncula (Murina, 1961, 1971, 1993; Cutler, 1973), Priapulidae (Murina and Starobogatov, 1961; Murina, 1975), Tanaidacea (Kudinova-Pasternak, 1993), Pantopoda (Turpaeva, 1974, 1975), Brachiopoda (Zezina, 1965, 1993) and Ophiuroidea (Baranova and Kuntsevitch, 1969). All these groups of invertebrates have a number of eurybathic species in the abyssal fauna. Such groups are more widely distributed in the abyssal than those with a prevalence of stenobathic abyssal species (Vinogradova, 1969a). In the Pacific, such groups as Cnidaria, Sipuncua and Cirripedia, where true abyssal species comprise 28.1, 36.4 and 37.7% of the total number of deep-sea species, respectively, include 43.8, 35 and 53.5% of endemic Pacific species. In Porifera, Isopoda, Pogonophora and Asteroidea, on the contrary, the former figure is 86.8, 88.7, 84.5 and 66%, respectively, and the latter 90, 93.2, 100 and 75% (Vinogradova, 1969a).

More recent studies indicate changing viewpoints on abyssal distribution. Monniot and Monniot (1975) note that the ranges of deep-sea ascidians continue to grow as new data accumulate. Kirkegaard (1995) has shown that supposed restricted ranges in four widely eurybathic polychaete species have now been expanded to worldwide. For Amphipoda, Barnard (1962) stated that strict regional endemism is confirmed from the discovery of a number of new species in each deep-water area studied, with a low frequency of previously known species. Nevertheless, conclusions about wide geographic ranges often result from the analysis of only a small portion of the fauna, and this portion is usually the most easily determinable, best studied, and belongs to the commonest genera. There are many examples of wide distribution among abyssal species, but this is not a reason to judge the fauna as a whole. For instance, Hansen (1967) analysed the distribution of 12 holothurian species from widespread abyssal genera and found that they were more widely distributed in the abyssal than he had suspected. He concluded that circumglobal distribution is bound to be the rule for deep-sea animals. In his fundamental work on deep-sea Elasipoda, Hansen (1975) emphasized that surprisingly few species are cosmopolitan, despite the expansion of the known ranges of many species due to recent synonymization. Nevertheless, he and other Danish investigators have generally held the opinion that wide geographic distribution of species from many groups is characteristic of the abyssal zone (Kirkegaard, 1954; Madsen, 1961; Knudsen, 1970; Hansen, 1975).

A second group of authors cite data in favour of greater geographic isolation of the fauna of intra-oceanic basins, and also of hadal trenches. For example, Parker (1963) described a peculiar fauna in the near-shore abyssal off the Pacific coast of Central America. Similar examples can be found in deep-sea basins near the coast of California (Hartman and Barnard, 1958). From data on abyssal molluscs (excluding Cephalopoda) of the World Ocean, Clarke (1962a) pointed out that each individual oceanic basin has its own endemic fauna. Of 1087 molluscan species in depths > 1800 m any one species is not found in more than two basins, and then in 90% of the cases these two basins are incompletely isolated from each other. This point of view is supported by Sanders (1965). Birstein (1963) found a high percentage of endemic abyssal species among deep-sea isopod crustaceans in the northwestern part of the Pacific. Menzies (1962, 1965) noted that only 14% of 158 species of Isopoda in the neighbouring Argentine and Cape basins were recorded from both basins, and none of the species was known from another ocean. Later (Menzies *et al.*, 1973), stated that a high degree of endemism is generally characteristic for each oceanic basin. Similar data were obtained for Cumacea (Jones and Sanders, 1972).

The third group of zoogeographers believe, with Ekman (1935, 1953), that there are rather large, but still distinct, deep-sea zoogeographic regions in the World Ocean, and the number of cosmopolitan species is not so large as to level the differences in composition of their faunas. Ekman (1954) has calculated that only 15% of 375 echinoderm species recorded from depths exceeding 2000 m are found in more than one ocean, and only 2.7% occur in three oceans. These ratios may vary in other groups. Vinogradova (1959b) analysed more than 1000 species of the abyssal bottom fauna and found that about 85% of species, including eurybathic forms, occur in only one ocean, and of the 15% with a wider distribution, only 4% are found in the Pacific, Atlantic and Indian Oceans. Similar data are given for abyssal Isopoda (Wolff, 1962), Tunicata (Millar, 1970), Spongia (Levi, 1964), and many other groups. Among the Echiura, 81% of 27 Pacific abyssal species are not known outside the Pacific and all these belong to the family Bonelliidae (Zenkevitch and Murina, 1976; Murina, 1978). Most species of deep-sea holothurians of the family Elpidiidae (67%) are endemics of a single ocean (Gebruk, 1990) and there are no species known from all three oceans.

Deep-sea bottom-living fish tend to be limited to one ocean; that is, about 75–85% of species (Grey, 1956). Following work by Grey (1956), Nybelin (1954) and Nielsen (1966), Briggs (1974) has suggested that each large ocean possesses its own fauna of bottom fish, though data are still insufficient for an ultimate judgement.

Future investigations will undoubtedly show the need for corrections, but we are certain that the material already in hand will reveal the general trends of distribution and allow the development of a working scheme of zoogeographic division, in spite of the fact that the great oceanic depths are still rather unevenly known.

In the late 1960s, Russian marine biologists carried out a detailed analysis of the geographic distribution of the deep-sea bottom fauna of the World Ocean (Vinogradova, 1956a, 1959a,b, 1962, 1969b, 1977a, 1979; Belyaev, 1966, 1972, 1989). By that time there were good data sets on geographic distribution of more than 1500 species of benthic invertebrates from abyssal and hadal depths, including the major groups of benthic Metazoa: Porifera, Cnidaria, Polychaeta, Mollusca, Crustacea, Echinodermata, as well as other, less species-rich groups.

Data on their distribution were extracted from the worldwide literature, covering samples of various deep-sea expeditions, both Russian and other nations, since the middle of 19th century. The extensive collections of deep-sea animals obtained by the "Vityaz" and other Russian research vessels in the Pacific, Indian and Atlantic oceans and around the Antarctic were also analysed. Leading taxonomists ensured that these data were critically estimated from the viewpoint of current systematics.

The zoogeographic analysis was based on the study of species ranges. Genera of all groups including fish are mostly very widely distributed, and cosmopolitan distribution is the rule for families. One cannot agree with the viewpoint of Menzies (Menzies *et al.*, 1973) that zoogeographic analysis is best performed at the genus level, unless it is only historical aspects that are in question. It is absolutely inappropriate to mix species and genus levels in zoogeographic analysis. It should also be realized that different taxonomic groups of animals may show different patterns of both vertical and horizontal distribution.

2.2. Relationship Between Geographic Distribution of Deep-sea Bottom Fauna and the Degree of its Eurybathy and Vertical Zonation

The patterns of geographic distribution of animals are different at different depths. Therefore, it is necessary to perform a separate zoogeographic analysis for each vertical faunistic unit, as noted by Ekman (1935, 1953), Zenkevitch (1956) and Birstein (1963), and subsequently supported by Briggs (1974).

Quantitative estimation was introduced into the analysis of deep-sea bottom fauna distribution by Vinogradova for the Pacific (1969a,b), Indian (1956b), and Southern (1960) Oceans and the entire World Ocean (1956a, 1957, 1959a,b, 1977a, 1979). The scheme of vertical biological zonation of the ocean, developed by Russian biological oceanographers (Vinogradova, 1956c, 1958a, 1969a, 1977b; Belyaev *et al.*, 1959; Vinogradova *et al.*, 1959; Belyaev, 1966, 1977), comprises the following zones of vertical distribution: *abyssal zone* (3000–6000 m), with a transitional horizon between bathyal and abyssal at 2500–3500 m *upper abyssal subzone* (3000–4500 m) and *lower abyssal subzone* (4500–6000 m). Deeper than 6000 m there is the *ultra-abyssal* (*hadal*) zone.

Two groups of species are clearly distinguishable among deep-sea animals by their vertical distribution: (i) eurybathic species with wide depth ranges, reaching far beyond the abyssal zone; and (ii) stenobathic abyssal species not recorded at less than 2500–3000 m depth. Among the latter group, one can distinguish upper abyssal species with an upper limit of occurrence in the upper abyssal subzone (3000–4500 m) and lower abyssal species, which are not found above the lower abyssal subzone (4500–6000 m). The extent of geographical distribution may be expected to differ in these two groups. More eurybathic species, which are

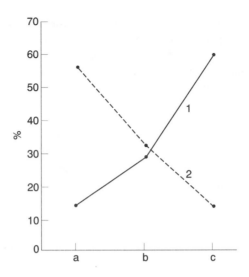

Figure 1   The vertical distribution of abyssal species that can extend above the abyssal zone, shown as percentage ratio of the total number of deep-sea species. 1, Eurybathic species; 2, stenobathic species. a, Species restricted to a relatively small region of the Pacific Ocean; b, those ranging over the whole Pacific Ocean; c, those occurring beyond the Pacific Ocean (after Vinogradova, 1969a).

basically more eurybiontic, should be distributed geographically more widely than the stenobathic species. Their distribution is primarily affected by surface water characteristics, principally climatic zonation.

In an analysis of the relationship between the vertical distribution of animals and their geographic distribution (Vinogradova, 1969b) separate consideration was given to species occurring deeper than 2500 m, but ascending to shallower depths, and to true abyssal species not found shallower than 2500 m. In the latter group the distributions of upper and lower abyssal species were also considered separately. Excluding species known from single records, 664 widely eurybathic deep-sea species and 147 true abyssal species were included in the analysis. Species with vertical ranges of less than 1000 m were considered stenobathic and those with ranges exceeding 1000 m were considered eurybathic. Ranges were divided into three types (Figure 1).

For both eurybathic and stenobathic species groups, it was found that the wider the range of vertical distribution of a species the wider is its geographic distribution. This rule was demonstrated by various animal groups for the entire World Ocean (Vinogradova, 1958a). The same relationship was noted earlier, without quantitative analysis, by some other authors (Ekman, 1935, 1953; Briggs, 1974). New data collected for various groups of deep-sea benthos confirm this rule

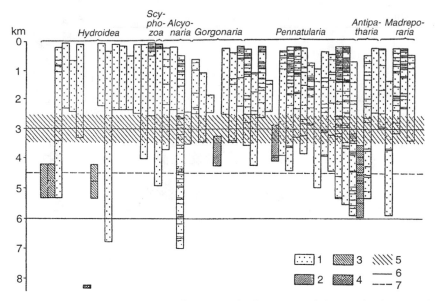

Figure 2   Vertical distribution of deep-sea Coelenterata, omitting species known only from single records. The columns show the vertical range of the distribution of individual species; the lines separately record: 1, eurybathic species; 2, upper-abyssal species; 3, lower-abyssal species; 4, hadal species; 5, transitional horizon; 6, upper borders of the abyssal and hadal zones; 7, border between the subzones of the abyssal zone (after Vinogradova, 1969a).

(Madreporaria – Keller, 1990; polychaetes – Detinova, 1982; cirripede crustaceans – Zevina, 1990; Shreider, 1994; Tanaidacea – Kudinova-Pasternak, 1990; Kudinova-Pasternak and Pasternak, 1981; starfish – Galkin and Korovchinsky, 1984; sea urchins – Mironov, 1987).

The role of stenobathic and eurybathic species in the abyssal fauna varies in the different taxonomic groups. For example, almost all the deep-sea cnidarians are widely eurybathic forms (Figure 2). The vast majority are found at sublittoral/shelf or bathyal depths, and only a few (*Stephanoscyphus simplex* and *Umbellula thompsoni*) occur more frequently in the abyssal zone. Only six species of true abyssal Cnidaria are known, with 11 additional species that have been recorded once in the abyssal zone. The Echinoidea, Polychaeta, Sipuncula, Cirripedia, Decapoda and Ophiuroidea are very similar in their vertical distribution to the Cnidaria. Deep-sea representatives of these groups have their main occurrence on the continental slope or shelf. On the other hand, the number of deep-sea species living on the ocean floor and not found in shallower zone from such groups as Isopoda, Echiura, Porifera, Pogonophora, Phanerozonia (Asteroidea) and Elasipoda (Holothuroidea) is much greater and sometimes exceeds

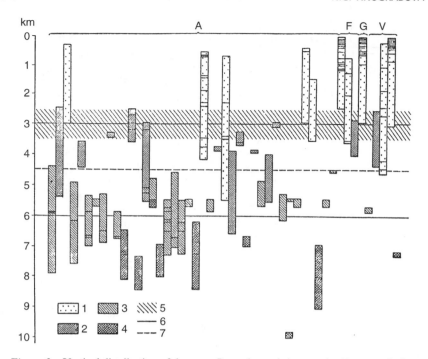

Figure 3 Vertical distribution of deep-sea Isopoda, omitting species known only from single records. The columns show the vertical range of the distribution of individual species; the lines separately record: 1, eurybathic species; 2, upper-abyssal species; 3, lower-abyssal species; 5, transitional horizon; 6, upper borders of the abyssal and hadal, zones; 7, border between the subzones of the abyssal zone; A, Asellota; F, Flabellifera; G, Gnathiidae; V, Valvifera (after Vinogradova, 1969a).

the number of species that have a wide vertical distribution (Figure 3).

When one considers the relative importance of eury- and stenobathic species below 2500 m (i.e. deeper than the upper limit of the transitional zone between the bathyal and abyssal zones) the widely eurybathic species penetrating the abyssal represent 67.3% of the total species number, omitting single species records (Table 1).

Two main categories of true deep-sea species can therefore be distinguished. The first includes taxonomic groups where true abyssal species comprise less than 50%, and the second those groups with more than 50% of true abyssal species. Both groups of taxa are about equally represented in the number of species in the abyssal fauna. In all taxa of the former group, the abyssal fauna is only a gradually diminishing remnant of mainly bathyal species. Individual species of these groups descend to the abyssal and have the lower limit of their distribution there. The second group includes an independent fauna represented by true deep-sea species,

Table 1 The number of stenobathic abyssal species in different animal groups in the deep-sea bottom fauna of the Pacific, as percentage of total number of deep-sea species (after Vinogradova, 1969a).

Group	Including species existing as single records	Omitting species existing as single records
Isopoda	88.7	76.5
Spongia	86.8	63.9
Echiura	84.6	80.0
Pogonophora	84.5	60.9
Cumacea	82.4	0
Ascidiae	77.8	60.0
Tanaidacea	75.0	57.2
Amphipoda	67.2	23.5
Monoplacophora	66.7	50.7
Asteroidea	66.0	45.0
Gastropoda	62.4	31.0
Bivalvia	59.6	37.7
Pantopoda	59.3	42.2
Holothuroidea	54.4	40.7
Priapula	50.0	50.0
Crinoidea	50.0	25.0
Polychaeta	42.8	18.2
Amphineura	41.6	12.5
Cirripedia	37.7	4.4
Echinoidea	37.5	28.6
Brachiopoda	36.4	12.5
Sipuncula	36.4	26.4
Decapoda	32.5	24.3
Bryozoa	31.0	4.6
Coelenterata	28.1	9.4
Ophiuroidea	28.6	11.3
Scaphopoda	9.1	9.1

adapted to life only at great depths and clearly delimited from the slope fauna. The classes Porifera, Isopoda, Asteroidea and Holothuroidea include large orders with a majority of species that are true abyssal forms (Hexactinellidae, Asellota, Phanerozonia and Elasipoda). Just this second group allows us to speak about specificity of deep-sea bottom fauna of the World Ocean.

These two categories probably reflect to some extent the history of colonization of the abyssal and may be termed secondarily deep-sea and primarily deep-sea, (or ancient deep-sea) animal groups (Andriyashev, 1953; Rass, 1959).

Figure 4 summarizes data showing the relationship between the extent of distribution of various animal groups in the World Ocean and the degree of isolation of their abyssal fauna. Despite some scatter, the relationship is clearly

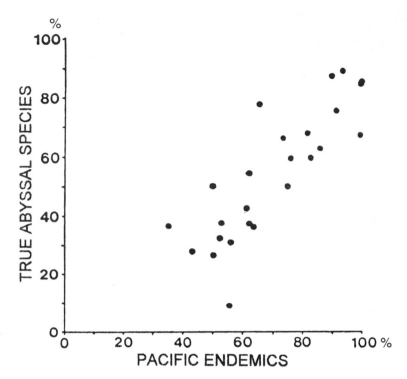

Figure 4 Relation of the degree of isolation of abyssal species, as % ratio to the total number of deep-sea species in the group. Examples from the Pacific Ocean: vertical scale, percentage of true abyssal species in different groups; horizontal scale, percentage of Pacific endemics in the same groups. Dots indicate different groups of animals, as shown in Table 1 (after Vinogradova, 1969b)

outlined. The degree of endemism of a group is higher when its "deep-sea nature" is better expressed, the number of true abyssal species in its composition is greater, and its taxonomic isolation in the abyssal zone is stronger.

2.3. Similarity of Abyssal Faunas in Different Regions of the World Ocean

The percentage of taxonomic similarity has been calculated for the bottom fauna of the Antarctic regions of the Pacific, Atlantic and Indian Oceans, south of 40°S, and the fauna of the remainder of each ocean. This index was obtained both for species extending to depths < 2000 m and those occurring only at depths > 3000 and 4000 m (Figure 5). It appears that in the antarctic region of the Atlantic there

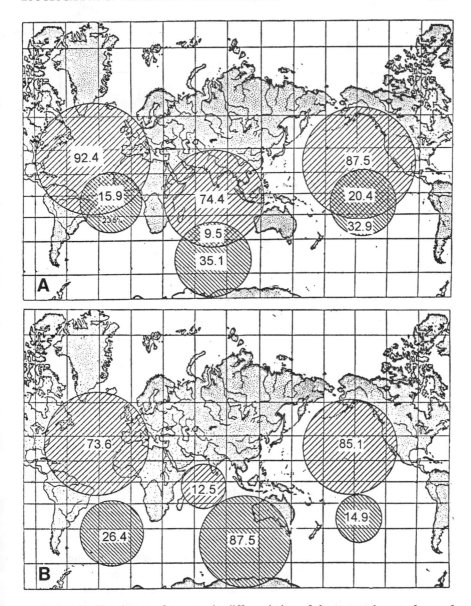

Figure 5 The degree of taxonomic differentiation of deep-water bottom fauna of Antarctic sectors of the Pacific, Indian and Atlantic Oceans. A, Those species rising to depths less than 2000 m; B, Species not found above 4000 m. The areas of circles are related to the number of species of deep-water invertebrates; the area common for intersecting circles is proportional to the number of common species in the areas compared. As per cent number of species out of the total number of species in each ocean (after Vinogradova, 1957).

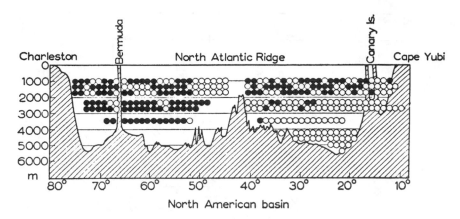

Figure 6 Vertical cross-section of the North Atlantic to show the distribution of species of bottom invertebrates of the North American and North African basins at various depths. The black circles represent the species of the North American basin, the open circles the species of the North African basin (after Vinogradova, 1959b).

are about 70% of species in common with the fauna of the rest of the ocean which also occur at depths < 2000 m. Species not extending above the 3000 m isobath comprised only 6% of species in common, and there are none among species which do not extend above the 4000 m isobath. The respective figures for the Pacific are 60, 10 and 0%. In other words, whereas the bottom fauna of Antarctic regions of the Atlantic and Pacific is very similar to that of more northern parts of the oceans at depths of about 2000 m and somewhat less, it becomes increasingly isolated with increasing depth, and species in common between the northern and southern parts of these oceans are absent from depths exceeding 4000 m. In the Indian Ocean, the percentage of benthic species in common between the northern and southern regions, for those ascending to depths < 2000 m, is 27%, for animals not ascending > 3000 m it is 2.5%, and for animals occurring only deeper than 4000 m it is 0%.

Between the east and west sides of the Atlantic there is the same increasing isolation of the deep-sea fauna with depth of species occurrence. Among animals living at depths < 2000 m, 49% of species are shared between east and west. For those not extending above the 3000 m isobath, this figure is 7.2%, and for species not extending above 4000 m it is 0%. A similar isolation is observed between the western and eastern parts of the North Pacific, where the respective figures are 49, 15.6 and 1.2% (see Fig. 7).

This increase in isolation of the deep-sea fauna with increasing depth cannot be explained simply by unequal sampling, though this must taken into account. The phenomenon is found both in oceanic regions where the deep-sea fauna is better known, for example the northern Atlantic (Figure 6) and in regions where

Table 2 Number of species of deep-sea benthic invertebrates occurring in the Pacific and beyond in different vertical zones, as % of species number in each zone, omitting species existing as single records (after Vinogradova, 1964b).

Species	Number of species	Pacific only	Pacific and Atlantic	Pacific and Indian Oceans	Pacific, Indian and Atlantic Oceans
Ascending above abyssal zone	572	44.6	19.9	15.7	19.8
Upper abyssal	175	70.3	5.7	13.7	10.3
Lower abyssal	54	92.6	—	7.4	—

it is less well known. Calculations for the entire deep-sea fauna of the Pacific compared with other oceans (Table 2) confirm that, as the depth increases, the degree of deep-sea faunal isolation of each ocean also increases, as is the case for different parts of an ocean. In the Pacific, 85% of deep-sea species found outside it are classed as eurybathic species, 14% are upper abyssal and only 1% lower abyssal. This example clearly demonstrates the importance of separate zoogeographic analysis for faunas of different vertical divisions even within a single vertical zone. A major factor affecting the isolation of the deep-sea fauna in different parts of an ocean is probably the ocean floor relief, especially the macrorelief. In the Atlantic the Mid-Atlantic Ridge separates the eastern and western basins. In the Pacific the Hawaiian Ridge lies between the eastern and western regions.

Analysis of geographic distribution of deep-sea benthic animals in the Pacific (Vinogradova, 1969b) shows that the fauna north of 35°N differs sharply in species composition from that of more southern areas (35°N to 30–40°S), and the latter differs from fauna of regions south of 40°S. The species composition of faunas of the eastern and western parts of these three latitudinal zones is also different. The number of species distributed over the entire Pacific is small, only a small percentage of the total number of Pacific species. Six natural regions of the Pacific designated by Vinogradova (1969a) are: northwestern, northeastern, western tropical, eastern tropical, southwestern, and southeastern. The northwestern region includes the northwestern Pacific, extending to southern Japan, and has been well studied by the R/V "Vityaz", as have the adjacent deep-sea areas of the Bering and Okhotsk seas. The Japan Sea was not included in the analysis due to the absence of true abyssal fauna. The deep ocean off the coast of North America to California is included in a northeastern region. The western tropical region is centred on the Indo-Malayan archipelago and extends from Japan in the north to

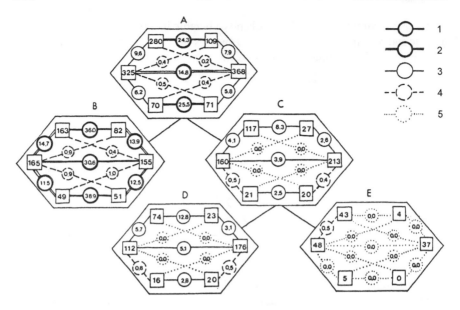

Figure 7 Similarity of the fauna of deep-sea benthic invertebrates in different regions of the Pacific (% ratio of the total number of species in the two compared regions). A, The entire fauna of the abyssal zone; B, species ascending above the abyssal zone; C, true abyssal species in general; D, species of the upper-abyssal subzone; E, species of the lower-abyssal subzone. Large hexagons, Pacific Ocean. Regions considered: northwest, northeast, tropical western, tropical eastern, southwest and southeast. In the squares are indicated the number of species in each region, in the circles the percentages of species in common. 1, More than 20% of common species; 2, from 10 to 20%; 3, from 1 to 10%; 4, less than 1%; 5, no common species (after Vinogradova, 1969b).

the Hawaiian Is in the east. The eastern tropical region covers the Gulf of Panama and the eastern part of the ocean, from California in the north to approximately the Juan-Fernandez Is in the south, and extends to the Hawaiian Is in the west. Western and eastern southern regions are adjacent to Australia and New Zealand or to South America, respectively. The degree of similarity between faunas of different regions and in different vertical faunistic groups varies (Figure 7). Faunas of the western and eastern parts of the ocean north of 35°N and in the central area, 35°N to 30–40°S, have many more species in common than do those of the northern and central regions in the eastern or the western parts of the Pacific. Many of the widespread species are widely eurybathic.

However, if we consider only the true abyssal species (Figure 7C), the number of species common to all these regions would drastically decrease. A faunistic break exists. The number of deep-sea species in common between regions among all those considered is approximately the same, 2.5–8.3%, with one exception.

Table 3 Coefficients of similarity between faunas in compared regions of the Pacific, as calculated for the whole deep-sea fauna (1029 species) (after Vinogradova, 1969b).

Number of species	Region	I	II	III	IV	V	VI
280	I						
109	II	0.43					
325	III	0.12	0				
368	IV	0	0.12	0.20			
70	V	—	—	0.09	0		
71	VI	—	—	0	0.10	0.34	—

There are only 0.5% (in the west) or 0.4% (in the east) species in common between the Antarctic regions of the Pacific south of 40°S and the tropical region. The eastern and western Antarctic regions are more closely related to each other (2.5% species in common). Thus, the Antarctic deep-sea fauna is isolated to a considerable extent and seems to be divorced from other Pacific fauna. With increasing depth, the difference between faunas of different regions of the Pacific becomes greater, and the lower abyssal faunas include very few species in common. However, this part of the fauna is rather poorly known.

To judge the degree of independence of fauna in regions considered, Preston's (1962a,b) index has been used for estimation of similarity between faunas of different areas (Vinogradova, 1969b). Table 3 and Figure 8A present the calculated coefficients of similarity between faunas of six regions of the Pacific; including both eurybathic and stenobathic species. The data show that regions I and II (i.e. northwestern and northeastern) as well as V and VI (i.e. southwestern and southeastern) have the highest similarity coefficients, 0.43 and 0.34, respectively. Faunas of all other regions are much more independent in respect to species. The fauna of Antarctic regions of the Pacific (regions V and VI) is the most isolated, as already mentioned, from the fauna of other parts of this ocean. The coefficients of similarity between regions V and III and VI and IV are only 0.09 and 0.10. A similar calculation was also performed separately for true abyssal species (Table 4, Figure 8B) and this indicates even greater divergence between faunas of different regions.

Different values of similarity coefficient should characterize the zoogeographic ranking of the regions compared. The greater the difference (the lower the value of $1 - z$), the higher a zoogeographic rank should be given to the regions compared, as supported by Kussakin (1973). The following grades of similarity

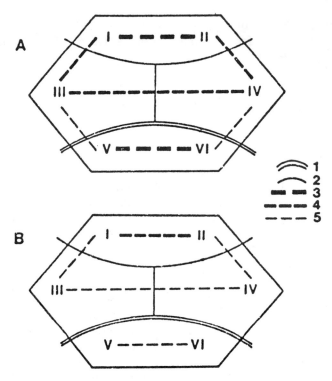

Figure 8 Scheme of co-ordination of the geographical affinities of bottom fauna in different deep-sea regions of the Pacific Ocean, based on Preston's index of similarity. A, For the entire fauna of the abyssal zone; B, for true abyssal species. I–VI. Regions compared: I, northwest; II, northeast; III, tropical western; IV, tropical eastern; V, southwest; VI, southeast. 1. Boundaries between zoogeographic regions; 2, boundaries between zoogeographic provinces; 3, index of similarity $(1 - z)$, on a species level, higher than 0.20; 4, $(1 - z)$ ranging from 0.20 to 0.10; 5, $(1 - z)$ less than 0.10. At $(1 - z)$ equal to zero there are no connecting lines (after Vinogradova, 1969b).

coefficients can be adopted for the whole fauna of the abyssal zone: >0.20; 0.20–0.10; <0.10; and, finally, 0. These values distinguish natural regions inhabited by bottom fauna defined on the basis of purely faunistic comparison.

The values of the coefficient of similarity between faunas, reflecting the closeness of zoogeographic relations, are shown in Figure 8 as lines of different pattern and thickness. It can be seen that the isolation of the fauna of southern regions of the Pacific (regions V and VI) is very great. The $1 - z$ value is less than 0.10 for the Antarctic when its entire deep-sea fauna is compared with that of other parts of the Pacific. In this case, the similarity coefficient is equal to 0 for the purely abyssal group of species, which indicates complete independence of the Antarctic abyssal fauna in regard to species. At the same time, if western and eastern

Table 4 Coefficients of similarity in faunistic regions of the Pacific, as calculated for those species living deeper than 2500–3000 m (532 species) (after Vinogradova, 1969b).

Number of species	Region	I	II	III	IV	V	VI
117	I						
27	II	0.18					
160	III	0.06	0				
213	IV	0	0.06	0.06			
21	V	—	—	0	0		
20	VI	—	—	0	0	0.03	

Antarctic regions are compared latitudinally, the value of $1-z$ appears to be 0.34 for the whole fauna of the abyssal zone and 0.03 for purely abyssal species. In spite of this extremely low coefficient for purely abyssal species, the value of $1-z$ for the whole abyssal faunas of these two regions indicates their relatively higher similarity within the Pacific sector of the Antarctic.

The Antarctic regions south of 40°S can be grouped as one large zoogeographic division, whereas the deep-sea faunas of other parts of the Pacific are much more closely interrelated. However, even there, within the limits of the subtropical–tropical and boreal regions of the surface zone, three regions of lower zoogeographic rank may be designated in the abyssal zone. These are the first (northwestern) and the second (northeastern) regions characterized by the highest $1-z$ value obtained: 0.43 for the whole fauna and 0.18 for its purely abyssal part. Thus, there is a single, relatively uniform region north of 35°N. Inside it the uniformity of the fauna is similar to that in the Antarctic region. Both in the west and in the east, the northern region is clearly delimited meridionally from the more southern regions. The coefficient of similarity between faunas of this region and regions III and IV is 0.12 for all deep-sea animals, and 0.06 for purely abyssal species; that is, it is three times lower than for the western and eastern parts of the northern region.

Regions III and IV, western and eastern central Pacific between 35°N and 40°S, have similarity coefficients of 0.20 and 0.06, respectively. The degree of similarity between faunas of these two regions is shown by values of the same order of magnitude as their similarity to the fauna of the northern regions. The central–western and central–eastern regions may therefore be considered as relatively independent and having equal zoogeographic rank, equal to that of the common northern region. Three of the regions identified within the Pacific on the

Table 5 Number of eurybathic and stenobathic abyssal species in different regions of
the Pacific that are also found in the Atlantic and Indian oceans. As % of number of species*
(after Vinogradova, 1969b).

Region of the Pacific	Species	Total number of species	Shared with Atlantic Ocean fauna	Shared with Indian ocean fauna
Northern (to north of 35°N)	Eurybathic	180	44.5	29.5
	Abyssal	133	3.0	7.5
	Whole fauna	313	26.8	19.8
Tropical (35°N–40°S)	Eurybathic	245	37.1	43.7
	Abyssal	359	2.8	5.0
	Whole fauna	604	16.7	20.7
Southern (to south of 40°S)	Eurybathic	72	73.6	80.5
	Abyssal	40	5.0	22.5
	Whole fauna	112	45.0	55.0

*Circum-Antarctic and circum-tropical species were counted twice in calculating Pacific
species in common with both Atlantic and Indian ocean fauna.

basis of Preston's index fit the zoogeographic scheme proposed by Vinogradova
(1956a, 1959b) and correspond to the three provinces of the Pacific subregion,
namely North, West and East Pacific (see below).

Relationships between the deep-sea bottom fauna of the different regions of the
Pacific and those of other oceans have been calculated (Vinogradova, 1969b)
(Table 5). The fauna of the northern region of the Pacific is generally more closely
related to that of the Atlantic (primarily the north) than the fauna of the central
region of the Pacific. This is clearly seen among eurybathic species, because
northern temperate regions of the Pacific harbour a number of amphiboreal and
Arctic–boreal widely ranging species which penetrate the abyssal zone. However,
in the true abyssal fauna even in this region, the number of species in common with
the abyssal fauna of the Indian Ocean is greater than those in common with the
Atlantic fauna.

In the central region of the Pacific the relationship with the Indian Ocean fauna
is notable at all depths. The southern Pacific has been insufficiently studied, but
it seems that the number of species in common with the fauna of the Indian Ocean
is rather large there, especially among forms living at > 3000 m depth.

Studies of the geographical distribution of the Indian Ocean deep-sea bottom
fauna (Vinogradova, 1956b, 1962), included about 260 species of Porifera,
Cnidaria, Isopoda, Decapoda, Cirripedia, Pantopoda, Asteroidea, Echinoidea,
Holothuroidea and Crinoidea. This analysis revealed that deep-sea fauna of the

Indian Ocean is less isolated (about 50% endemics) than those of the Pacific and Atlantic, where 73.2% and 76.0% of deep-sea species, respectively, are endemics. The fauna of northern areas of the Indian Ocean (north of 35–40°S) differs sharply from that of southern regions. Only 2.4% species in common were recorded at depths > 3000 m. Closer faunistic relationships with the Pacific were reported in the northern Indian Ocean – 47.7%; 33.3% are in common with those deep-sea species of central and northern parts of the Pacific which migrate through deep-water seas and straits of the Malayan Archipelago. There are only 4% of species in common between the northern Indian Ocean and the Atlantic.

The main relationship of the deep-sea fauna of the southern Indian Ocean is to Antarctic species (77%). The proportion of species in common with the deep-sea fauna of the Atlantic increases (10%), whereas species in common with the Pacific decrease (4%), compared with the situation in the northern Indian Ocean. The northern part of the Indian Ocean is termed the North–Indian subregion of the Pacific–North–Indian deep-sea region, whereas the southern part is termed the Indian province of the Antarctic–Indian–Pacific subregion of the Antarctic deep-sea region (see below). The duality of the deep-sea fauna of the Indian Ocean can probably be explained by its paleogeographic history.

The deep-sea fauna of Antarctic regions of the Pacific, Atlantic and Indian oceans was shown to have a high degree of species endemism (Vinogradova, 1957), with only 4.5% of species in common with the Pacific and Indian Oceans, and 8.4% with the Atlantic (see above). The use of Preston's index also indicated the independence of deep-sea bottom fauna of the Southern Ocean. A further zoogeographic analysis of deep-sea bottom fauna of the Antarctic south of 40° S (Vinogradova, 1960) was based on data on the distribution of 274 benthic species living at depths exceeding 2000 m, collected by Antarctic expeditions of various countries. The material contained representatives of Porifera, Cnidaria, Isopoda, Decapoda, Crinoidea, Asteroidea, Echinoidea, and Holothuroidea.

In the Antarctic regions of the Atlantic, deep-sea species that extend up into depths less than 2000 m include 67.8% in common with the fauna of other parts of the Atlantic; those species not extending above 2000 m include 49.6% in common; species not extending above 3000 m include 5.9% in common; and there are none in common among the species that are restricted to depths below 4000 m. For the Pacific these figures are 62, 14.2, 9.9 and 0% respectively; those for the Indian Ocean are 18.5, 4.4, 2.5 and 0%.

From this analysis, it is evident that the deep-sea fauna of the Antarctic regions of the Pacific and Atlantic shows increasing taxonomic isolation with increasing depth, and there are two independent faunas at depths > 3000 m in these regions. In regard to deep-sea fauna of northern and southern parts of the Indian Ocean, the taxonomic heterogeneity manifests itself there already at the depth of 2000 m. Barthel and Tendal (1994) state that 23 species of sponges of the family Hexactinellidae are known from the abyssal of the Antarctic, but only five of them are also found in other oceans. However, of the 23 species, 16 are known from single records. All the evidence confirms the Antarctic deep-sea region south of 40°S as a separate zoogeographic region with a high degree of species endemism.

The pattern of distribution of deep-sea animals within the Antarctic region varies with depth. Many species extending above 2000 m are widely distributed here and widely eurybathic. Some of them are circum-Antarctic (e.g. the starfish *Psylaster charcoti* and the sea urchin *Cerathophysa cerathopyga*). Among echinoderms extending above 2000 m, there are a number of species in common between the Antarctic regions of the Indian and Atlantic, Indian and Pacific, and also Pacific and Atlantic Oceans: 50, 33.3 and 22.2% of the whole Antarctic fauna, respectively. The first group of species is even more numerous than the second, while species shared by the Pacific and Atlantic are fewest. The South-Atlantic Ridge, separating the great depths of the southern regions of the Pacific and Atlantic, is an almost impassable barrier for many deep-sea species, even widely eurybathic ones.

Another relationship between faunas is observed at depths >3000 m. Here, there are no species in common between the fauna of the Atlantic, and both the Pacific and Indian oceans, but the faunas of the Pacific and Indian Oceans are closely related, and, for example, there are 37.5% of species in common among echinoderms. The distribution of other animal groups is similar.

Based on the above evidence, the Antarctic deep-sea region has been divided into two subregions: (i) Antarctic–Atlantic; and (ii) Antarctic–Indian–Pacific. The latter subregion in its turn consists of two provinces: Indian and Pacific. The border between the two subregions of the Antarctic deep-sea region passes along the South-Atlantic Ridge and from the Cape of Good Hope through Crozet Is to Enderby Land. Initially this border was established (Vinogradova, 1958b) on the basis of purely faunistic data. After an expedition on the R/V "Ob" discovered a submarine ridge up to 3000 m high passing from Crozet Is to Enderby Land (Zhivago and Lisitzin, 1958), the position of the border was supported by geomorphological evidence. The border between the two provinces of the Antarctic–Indian–Pacific subregion passes from New Zealand to King George V Land. The Argentine Basin and the Agulhas Basin have common deep-sea faunas, though some differences between their composition may be found.

It should be noted that the possible independence of an Antarctic deep-sea zoogeographic region was suggested by a number of investigators, beginning with Ekman (1935), who gave examples of genera of Hyalospongiae, *Aulorossela* (six species), *Gymnorossela* (two species), *Rossella* (16 species and subspecies), and six other genera with few species distributed only in the Antarctic. Fisher (1940) recorded 23 true abyssal starfish species of many Antarctic genera from the "Discovery" collections. Kirkegaard (1954) listed 21 polychaete species of the Antarctic region, 16 occurring deeper than 2000 m. A number of Antarctic abyssal endemics belonging to various benthic groups were also found by a recent Russian Antarctic expedition on the R/V "Dmitri Mendeleev" in 1989. These include representatives of Cirripedia Thoracica (Zevina, 1993), Tanaidacea (Kudinova-Pasternak, 1993), Pantopoda (Turpaeva, 1993), Echinoidea (Mironov, 1993) and Holothuroidea Elasipoda (Gebruk, 1993). Ekman stated that the existence of endemic Antarctic species is not surprising, because the Southern Ocean is hydrologically a separate unit, and that the independence of an Antarctic deep-sea fauna could be expressed in the absence of topographic isolation.

The Atlantic deep-water bottom fauna and its zoogeographic relations with those of the Pacific and Indian oceans have been discussed by Vinogradova (1957). Data on divergence of the faunas of some Atlantic basins with increase in depth are summarized in Figure 6. About 50% of widely eurybathic abyssal species living in the northern Atlantic east and west of the Mid-Atlantic Ridge are shared by both regions. Shared stenobathic abyssal species living deeper than 3000 and 4000 m constitute only 7.2 and 0%, respectively, of the faunas.

2.4. Schemes of Zoogeographic Zonation of the Abyssal Zone

The following scheme was proposed by Vinogradova (1956a, 1959b, 1969b, 1979) for the zoogeographic regionation of the abyssal, including three regions, six subregions and eight provinces (Figure 9):

Pacific–North-Indian deep-sea region
    Pacific subregion
        North-Pacific province
        West-Pacific province
        East-Pacific province
    North-Indian subregion
Atlantic deep-sea region
    Arctic subregion
    Atlantic subregion
        North-Atlantic province
        West-Atlantic province
        East-Atlantic province
Antarctic deep-sea region
    Antarctic–Atlantic subregion
    Antarctic–Indian–Pacific subregion
        Indian province
        Pacific province

This scheme is close to that proposed by Ekman in 1935. He separated four deep-sea regions: Atlantic, Indo-Pacific, Antarctic and Arctic, the last regarded by Vinogradova as a subregion of the Atlantic deep-sea region. A similar division of the abyssal was given by Kirkegaard (1954) for polychaetes.

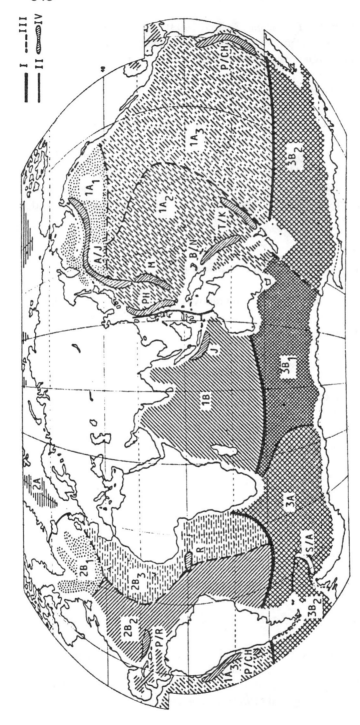

Figure 9 Zoogeographic demarcation of the abyssal and hadal zones of the ocean. Boundaries of regions (I), subregions (II), abyssal provinces (III) and hadal provinces (IV). Pacific–North-Indian deep-sea region: 1A, Pacific subregion. 1A₁, North-Pacific abyssal province; 1A₂, West-Pacific abyssal province; 1A₃, East Pacific abyssal province; A/J Aleutian–Japan hadal province; PH, Philippine; M, Mariana; B/N, Bougainville – New Hebrides; T/K Tonga–Kermadec; P/CH, Peru–Chile hadal provinces. 1B, North-Indian subregion; J, Java hadal province. Atlantic deep-sea region: 2A, Arctic subregion; 2B, Atlantic subregion; 2B₁, North Atlantic abyssal province; 2B₂ West Atlantic abyssal province; 2B₃ East-Atlantic abyssal province; P/R, Puerto-Rico hadal province; R, province of the Romanche Trench. Antarctic deep-sea region: 3A, Antarctic–Atlantic subregion; S/A, South Antilles hadal province; 3B, Antarctic–Indian-Pacific subregion: 3B₁, Indian Ocean abyssal province; 3B₂, Pacific abyssal province. (Zonation of the abyssal after Vinogradova, 1959a; hadal after Belyaev, 1974).

Publications on various animal groups, both invertebrates and fish, confirm the scheme suggested. For example, Nybelin (1954) gives the following analysis of geographic distribution of deep-sea fish of the family Brotulidae, which includes benthic forms: of 44 species of these fish, excluding single records, 21 species are known only from the Atlantic, five only from the Indian Ocean, and 15 only from the Pacific. Two species were found in both the Pacific and Atlantic, one species in both the Pacific and Indian Oceans, and none of the Brotulidae was found in both the Atlantic and Indian Oceans or in all three oceans. A similar distribution of benthic fish was reported by Grey (1956) and Rass (1959). Data on various invertebrate groups (Filatova, 1976 – abyssal bivalve molluscs of the genus *Spinula*; Millar, 1970 – ascidians; Zezina, 1985 – brachiopods) fit well in this zoogeographic scheme.

A very similar scheme of zoogeographic zonation of oceanic depths is given by Kussakin (1971a, 1973), based on the distribution of abyssal Isopoda:

Austral deep-sea region
  Andean austral province
  Gondwanian austral province
Indo-Pacific deep-sea region
  Indian province
  West-Pacific province
  East-Pacific province
  North-Pacific province
Atlantic deep-sea region
  West-Atlantic province
  East-Atlantic province
  North-Atlantic province
  Arctic province

The zoogeographic rank of the Indian and Arctic units is lowered from subregion to province in Kussakin's scheme, but apart from this, it is very close to that of Vinogradova.

Confirmation of a border between the two Antarctic subregions (Vinogradova, 1959b) has been found in the Scotia Sea area (Filatova, 1974; Filatova and Vinogradova, 1974; Filatova *et al.*, 1975; Pasternak, 1975a). The northern border of the Antarctic region in the western Atlantic and eastern Pacific was also confirmed. In areas south of Australia, New Zealand and Africa, the northern border of the Antarctic region is 5–15° south of the border suggested in the scheme of 1956 (Vinogradova, 1956b), according to the distribution of sea urchins and isopod crustaceans (Kussakin, 1973; Mironov, 1974). The data on the Indian Ocean boundaries have been confirmed by analysis of the distribution of stalked barnacles (Cirripedia, Lepadomorpha) (Shreider, 1994).

However, there is no common viewpoint on the zoogeographic relationships

Table 6 Number of deep-sea echinoderm species with wide geographic distribution (after Madsen, 1954).

	Animal groups		
Character of range	Starfishes	Ophiurans	Holothurians
Species from deeper than 3000 m	139	98	135
Cosmopolitan species	4	11	10
Species found in the Atlantic and Indian oceans but not in the Pacific	5	6	4
Species found in the Pacific and Indian oceans but not in the Atlantic	1	0	3

of the deep-sea bottom fauna of different regions of the World Ocean. In particular, there is a lack of agreement about the links between faunas of the Pacific and Indian oceans, on the one hand, and Indian and Atlantic oceans, on the other hand. Madsen (1954, 1961), having studied the echinoderms, came to the conclusion that deep-sea faunas show the closest relationship between the Indian and Atlantic oceans, whereas the fauna of the eastern Pacific is more isolated (Table 6). The zoogeographic regions recognized by Madsen in the abyssal correspond in area and rank to those discussed above but his ideas of the main faunistic relationships were different. He divided the fauna of great oceanic depths into two main regions, a Pacific region and a region embracing the Atlantic and Indian oceans, pointing out that his contrasted to the usual division of the shelf seas, where the Atlantic is distinguished from an Indo-West Pacific region (Madsen, 1954). Later, after a full analysis of the "Galathea" collections the same author added a third group, the Antarctic (Madsen, 1961).

Knudsen (1970) and Hansen (1975), in discussing the distribution of bivalve molluscs and holothurians, respectively, also noted the zoogeographic isolation of the eastern regions of the Pacific. Data on deep-sea ascidians collected by the SAFARI expedition also indicate strong affinities between the northern part of the Indian Ocean and the north of the Atlantic. More than 50% of species are shared by these regions (Monniot and Monniot, 1985a).

However, Vinogradova (1977a) reported a small number of species in common, from depths > 2000 m, for the Atlantic and Indian oceans (Table 7). Gebruk (1990) found only one species in both the Atlantic and Indian oceans among 74 deep-sea species of holothurians of the family Elpidiidae. Shreider (1994) listed 56 species of cirripedes from the Indian Ocean in > 3000 m, of which 50% have an Indo-Pacific distribution. Only three species are shared with the North–West Atlantic fauna, and all of those were found only in the northwestern part of the Indian Ocean.

Table 7 Quantity of common species known from depths exceeding 200 m in the different parts of the world ocean.

Ocean	% of species
Only Atlantic	32.5
Only Pacific	35.6
Only Indian	16.8
Atlantic and Pacific	4.0
Pacific and Indian	4.8
Atlantic and Indian	2.1
Atlantic, Pacific and Indian	4.2

Further development of ideas about distribution of the deep-sea fauna appeared in the 1960s (Menzies *et al.*, 1973). Having analysed the results of new studies, Menzies and his co-workers proposed smaller zoogeographic units in the abyssal, with a very high degree of endemism in each individual oceanic basin. In their scheme of zoogeographic division of the World Ocean at depths >4000 m, five large deep-sea zones are recognized: Pacific, Arctic, Atlantic, Indian and Antarctic. These zones are divided into 13 provinces and 17 regions and subregions (Figure 10). The divisions were based on topography of the ocean floor and hydrological characteristics of the bottom waters, essentially the distribution of temperature at depths >3000–4000 m. This scheme is not yet confirmed by faunistic material. Such an approach to zoogeographic constructions, based on changes in environmental factors but not on studying ranges of species, seems to be incorrect. The position of the proposed borders, especially the northern border of the Antarctic deep-sea region placed off northern Brazil, close to India, and about 20°N in the central area of the Pacific, is very unusual and conflicts with existing hydrological data.

Several invertebrate groups demonstrate quite narrow species ranges in the lower abyssal and show high endemism of the fauna of certain intra-oceanic basins. In addition to isopod crustaceans, Menzies (1965) considers such distribution to be characteristic of Bivalvia (Clarke, 1962a,b). Works by American investigators on the fauna of various East-Pacific basins (Hartman and Barnard, 1958; Barnard, 1961; Parker, 1963) have already been mentioned. Other examples indicate the relative isolation of the fauna of some deep-sea oceanic areas (Barnard, 1962; Jones and Sanders, 1972; Kudinova-Pasternak, 1975; Monniot and Monniot, 1975; Zevina, 1975a,b; Tabachnik, 1994). The Monniots correctly point out that each investigation initially finds a lot of new species in a region then the known limits of ranges of these species later begin to extend, as more material accumulates. One cannot agree with their conclusion that this will finally lead to deep-sea cosmopolitanism. The zoogeographic analysis of deep-sea Atlantic ascidians collected by a number of French expeditions may produce an interesting result. Monniot and Monniot (1985b) gave localities of 80 species of ascidians, and many were recorded in the Atlantic for the first time.

Hydrological characteristics and near-bottom currents are not without importance as factors determining the geographic distribution of animals. However, most data have been obtained 300–500 m above the bottom. At such a distance, differences between water masses cannot directly affect the distribution of life on the bottom. It is known that the benthic boundary layer, which is in direct contact with benthic animals, differs in its characteristics from the rest of the water column, but instrumental observations in the benthic boundary layer over the World Ocean are still incomplete.

The role of mid-oceanic ridges as zoogeographic barriers has been emphasized, because the direction of the main zoogeographic borders corresponds to macroscale disposition of the ridges (Vinogradova, 1957, 1959a; Belyaev *et al.*, 1973). Menzies *et al.* (1973) took into account the effect of relief on the distribution of fauna to an even greater extent. In this connection, special consideration should be given to the works of the Monniots on the deep-sea ascidians of the Atlantic, because they contain criticisms of earlier concepts and give new data on zoogeography at great depths in this ocean (Monniot and Monniot, 1973, 1974, 1975, 1978; Monniot, 1979). Their main conclusion, based on new material, was that the Mid-Atlantic Ridge and, probably, the Walvis Ridge are not zoogeographic barriers for ascidians, because water masses rise over the ridges and, therefore, the fauna can ascend to lesser depths in areas near the ridges. However, the accompanying map of the distribution of this group in the Atlantic has both similarities to and differences from the general scheme of zoogeographic division of Atlantic abyssal, proposed earlier by Vinogradova (as well as from Menzies' scheme for the North Atlantic). The Monniots consider ridge areas as

---

Figure 10   A hypothetical scheme of lower abyssal zoogeographic regions, provinces and areas. A, Pacific Deep-Water Region: A-1, Northwest Pacific province; A-2, Central Pacific province: A-2a, Northern Mid-America trench area; A-2b, Southern Mid-America trench area; A-2c, Peruvian area; A-2d, Easter Is area; A-2e, Tuamoto–Marquesas area; A-2f, Northern New Zeland area, A-2g, New Guinea–Borneo–Philippine area; A-2h, China area. B. Arctic Deep-Water Region; B-1, Norwegian province; B-2, Greenland–Fram province; B-3, Eurasian province; B-4, Siberian province; B-5, Canadian province. C. Atlantic Deep-Water Region: C-1, Northwestern Atlantic province; C-2, North–South Eastern Atlantic province; C-3, Caribbean–Gulf province; C-4, Mediterranean province. D. Indian Deep-Water Region: D-1, Andaman province; D-1a, Southern India area; D-1b, Arabian area; D-1c, Afro-Indian area. E. Antarctic Deep-Water Region: E-1, Antarctic circumpolar province; E-1a, Atlanto-Indian Antarctic area; E-1a(1), Eastern South Atlantic subarea; E-1a(2), western South Atlantic subarea; E-1a(3), Southeastern Indian subarea; E-1b, Austro-Indian Antarctic area; E-1b(1), Southwestern Indian subarea; E-1b(2), Eastern Australian subarea; E-1c, Southeastern Pacific Antarctic area; E-1c(1), South Central Pacific subarea (after Menzies *et al.*, 1973).

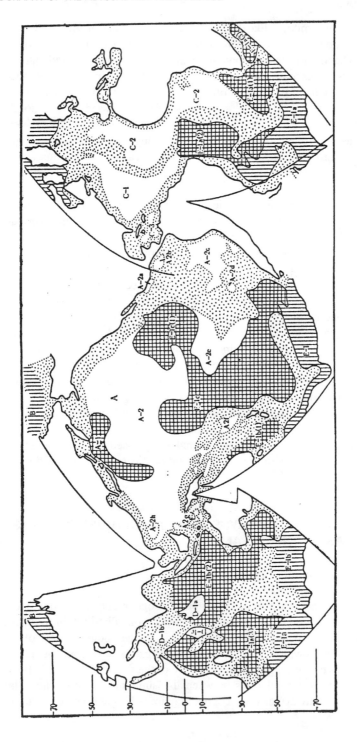

a zone of faunal mixing, like the North-Atlantic province of Vinogradova, the latter province, however, having 4% endemic species. This is not a contradiction, because the mixing areas are boundary areas. The Monniots noted that the scheme of Menzies seems more convincing, because, in their opinion, main zoogeographic barriers are defined by the hydrology of the near-bottom boundary layer but not by relief. However, Menzies' scheme is primarily based on geomorphological structure of the ocean floor. The hydrology of the boundary layer is very poorly known, as mentioned above. Knudsen also thought that mid-ocean ridges did not limit the distribution of any species of deep-sea bivalve molluscs, but all his examples, except for two species, are eurybathic bathyal–abyssal forms (Knudsen, 1970).

Nevertheless, there are examples where the existence of zoogeographic barriers in the form of large underwater ridges or other geomorphological formations has been predicted from the known distribution of benthic organisms. This was first done by Guryanova (1948) for the then unknown large Lomonosov Ridge in the Arctic. Later, she compared peculiarities of the abyssal fauna of the Pacific and Atlantic basins of the Arctic Ocean and confirmed the great isolating role of the Lomonosov Ridge. She gave about 15 examples of benthic invertebrate species living in one or the other basin (Guryanova, 1957). Similarly, on the basis of faunistic data, Vinogradova (1956a) revealed the border between two subregions of the Antarctic deep-sea region in the area of Crozet Is – Enderby Land, where a vast rise was subsequently discovered (p. 346).

The causes determining animal distribution on the immense spaces of ocean floor are much more complex than early workers realized. Many of the boundaries separating different zoogeographic regions are latitudinal; that is, they do not coincide with the direction of mid-oceanic ridges. This is connected with the latitudinal climate zones of the world, and is expressed through trophic factors that link the sea surface with the great oceanic depths (see Sokolova, this volume).

3. PATTERN OF THE ABYSSAL FAUNA DISTRIBUTION IN SOME AREAS OF THE OCEAN

3.1. Arctic Ocean

Having considered the general questions of zoogeography of the abyssal zone, this review moves on to peculiarities of deep-sea fauna distribution in distinct areas. Primarily, this concerns a detailed analysis of the zoogeography of the Arctic Basin, a topic to which many Russian investigators have contributed.

Problems concerning the independence of the fauna of the Arctic Basin and its individual parts, the relations between its fauna and those of the Atlantic and Pacific, and pathways of faunal spreading are still under active discussion at the present day. Among Russian authors, Gorbunov (1946) estimated the degree of endemism of the Arctic Basin abyssal fauna as very high (71% endemic species). Zenkevitch (1947) distinguished the abyssal of the Arctic Basin as an "abyssal Arctic subregion", covering all deeps of the Arctic Basin and differing from the abyssal of the Atlantic itself. A similar viewpoint was expressed by Filatova (1957), based on the analysis of the bivalve fauna. She considered there was a common abyssal fauna in the Arctic Ocean and the Greenland and Norwegian seas consisting of a complex of deep-sea species with a number of endemics. Some species were related to the North Atlantic abyssal fauna, whereas Pacific elements were absent. Kussakin (1971a) also emphasized the presence of true Arctic abyssal species of Isopoda, and noted the influence of deep-sea North Atlantic fauna.

Guryanova (1938, 1951, 1957, 1970, etc.) made important studies on the formation of the Arctic deep-sea fauna and the problem of the age of the Arctic Basin abyssal depths. She considered that the origin of the deep-sea fauna was related to geological and hydrophysical peculiarities, including transgressions and regressions of the sea, periods of warming and cooling, and fluctuations in salinity of the surface waters. Salinity oscillations forced relatively shallow-water species to descend deeper, thus forming deep-sea species. In Guryanova's opinion, the Arctic deep-sea fauna is relatively young, of Pleistocene age.

Guryanova used all available data on the Arctic abyssal benthos collected by various European and Russian expeditions. Extensive lists of deep-sea benthic animal species collected by Russian expeditions are given by Gorbunov (1946), Koltun (1964), Guryanova (1970). An analysis based on the best-studied benthic groups, Crustacea (Amphipoda and Isopoda) and Echinodermata, has reconfirmed the high degree of species endemism (from 72% in the Central Basin to 56% in the near-Pacific deep), the absence of families and genera endemic for the Arctic, and a noticeable admixture of widely eurybathic species in the deep-sea fauna.

The western sector of the Arctic (Baffin and Scandian–Norwegian and Greenland Basins) possesses a common fauna. The fauna of these areas is distinguished by much higher diversity in comparison with the bottom fauna of the central part of the Arctic. This is not due to an increase in the number of truly abyssal species but to descent (submergence) of bathyal and shelf species to greater depths in these regions.

In the Scandian and Central basins, Guryanova recorded more than 167 species in the three invertebrate groups analysed. Of these, 71.8% are common to both basins (Guryanova, 1970). Guryanova (1934, 1936, 1938) suggested that there is an ongoing invasion of North Atlantic deep-sea fauna to high latitudes of the Arctic, up to the New Siberian Is, and that this affects the fauna of the western sector of the Arctic. According to Gorbunov (1946), the deep waters of the Arctic are inhabited by an autochthonous fauna. This controversial question was also taken up by Koltun (1959, 1964), who was of the same opinion as Gorbunov and pointed out that North Atlantic deep-sea species were not recorded farther north than Franz Josef Land or even the region northward of Spitzbergen.

It seems that the deep-sea Arctic fauna was formed largely as a consequence of repeated descents of the shelf fauna under the influence of temperature and salinity oscillations, resulting in splitting off of deep-sea sibling species. This is well shown for deep-sea fish (Andriyashev, 1953) and was also reported for sponges (Koltun, 1964), bivalve molluscs (Gorbunov, 1946), crustaceans (Guryanova, 1946, 1970) and echinoderms (Djakonov, 1945; Baranova, 1964). In other words, the basins of the Arctic are populated by pseudo-abyssal rather than abyssal species. Koltun (1964) suggested that the bathyal zone of the Kara and Laptev seas was inhabited not by boreal North Atlantic species, but by morphologically similar (their derivatives) Arctic bathyal and sub-Arctic species, which penetrated here before the establishment of the present hydrological regime in the Arctic.

Comparison of the deep-sea benthos of the near-Pacific Basin and that of the rest of the Arctic, separated by the Lomonosov Ridge, shows that they are basically similar in species composition, despite some differences arising from the isolating role of the ridge (Guryanova, 1934; Brodsky, 1956; Koltun, 1964). Koltun (1964) noted that 63.3% of species were common to both areas. He stated that in the Lomonosov Ridge area, at least down to 2000 m (depending on hydrographic conditions), there is an exchange in deep-sea faunas between basins of the Central Arctic.

Subsequent investigations have confirmed the similarity in the species composition of Baffin, Scandian and near-Atlantic basins of the Arctic, as well as the Central and near-Pacific basins, described by Guryanova (1957). At the same time, it was established that there was little connection between the recent abyssal fauna of the near-Pacific basin of the Arctic and the abyssal fauna of the Pacific.

The old names of the basins cited in the Russian publications noted above should be put into correspondence with the more recently discovered complex bottom relief of the Arctic. The Scandian Basin corresponds to the Norway and Greenland basins, the Central Basin to the Nansen and Amundsen basins (or the Fram and Eurasia basins, according to Menzies, 1963) and the near-Pacific Basin corresponds to the Makarov (or Siberian) and Canadian basins.

In the 1960–70s, American investigators carried out extensive studies on the fauna of the great depths of the Arctic (Menzies, 1963; Menzies et al., 1973) and suggested the Arctic Basin abyssal fauna should be regarded as a separate zoogeographic region. Menzies (1963) analysed the distribution of 16 species of Arctic abyssal Isopoda and recorded 87% of endemic species among benthic forms. He stated that at least a half of the Arctic abyssal fauna is endemic at the species level. In his opinion, the faunistic relationship with the depths of the Atlantic is characteristic primarily of abyssopelagic species. *Mesidotea* is the most distinctive and the only endemic of the Arctic at the genus level. In this connection, Menzies was even inclined to regard the Arctic abyssal as the Mesidotea Abyssal region. According to Menzies (1963), only one species of isopod is common to the Norway and Eurasia basins. Based on this fact, he suggested the presence of two faunal provinces in the Arctic abyssal at depths exceeding 2000 m: Norwegian and Eurasian. Regarding the complex bottom relief of the Arctic and the presence of several more or less separate basins, he suggested that, as new data become available, each basin in the Arctic abyss would be found to have its own typical fauna. In developing these concepts, five provinces were distinguished for the Arctic deep-sea region: Norwegian, Eurasian, Greenland–Fram, Siberian and Canadian (Menzies *et al.*, 1973).

A number of trawl and grab samples were taken by American and Russian investigators at drifting ice stations in the 1970s. American expeditions on Fletcher's Ice Is (T3) obtained 75 quantitative and 28 qualitative samples of bottom fauna at 1000–2500 m in the area of the Mendeleev Ridge in 1969–1972 (Paul and Menzies, 1974). However, their strategy was targeted to the study of ecological aspects and species diversity of bottom biocenoses. As far as we know, no zoogeographic conclusions were drawn up on the basis of the newly obtained material.

In 1976 and 1977, the Russian drifting stations "North Pole 22" (Afanasjev, 1978) and "North Pole 23" (Melnikov and Tzinovskyi, 1978) obtained 50 samples of bottom fauna from the Canadian Basin at 2600–3550 m. By that time, 146 species of bottom animals were recorded in the abyssal of the central part of the basin (Vinogradov and Melnikov, 1980). Afanasjeva and Filatova (1980) listed 42 species of Foraminifera, Spongia, Actiniaria, Polychaeta, Brachiopoda, Bivalvia, Gastropoda, Cephalopoda, Amphipoda, Holothuroidea, Crinoidea and Ophiuroidea. Only one general conclusion was made about the distribution of the collected animals: Scandian–North-Atlantic and panoceanic species prevailed, and there were some endemic Arctic species. Material collected by Russian expeditions has formed the basis for several studies. For example, Zezina (1980) recorded an inarticulate brachiopod *Pelagodiscus atlanticus* in the abyssal of the Arctic, a species previously known from the bathyal and abyssal of all oceans except the Arctic. Kamenskaya (1980) noted that of eight amphipod species at abyssal depths in the Canadian Basin, half are endemics of the Arctic; three others are widely eurybathic species in the North Atlantic, while one species, *Leucothoe uschakovi*, is known from the North Pacific.

Quite different zoogeographic results were obtained for Polychaeta by Zhirkov (1980, 1982). Having analysed the Arctic abyssal fauna he wrote that at the genus level there is a more pronounced relationship to the Pacific fauna than to the Atlantic one. He came to the conclusion that "the Arctic abyssal and any other abyssal region of the World Ocean have neither common stenobathic nor even eurybathic species of polychaetes". Therefore, he believed that, as regards to the polychaete fauna, "the rank of the Arctic abyssal is higher than that of any of other regions and is equal to the rank of the abyssal of the World Ocean as the whole"

(Zhirkov, 1980, p. 235). This view of the high zoogeographic status of the Arctic abyssal was not supported by other authors. Kussakin (1979), for example, regarded the deep waters of the Arctic as only a province of the Atlantic deep-sea region. He found a low value of Preston's index $(1 - z = 0.81)$ for comparison of the faunas of the North Atlantic and Arctic.

Among the few general works of the same period, Bouchet and Warén (1979) carried out a zoogeographic analysis of the molluscan abyssal fauna of the Norwegian Sea, summarizing material from various museums round the world. Essentially, these authors confirm the concept of the Arctic abyssal, including the Norwegian Sea, as a common zoogeographic region. They estimated the endemism of the Arctic abyssal molluscan fauna as 83% at the species level, but noted that there were no endemic genera. Seven species of 26 molluscs of the Norwegian Sea abyssal region are common to the North Atlantic fauna, and 10 species are common to deep-sea fauna of the central Arctic. No species are known in common with the North Pacific fauna.

In an analysis of the distribution of families and genera of deep-sea Isopoda of the suborder Asellota, presented at the 6th Deep-Sea Biology Symposium in Copenhagen, Stromberg and Svavarsson (1991) found a surprisingly rich fauna of Asellota in the Arctic: representatives of 13 deep-sea families were recorded in the area of the Nansen Basin, Norwegian and Greenland seas. The endemism is rather high at the species level, 44%, but there is only a single endemic Arctic genus, whereas endemic families are absent. It was noted that this fauna is weakly related to the fauna of the Pacific, whereas North Atlantic species are common in Arctic depths.

Investigations of the Arctic were made during the project "German–Russian Investigations of the Ecology of Shelf of Eurasian Arctic Seas" (GRYMSEA) in recent years. Publications on the project have concentrated on quantitative and ecological approaches to the distribution and composition of marine biota (Hinz and Schmid, 1995). In 1993 an expedition aboard R/V "Polarstern" investigated depths exceeding 3000 m in the Nansen and Amundsen basins (Laptev Sea). A biocenosis dominated by the Arctic holothurians *Elpidia glacialis* and *Kolga hyalina* and the sea urchin *Pourtalesia jeffreysi* was recorded at depths of 2000–2500 m and more (Petryashev *et al.*, 1994; Sirenko *et al.*, 1995). It is known that both holothurian species are widely distributed in the Arctic Ocean, as far as the central areas (Gebruk, 1990).

Finally, mention should be made of the most recent Russian expeditions on the R/V "Akademik Mstislav Keldysh" in the northeastern part of the Norwegian Sea, in the area of the wreck of the nuclear submarine "Komsomolets", at 1500–2000 m depth, in 1989–1995 (Vinogradov *et al.*, 1996). Besides the description of bottom biocoenoses and quantitative characteristics of the benthic fauna, there is an extensive list, with brief ecological and zoogeographic characteristics, of benthic animal species from the main high-level taxa, including the Komokiacea and other large foraminiferans.

3.2. Caribbean Sea

The Caribbean region is geologically an area that had previously belonged to the Pacific. The separation of the Caribbean Sea from the Pacific and its union with the Atlantic occurred about 7–10 million years ago (Khain, 1975). As a consequence, the faunistic relationships of its deep-sea animals might be expected to be interesting (Rass, 1975). However, data on the deep-sea benthic animals are contradictory. Dahl (1954) thought that some groups of deep-sea crustaceans were completely different on either side of the Panama isthmus; the Atlantic deep-sea fauna was closer to that of the Indian Ocean, whereas shallow-water Atlantic forms were related to Pacific ones. Menzies and Frankenberg (1967) found that among Isopoda of the genera *Mesosignum* and *Storthyngura*, the Caribbean abyssal forms are close to those of the East Pacific deep-sea fauna. However, newer information about deep-sea Caribbean Isopoda does not confirm this view. An analysis of distribution of closely related forms in 19 good species of deep-sea Caribbean Isopoda shows that the number of related forms inhabiting the Pacific, on one hand, and living in the Atlantic–Arctic region, on the other hand, is essentially the same (Wolff, 1975). A study of near-bottom deep-sea fish of the same region indicates their close zoogeographic relations to the Pacific fauna. This is confirmed by the example of a new hadal species of the genus *Leucicorus*, found in the Cayman trench in the Caribbean Sea at 5800–6800 m depth. Another closely related species is known from the eastern Pacific. In contrast, the neighbouring Puerto-Rico Trench is inhabited by another genus and species, *Bassogigas profundissimus*, known also from trenches of the central parts of the Atlantic and Indian Oceans (Rass, 1974, 1975; Rass *et al.*, 1975). A close relationship between faunas of the southeastern Pacific and the Caribbean region is suggested for deep-sea pelagic fish (Parin *et al.*, 1973). Based on data from various authors including Chesher (1972), Mironov (1975) pointed out that the similarity between sea urchin deep-sea faunas of the West Indies and the Pacific coast of America was repeatedly mentioned in the literature. According to him, new data on the distribution of sea urchins make this similarity even more obvious. The same tendency is seen in the distribution of pennatularians of the genus *Umbellula* (Pasternak, 1975a,b). Two species, *U. magniflora* and *U. durissima*, occur in the Caribbean Basin in the bathyal and upper abyssal subzone. They are widely distributed in the Pacific and Indian Oceans and in the Antarctic. Pasternak believes that they most probably appeared in the Caribbean Sea through the then deep-water Panama Strait, 10 million years ago. Kudinova-Pasternak and Pasternak (1978) came to similar conclusions when analysing the distribution of the deep-sea Tanaidacea of this region.

The validity of these conclusions about the Pacific origin of abyssal bottom fauna of the Caribbean region requires additional confirmation from other groups. Groups such as the brachiopods (Zezina, 1975, 1990), ophiurans (Litvinova, 1975), cirripeds (Zevina, 1975a) do not show close faunistic relations between the eastern Pacific and the Caribbean. The East Pacific brachiopod fauna is considered by many authors to contain endemic forms and different genera and even families in comparison with the Caribbean Basin. At the same time, all deep-sea brachiopods of the Caribbean region are eurybathic. Three of them are known only from the Atlantic, and one species, *Pelagodiscus atlanticus*, has a worldwide distribution. Cnidarians provide rather ambiguous results on this question (Keller *et al.*, 1975).

Thus, this very interesting theoretical question still remains open. Apparently, different species could appear in many groups during 7–10 million years of separate existence of these regions, while the distribution of genera often appears to be much wider than deep-sea areas on both sides of the Panama isthmus. Zenkevitch (1956) believed that tens and hundreds of thousand years are necessary for species formation, millions and tens of million years for new genera, and much longer periods are required for separation of families.

3.3. Mediterranean Sea

There has been some argument about the existence of a truly deep sea fauna in the Mediterranean, due to the extremely complex climatic and geological history of the basin (George and Menzies, 1968). In the opinion of these authors, the last stagnation of near-bottom water and the hydrogen sulphide contamination, preventing deep-sea basins from being inhabited, occurred only 7000 years ago, and the bottom fauna could not colonize abyssal depths in such a short time. Pérès (1967) noted that the problem of existence of a true abyssal fauna in the Mediterranean was far from being resolved. However, as Pasternak (1982) pointed out, recent data show that the process of abyssal species origin from shallow-water ancestors may be relatively rapid and may require less than 7000 years. In his opinion, this evolution may have occurred in the deeps of the American Mediterranean Sea, and a number of Pacific abyssal species of Cnidaria, Tanaidacea and Echinodermata passed to the Caribbean Sea before its final separation from the Pacific by the isthmus of Panama.

However, a diverse bottom fauna has been collected down to 3670 m in the western part of the Mediterranean by the expedition on board of the R/V "Vityaz" (65th cruise) in 1979 (Vinogradova *et al.*, 1982). Kudinova-Pasternak (1982) has shown that oceanic Atlantic species of deep-sea Tanaidacea are absent from the western Mediterranean, and that the depths of the Mediterranean are inhabited by local abyssal forms. Pasternak (1982) recorded four species of abyssal Isopoda from 2847–3650 m and the "Jean Charcot" expeditions also found numerous Isopoda at maximum depths of the Mediterranean abyssal (Chardy *et al.*, 1973a,b). There is one abyssal Mediterranean endemic species of Echiura, which was recorded at 3650 m (Murina, 1984). Polychaeta (Laubier and Ramos, 1973) and Cumacea (Reyss, 1972) also possess abyssal autochthonous species. Monniot and

Monniot (1978) report only a single deep-sea eurybathic ascidian, *Agnesiopsis translucida*, from depths of 400–4000 m in the Mediterranean, but this species is also found in the Bay of Biscay. Bellan-Santini (1990) has reviewed the data about deep-sea Amphipoda in the Mediterranean, and although there are many eurybathic forms, below 2000 m there are a few endemic species, belonging to genera otherwise found in the Atlantic and the Pacific. It seems possible that the Mediterranean fauna began to differentiate after the isolation of the basin. Species that evolved in the Mediterranean abyssal thus have taxonomic importance, though a substantial part of the fauna of the deep-sea basins is represented by eurybathic species.

3.4. Circular Ranges

Trophic conditions are important among the various environmental factors that determine the distribution of the bottom fauna in the abyssal zone. Trophic relations affect not only ecological aspects and quantitative distribution of benthic organisms, but also zoogeographic issues. Sokolova (1986, see also this volume) has shown that the nutrition of deep-sea benthic organisms under abyssal conditions, where food resources are scarce, is primarily connected with the presence of organic matter in bottom deposits. The distribution of organic matter in these deposits is in turn affected by various processes in the surface zone of the ocean, which depend on the climatic zonation of Earth, on the one hand, and on the influence of shores or circumcontinental zonality, on the other hand. Sokolova has distinguished eutrophic and oligotrophic regions in the ocean. There are three eutrophic regions: Near-continental Eutrophic, encircling the periphery of every ocean; Oceanic Eutrophic, adjacent to the Near-continental; Equatorial Eutrophic, which more or less completely crosses the oceans in the equatorial zone; and two oligotrophic (central) regions: Northern and Southern, situated under anticyclonic gyres (Sokolova, 1977, 1979, 1981a,b). These areas are expressed in full measure in the Pacific. Only one Southern Oligotrophic Region is present in the Indian Ocean. According to the most recent views (Sokolova, 1979), oligotrophic regions are not expressed in the Atlantic at all.

Sokolova (1978, 1979, 1984) performed an analysis of macrobenthic species distribution in the trophic regions. The analysis was based on a set of 132 animal species. It revealed three types of distribution, associated with life in different trophic conditions, from the most favourable to the worst. The first type is near-continental. It includes species (42% of all analysed) with ranges limited by the Near-continental trophic region. The second type is oceanic (44%). Its species ranges cover both eutrophic regions, Oceanic and Near-continental. In most cases such species enter the Near-continental Region, but oligotrophic conditions appear to be unacceptable for them. Finally, the third type is panthalassic (14% of species). The most eurybiontic species belong to this type and are able to inhabit both eutrophic and oligotrophic regions. Generally, eurybathic forms prevail among near-continental species (63%), and abyssal forms dominate among oceanic and panthalassic species (57% and 66% respectively). The fauna of the Near-continental Eutrophic Region is the richest in number of macrobenthic species.

The relative distribution of the antipatharians *Bathypates patula* and *B. lyra* in eutrophic and oligotrophic regions of the World Ocean was reported by Pasternak (1977). The former species has a worldwide distribution within eutrophic regions, whereas the latter also enters oligotrophic regions.

Data on the existence of deep-sea animal ranges, associated with offshore oceanic areas to one or another extent, have long been accumulated. Ekman (1953) called these "offshore-deep-sea species". Rass (1959) distinguished a group of continental deep-sea fish, mainly including near-bottom species. Using the pennatulid genus *Kophobelemnon* and others, Pasternak (1961, 1973, 1975a) showed the existence of ranges of deep-sea benthic invertebrates, which narrowly border the coasts. Among abyssal Polychaeta, the offshore-deep-sea type of range was recorded for *Onuphis lepta* (family Onuphidae) by Kucheruk (1975). This author described a somewhat different range for other representatives of Onuphidae: a near-slope deep-sea type of linear distribution of benthic organisms, when species extend outside the Near-continental Region to the Oceanic Eutrophic Region along underwater ridges (Kucheruk, 1978). Detinova (1982) noted that all deep-sea species of the genus *Maldanella* (Polychaeta, Maldanidae) of the Pacific (five species) occur only in the Near-continental Eutrophic Region, whereas *M. antarctica* lives also in the open ocean like other Onuphidae, but only near underwater mountains and ridges. She regards such near-slope deep-sea type of distribution as a specific form of near-continental type. Findings of *M. antarctica* along underwater ridges in the sub-Antarctic confirm the suggestion that underwater ridges are not only zoogeographic barriers, but also pathways for spread of the near-continental abyssal fauna (Kucheruk, 1979).

A review of geographic distribution of crustaceans of the order Tanaidacea in the Southern Ocean (Kudinova-Pasternak and Pasternak, 1981) demonstrates that these animals do not move far from the base of the continental slope, though different species are characterized by different linear extent of their ranges. Among deep-sea starfishes of the genus *Freyella*, Galkin and Korovchinsky (1984) found three species with extended ranges in offshore areas. Most species of sea urchins are limited in their distribution to near-continental areas (Mironov and Sagaidachny, 1984). Gebruk (1990) reported that holothurians of the family Elpidiidae, the most characteristic representatives of the deep-sea fauna, are associated with near-continental areas of eutrophic oceanic regions, and this is one of the regularities of their distribution. This is also characteristic of most species of the family Myriotrochidae (Belyaev and Mironov, 1982). Chindonova (1981) recorded such distribution in deep-sea Mysidacea of the subfamily Mysinae. Southward (1979) described the wide band-shaped distribution of lower slope Pogonophora in the Atlantic.

Kucheruk (1979; Kucheruk *et al.*, 1981) follows Sokolova (1978, 1979, 1981a) in recognizing three types of geographic ranges in the abyssal, which correspond to the three types of trophic distribution. He makes separate divisions for near-continental, oceanic and panthalassic species; that is, only within the zone of occurrence of species with one type of trophic distribution. On this basis it is impossible to construct a single zoogeographic scheme for the abyssobenthal division; species with different types of trophic distribution are eurybiontic to a different extent, and, therefore, are characterized by different scales of the three schemes of division. These considerations are undoubtedly of interest, but unfortunately they are in no way supported by actual data. Kucheruk *et al.* (1981) give only six examples of species with near-continental range, five with oceanic and six with panthalassic, without an indication of their vertical range. Moreover, all these examples are from the Pacific. This point of view is reasonable and has its logic, because at present it is impossible to neglect the role of trophic factors in the distribution of deep-sea bottom fauna, as was already widely discussed earlier. Besides the works already mentioned, the question has been discussed by Birstein and Vinogradov (1971), Vinogradova (1969a,b, 1977a, 1979) and Belyaev *et al.* (1973) among others.

However, the point of view that it is impossible to construct a single scheme of zoogeographic division of the abyssobenthal is now in doubt because there are insufficient calculations or analyses of the composition of the entire known fauna. Also because such an approach requires development of different schemes of the abyssobenthal division in the case of, for example, species with different modes of reproduction (presence or absence of pelagic larva), or species with different ranges of vertical distribution. This was essentially proposed by Menzies (Menzies *et al.*, 1973) and Zhirkov (1980), who took into account only the truly abyssal part of the benthos in constructing their schemes of zoogeographic division.

Both upper and lower abyssal subzones are also inhabited by widely eurybathic deep-sea species. Their role is quite significant at all depths of the abyssal, and they generally reveal the faunal divergence between different areas, which appears in stenobathic species, the most deep-sea part of the fauna, as the depth increases. In other words the eurybathic species are conceived as determining the united pattern of zoogeographic division of the deep-sea bottom fauna at all horizons of the oceanic abyssal zone.

Some workers (Mironov, 1981, 1983; Mironov and Sagaidachny, 1984), discussing the distribution of sea urchins, state that "within certain vertical zones there are faunistic borders, which extend circumcontinentally and are determined mainly (in the abyssal – exclusively) by the impoverishment of the taxonomic composition of the fauna at increasing distance from the continents" (Mironov and Sagaidachny, 1984, p. 201). According to Mironov (1983), the sublittoral, bathyal and abyssal should be divided into two faunistic zones: circumcontinental subcontinental and oceanic. This concerns all three oceans – the Pacific, Atlantic and Indian. Mironov (1987) recognizes five faunistic regions in near-continental

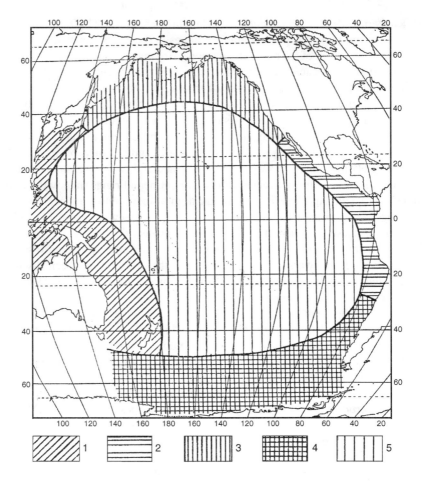

Figure 11 Faunistic regions of the Pacific ocean floor. 1, Western; 2, Eastern; 3, Northern; 4, Antarctic; 5, Central (after Mironov, 1987).

areas of the ocean floor: Western, Eastern, Northern, Antarctic and Central, whose borders coincide rather precisely with previously known faunistic borders, established on the basis of data on various groups of benthic animals. The central–oceanic areas of the Pacific are regarded by Mironov as a separate central region (Figure 11). Mironov mentioned that only seven species out of the 41 known Pacific deep-sea Echinoidea were recorded at a considerable distance from continents, and that these sea urchins include very few species with a distribution limit passing near the ocean mid-line.

Biogeographic circular structures, including the abyssal zone, were also described by Semenov (1981). He recognized oceanic systems in the abyssal that correspond to the trophic regions of Sokolova (1979).

According to Springer (1982), a faunistic border passes between the near-continental and oceanic regions of the Pacific. Its existence is determined by the impoverishment of taxonomic composition of the fauna in oceanic regions, as well as by the high degree of faunal endemism. Springer stated that these borders are connected with geomorphological peculiarity of the Pacific lithospheric plate, which has long been the centre of origin of many taxa. However, the above-mentioned structures are observed also in other oceans, so that Springer's statement is in doubt.

Sokolova (1984) has shown that there are no endemic species in oligotrophic regions of the Pacific remote from the coasts, and only 5% of oceanic species occur also in the Near-continental Region.

It is evident that the problem of isolation of zoogeographic regions on the basis of impoverishment in species composition is rather controversial. Vinogradova (1969b) has shown the transitory character of the deep-sea fauna in the central part of the Pacific, near the Hawaiian Is and southward; that is, between West- and East-Pacific provinces of the Pacific deep-sea subregion. In the Pacific, there is a clear relationship between the most pronounced borders, based on change in faunal composition, and the borders of trophic regions, at which the feeding conditions alter dramatically. Within the central part of the ocean abyssal, the eutrophic and oligotrophic regions extend latitudinally, but not circumcontinentally (Sokolova, 1969; Vinogradova, 1969b).

All the above conclusions are relevant to the basic problems of the biological structure of the ocean. As shown first by the Russian investigators, Zenkevitch and Bogorov, and their successors, the distribution of life in the ocean ultimately depends on three types of zonality: vertical, latitudinal (climatic) and circum-oceanic (circumcontinental), and also on historical factors.

3.5. Bipolar Distributions

The existence of circumoceanic ranges in deep-sea animals is connected with the widely discussed question of bipolar and amphioceanic species in the abyssal bottom fauna. These types of ranges have been disputed by some investigators, who think this phenomenon is only the result of insufficient knowledge of the deep-sea bottom fauna (Menzies, 1959; Murina, 1961, 1978; Knudsen, 1970; Pasternak, 1973). Based on the idea of wide distribution of true abyssal species within trophic zones and the absence of any physical barriers, Kudinova-Pasternak and Pasternak (1981) said that, unlike eurybathic species, the abyssal ones cannot have disrupted ranges. Therefore, according to these authors, one should accept the statement that bipolar, amphi-Pacific and other disrupted ranges do not occur in purely abyssal species (Kucheruk et al., 1981). Nevertheless, examples of bipolar distribution still remain in various groups. For instance, Birstein (1960) noted the bipolar species *Stylomesus inermis*, was found in the Antarctic in 2450 m and in the North Pacific in 3575 m. According to his data, almost the whole family Ischnomesidae are bipolar. There are five deep-sea bipolar species in the Tanaidacea (Kudinova-Pasternak, 1978, 1990, 1993; Kudinova-Pasternak and Pasternak, 1981).

Gebruk (1990) found that in holothurians of the family Elpidiidae the closest relationship occurs between faunas of the Southern Ocean and the Pacific, which have 14 common species. Ten of them inhabit both oceans. He noted that holothurians of the family Elpidiidae possess four major types of geographic distribution at the genus level, and *Peniagone*, *Kolga*, *Elpidia* and *Ellipinion* inhabit both the Antarctic and temperate and high latitudes of the northern hemisphere (Gebruk, 1994). Mironov (1993) reported that the ranges of three genera of sea urchins (*Echinosigra*, *Plexechinus* and *Urechinus*) of deep-sea Antarctic origin are widespread in middle latitudes of the Atlantic, from 53–33°S to 17–47°N. Other examples include: Ophiuroidea (Belyaev and Ivanov, 1961), Priapulida (Murina and Starobogatov, 1961; Murina, 1964), Sipuncula (Murina, 1993), gastropods of the genus *Tacita* (Lus, 1975; Sysoev and Kantor, 1987), Cirripedia (Zevina, 1976, 1993), and Pantopoda (Turpaeva, 1993).

Vinogradova (1969b) said that about 20 bipolar deep-sea species are known for the Pacific, excluding widely eurybathic species. This phenomenon is attracting renewed attention, because the study of the hadal fauna of deep-sea trenches of the Southern Ocean in very recent years has revealed zoogeographic relationships to the fauna of North Pacific trenches (see p. 371).

Vinogradov (1968) noted that there are a number of deep-sea bipolar species belonging to purely deep-sea genera. In this case, the concept of their origin from a common bipolar shallow-water ancestor (Birstein, 1963) is inapplicable. When analysing the phenomenon of bipolar distribution of deep-sea pelagic forms, Vinogradov paid great attention to the importance of inflow of cold near-bottom waters of polar origin, which passes meridionally, either from north to south or from south to north, in different areas of the World Ocean (Vinogradov, 1959, 1962, 1968). Meridional deep flows are believed to be stronger in periods of climate cooling, and may have formed bridges for migration of animals from temperate and cold zones of one hemisphere to another. In warmer periods the intensity of these flows appears to have decreased, thus isolating the deep-sea polar ranges. The effect of deep currents on the deep-sea fauna distribution may be well illustrated by the following consideration. Meridional deep currents are at present much more pronounced in the Atlantic than in the Pacific. Accordingly, deep-sea pelagic species are bipolar in the Pacific, whereas in the Atlantic they cross the tropical zone with the help of deep currents along the eastern border of America. A similar but less pronounced bridge also exists in the Pacific and passes along the western coast of America. Similar views were held by Zarenkov (1965) who pointed out that bipolar ranges of Antarctic decapod crustaceans, including the abyssal *Munidopsis antoni* (2520–3900 m), have formed due to deep-sea migrations along continental margins, primarily along the western coast of America. *Priapulus tuberculatospinosus* (Murina and Starobogatov, 1961; Murina, 1975) and the abyssal Polychaeta may include other examples. Among 16 species of the latter group found in the Orkney Trench (Detinova, 1993) there are 14 species in common with the fauna of the northwestern Pacific. These species are widely distributed in the Antarctic abyssal and pass northward along the Pacific coast of America. There are only six species in common with the North Atlantic fauna. Detinova concluded that faunistic connections between polychaete faunas of abyssal depths of the Antarctic and North Pacific are much closer than those between the Antarctic and North Atlantic. Therefore, the problem of bipolar distribution of abyssal species still remains open.

4. GEOGRAPHIC DISTRIBUTION OF ULTRA-ABYSSAL (HADAL) FAUNA

Data on life in deep-sea trenches of the World Ocean, on the composition of their bottom fauna, collected by expeditions of several different countries and scattered in numerous worldwide periodicals, were summarized by Belyaev (1966, 1972, 1989) and Wolff (1960, 1970). Belyaev (1989) compiled lists of animals living at depths exceeding 6000 m (no less than 700 species of Metazoa) and indicated the vertical and geographic distribution of each species, with references. He also revealed regularities of vertical and geographic distribution of hadal fauna. This review follows Belyaev, with the addition of new data, obtained after publication of his last monograph.

At present, 37 deep-sea trenches are known. Most of them "frame" to some extent the periphery of the oceans, extending along frontal edges of geotectonic plates and island arcs. These trenches also include transform fracture zone trenches situated far from coasts, in areas of rift zones of mid-oceanic ridges. However, Belyaev excluded geomorphological structures corresponding to the concept of a deep-sea trench, if their depth was less than 6000 m. This is because at depths >6000 m there is believed to be a special zone of life inhabited by a specific ultra-abyssal (Zenkevich et al., 1955) or hadal (Bruun, 1956) fauna.

Most of the known trenches are situated in the Pacific, 28, including the nine deepest, with depths of about 9–11 km. Five trenches are known from the Atlantic and four from the Indian Ocean. Specific environmental conditions are responsible for the ecological isolation of hadal depths from surrounding areas of ocean floor, and these conditions define high specificity of the hadal fauna as a whole. In the whole hadal fauna, endemics comprise 56.4% of the species, though only 10% at the genus level. Although the hadal zone is characterized by considerable faunistic isolation from the zones above it, Belyaev (1989) believed that there was no reason to regard it as a specific zoogeographic region.

The main reserve for population of the deep-sea trenches was the abyssal fauna of adjacent (to each trench) areas of ocean floor. Therefore, in species inhabiting hadal depths but not endemic there (i.e. having a wider vertical distribution), on average about half are species with a local distribution. Their ranges do not extend outside the limits of the abyssal province or subregion adjacent to the given trench, assuming use of the schemes of zoogeographic division of the oceanic abyssal proposed by Vinogradova (1956a) (see p. 347). Depths exceeding 6 km cannot be considered as a single unit in a zoogeographic sense also because they are separated from each other by vast distances of the abyssal plains. Correspondingly, individual trenches or groups of trenches possess a distinct fauna at the species level. Using more than 600 identified species, Belyaev (1989) provided the following data on the character of distribution of bottom and near-bottom fauna in the hadal zone.

Type of distribution	% of all hadal forms
Species endemic to the hadal	56.4
Found in a single trench	47
Found in two or several neighbouring trenches	6.4
Found in two or several widely spaced trenches	3
Species found also at lesser depths	43.6
Only in the region of trench location (within the same zoogeographic province)	22
Known from different regions of the same ocean	4.6
Known from two oceans	11
Known from three oceans	6

Therefore, the most characteristic feature of the hadal bottom fauna is geographic confinement of species to certain local zones of one ocean. Among hadal endemics, 95% live in only one trench or in a group of neighbouring trenches. This part of the fauna comprises 53.5% of the whole hadal fauna. Belyaev's view of the endemism of the fauna of deep-sea trenches was followed by Angel (1982).

Of the species found also at lesser depths, half do not occur outside the ocean region where the trench is situated. That is, three-quarters of the entire trench fauna are species with local distribution. For example, Gebruk (1990) demonstrated that almost all hadal holothurians of the family Elpidiidae are known from only one trench, and mainly from the Pacific. Outside the Pacific there are only three known true hadal species of *Elpidia*, and two of them – *E. lata* and *E. ninae* – occur only in the South Sandwich Trench.

On the other hand, Belyaev noted also a considerable faunistic similarity between some groups of trenches, forming united chains. The Aleutian, Kurile–Kamchatka, Japan, and Izu–Bonin trenches have 26% of hadal species in common between two, three or four trenches. Among new findings not included by Belyaev (1989), a new species *Chevroderma hadalis* (Caudofoveata, Aplacophora) may be mentioned. This species was described from the Aleutian and Kurile–Kamchatka trenches, in 7250–8390 m (Ivanov, 1996). Among non-endemic species, 46% do not leave the limits of the North-Pacific Province.

The Tonga and Kermadec trenches share two new species of Bivalvia, identified by Z.A. Filatova but left undescribed. The starfish *Lethmaster rhipidophorus* was found only in the Ryukyu and Philippine trenches in 6460–7880 m (Belyaev, 1969, 1985). The last group of trenches consists of Volcano, Mariana, Yap and Palau trenches, where six of the 12 hadal endemics of the Mariana Trench were also found in neighbouring trenches. Belyaev explained such similarity between hadal faunas of trench chains by their possession of a common geological history and the fact that earlier the trenches were linked at depths of 6.5–7 km. In Belyaev's opinion, the thresholds between trenches appeared later and are rather young. The separation into individual trenches most probably took place, as he believed, in Quaternary time.

There are 20 species of hadal benthic animals living in widely separated trenches. Seven species are active Amphipoda, apparently of nectobenthic habit and able to ascend to the abyssal zone, though not yet found there. The wide distribution of such forms can be explained by their active mode of life. Kamenskaya (1981a,b) analysed the distribution of Pacific hadal Amphipoda and found 34 species, 79.4% being endemics of the hadal zone. Among the latter group, she recognized two types of species distribution in trenches: (i) amphipods that construct shelters, and other less-active forms with weakly developed limbs and musculature are usually associated with one trench or a group of neighbouring trenches; and (ii) active nectobenthic forms with a much wider distribution. For example, benthopelagic *Scopelocheirus schellenbergi* was considered a typical hadal species. It was recorded in the hadal of six trenches of the West Pacific (Kamenskaya, 1981b), in the Java Trench (Birstein and Vinogradov, 1964) and was found for the first time at abyssal depth in the Orkney Trench in the South Atlantic (Vinogradov and Vinogradov, 1993). It should be noted, however, that an ascent of deep-sea fauna to lesser depths is characteristic of cold-water polar regions.

Some less active hadal species also can have wide distributions. The gastropod *Bonus petrochenkoi* was recorded in the Kurile–Kamchatka and Tonga trenches (8240–9530 m) (Moskalev, 1973); the polychaetes *Macellicephaloides grandelythris* in the South Sandwich and Macquarie trenches (Levenstein, 1975) and *Bathykermadeca hadalis* in the Kermadec, Yap, Japan, Banda and Philippine trenches (Kirkegaard, 1956; Pettibone, 1976; Levenstein, 1978, 1982); the bivalve *Spinula vityazi* (6450–9335 m) in the Aleutian, Kurile–Kamchatka, Japan and Izu–Bonin trenches (Filatova, 1976); the holothurian *Peniagone heronardi* in the South Sandwich and Puerto Rico trenches (7694–8100 m) (Gebruk, 1990).

As one possible explanation of distribution of such species, Belyaev (1989) suggested that they have colonized the hadal depths in different parts of the ocean relatively recently, and that they may have evolved in parallel in different trenches, from a common and widely spread ancestor. The populations of such species in different trenches may possibly be different biological species, which are currently indistinguishable at morphological level. For example, the hadal lysianassoid amphipod *Hirondella gigas*, is an active planktonic form living in the near-bottom water layer but capable of descending to the bottom for food. This species has been found in eight trenches of the West Pacific: Kurile–Kamchatka, Japan, Izu-Bonin, Volcano, Yap, Mariana, Palau and Philippine. A statistical study (France, 1993) of *H. gigas* from the latter three trenches shows some morphological differences, so France suggests that the geographically isolated populations may have reduced levels of gene flow between them, causing divergence.

However, it is also possible that the distributions of some groups of the deep-sea fauna are not fully known. For example, the Galatheanthemidae had been considered typically hadal actinians, but then were found to be widely distributed in depths of >4 km in Antarctic and Subantarctic waters (Vinogradova *et al.*, 1974). The errant polychaete *Bathyeliasona kirkegaardi* was originally known from nine trenches in the West Pacific and from the Java and Banda trenches. Later it was found at 5275 m in the Atlantic (Pettibone, 1976), and in 5525 m in the Pacific (Levenstein, 1978).

The degree of isolation of the trench fauna from adjacent abyssal areas of the ocean floor determines the zoogeographic status of certain trenches or groups of trenches as separate zoogeographic units ranked as provinces in the united scheme of zoogeographic division of the abyssal and hadal (Belyaev, 1974; Belyaev and Vinogradova, 1977; Vinogradova, 1977a, 1979) (Figure 9). Each of these provinces is included in the deep-sea region or subregion, within which the respective trenches or groups of trenches are situated. Belyaev (1989) recognized the following separate hadal provinces:

- *The Pacific–North-Indian deep-sea region.* The Pacific subregion includes: (i) Aleutic–Japan Province (Aleutian, Kurile–Kamchatka, Japan and Izu-Bonin trenches); (ii) Philippine (Philippine and Ryukyu trenches); (iii) Mariana (Volcano, Mariana, Yap and Palau trenches);* (iv) Bougainville–New-Hebrides (New Britain, Bougainville, Santa-Cruz and New Hebrides trenches); (v) Tonga–Kermadec; and (vi) Peru-Chilean.** The North-Indian subregion includes: (vii) Java hadal province.
- *The Atlantic deep-sea region.* The Atlantic subregion includes: (i) Puerto Rico Province; and (ii) Romanche Province.

- *The Antarctic deep-sea region*. The Antarctic–Atlantic subregion includes: (i) South Antillean hadal province. Earlier, this consisted only of the South Sandwich Trench (Belyaev, 1974).

New data obtained in the hadal zone of the southern oceans by more recent Russian expeditions using trawls (Filatova and Vinogradova, 1974; Vinogradova *et al.*, 1974, 1978, 1993; Vinogradova, 1996); allow inclusion of the Orkney Trench. For example, *Elpidia decapoda* was recorded as a common species in the Orkney Trench in 6420 m, and *Elpidia gracilis* was also found. The latter species was listed by Belyaev (1975) from the South Sandwich, Endurance (a fracture in the Scotia Sea), and Lori (former name – South Orkney) trenches. Gebruk (1993) indicates that these species are not known from other oceanic regions, except for abyssal areas adjacent to the above-listed trenches. A high similarity between faunas of these trenches was also confirmed for Polychaeta (Levenstein, 1975). A new gastropod mollusc, *Tromina abyssalis*, of the family Buccinidae was discovered in the Orkney and Lori trenches (Lus, 1993).

In spite of the high degree of endemism of the bottom fauna of trenches of the Banda and Cayman seas (43% and 47%, respectively), they have not been recognized as separate provinces, because their fauna is still insufficiently known to resolve this question.

Belyaev believed that the status of separate hadal provinces should not be assigned to deeps of some oceanic basins, with depth slightly exceeding 6 km, including the Hjort Trench and, probably, the Macquarie Trench, as well as the Endurance Fracture and the Lori Trench in the South Atlantic. Their fauna differs only slightly from that of adjacent abyssal regions (20–28% of endemics). The intermediate character of the fauna of such trenches was also mentioned by Vinogradova *et al.* (1974, 1978).

A number of authors have noted the faunal connections between the hadal Antarctic fauna and the fauna of some other remote trenches, primarily the trenches of the North Pacific. These connections are seen mainly at the genus level, among closely related species, more rarely at the species level. Belyaev (1974, 1975, 1989) exemplified such connections among various species of holothurians of the genus *Elpidia* (family Elpidiidae), based on an assumption that colonization of Pacific trenches proceeded along the west coast of America. The colonization of the Arctic occurred through abyssal depths of the Atlantic (Belyaev, 1989).

---

*Belyaev did not exclude the possibility that further studies of the fauna of trenches around the Philippine Sea would result in unification of the Philippine and Mariana hadal provinces. This is seemingly confirmed by the distribution of the hadal bivalve mollusc genus *Parayoldella* (family Ledellidae), which includes nine species inhabiting only the West Pacific trenches. For example, *P. inflata* lives in the Mariana, Yap, Palau and Philippine trenches in 7340–9990 m (Filatova and Schileyko, 1985).

**Based on the distribution of benthic Mysidacea and Cumacea, as well on data of some other authors, Bacescu (1981) follows Belyaev in recognizing the faunistic independence of the Peru–Chile Trench and in its isolation as a separate zoogeographic unit.

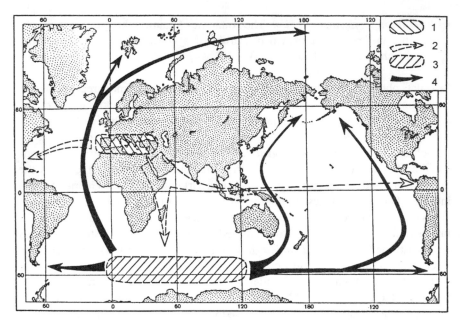

Figure 12. The probable path of distribution of the recent Elasipoda. 1, The primary centre of origin and distribution; 2, the direction of distribution from the primary centre; 3, centre of the deep-sea fauna formation; 4, pathways of distribution of the deep-sea fauna (after Gebruk, 1990).

Similar views on the centre of formation of the deep-sea species of *Elpidia* and pathways for spreading of this genus were expressed by A.V. Gebruk (1990, 1994*) (Figure 12).

Spreading along the Pacific coast of America is known for deep-sea sea urchins (Mironov, 1982); near-bottom-pelagic mysids *Birsteiniamysis* (Chindonova, 1981, 1993); and polychaetes of the genus *Maldanella*, including *M. antarctica* and other forms (Uschakov, 1975; Detinova, 1982, 1993). A new species and genus of Isopoda of the family Arcturidae, *Globarcturus angelikae*, was found in the South Sandwich Trench in 6766–7212 m. This was the first discovery of hadal Arcturidae in the southern hemisphere (Kussakin and Vasina, 1993). Four species of this family were previously known from only the hadal of the northwestern Pacific (Birstein, 1963; Kussakin, 1971b; Mezhov, 1980). The near-bottom amphipod *Alexandrella carinata* found recently in the Orkney Trench (Vinogradov and Vinogradov 1993, 1996) was described originally from the hadal of the Kurile–Kamchatka Trench (Birstein and Vinogradova, 1960). There are

*Authors of this work were erroneously printed as Gebruk, A.V. and Shirshov, P.P.

other examples of similar ranges in the Tanaidacea (Kudinova–Pasternak and Pasternak, 1981), for example, *Leptognathia robusta* is reported from the South Sandwich and the Kurile–Kamchatka trenches. Two new species of the aberrant deep-sea ascidian of the genus *Situla* were described from the South Sandwich Trench (Vinogradova, 1975). This genus was initially described from the Kurile–Kamchatka Trench at 5035–8430 m (Vinogradova, 1969c), and an additional species was subsequently found in the North Atlantic in 2115–4690 m (Monniot and Monniot, 1973). Two different species of pogonophorans of the genus *Spirobrachia* were found in the South Sandwich and Orkney trenches (Gureeva, 1975; Vinogradova *et al.*, 1993). Two other species of this genus were described earlier (Ivanov, 1952) from the Bering Sea and from the hadal of the Kurile–Kamchatka Trench.

The number of similar examples of discontinuous distribution may increase, and they should be discussed in connection with the apparent bipolarity of bottom fauna distribution at great depths of the World Ocean. Many investigators believe that Antarctic and Subantarctic regions were the centre of formation of at least some species of deep-sea genera of various invertebrate groups. Starting from there, these species colonized more northern deep-sea oceanic areas and deep-sea trenches, where endemic species were formed. This was facilitated by a penetration of deep Antarctic waters far to the north. Besides the spreading pathways already noted, deep-sea animals may have spread from the Antarctic to the North Pacific along the Pacific margin of America; and a pathway along the western margin of the Pacific has also been suggested (Belyaev, 1974; Mironov, 1982, 1983; Gebruk, 1990). Pasternak (1973) discussed the possible spreading of *Kophobelemnon biflorum* (Pennatularia) from the North Pacific to the area of the Scotia Sea and South Sandwich Trench along the continental slope of America and around its southern extremity. A similar, but reverse, pathway was suggested for some bivalve molluscs of the deep-sea family Ledellidae (Protobranchia) (Filatova and Schileyko, 1984). In this section only a few studies by Russian authors have been selected out of numerous publications on the formation and spreading of deep-sea bottom fauna of the World Ocean. These were selected because their data have been used in other sections of this review. This interesting and important area of historical zoogeography deserves a separate review.

REFERENCES

Afanasjev, I.F. (1978). Studies of the deep-sea bottom fauna in the central part of the Arctic Ocean. *Okeanologia* **18**(5), 950–951. (In Russian).
Afanasjev, I.F. and Filatova, Z.A. (1980). Investigations of the bottom fauna in the Canadian Basin of the Polar Ocean. *In* "Biology of the Central Arctic Basin" (M.E. Vinogradov and I.A. Melnikov, eds), pp. 219–229, Nauka, Moscow. (In Russian).

Andriyashev, A.P. (1953). Ancient deep-water and secondary deep-water fishes and their significance in a zoogeographical analysis. *In* "Notes on Special Problems in Ichthyology" pp. 57–64. Izdatelstvo Akademii Nauk SSSR, Moscow. (In Russian).

Angel, M.V. (1982). Ocean trench conservation. *Environmentalist* **2** (Suppl. 1), 1–17.

Bacescu, M. (1981). Contribution to the knowledge of some Mysidacea from the Peru-Chile Trench, Californian coast and Philippinian sea. *In* "Biology of the Pacific Ocean Depths" (N.G. Vinogradova, ed.), pp. 34–39. Proceedings of the XIV Pacific Science Congress, Khabarovsk, August 1979. Section Marine Biology, Issue I. Vladivostok. (In Russian).

Baranova, Z.I. (1964). Echinoderms (Echinodermata) collected by the r/v "F. Litke" expedition in 1955. *Trudy Arcticheskogo i Antarcticheskogo nauchno- issledovayelskogo Instituta* **259**, 355–372. (In Russian).

Baranova, Z.I. and Kuntsevich, Z.V. (1969). Taxonomic composition of deep water benthic fauna: Ophiuroidea. *In* "The Pacific Ocean. Biology of the Pacific Ocean, 2. The Deep-sea Bottom Fauna; Pleuston" (L.A. Zenkevich, ed.), pp. 115–124. Nauka, Moscow. (In Russian).

Barnard, J.L. (1961). Gammaridean Amphipoda from depths of 400 to 6000 m. *Galathea Report* **5**, 23–128.

Barnard, J.L. (1962). South Atlantic abyssal amphipods collected by R/V "Vema". *Vema Research Series* **1**, 1–78.

Barthel, D. and Tendal, O.S. (1994). Antarctic Hexactinellidae. *Synopses of the Antarctic Benthos* **6**, 1–154.

Bellan-Santini, D. (1990). Mediterranean deep-sea amphipods: composition, structure and affinities of the fauna. *Progress in Oceanography* **24**, 275–387.

Belyaev, G.M. (1966). "Bottom fauna of the Ultraabyssal Depths of the World Ocean". Nauka, Moscow. (In Russian).

Belyaev, G.M. (1969). New starfish species from the abyssal and ultraabyssal depths of the Pacific. *Byulleten Moskovskogo Obshchestva ispytatelei prirody. Otdel biologicheskij* **74**(3), 5–26. (In Russian).

Belyaev, G.M. (1972). "Hadal Bottom Fauna of the World Ocean". Israel Program for Scientific Translations, Jerusalem.

Belyaev, G.M. (1974). On the age of deep-sea fauna of the ocean and ultra-abyssal fauna of trenches. *Byulleten Moskovskogo Obshchestva ispytatelei prirody, Otdel biologicheskij* **79**(5), 94–112. (In Russian).

Belyaev, G.M. (1975). New species of holothurians of the genus *Elpidia* from the south part of Atlantic Ocean. *Trudy Instituta Okeanologii Akademii Nauk SSSR* **103**, 259–280. (In Russian with English summary).

Belyaev, G.M. (1977). Ultraabyssal (hadal) fauna. *In* "Oceanology; Biology of the Ocean, 1" (M.E. Vinogradov, ed.), pp. 198–204. Nauka, Moscow. (In Russian).

Belyaev, G.M. (1985). New findings of starfishes of the family Porcellanasteridae in the ultraabyssal zone. *Zoologicheskii Zhurnal* **64**(4), 538–548. (In Russian).

Belyaev, G.M. (1989). "Deep-sea Trenches and their Fauna". Nauka, Moscow. (In Russian).

Belyaev, G.M. and Ivanov, B.G. (1961). On the bipolarity of ophiuroids of the genus *Toporkovia* Djakonov. *Zoologicheskii Zhurnal* **40**(8) 1258–1260. (In Russian).

Belyaev, G.M. and Mironov, A.N. (1982). The holothurians of the family Myriotrochidae (Apoda): composition, distribution and origin. *Trudy Instituta Okeanologii Akademii Nauk SSSR* **117**, 81–120. (In Russian with English summary).

Belyaev, G.M. and Vinogradova, N.G. (1977). Zoogeography of abyssal and ultraabyssal depths. *In* "I sjezd sovetskich okeanologov. Biologia i Chimia Okeana, 2", pp. 47–48. Nauka, Moscow. (In Russian).

Belyaev, G.M., Birstein, J.A., Bogorov, V.G., Vinogradova, N.G., Vinogradov, M.E. and Zenkevitch, L.A. (1959). A diagram of the vertical biological zonality of the ocean. *Doklady Akademii Nauk SSSR* **129**(3), 658–661. (In Russian).

Belyaev, G.M., Vinogradova, N.G., Levenstein, R.Ya., Pasternak, F.A., Sokolova, M.N. and Filatova, Z.A. (1973). Distribution laws of the deep-sea bottom fauna in the light of the idea of the biological structure of the ocean. *Okeanologia* **13**(1), 149–197. (In Russian).

Birstein, Ja.A. (1960). The family Ischnomesidae (Crustacea, Isopoda, Asellota) in the north-western Pacific and the problem of amphiboreal and bipolar distribution of deep-sea fauna. *Zoologicheskii Zhurnal* **39**(1), 3–28. (In Russian).

Birstein, Ja.A. (1963). "Deep-sea Isopods in the North-west Pacific". Isdatelstvo Akademii Nauk SSSR, Moscow. (In Russian).

Birstein, Ja.A. and Vinogradov, M.E. (1964). The pelagic amphipods gammarids of the northern part of the Indian ocean. *Trudy Instituta Okeanologii Akademii Nauk SSSR* **65**, 152–196. (In Russian).

Birstein, Ja.A. and Vinogradov, M.E. (1971). Role of the trophic patterns in the taxonomic discreteness of marine deep-sea fauna. *Byulleten Moskovskogo obshchestva ispytatelei prirody, Otdel biologicheskij* **76**(3), 59–92. (In Russian).

Birstein, Ja.A. and Vinogradova, N.G. (1960). The bottom ultraabyssal gammarids from the north-western Pacific, I. *Trudy Instituta Okeanologii Akademii Nauk SSSR* **34**, 147–164. (In Russian).

Bouchet, P. and Warén, A. (1979). The abyssal molluscan fauna of the Norwegian sea and its relation to other faunas. *Sarsia* **64**, 211–243.

Briggs, J.C. (1974). "Marine Zoogeography". McGraw-Hill, New York.

Brodsky, K.A. (1956). The pelagic life in the Polar basin. *Priroda* 5, 41–48. (In Russian).

Bruun, A.F. (1956). The abyssal fauna: its ecology, distribution and origin. *Nature* **177**, 1105–1108.

Bruun, A.F. (1957). Deep-sea and abyssal depths. *Geological Society of America Memoir* **67**(1), 641–672.

Bruun, A.F. and Wolff, T. (1961). Abyssal benthic organisms: nature, origin, distribution and influence on sedimentation. "Oceanography" pp. 391–397. American Association for the Advancement of Science, Publ. 67, Washington DC.

Chardy, P., Laubier, L., Reyss, D. and Sibuet, M. (1973a). Données préliminaires sur les résultats biologiques de la campagne Polymede, Part 1. Dragages Profonds. *Rapports et Procès-verbaux des Réunions. Commission Internationale pour l'Exploration Scientifique de la Mer Mediterranée* **21**(9), 621–625. Paris.

Chardy, P., Laubier, L. Reyss, D. and Sibuet, M. (1973b). Dragages profonds en mer Ionien, données préliminaires. *Rapports et Procès-verbaux des Réunions. Commission International pour l'Exploration Scientifique de la Mer Mediterranée* **22**(4), 103–105.

Chesher, R.H. (1972). The status of knowledge of Panamian echinoids, 1971, with comments on other Echinoderms. *Bulletin of the Biological Society of Washington* **2**, 139–158.

Chindonova, Ju.G. (1981). New data on the systematic position of some deep-sea mysids (Mysidacea, Crustacea) and their distribution in the World Ocean. *In* "Biology of the Pacific Ocean Depths" (N.G. Vinogradova, ed.), pp. 24–33. Proceedings of the XIV Pacific Science Congress, Khabarovsk, August 1979, Section Marine Biology, Issue 1. Vladivostok. (In Russian).

Chindonova, Ju.G. (1993). The distribution of the deep-sea near-bottom pelagic Mysidacea (Crustacea) in the World Ocean and the relationship between high taxa in this group of animals. *Trudy Instituta Okeanologii Rossijskoi Akademiii Nauk* **127**, 147–158. (In Russian).

Clarke, A.H. (1962a). On the composition, zoogeography, origin and age of the deep-sea mollusk fauna. *Deep-Sea Research* **9**(5) 291–306.

Clarke, A.H. (1962b). Annotated list and bibliography of the abyssal marine molluscs of the World. *Bulletein of the National Museum of Canada* **181**, 1–114.

Cutler, E.B. (1973). Sipuncula of the western North Atlantic. *Bulletin of the American Museum of Natural History,* **152**, 107–204.

Dahl, E. (1954). The distribution of deep-sea Crustacea. *In* "On the Distribution and Origin of the Deep-Sea Bottom Fauna", pp. 43–48. *International Union of Biological Sciences, Series B* **16**. Naples.

Detinova, N.N. (1982). Deep-water Maldanidae (Polychaeta) of the Pacific Ocean. 1. The genus *Maldanella*. *Trudy Instituta Okeanologii Akademii Nauk SSSR* **117**, 63–75. (In Russian).

Detinova, N.N. (1993). The polychaetous worms of the Orkney Trench. *Trudy Instituta Okeanologii Rossijskoi Akademii Nauk* **127**, 97–106. (In Russian).

Djakonov, A.M. (1945). The relationship between the Arctic and Pacific sea fauna based on zoogeographical analysis of the echinoderms. *Zhurnal obtschei biologii* **6**(2), 125–155. (In Russian).

Ekman, S. (1935). "Tiergeographie des Meeres". Akademisches Verlag, Leipzig.

Ekman, S. (1953). "Zoogeography of the Sea". Sidgwick & Jackson, London.

Ekman, S. (1954). Betrachtungen über die Fauna der abyssalen Ozeanboden. *In* "On the Distribution and Origin of the Deep-Sea Bottom Fauna", pp. 5–11. *International Union of Biological Sciences, Series B* **16**. Naples.

Filatova, Z.A. (1957). Zoogeographical zonation of northern seas from the occurrence of bivalve molluscs). *Trudy Instituta Okeanologii Akademii Nauk SSSR* **23**, 195–215. (In Russian).

Filatova, Z.A. (1974). On the bivalves of the deep-ocean trenches of the southern part of the Atlantic Ocean. *Trudy Instituta Okeanologii Akademii Nauk SSSR* **98**, 270–276. (In Russian).

Filatova, Z.A. (1976). Structure of the deep-sea genus of bivalve mollusca *Spinula* (Dall, 1908, Malletiidae) and their distribution in the Ocean. *Trudy Instituta Okeanologii Akademii Nauk SSSR* **99**, 219–240. (In Russian with English summary).

Filatova, Z.A. and Schileyko, A.A. (1984). Size, structure and distribution of the deep-sea Bivalvia of the family Ledellidae (Protobranchia). *Trudy Instituta Okeanologii Akademii Nauk SSSR* **119**, 106–144. (In Russian with English summary).

Filatova, Z.A. and Schileyko, A.A. (1985). The composition, morphology and distribution of the ultraabyssal genus *Parayoldiella* (Bivalvia). *Trudy Zoologicheskogo Instituta* **135**, 76–94. (In Russian).

Filatova, Z.A. and Vinogradova, N.G. (1974). Bottom fauna of the South Atlantic deep-sea trenches. *Trudy Instituta Okeanologii Akademii Nauk SSSR* **98**, 141–156. (In Russian).

Filatova, Z.A., Vinogradova, N.G. and Moskalev, L.I. (1975). The mollusc *Neopilina* (Monoplacophora) in the Antarctic. *Okeanologia* **15**(1), 143–145. (In Russian).

Fisher, W.K. (1940). Asteroidea. *Discovery Reports* **20**, 69–306.

France, S.C. (1993). Geographic variation among three isolated populations of the hadal amphipod *Hirondella gigas* (Crustacea: Amphipoda: Lysianassoidea). *Marine Ecology Progress Series* **92**, 277–287.

Gage, J.D. and Tyler, P.A. (1991). "Deep-Sea Biology: a Natural History of Organisms at the Deep-sea Floor". Cambridge University Press, Cambridge.

Galkin, S.V. and Korovchinsky, N.M. (1984). Vertical and geographical distribution of the starfishes of the genus *Freyella* (Brisingidae) with some remarks on their ecology and evolution. *Trudy Instituta Okeanologii Akademii Nauk SSSR* **119**, 164–178. (In Russian with English summary).

Gebruk, A.V. (1990). "Deep-Sea Holothurians of the Family Elpidiidae". Nauka, Moscow. (In Russian).

Gebruk, A.V. (1993). New data on elasipod fauna of the South Atlantic. *Trudy Instituta Okeanologii Rossijskoi Akademii Nauk* **127**, 228–244. (In Russian with English summary).

Gebruk, A.V. (1994). Two main stages in the evolution of the deep-sea fauna of elasipodid holothurians. *In* "Echinoderms through Time" (B. David, A. Guille, J.P. Feral and M. Roux, eds.), pp. 507–514. Balkema, Rotterdam.

George, R.Y. and Menzies, R.J. (1968). Additions to the Mediterranean deep-sea isopod fauna (Vema-14). *Revue Roumaine de Biologie, Serie Zoologie* **13**(6), 367–383.

Gorbunov, G.P. (1946). The bottom fauna of the Novosibirsk Islands shoal and the central part of the Arctic ocean. *Trudy dreifujutschei expedicii na ledokolnom parahode "G. Sedov"* *1937–1940* **3**, 30–138. (In Russian).

Grey, M. (1956). The distribution of fishes found below a depth of 2000 meters. *Fieldiana: Zoology* **36**(2), 75–337.

Gureeva, M.A. (1975). A new *Spirobrachia* (Pogonophora) from the South-Sandwich Trench. *Trudy Instituta Okeanologii Akademii Nauk SSSR* **103**, 307–312. (In Russian).

Guryanova, E.F. (1934). The crustaceans of the Kara Sea and the way of penetration of the Atlantic fauna in the Arctic. *Doklady Akademii Nauk SSSR* **1**(2), 91–96. (In Russian).

Guryanova, E.F. (1936). Zoogeography of Kara Sea. *Izvestija Akademii Nauk SSSR (matematicheskie i estestwennije nauki)* **2**(18), 565–598. (In Russian).

Guryanova, E.F. (1938). On the question of the composition and origin of the abyssal fauna of the Polar basin. *Doklady Akademii Nauk SSSR* **20**(4), 330–336. (In Russian).

Guryanova, E.F. (1946). Individual and age variability of the marine assel *Mesidothea entomon* and its importance in evolution of the genus *Mesidothea* Rich. *Trudy Zoologicheskogo Instituta Akademii Nauk SSSR* **8**(1), 105–144. (In Russian).

Guryanova, E.F. (1948). The Arctic fauna and its exchange of fauna with the neighbouring regions of the World Ocean. *Trudy 2–go vsesojusnogo geographicheskogo sjezda, Moskva, 1947* **2**, 42–44. (In Russian).

Guryanova, E.F. (1951). "Gammaridea of the Seas of the U.S.S.R. and Adjacent Waters." *Opredeliteli po faune SSSR* **41**. Isdatelstvo Akademii Nauk SSSR, Moscow/Leningrad. (In Russian).

Guryanova, E.F. (1957). On zoogeography of the Arctic Basin. *In* "Data of investigations of scientific ice-drifting stations "North Pole-3, North Pole-4" in the years 1954–1956" Vol. **1**, pp. 343–355. Isdatelstvo "Morskoi transport", Leningrad. (In Russian).

Guryanova, E.F. (1970). The particularities of the Arctic fauna and its significance for the interpretation of the history of its formation. *In* "Severnyi Ledovitii okean i ego pobereje v kainozoe", pp. 126–161. Izdatelstvo "Gidrometeoizdat", Leningrad. (In Russian).

Hansen, B. (1967). Taxonomy and zoogeography of the deep-sea holothurians in their evolutionary aspects. *Studies in Tropical Oceanography* **5**, 480–501.

Hansen, B. (1975). Systematics and biology of the deep-sea holothurians, pt 1. Elasipoda. *Galathea Report* **13**, 1–262.

Hartman, O. and Barnard, J.L. (1958). The benthic fauna of the deep basins off Southern California, 1. *Allan Hancock Pacific Expeditions* **22**(1), 1–67.

Hinz, K. and Schmid, M.K. (1995). Studies on benthic communities in the Laptev Sea. *In* "Modern State and Perspectives of the Barents, Kara and Laptev Seas Ecosystems Researches". Abstracts of papers submitted to the International Conference,10–15 October 1995, Murmansk, pp. 92–93. (In Russian).

Ivanov, A.V. (1952). New Pogonophora from the Far Eastern Seas. *Zoologicheskii Zhurnal* **31**(3), 372–391. (In Russian).

Ivanov, D.L. (1996). *Chevroderma hadalis*, a new species of Prochaetodermatidae (Caudofoveata, Aplacophora) from the north-west Pacific. *Ruthenica* **6**(1), 83–84.

Jones, N.S. and Sanders, H.L. (1972). Distribution of Cumacea in the deep Atlantic. *Deep-Sea Research* **19**(10), 737–745.

Kamenskaya, O.E. (1980). Deep-Sea amphipods (Amphipoda, Gammaridea) collected by the drift-ice research unit "North Pole-22" expedition. *In* "Biology of the Central Arctic Basin" (M.E. Vinogradov and I.A. Melnikov, eds.), pp. 241–251. Nauka, Moscow. (In Russian).

Kamenskaya, O.E. (1981a). Ultraabyssal (hadal) amphipods from the trenches of the Pacific Ocean. *In* "Biology of the Pacific Ocean Depths" (N.G. Vinogradova, ed.), pp. 40–43. Proceedings of the XIV Pacific Science Congress, Khabarovsk, August 1979, Section Marine Biology, Issue 1. Vladivostok. (In Russian).

Kamenskaya, O.E. (1981b). The amphipods (Crustacea) from deep-sea trenches in the western part of the Pacific Ocean. *Trudy Instituta Okeanologii Akademii Nauk SSSR* **115**, 94–107. (In Russian).

Keller, N.B. (1990). About the distribution of the madreporarian corals in Antarctic and Subantarctic basins. *Trudy Instituta Okeanologii Akademii Nauk SSSR* **126**, 74–79. (In Russian with English summary).

Keller, N.B., Naumov, D.V. and Pasternak, F.A. (1975). Bottom deep-sea Coelenterata from the Gulf of Mexico and Caribbean Sea. *Trudy Instituta Okeanologii Akademii Nauk SSSR* **100**, 147–159. (In Russian).

Khain, V.E. (1975). Main stages of the geological development of the Mexican–Caribbean region. *Trudy Instituta Okeanologii Akademii Nauk SSSR* **100**, 25–46. (In Russian).

Kirkegaard, J.B. (1954). The zoogeography of the abyssal Polychaeta. *In* "On the Distribution and Origin of the Deep-sea Bottom Fauna", pp. 40–43. *International Union of Biological Sciences, Series B* **16**. Naples.

Kirkegaard, J.B. (1956). Benthic Polychaeta from depths exceeding 6000 meters. *Galathea Report* **2**, 63–78.

Kirkegaard, J.B. (1995). Bathyal and abyssal Polychaeta (errant species). *Galathea Report* **17**, 7–56.

Knudsen, J. (1961). The bathyal and abyssal *Xylophaga* (Pholadidae, Bivalvia). *Galathea Report* **5**, 163–209.

Knudsen, J. (1970). The systematics and biology of abyssal and hadal Bivalvia. *Galathea Report* **11**, 7–236.

Koltun, V.M. (1959). Abyssal fauna of the central Polar Basin. *Doklady Akademii Nauk SSSR* **129**(3), 662–665. (In Russian).

Koltun, V.M. (1964). About the study of the bottom fauna of the Greenland Sea and the central part of the Arctic basin. *Trudy Arcticheskogo i Antarkticheskogo nauchno-issledovatelskogo Instituta* **259**, 13–78. (In Russian).

Kucheruk, N.V. (1975). On the distribution of an abyssal eunicid species (Polychaeta). *Trudy Instituta Okeanologii Akademii Nauk SSSR* **103**, 151–153. (In Russian with English summary).

Kucheruk, N.V. (1978). Deep-water Onuphidae (Polychaeta) from the collections of 16th cruise of r/v "Dmitry Mendeleev" (to the generic diagnosis of family Onuphidae). *Trudy Instituta Okeanologii Akademii Nauk SSSR* **113**, 88–106. (In Russian with English summary).

Kucheruk, N.V. (1979). Zoogeographical zonation of the abyssobenthal. *Byulleten Moskovskogo obshchestva ispytatelei prirody, Otdel biologicheskij biologicheskaja* **5**, 59–66. (In Russian).

Kucheruk, N.V., Keller, N.B., Kudinova-Pasternak, R.K., Levenstein, R.Ja., Pasternak, F.A. and Filatova, Z.A. (1981). New data on the distribution of some groups of bottom invertebrates and the zoogeographical division of the abyssobenthal of the Pacific Ocean. *In* "Biology of the Pacific Ocean depths" (N.G. Vinogradova, ed.), pp. 15–19. Proceedings of the XIV Pacific Science Congress, Khabarovsk, August 1979, Section Marine Biology, Issue 1. Vladivostok. (In Russian).

Kudinova-Pasternak, R.K. (1975). Tanaidacea (Crustacea, Malacostraca) from the Atlantic sector of Antarctic and Subantarctic. *Trudy Instituta Okeanologii Akademii Nauk SSSR* **103**, 194–229. (In Russian).

Kudinova-Pasternak, R.K. (1978). Gigantapseudidae fam. n. (Crustacea, Tanaidacea) and composition of the suborder Monokonophora. *Zoologicheskii Zhurnal* **57**(8), 1150–1161. (In Russian with English summary).

Kudinova-Pasternak, R.K. (1982). Deep-sea Tanaidacea (Crustacea, Malacostrata) from the Mediterranean Sea. *Trudy Instituta Okeanologii Akademii Nauk* **117**, 151–162. (In Russian).

Kudinova-Pasternak, R.K. (1990). Tanaidacea (Crustacea; Malacostraca) of the south eastern part of Atlantic ocean and the region to the North of Mordvinov (Elephant) Island. *Trudy Instituta Okeanologii Akademii Nauk SSSR* **126**, 90–107. (In Russian).

Kudinova-Pasternak, R.K. (1993). Tanaidacea (Crustacea, Malacostrata) collected in the 43rd cruise of the r/v "Dmitry Mendeleev" in the south-western Atlantic and the Weddell Sea. *Trudy Instituta Okeanologii Rossijskoi Akademii Nauk SSSR* **127**, 134–146. (In Russian with English summary).

Kudinova-Pasternak, R.K. and Pasternak, F.A. (1978). Deep-sea Tanaidacea (Crustacea, Malacostraca) collected in the Caribbean Sea and Puerto-Rico Trench during the 16th cruise of r/v "Akademik Kurchatov" and the resemblance between fauna of deep-sea Tanaidacea of the Caribbean region and the Pacific Ocean. *Trudy Instituta Okeanologii Akademii Nauk SSSR* **113**, 178–197. (In Russian with English summary).

Kudinova-Pasternak, R.K. and Pasternak, F.A. (1981). Tanaidacea (Crustacea, Malacostrata) collected by the Soviet Antarctic Expedition during the years 1955–1958 and the correlation of the ranges of the Tanaidacea obtained in the South Ocean and their bathymetrical distribution. *Trudy Instituta Okeanologii Akademii Nauk SSSR* **115**, 108–125. (In Russian with English summary).

Kussakin, O.G. (1971a). "Some particularities of the horizontal and vertical distribution of the isopods in the cold and temperate waters of the World Ocean". Doctoral Thesis, Zoological Institute Russian Academy of Sciences, Leningrad, SSSR. (In Russian).

Kussakin, O.G. (1971b). Additions to the fauna of isopods (Crustacea, Isopoda) of the Kurile–Kamchatka Trench. III. Flabellifera and Valvifera. *Trudy Instituta Okeanologii Akademii Nauk SSSR* **92**, 239–273. (In Russian).

Kussakin, O.G. (1973). Peculiarities of the geographical and vertical distribution of marine isopods and the problem of deep-sea fauna origin. *Marine Biology* **23**, 19–34.

Kussakin, O.G. (1979). "Marine and Brackish Water Isopods, Suborder Flabellifera". Nauka, Leningrad. (In Russian).

Kussakin, O.G. and Vasina, G.S. (1993). Description of *Globarcturus angelikae* gen.et sp.n., the first Antarctic hadal arcturid from the South Sandwich Trench (Crustacea, Isopoda: Arcturidae). *Zoosystematica Rossica* **2**(2), 241–245.

Laubier, L. and Ramos, J. (1973). Paraonidae (Polychètes sédentaires) de Méditerranée. *Bulletin du Muséum National d'Histoire Naturelle, Paris, (Sér. 3) Zoologie* **113**(168), 1097–1148. Paris.

Levenstein, R.Ja. (1975). The polychaetous annelids of the deep-sea trenches of the Atlantic sector of the Antarctic Ocean. *Trudy Instituta Okeanologii Akademii Nauk SSSR* **103**, 119–142. (In Russian).

Levenstein, R.Ja. (1978). Polychaetes of the family Polynoidae (Polychaeta) from the deep-water trenches of the western part of the Pacific. *Trudy Instituta Okeanologii Akademii*

Nauk SSSR **112**, 162–173. (In Russian).

Levenstein, R.Ja. (1982). On the polychaete fauna (fam. Polynoidae) from the Japan Trench. *Trudy Instituta Okeanologii Akademii Nauk SSSR* **117**, 59–62. (In Russian).

Levi, C. (1964). Spongiaries des zones bathyale, abyssale et hadale. *Galathea Report* **7**, 63–112.

Litvinova, N.M. (1975). Ophiuroids of the Caribbean and Gulf of Mexico collected during the 14th cruise of the r/v "Akademik Kurchatov". *Trudy Instituta Okeanologii Akademii Nauk SSSR* **100**, 196–204. (In Russian).

Lus, V.Ja. (1975). New species of mollusc – *Tacita zenkevitchi* (Buccinidae) from the lower-abyssal zone of Peru–Chile trench region with description of the egg capsules and some stages of ontogenesis. *Trudy Instituta Okeanologii Akademii Nauk SSSR* **103**, 162–178. (In Russian).

Lus, V.Ja. (1993). New species of *Tromina* (Neogastropoda, Buccinidae, Tromina) from the lower abyssal zone of Lori and Orkney trenches (Antarctica). *Trudy Instituta Okeanologii Rossijskoi Akademii Nauk SSSR* **127**, 176–197. (In Russian).

Madsen, J.F. (1954). Some general remarks on the distribution of the echinoderm fauna of the deep-sea. "On the Distribution and Origin of the Deep-sea Bottom Fauna", pp. 30–37. *International Union of Biological Sciences, Series B* **16**. Naples.

Madsen, J.F. (1961). On the zoogeography and origin of the abyssal fauna in view of the knowledge of the Porcellanasteridae. *Galathea Report* **4**, 177–218.

Melnikov, I.A. and Tzinovskyi, V.D. (1978). Hydrobiological studies in the Arctic Ocean (May–October, 1977). *Okeanologia* **18**(2), 378–379. (In Russian).

Menzies, R.J. (1959). *Priapulus abyssorum*, new species, the first abyssal priapulid. *Nature* **184**, 1585–1586.

Menzies, R.J. (1962). The isopods of abyssal depths in the Atlantic ocean. *Vema Research Series* **1**, 79–206.

Menzies, R.J. (1963). The abyssal fauna of the floor of the Arctic Ocean. *In* "Proceedings of the Arctic Basin Symposium, 1962", pp. 46–66. Arctic Institute of North America, Washington DC.

Menzies, R.J. (1965). Conditions for the existence of life on the abyssal sea floor. *Oceanography and Marine Biology Annual Review* **3**, 195–210.

Menzies, R.J. and Frankenberg, D. (1967). Systematics and distribution of the bathyal-abyssal genus *Mesosignum* (Crustacea: Isopoda). *Antarctic Research Series, American Geophysical Union* **11** (Biology of the Antarctic Seas III), 113–140.

Menzies, R.J., George, R.Y. and Rowe, G.T. (1973). "Abyssal Environment and Ecology of the World Ocean". Wiley, New York.

Mezhov, B.V. (1980). On the fauna of Isopoda (Crustacea) of the Japanese and Idzu–Bonin Troughs of the Pacific. *Zoologicheskii Zhurnal* **59**(6), 818–829. (In Russian).

Millar, R.H. (1970). Ascidians, including specimens from the deep-sea, collected by the R.V. Vema and now in the American Museum of Natural History. *Zoological Journal of the Linnaean Society* **49**(2), 99–159.

Mironov, A.N. (1974). Pourtalesiid sea urchins of the Antarctic and Subantarctic (Echinoidea: Pourtalesiidae). *Trudy Instituta Okeanologii Akademii Nauk SSSR* **98**, 240–252. (In Russian with English summary).

Mironov, A.N. (1975). Deep-sea urchins (Echinodermata, Echinoidea) collected during the 14th cruise of the r/v "Akadevik Kurchatov". *Trudy Instituta Okeanologii Akademii Nauk SSSR* **100**, 205–214. (In Russian).

Mironov, A.N. (1981). On principles of zonal regionalization of the benthal zone on the faunistic basis. *Zoologicheskii Zhurnal* **60**(8), 1125–1128. (In Russian).

Mironov, A.N. (1982). The role of Antarctic in formation of the deep-sea fauna of the World Ocean. *Okeanologia* **22**, 486–491. (In Russian).

Mironov, A.N. (1983). The particularities of the distribution of sea urchins in the central-oceanic regions. *In* "Sravnitelnaja morphologia, evolutia i raspostranenie sovremennich i vimershih iglokojih. Tezisi dokladov V vsesojuznogo simposiuma po iglokozhim", pp. 47–48. Lvov. (In Russian).

Mironov, A.N. (1987). "Echinoidea of the World Ocean and the formation of recent faunistic complexes". Doctoral thesis. Institut Okeanologii, Moscow. (In Russian).

Mironov, A.N. (1993). Deep-sea Echinoids (Echinodermata: Echinoidea) of the South Atlantic. *Trudy Instituta Okeanologii Rossijskoi Akademii Nauk* **127**, 218–227. (In Russian with English summary).

Mironov, A.N. and Sagaidachny A.Yu. (1984). Morphology and distribution of the recent echinoids of the genus *Echinocyamus* (Echinoidea: Fibulariidae). *Trudy Instituta Okeanologii Akademii Nauk SSSR,* **119**, 179–204. (In Russian with English summary).

Monniot, C. and Monniot, F. (1973). Ascidies abyssales recoltées en course de la campagne oceanographique Biacores par le "Jean-Charcot". *Bulletin du Muséum National d'Histoire Naturelle, Paris (Sér. 3) Zoologie* **93**, 389–475.

Monniot, C. and Monniot, F. (1974). Ascidies abyssales de l'Atlantique recoltées par le "Jean Charcot" (Campagnes Noratlante, Walda, Polygas A.). *Bulletin du Muséum National d'Histoire Naturelle, Paris (Ser. 3) Zoologie* **154**, 721–786.

Monniot, C. and Monniot, F. (1975). Abyssal tunicates: an ecological paradox. *Annales de l'Institute Océanographique* **51**(1), 99–120.

Monniot, C. and Monniot, F. (1978). Recent work on the deep-sea Tunicates. *Oceanography and Marine Biology Annual Review* **16**, 181–228.

Monniot, C and Monniot, F. (1985a). Tuniciers profonds de l'ocean Indien: Campagne SAFARI du "Marion Dufresne". *Bulletin du Muséum National d'Histoire Naturelle, Paris. (Ser. 4) Section A* **7**, 279–308.

Monniot, C and Monniot, F. (1985b). Nouvelles recoltés de Tuniciers benthiques profonds dans l'Océan Atlantique. *Bulletin du Muséum National d'Histoire Naturelle, Paris (Ser. 4) Section A,* **7**, 5–37.

Monniot, F. (1979). Faunal affinities among abyssal Atlantic basins. *Sarsia* **64**, 93–95.

Moskalev, L.I. (1973). Pacific Bathysciadiidae (Gastropoda) and related forms. *Zoologicheskii Zhurnal* **52**(9), 1297–1303. (In Russian).

Murina, V.V. (1961). On the geographical distribution of the abyssal sipunculid *Phascolion lutense* Selenka. *Okeanologia* **1**(1), 140–142. (In Russia).

Murina, V.V. (1964). On the problem of bipolar distribution of Priapulidae. *Okeanologia* **4**(5), 873–875. (In Russian).

Murina, V.V. (1971). Geographical distribution of marine worms of the phylum Sipuncula in the World Ocean. *Zoologicheskii Zhurnal* **50**(2), 184–192. (In Russian).

Murina, V.V. (1975). On the taxonomic rank and geographical distribution of abyssal Priapulids. *Trudy Instituta Okeanologii Akademii Nauk SSSR* **103**, 154–161. (In Russian with English summary).

Murina, V.V. (1978). New and rare echiurids of the family Bonellidae. *Trudy Instituta Okeanologii Akademii Nauk SSSR* **113**, 107–119. (In Russian).

Murina, V.V. (1984). The composition and distribution of the echiurans of the Mediterranean. *Trudy Instituta Okeanologii Akademii Nauk SSSR* **119**, 82–98. (In Russian).

Murina, V.V. (1993). New data of the fauna of Sipuncula, Echiura and Priapulida from Atlantic Ocean. *Trudy Instituta Okeanologii Rossijskoi Akademii Nauk* **127**, 107–121. (In Russian).

Murina, V.V. and Starobogatov, Ja.I. (1961). Classification and zoogeography of priapulids. *Trudy Instituta Okeanologii Akademii Nauk SSSR* **46**, 179–200. (In Russian).

Murray, J. and Hjort, J. (1912). "The Depths of the Ocean". Macmillan, London.
Nielsen, J.G. (1966). Synopsis of the Ipnopidae (Pisces, Iniomi). *Galathea Report* **8**, 49–75.
Nybelin, O. (1954). Sur la distribution géographique et bathymetrique des brotulides, trouvés au-dessus de 2000 m de profondeur. *In* "On the Distribution and Origin of the Deep-Sea Bottom Fauna", pp. 65–71. *International Union of Biological Sciences, Series B* **16**. Naples.
Ortman, A. (1896). "Grundzuge der marinen Tiergeographie". Gustav Fischer, Jena.
Parin, N.V., Bekker, V.E., Borodulina, O.D. and Chuvasov, V.M. (1973). Deep-sea pelagic fishes of the south-eastern Pacific Ocean and adjacent waters. *Trudy Instituta Okeanologii Akademii Nauk SSSR* **94**, 71–159. (In Russian with English summary).
Parker, R.H. (1963). Zoogeography and ecology of some macroinvertebrates, particularly Mollusks, in the Gulf of California and the continental slope of Mexico. *Videnskabelige Meddelelser fra Dansk Naturhistorisk Forening i København* **126**, 1– 178.
Pasternak, F.A. (1961). Some new data on the species composition and the distribution of deep-sea Pennatularia, genus *Kophobelemnon* in Northern Pacific. *Trudy Instituta Okeanologii Akademii Nauk SSSR* **45**, 240–258. (In Russian).
Pasternak, F.A. (1964). The deep-sea Pennatularians and Antipatharians obtained by R/V "Vityaz" in the Indian Ocean and the resemblance between the faunas of the Pennatularians of the Indian Ocean and Pacific. *Trudy Instituta Okeanologii Akademii Nauk SSSR* **69**, 183–215. (In Russian).
Pasternak, F.A. (1973). The deep-water sea pens (Octocorallia, Pennatularia) of the Aleutian Trench and Gulf of Alaska. *Trudy Instituta Okeanologii Akademii Nauk SSSR* **91**, 108–127. (In Russian).
Pasternak, F.A. (1975a). New data on the specific composition and distribution of the deep-sea Pennatularians (Octocorallia, Pennatularia) of the Peru–Chile region and South Atlantic. *Trudy Instituta Okeanologii Akademii Nauk SSSR* **103**, 101–118. (In Russian with English summary).
Pasternak, F.A. (1975b). Deep-sea pennatularians of the genus *Umbellula* from the Caribbean Sea and Puerto-Rican Trench. *Trudy Instituta Okeanologii Akademii Nauk SSSR* **100**, 160–173. (In Russian with English summary).
Pasternak, F.A. (1977). Antipatharia. *Galathea Report* **14**, 157–164.
Pasternak, F.A. (1982). Composition, origin and peculiarities of distribution of the Mediterranean deep-sea Isopod fauna. *Trudy Instituta Okeanologii Akademii Nauk SSSR* **117**, 163–177. (In Russian with English summary).
Paul, A.Z. and Menzies, R.J. (1974). Benthic ecology of the high Arctic deep-sea. *Marine Biology* **27**(3), 251–262.
Pérès, J.M. (1967). The Mediterranean benthos. *Oceanography and Marine Biology. Annual Review* **5**, 449–533.
Petryashev, V.V., Sirenko, B.I., Rachor, E. and Hinz, K. (1994). Distribution of macrobenthos in the Laptev Sea from materials of the expeditions of r/v "Ivan Kireev" and ice-breaker r/v "Polarstern" in 1993. "Scientific Results of the Expedition LAPEX-93", pp. 277–288. St Petersburg. (In Russian).
Pettibone, M.H. (1976). Revision of the genus *Macellicephala* McIntosh and the subfamily Macellicephalinae Hartman–Schroder (Polychaeta: Polynoidae). *Smithsonian Contributions to Zoology* **229**, 1–71.
Preston, F.M. (1962a). The canonical distribution of commoness and rarity. Part I. *Ecology* **43**, 185–215.
Preston, F.M. (1962b). The canonical distribution of commonness and rarity. Part II. *Ecology* **43**, 410–432.
Rass, T.S. (1959). Deep-sea fishes. *In* "Itogi Nauki. Dostizheniya okeanologii" (L.A.

Zenkevitch, ed.), 1, pp. 285–315. Isdatelstvo Akademii Nauk SSSR, Moscow. (In Russian).

Rass, T.S. (1974). Fishes of the greatest depths of the ocean. *Doklady Akademii Nauk SSSR* **217**(1), 209–212. (In Russian).

Rass, T.S. (1975). Scientific problems of the American Mediterranean Sea study and some general outlines of the recent new data. *Trudy Instituta Okeanologii Akademii Nauk SSSR* **100**, 7–24. (In Russian with English summary).

Rass, T.S., Grigorash, V.A., Spanovskaya, V.D. and Shcherbachev, Yu.N. (1975). Deep-sea bottom fishes caught during 14th cruise of the R/V "Akademik Kurchatov". *Trudy Instituta Okeanologii Akademii Nauk SSSR* **100**, 337–347. (In Russian).

Reyss, D. (1972). Résultats scientifiques de la campagne Polymede du N.O. "Jean Charcot" en Méditerranée occidentale, mai-juin-juillet 1970. Cumacés. *Crustaceana* (Suppl. 3), 362–377.

Sanders, H.L. (1965). Time, latitude and structure of marine benthic communities. *Revista de la Academia Brasileira de Ciencias* **37** (Suppl.), 83–86.

Semenov, V.N. (1981). The ring structure in the biogcography of the sea. *In* "IV sjesd Vsesojuznogo Gidrobiologicheskogo obtschestva, 1"", pp. 38–39. Izdatelstvo "Naukova Dumka". Kiev. (In Russian).

Shreider, M.Yu. (1994). The benthic Lepadomorpha (Cirripedia, Thoracica) from the western part of the Indian ocean. *Trudy Instituta Okeanologii Rossijskoi Akademii Nauk* **129**, 156–164. (In Russian).

Sirenko, B.I., Petryashev, V.V., Rachor, E. and Hinz, K. (1995). Bottom biocoenoses of the Laptev Sea and adjacent areas. *Berichte zur Polarforschung* **176**, 211–221.

Sokolova, M.N. (1969). The distribution of deep-sea benthic invertebrates in relation to their methods and conditions of feeding. *In* "The Pacific Ocean. Biology of the Pacific Ocean 2. The Deep-sea Bottom Fauna, Pleuston" (L.A. Zenkevitch, ed.), pp. 182–201, Nauka, Moscow. (In Russian).

Sokolova, M.N. (1977). Trophic structure of the deep-sea bottom fauna. *In* "Okeanologia. Biologia Okeana 2" (M.E. Vinogradov, ed.), pp. 176–183, Nauka, Moscow. (In Russian).

Sokolova, M.N. (1978). The trophic classification of the types of distribution of the deep-sea macrobenthos. *Doklady Akademii Nauk SSSR* **241**(2), 471–474. (In Russian).

Sokolova, M.N. (1979). On the global spreading of trophic regions over the ocean bed. *Doklady Akademii Nauk SSSR* **246**(1), 250–252. (In Russian).

Sokolova, M.N. (1981a). Trophic large-scale regions of the World Ocean floor and characters of their population. *In* "Biology of the Pacific Ocean depths" (N.G. Vinogradova, ed.) pp. 8–14. Proceedings of the XIV Pacific Science Congress, Khabarovsk, August 1979, Section Marine Biology, Series 1. Vladivostok. (In Russian).

Sokolova, M.N. (1981b). On characteristic features of the deep-sea benthic eutrophic regions of the World Ocean. *Trudy Instituta Okeanologii Akademii Nauk SSSR* **115**, 5–13. (In Russian with English summary).

Sokolova, M.N. (1984). About species composition of the population of the deep-sea trophic regions. *Trudy Instituta Okeanologii Akademii Nauk SSSR* **119**, 33–46. (In Russian).

Sokolova, M.N. (1986). "Feeding and trophic structure in the deep-sea macrobenthos". Nauka, Moscow. (In Russian).

Southward, E.C. (1979). Horizontal and vertical distribution of Pogonophora in the Atlantic Ocean. *Sarsia* **64**, 51–55.

Southward, E.C. and Southward, A.J. (1967). The distribution of Pogonophora in the Atlantic Ocean. *Symposia of the Zoological Society of London* **19**, 145–158.

Springer, V.G. (1982). Pacific plate biogeography with special reference to shorefishes. *Smithsonian Contributions to Zoology* **367**, 183 pp.

Strömberg, J.O. and Svavarsson, J. (1991). Some aspects of the distribution of deep-sea asellote families and genera with special reference to Arctic basins. *The 6th Deep-Sea Biology Symposium, Copenhagen*, Abstracts 62–63.

Sysoev, A.V. and Kantor, Ju.I. (1987). Deep-sea gastropods of the genus *Aforia* (Turridae) of the Pacific: species composition, systematics and functional morphology of the digestive system. *Veliger* **30**, 105–126.

Tabachnick, K.R. (1994). Distribution of recent Hexactinellida. *In* "Sponges in Time and Space. Biology, Chemistry, Paleontology" (R.W.M. Van Soest, T.M.G. Van Kempen, J.-C. Braekman, eds), pp. 225–232. Balkema, Rotterdam.

Turpaeva, E.P. (1974). The pycnogonids of the Scotia Sea and the surrounding waters. *Trudy Instituta Okeanologii Akademii Nauk SSSR* **98**, 277–305. (In Russian).

Turpaeva, E.P. (1975). On some deep-water Pantopods (Pycnogonida) collected in north-western and south-eastern Pacific. *Trudy Instituta Okeanologii Akademii Nauk SSSR* **103**, 230–246. (In Russian).

Turpaeva, E.P. (1993). Pycnogonida collected during "Dmitry Mendeleev" 43rd cruise to the South Atlantic basins. *Trudy Instituta Okeanologii Rossijskoi Akademii Nauk SSSR* **127**, 159–175. (In Russian).

Uschakov, P.V. (1975). Deep-water Phyllodocidae (Polychaeta) from the South-Sandwich Trench collected by the R/V "Akademik Kurchatov" in 1971. *Trudy Instituta Okeanologii Akademii Nauk SSSR* **103**, 143–150. (In Russian).

Vervoort, W. (1966). Bathyal and abyssal hydroids. *Galathea Report* **8**, 82–160.

Vinogradov, M.E. (1959). About quantitative distribution of deep-sea plankkton in the Western Pacific Ocean and its relation to deep-water circulation. *Doklady Akademii Nauk SSSR* **127**, 877–880. (In Russian).

Vinogradov, M.E. (1962). Quantitative distribution of deep-sea plankton in the western Pacific and its relation to deep-water circulation. *Deep-Sea Research* **8**, 251–258.

Vinogradov, M.E. (1968). "Vertical distribution of the oceanic zooplankton". Nauka, Moscow. (In Russian; English translation (1970)) Israel Program for Scientific Translation, Jerusalem.

Vinogradov, M.E. and Melnikov, I.A. (1980). Investigations of the ecosystem of the central Arctic basin. *In* "Biology of the Central Arctic Basin" (M.E. Vinogradov and I.A. Melnikov, eds), pp. 5–14. Nauka, Moscow. (In Russian).

Vinogradov, M.E., Sagalevitch, A.M. and Khetogurov, S.V. (eds.) (1996). "Oceanographic Research and Underwater Technical Operations on the Site of Nuclear Submarine "Komsomolets" Wreck. Nauka, Moscow, Russia. (In Russian with English abstract).

Vinogradov, M.E. and Vinogradov, G.M. (1993). Notes about pelagic and benthopelagic gammarids in the Orkney Trench. *Trudy Instituta Okeanologii Akademii Nauk SSSR* **127**, 129–133. (In Russian with English summary).

Vinogradov, M.E. and Vinogradov, G.M. (1996). Finding of hadal Pacific gammarid *Vitjaziana gurjanovae* (Crustacea, Amphipoda) in the Atlantic Ocean and a problem of endemism in abyssopelagic animals. *Zoologicheskii Zhurnal* **75**(1), 45–51. (In Russian).

Vinogradova, N.G. (1956a). Zoogeographical subdivision of the abyssal of the World Ocean. *Doklady Akademii Nauk SSSR* **111**, 195–198. (In Russian).

Vinogradova, N.G. (1956b). On the zoogeography of the abyssal fauna of the Indian Ocean. *Doklady Akademii Nauk SSSR* **111**, 459–461. (In Russian).

Vinogradova, N.G. (1956c). Some regularities of the vertical distribution of the abyssal bottom fauna in the World Ocean. *Doklady Akademii Nauk SSSR* **110**(4), 684–687. (In Russian).

Vinogradova, N.G. (1957). Peculiarities in the distribution of the deep-water bottom fauna of the ocean. *Priroda* 6, 93–96. (In Russian).

Vinogradova, N.G. (1958a). The vertical distribution of the deep-sea bottom fauna in the ocean. *Trudy Instituta Okeanologii Akademii Nauk SSSR* 27, 86–122. (In Russian).

Vinogradova, N.G. (1958b). On the question of the geographical distribution of the deep-sea bottom fauna of the Antarctic. *Byulleten Sovestskoi Antarcticheskoi expedicii* 3, 45–46. (In Russian).

Vinogradova, N.G. (1959a). The zoogeography of the abyssal zone of the ocean (bottom fauna). *In* "Itogi Nauki. Dostizheniya Okeanologii", Vol. 1 (L.A. Zenkevitch, ed.) pp. 148–165. Isdatelstvo Akademii Nauk SSSR, Moscow. (In Russian).

Vinogradova, N.G. (1959b). The zoogeographical distribution of the deep-water bottom fauna in the abyssal zone of the ocean. *Deep-Sea Research* 5, 205–208.

Vinogradova, N.G. (1960). On the problem of geographic distribution of the deep-water bottom fauna in the Antarctic waters. *Okeanologicheskie issledovaniya* 2, 108–111. (In Russian).

Vinogradova, N.G. (1962). Some problems of the study of deep-sea bottom fauna. *Journal of Oceanographical Society of Japan*. 20th Anniversary Volume, 724–741.

Vinogradova, N.G. (1969a). The vertical distribution of the deep-sea bottom fauna. *In* "Pacific Ocean, Biology of the Pacific Ocean, 2. The Deep-sea Bottom Fauna, Pleuston" (L.A. Zenkevitch, ed.), pp. 129–153. Nauka, Moscow. (In Russian).

Vinogradova, N.G. (1969b). The geographical distribution of the deep-sea bottom fauna. *In* "Pacific Ocean. Biology of the Pacific Ocean, 2. The deep-sea bottom fauna, Pleuston" (L.A. Zenkevitch, ed.), pp. 154–181. Nauka, Moscow. (In Russian).

Vinogradova, N.G. (1969c). On the finding of a new aberrant ascidian in the ultra-abyssal of the Kurile–Kamchatka Trench. *Byulleten Moskovskogo obtschestva Ispitatelly Prirody, Otdel Biologicheskii* 74(3), 27–43. (In Russian).

Vinogradova, N.G. (1975). On the discovery of two new species of an aberrant deep-water Ascidiacean genus *Situla* in the South Sandwich Trench. *Trudy Instituta Okeanologii Akademii Nauk SSSR* 103, 289–306. (In Russian).

Vinogradova, N.G. (1977a). Abyssal and ultraabyssal bottom fauna. *In* "Oceanology. Biology of the Ocean, 1" (M.E. Vinogradov, ed.), pp. 281–298. Nauka, Moscow. (In Russian).

Vinogradova, N.G. (1977b). The fauna of the shelf, slope and abyssal. *In* "Oceanology. Biology of the Ocean, 1" (M.E. Vinogradov, ed.), pp. 178–197. Nauka, Moscow. (In Russian).

Vinogradova, N.G. (1979). The geographical distribution of the abyssal and hadal (ultraabyssal) fauna in relation to the vertical zonation of the ocean. *Sarsia* 64, 41–50.

Vinogradova, N.G. (1996). Hadal bottom fauna of the Antarctic trenches. *Deep-Sea Newsletter* 24, 9–12.

Vinogradova, N.G., Birstein, Ja.A. and Vinogradov, M.E. (1959). Vertical zonation in the distribution of the deep-sea fauna. *In* "Itogi Nauki. Dostizhenia Okeanologii, 1" (L.A. Zenkevitch, ed.), pp. 166–187. Isdatelstvo Akademii Nauk SSSR, Moscow. (In Russian).

Vinogradova, N.G., Kudinova-Pasternak, R.K., Moskalev, L.I., Muromtseva, T.L. and Fedikov, N.F. (1974). Some regularities of the quantitative distribution and trophic structure of the bottom fauna of the Scotia Sea and the deep-sea trenches of the Atlantic sector of the Antarctic. *Trudy Instituta Okeanologii Akademii Nauk SSSR* 98, 157–182. (In Russian with English summary).

Vinogradova, N.G., Zezina, O.N. and Levenstein, R.Ja. (1978). Bottom fauna of deep-sea trenches of the Macquarie complex. *Trudy Instituta Okeanologii Akademii Nauk SSSR*

112, 174–192. (In Russian with English summary).

Vinogradova, N.G., Zezina, O.N., Levenstein R.Ja., Pasternak, F.A. and Sokolova, M.N. (1982). Studies of deep-water benthos of Mediterranean Sea. *Trudy Instituta Okeanologii Akademii Nauk SSSR* **117**, 135–146. (In Russian with English summary).

Vinogradova, N.G., Belyaev, G.M., Gebruk, A.V., Zhivago, A.V., Kamenskaya, O.E., Levitan, M.A. and Romanov, V.N. (1993). Investigations of Orkney Trench in the 43rd cruise of r/v "Dmitry Mendeleev". Geomorphology, sediments and benthos. *Trudy Instituta Okeanologii Rossijskoi Akademii Nauk* **127**, 9–32. (In Russian with English summary).

Wolff, T. (1960). The hadal community, an introduction. *Deep-Sea Research* **6**, 95–124.

Wolff, T. (1962). The systematics and biology of bathyal and abyssal Isopoda Asellota. *Galathea Report* **6**, 1–320.

Wolff, T. (1970). The concept of the hadal or ultra-abyssal fauna. *Deep-Sea Research* **17**(6), 983–1003.

Wolff, T. (1975). Deep-sea Isopoda from the Caribbean Sea and the Puerto-Rico Trench. *Trudy Instituta Okeanologii Akademii Nauk SSSR* **100**, 215–232. (In Russian).

Zarenkov, N.A. (1965). "The taxonomy and zoogeography of the decapods of the Antarctic and south temperate regions". Candidat thesis, Moscow State University, Moscow.

Zenkevitch, L.A. (1947). "Fauna and Biological Productivity of the Sea, 2". Isdatelstvo "Sovetskaja Nauka", Moscow. (In Russian).

Zenkevitch, L.A. (1956). Recent oceanographic research in the northwestern Pacific. *Izvestiya Akademii Nauk SSSR, seriya geograficheskaya* **4**, 26–37. (In Russian).

Zenkevitch, L.A. and Birstein, Ja.A. (1955). Study of the deep-sea fauna and connected questions. *Vestnik Moskovskogo Gosudarstvennogo Universiteta. Seria 16*, Biologia **4/5**, pp. 231–242. (In Russian).

Zenkevitch, L.A. and Murina, V.V. (1976). Deep-Sea Echiurida of the Pacific Ocean. *Trudy Instituta Okeanologii Akademii Nauk SSSR* **99**, 102–114. (In Russian).

Zevina, G.B. (1975a). Cirripedia Thoracica of the American Mediterranean. *Trudy Instituta Okeanologii Akademii Nauk SSSR* **100**, 233–258. (In Russian with English summary).

Zevina, G.B. (1975b). Cirripedia Thoracica collected by R/V "Akademik Kurchatov" in the Atlantic sector of the Antarctic. *Trudy Instituta Okeanologii Akademii Nauk SSSR* **103**, 183–193. (In Russian).

Zevina, G.B. (1976). Abyssal species of barnacles (Cirripedia Thoracica) of the North Atlantic. *Zoologicheskii Zhurnal* **55**(8), 1149–1157. (In Russian).

Zevina, G.B. (1990). Deep-sea Cirripedia Thoracica of the South Atlantic. *Trudy Instituta Okeanologii Akademii Nauk SSSR* **126**, 80–89. (In Russian with English summary).

Zevina, G.B. (1993). Abyssal species of Scalpellidae (Cirripedia, Thoracica) in the Atlantic region of the Antarctic. *Trudy Instituta Okeanologii Akademii Nauk SSSR* **127**, 122–128. (In Russian with English summary).

Zezina, O.N. (1965). On the distribution of the deep-sea brachiopod *Pelagodiscus atlanticus* (King). *Okeanologia* **5**(2), 354–358. (In Russian).

Zezina, O.N. (1975). Recent Caribbean deep-sea brachiopod fauna, the sources and the conditions of its formation. *Trudy Instituta Okeanologii Akademii Nauk SSSR* **100**, 188–195. (In Russian with English summary).

Zezina, O.N. (1980). About the deep-sea finding of the brachiopoda in the Arctic basin. *In* "Biology of the Central Arctic Basin" (M.E. Vinogradov and I.A. Melnikov, eds), pp. 240–241. Nauka, Moscow. (In Russian).

Zezina, O.N. (1985). "The Recent brachiopods and Problems of the Bathyal Zone of the World Ocean". Nauka, Moscow. (In Russian).

Zezina, O.N. (1990). Rare and new brachiopods from the Southern Atlantic. *Trudy Instituta*

Okeanologii Akademii Nauk SSSR **126**, 127–131. (In Russian with English summary).

Zezina, O.N. (1993). New findings of recent deep-sea brachiopods from the Weddell Sea and the nearby Orkney Trench. *Trudy Instituta Okeanologii Rossijskoi Akademii Nauk* **127**, 198–200. (In Russian with English summary).

Zhirkov, I.A. (1980). About the abyssal polychaete fauna in the Canadian basin. *In* "Biology of the Central Arctic Basin" (M.E. Vinogradov and I.A. Melnikov, eds), pp. 229–236. Nauka, Moscow. (In Russian).

Zhirkov, I.A. (1982). About the abyssal polychaete fauna in the Norwegian Sea. *Trudy Instituta Okeanologii Akademii Nauk SSSR* **117**, 128–134. (In Russian with English summary).

Zhivago, A.V. and Lisitzin, A.P. (1958). Relief and bottom sediments of the Southern Ocean. *Informacionnii Byulleten Sovetskoi Antarkticheskoi Expedicii* **3**, 21–22. (In Russian).

Biogeography of the Bathyal Zone

O.N. Zezina

*P.P. Shirshov Institute of Oceanology, Russian Academy of Sciences,
Nakhimovsky Prospekt 36, Moscow 117851, Russia*

ABSTRACT

The bathyal zone lies along the slopes of continents and on seamounts and underwater rises. It extends from the edge of the shelf to the beginning of the abyss and is a substantial part of the ocean, being larger than the shallow shelf zone, including the sublittoral. Some taxa of benthic animals attain their optimal number of species and abundance in the bathyal zone.

The distribution of the bathyal fauna is described on the basis of groups of species with comparable geographical range, termed geographic faunistic elements or types of range. The Brachiopoda, which have been thoroughly studied from a large database of samples and records, are used to establish clear biogeographical patterns in the bathyal zone of the world ocean.

There are depth-related changes within the limits of the bathyal zone: the number of species; the number of geographic faunistic elements; and the number of latitudinal (climatic) faunistic belts diminish with increasing depth. Correspondingly, there is a reduction in number of faunistic provinces. The

ADVANCES IN MARINE BIOLOGY VOL. 32
ISBN 0-12-026132-4

simplification in biogeographic structure of bottom fauna down the slopes is in accordance with the simplification in the structure of the water masses that are in contact with the bottom along the slopes. Food supply is also an important factor related to depth distribution of the macrobenthic animals.

The basic biogeographical divisions of the bathyal zone become asymmetric under the influence of the unequal distribution of land and water masses on the globe, and in relation to oceanic gyres that cause differences in productivity on the eastern and western sides of the oceans. This inequality results in faunistic differences especially in the number of species in different taxa.

The bathyal zone may have acted as a reserve of species for recolonization of the shelves and the abyss between periods of global changes in climate. Partly related to this, the bathyal zone contains many relict species, some of which are the most primitive extant members of their groups. The function of the bathyal zone as a reserve of species is challenged by the consequences of commercial exploitation of the non-sustainable fish and shellfish populations on the upper part of the continental slope and on seamounts.

1. INTRODUCTION

The bathyal zone is situated between the outer edge of the continental shelf and the beginning of the abyssal zone at the base of the continental slope (Table 1). Taking into account all oceans and seas, and including depths from 200 to 3000 m (Atlas, 1980), the bathyal zone occupies 17.8% of the World Ocean area, and this is 2.4 times larger than all the continental shelves, which occupy 7.3% of the World Ocean area. If we omit the smaller seas and confine ourselves to the major ocean basins these figures are respectively 11.9%, 4.2 times and 2.6%.

For a long time the bathyal zone was less studied than the other vertical zones, less so even than the abyssal and ultra-abyssal zones which attracted considerable scientific attention. But after 1958, during preparations for the International Agreement on 200-mile National Economic Zones, and after the signing of the treaty, interest in the bathyal zone increased. This increased interest was in great part a reflection of the productive potential of the upper slope, where conditions are to some extent similar to those on the shelves, and trophic resources are rich enough to permit commercial exploitation of fish and shellfish. Many systematic groups (e.g. bivalves, gastropods, stalked crinoids, brachiopods and fishes) attain their maximum species number in the bathyal zone. In terms of biomass of bottom fauna (Lukjanova, 1974) the bathyal zone takes second place (32% of total World Ocean bottom biomass) after the shelf zone (58% of total World Ocean bottom biomass).

Zenkevich (1968, 1973) encouraged Russian scientists to study the bathyal depths and posed numerous questions on the subject under the general description of "The bathyal zone problem". Later The Bottom Fauna Laboratory of the P.P. Shirshov Institute of Oceanology, which inherited the pupils and followed the

Table 1 Vertical distribution of the major benthic biological zones as they relate to macro-relief units in the ocean.

Depths, m	Macro-relief units	Biological zones
0–200	Shelves	Sublittoral
200–3000	Slopes of continents, seamounts and underwater rises	Bathyal
3000–6000	Floor of the ocean basins	Abyssal
>6000	Deep-sea trenches	Ultra-abyssal

scientific ideas of Zenkevitch, prepared four special volumes on the bathyal zone: Kuznetsov and Mironov (1981); Kusnetsov (1985); Mironov and Rudjakov (1990); and Kuznetsov and Mironov (1994).

Some aspects of the bathyal zone problem, especially biogeography, were analysed during studies of the vertical and geographic distribution of separate systematic groups such as: brachiopods (Zezina, 1976a, 1985a); madreporarians (Cairns and Keller, 1993); sponges (Tabachnick, 1994); and sea urchins (Mironov, 1994). These studies were based on collections gathered by trawls and grabs over many years during various Russian and other expeditions around the world.

Considerable progress in bathyal zone biology was also made as a result of more localized expeditions with epibenthic trawls and submersibles on special transects. For example, down the slope off the American east coast (Sanders *et al.*, 1965; Day and Pearcy, 1968; Griggs *et al.*, 1969; Cutler, 1975; Grassle *et al.*, 1975; Haedrich *et al.*, 1975, 1980; Jumars, 1975; Zezina, 1975; Gardiner and Haedrich, 1978; Rice *et al.*, 1979, 1982; Rowe and Haedrich, 1979; Rowe *et al.*, 1982; Carney *et al.*, 1983; Messing, 1985; Messing *et al.*, 1990); and in the northeast Atlantic Ocean (Gage, 1977; Laubier and Sibuet, 1977a,b; Conan *et al.*, 1981; Rex, 1983; Laubier and Monniot, 1985; Pfannkuche, 1985; Weston, 1985; Lampitt *et al.*, 1986, 1995; Hecker, 1990). Large collections, supported by photo- and video-materials, were also made along bathyal transects in the West Pacific (Levi and Levi, 1982, 1983a,b, 1988; Ohta, 1983; Colin *et al.*, 1986; Lambert and Roux, 1991; Richer de Forges, 1992) and on the northern slope of the Eurasian continent (Koltun, 1964; Golikov, 1990; Sirenko and Piepenburg, 1994; Hinz and Schmidt, 1995; Pogrebov *et al.*, 1995; Sirenko and Petryashev, 1995; Sirenko *et al.*, 1995; Sokolova *et al.*, 1996; Vinogradov *et al.*, 1996). Trawl and grab transects were also made in the bathyal zone of the southeast Pacific (Romanova, 1972; Romanova and Karpinsky, 1979, 1983; Karpinsky, 1985, 1996; Kucheruk and Savilova, 1985; Kucheruk, 1995) and in the eastern Atlantic (Thiel, 1978, 1981, 1982; Vinogradova *et al.*, 1990), in regions where upwelling disturbs the normal structure of bathyal benthic assemblages. Some additional new material was obtained recently from the Antarctic slope (Gutt, 1988; Gutt *et al.*, 1991; Zezina, 1993; Brey *et al.*, 1995).

All these materials helped to build up a picture of bathyal zone biogeography, which began to appear more complicated than suggested by the earlier studies.

2. THE TASK AND THE METHOD

The study of the biogeography of the bathyal zone is a relatively recent development, as already noted. The first objective of marine biogeography was the shallow-water regions, especially the littoral (intertidal) and the shelf zones, which were more accessible to investigation. When studies began on the abyssal fauna, it became evident that the shallow-water fauna and the deep-sea fauna had very different patterns of geographical distribution. In order to understand the nature of the causes of the change in biogeography of the benthos between the continental shelves and the ocean floor it was necessary to know the fauna of the continental slope and choose a comparable analytical method for classifying the vertical zonation of different groups of animals.

The most logical analytical procedure is that proposed and used in biogeographical studies of plankton and fish (Beklemishev, 1967, 1969, 1982; Parin, 1968; Beklemishev *et al.*, 1972, 1977). This method looks for natural faunistic boundaries (frontiers), compares them with areas of changing environmental factors, and seeks a causative relationship. The first step in this method is to define "types of ranges" in order to combine separate species with similar geographical ranges into groups which are called "geographic faunistic elements" (Beklemishev, 1967). Such a method of analysis is based on separate species only, without reference to the systematic position of the species or their genetic and cladistic relationships. The species-level approach allows detection of the recent (modern, contemporary) causes of distribution patterns and enables us to distinguish between actual and historical causes.

It is important to bear in mind (Beklemishev and Zezina, 1972; Zezina, 1982) that all taxa higher than species can show historical relationships between faunas but do not necessarily reflect the strength of contemporary biogeographical barriers. To illustrate this position we can compare the generic composition of the recent North Atlantic fauna with that of the Middle Atlantic. Many benthic groups and all planktonic groups have no endemics above species level (Beklemishev and Zezina, 1972). This means that the recent biogeographical boundary is real, but suggests the northern fauna is too young for endemics of higher taxa to have evolved. Ekman (1935, 1953) and Briggs (1974) preferred the species level in their biogeographical divisions without explaining the principles involved. Guryanova (1964, 1972) followed the "generic" way in her investigations of historical relations between local recent shelf faunas.

Some general principles of recent historical biogeogeography were reviewed at a special scientific meeting (Babin and Roux, 1982) where many different systematic groups of animals were examined through time. See especially Roux (1982) and Valentine and Jablonski (1982).

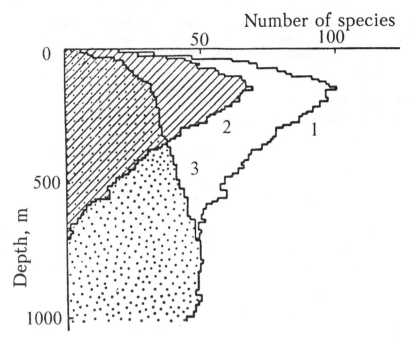

Figure 1 Change in number of recent brachiopod species in relation to depth. 1, Number of species at each 10-m interval down the slope; 2, a similar treatment, but for species that live at depths above 700 m; 3, the same, for species that can occur in depths over 700 m. (After Zezina, 1971, Figure 11, in part).

The brachiopods are usually regarded as a very conservative group that has persisted relatively unchanged through long geological periods, and they have proved very useful for studying the biogeography of the bathyal zone (Figure 1). The contemporary distribution of brachiopods shows (Zezina, 1970, 1976a) that the majority of species live on the outer edge of the shelves and on the slopes of continents, islands, seamounts and on underwater ridges. A large collection of recent brachiopods was gathered during expeditions of the research vessels "Vityaz", "Akademik Kurchatov", "Dmitri Mendeleev" and "Akademik Mstislav Keldysh". With the addition of examples collected by Fisheries Institutes' expeditions (VNIRO, PINRO, TINRO, AtlantNIRO and AzcherNIRO) and also by researches of Moscow University, the Institute of Marine Biology in Vladivostok and others who sent material for identification, it has been possible to build up a rich collection of recent brachiopods (about 12 000 specimens from 1275 stations) which is now deposited in the Shirshov Institute of Oceanology. Later, examples from additional research vessels, including "Albatross", "Chain" and "Panulirus" (USA), "Galathea" (Denmark) and from the BEN-THEDI project (France) were also identified (Zezina, 1975, 1981a,b, 1987). It thus became possible to define "geographic faunistic elements" for the bathyal zone of the World Ocean (Figure 2) and prepare biogeographical divisions, based on recent brachiopod distribution (Zezina, 1973, 1976b). These divisions were the first biogeographical schemes for the bathyal zone. For the first scheme all recent brachiopod species are considered that live at the shelf edge and on the upper parts of the slopes; the second scheme is based on species that reach depths over 700 m, where many species appear to reach the lower limit of their vertical range (see Figure 1). The schemes are shown in Table 2. In these schemes the similarity and the differences between the faunas of the regions were evaluated with Preston's Coefficient (Preston, 1962). The grades of subdivisions were taken in the same manner as for the scheme prepared by Vinogradova (1969, 1977) for the abyssal zone, which also was based on species distribution. A coefficient of similarity $1 - z$ less than 0.10 was accepted for the areas; $1 - z$ more than 0.10, but less than 0.20 for the subareas; and a coefficient $1 - z$ more than 0.20 for the provinces. So the schemes for the bathyal zone and for the abyssal zone became comparable and the two can now be associated together in the general species-level scheme (Vinogradova and Zezina, 1996), prepared for the slopes and the floor of the World Ocean. Scheme construction of this type is not the only aim of biogeographical analysis, but rather an important stage of the study, offering conceptual results that provide a convenient basis for comparison of different methods (Beklemishev, 1982). Among the most important results obtained from analysis of "geographic faunistic elements" applied to recent brachiopods of the bathyal zone are: (i) demonstration of the change of biogeographical zones with depth (a "simplification"); and (ii) interpretation of changes in the symmetry of the fauna in respect to the equator and in respect to the meridional axis of the ocean.

3. LATITUDINAL ZONE CHANGES RELATED TO DEPTH

Important changes occur in the faunistic structure of the oceanic benthos within the large vertical interval of the bathyal zone. The most obvious difference

---

Figure 2 Geographic elements of recent brachiopods as distributed in latitudinal faunistic belts (after Zezina, 1983). The latitudinal (climatic) faunistic belts: I, limits of tropical species distribution; II, limits for northern and southern subtropical species; III, limits for low boreal and antiboreal species; IV, limits for the most cold-water species. Geographic elements are shown as Arabic numerals: 1, Caribbean; 2, West African; 3, Indo-West Pacific together with Indo-oceanic and West Pacific; 4, East Pacific; 5, West Atlantic; 6, East Atlantic together with Mediterranean and Lusitano–Mauritano–Mediterranean; 7, Japanese; 8, Californian; 9, South Brazilian-Uruguayan; 10, South African; 11, New Amsterdamian; 12, South Australian; 13, West European; 14, Oregonian; 15, South American; 16, New Zealandian; 17, North Atlantic; 18, North Pacific; 19, Antarctic (including circumantarctic, local antarctic and antiboreal–antarctic).

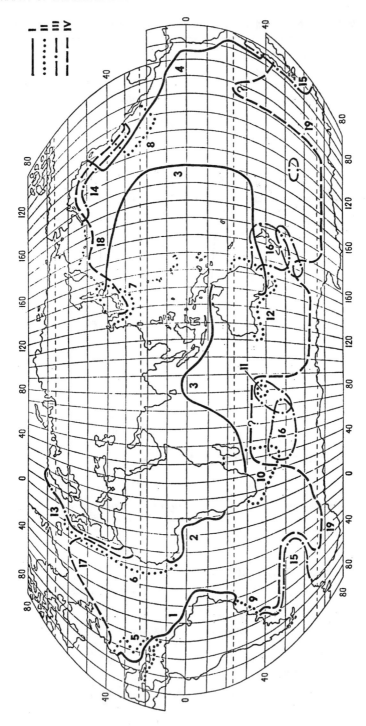

Table 2   Schemes for classifying the geographical distribution of the bathyal fauna.

A.   *For depths less than 700 m (Zezina, 1973)*

BOREAL–ARCTIC AREA

    North Pacific Subarea   *Asian–Aleutic Province*
                                     North-American Province
                                     Californian Province (subtropical)
    North Atlantic Subarea
    Arctic Subarea

AMPHIATLANTIC TROPICAL AREA

    Atlantic–Central American Subarea   *Caribbean Province (subtropical)*
                                       Brazilian Province
    Lusitano-Mauritanian Subarea (subtropical)
    Mediterranean Subarea (subtropical)

WEST INDO-OCEANIC TROPICAL AREA

INDO-WEST PACIFIC TROPICAL AREA

    Indomalayan Subarea
    Japanese Subarea (subtropical)

SOUTH BRAZILIAN–URUGUAYAN SUBTROPICAL AREA

SOUTH AFRICAN SUBTROPICAL AREA

SOUTH AUSTRALIAN SUBTROPICAL AREA
                                     Australian Province
                                     Tasmanian Province

NEW AMSTERDAMIAN ANTIBOREAL AREA

NEW ZEALANDIAN–KERGUELENIAN ANTIBOREAL AREA

    Newzealandian Subarea   *North New Zealandian Province*
                                       South New Zealandian Province
    Kerguelenian Subarea
    Macquarian Subarea

ANTARCTIC–SOUTHAMERICAN AREA

    Southamerican Subarea
    Antarctic Subarea

B.   *For depths 700–2000 m (Zezina, 1973, 1976b)*

BOREAL BATHYAL AREA
    Northatlantic Subarea
    Northpacific Subarea

Table 2 Continued

AMPHI-ATLANTIC BATHYAL AREA

                          Central Atlantic Province
                          Lusitano-Mauritano-Mediterranean Province
                          (transitional)

WEST INDO-OCEANIC BATHYAL AREA

WEST PACIFIC BATHYAL AREA   *Malayan Province*
                                   Japanese Province (transitional)

ANTARCTIC BATHYAL AREA

---

between shallow water and the deep sea is the reduction in number (simplification) of latitudinal or climatic belts, both in biomass and faunistic structure. Bogorov and Zenkevich and some of their pupils and disciples (Zenkevich, 1948; Bogorov, 1959, 1967; Bogorov and Zenkevich, 1966; Vinogradova, 1969, 1976; Belyaev *et al.*, 1973) regarded the simplified biogeographical division of the deep sea as "a washed-out or eroded projection" of the surface picture.

The manner of transformation from shallow-water biogeographic patterns to the deep-sea pattern is well exemplified by the distribution of recent brachiopods. The first scheme devised for recent brachiopods living at depths less than 700 m (Table 2A) is similar to the schemes suggested for the fauna of the shelves, based on many benthic groups (Ekman, 1935, 1953; Guryanova, 1964, 1972). It shows seven latitudinal belts: arcto boreal, low-boreal, northern subtropical, tropical, southern subtropical, antiboreal and antarctic. The second scheme based on recent brachiopods living deeper than 700 m (Table 2B) shows only three main latitudinal belts (northern, low latitudinal and southern), comparable to the scheme prepared by Vinogradova (1956, 1969, 1977) for the abyssal fauna occurring deeper than 3 km. The same three belts can be seen in the schemes of Sokolova (1976a,b, 1979, 1986) based on trophic structure of biocoenoses but connected with the species content of benthic samples (Sokolova, 1978). The changing number of latitudinal belts according to depth is illustrated in Figures 3 and 4.

At depths below the lower boundary of the intermediate zone there are smaller numbers of "geographic faunistic elements". For example, the subtropical faunistic elements (Japanese, Californian, South Australian, also South African and South Brazilian–Uruguayan), low-boreal (Oregonian) and antiboreal (South American and New Zealandian) are not found where deep-water masses exist. This phenomenon reflects a decrease of water mass diversity down the vertical scale (Sverdrup *et al.*, 1942; Stepanov, 1965, 1974, 1983). Deeper than 2000 m the differences between the physical characteristics of the water masses are very small,

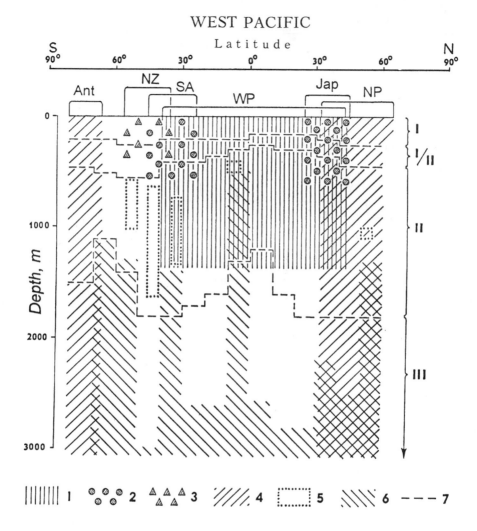

Figure 3 Scheme of the vertical distribution of the geographic faunistic elements in the western Pacific as shown by recent brachiopods (after Zezina, 1979, modified). Faunistic latitudinal belts: 1, tropical; 2, northern and southern subtropical; 3, low-boreal and antiboreal; 4, boreal and antarctic; 5, distribution of bathyal endemics; 6, distribution of deep-sea species; 7, borders of hydrological structural zones (after Stepanov, 1974): I, surface; II, intermediate; III, deep-sea; I/II, bordering layer. Geographic faunistic elements: Ant, Antarctic; NZ, New Zealandian; SA, South Australian; WP, West Pacific; Jap, Japanese, NP, North Pacific.

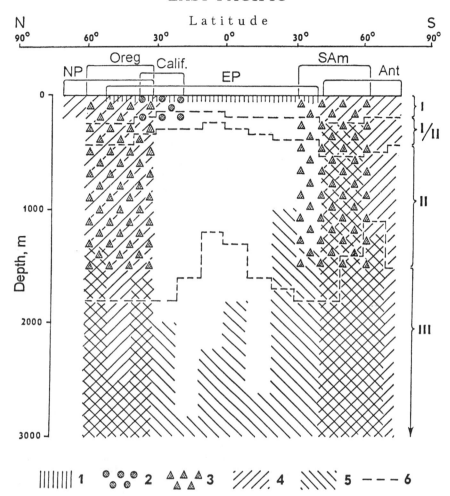

Figure 4 Scheme of vertical distribution of the geographic faunistic elements in the Eastern Pacific as shown by recent brachiopods (after Zezina, 1979, modified). Faunistic latitudinal belts: 1, tropical; 2, northern and southern subtropical; 3, low-boreal and antiboreal; 4, boreal and antarctical; 5, distribution of bathyal endemics; 6, distribution of deep-sea species; 7, borders of hydrological structural zones (after Stepanov, 1974): I, surface; II, intermediate; III, deep-sea; I/II, bordering layer. Geographic faunistic elements: NP, North Pacific; Oreg, Oregonian; Calif, Californian; EP, East Pacific; SAm, South American; Ant, Antarctic.

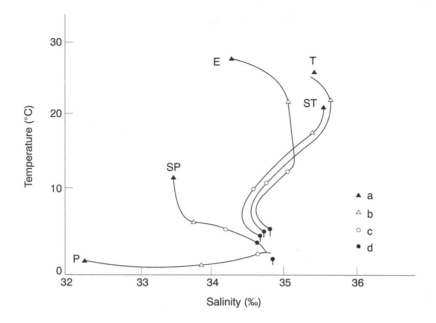

Figure 5 Temperature and salinity characteristics of the main types of water masses (after Stepanov, 1965, simplified). E, Equatorial; T, tropical; ST, subtropical; SP, subpolar structure; P, polar. Characteristics: a, near the surface; b, at 100 m deep; c, at 500 m; d, at 2000 m.

almost negligible (Figure 5), and hence the most important biogeographical differences lie in differences of food supply (Sokolova, 1986 and this volume; Zezina *et al.*, 1986, 1993). A similar simplification of geographical distribution related to water masses is seen in the plankton (Parin and Beklemishev, 1966; Beklemishev, 1969; Beklemishev *et al.*, 1972, 1977; Parin, 1979; Semina, this volume).

This relationship of benthos and plankton to water-mass distribution allows us to propose a universal scheme for the biogeographical division of the World Ocean based on the faunistic point of view (Beklemishev *et al.*, 1972; Zezina *et al.*, 1986) and on the landscape point of view (Zezina *et al.*, 1993).

4. SYMMETRY AND ASYMMETRY OF THE FAUNA

The symmetry of the latitudinally arranged faunistic belts of the continental shelves due to climatic zonation is shown also in the bathyal zone (Neyman *et al.*, 1977), but there are some asymmetrical features, however (Zezina, 1978). These asymmetrical features are evident in the data and are important when considering the history of the ocean fauna. The first feature of distortion in the symmetry in respect to the equator is the different degree of separation of faunistic latitudinal belts in the northern and in the southern hemispheres. These belts are disjunct in the southern hemisphere, but conjunct and even interfused (see Figures 2, 3 and 4) in the northern hemisphere, especially on the western sides of the oceans. This feature appears to be related to the asymmetry of the oceanic circulation which displaces the climatic equator to the north of the geographical equator. The asymmetry in the ocean circulation and in the biogeography of the faunas reflects the asymmetry of the distribution of land and ocean on earth.

With regard to the longitudinal axes of the Pacific and the Atlantic oceans, distortion of symmetry is shown by the differing widths of latitudinal faunistic belts and in the different degrees of their separation on the western and eastern sides of each ocean. This distortion is also related to the positions and directions of oceanic circulation, influenced by the earth's rotation (Coriolis force) and the configuration of the oceans. The global gyres of surface and intermediate waters have opposite directions in the north and in the south (Figure 6). The latitudinal faunistic belts are broader and more separated on the eastern side of an ocean and are narrower, conjunct and interfused at the western edges.

A third distortion in the symmetry is related to the difference in species number between the western and the eastern warm-water faunas (tropical and subtropical faunistic belts) in the Pacific, the Indian and the Atlantic oceans. In the west the warm-water faunas are richer in species than in the east. This relationship is known for many taxa, and is especially well shown for echinoderms (Djakonov, 1950; Mortensen, 1951; Mironov, 1983a,b, 1989), recent brachiopods (Zezina, 1976a, 1985a) and solitary madreporarians (Cairns and Keller, 1993). Among recent articulate brachiopods (Testicardines) the number of warm-water species on the western shelves and slopes greatly surpasses the number of species on the eastern shelves and slopes: 25 times in the Pacific, 14 times in the Atlantic and five times in the Indian Ocean (Figure 7). These differences depend on the direction of global water streams and on the historical disposition of centres of speciation. Asymmetry of the warm-water faunas related to the meridional axis reflects the hydrological asymmetry which is determined chiefly by westward movement of surface water at low latitudes and the constant upwelling near the eastern shores in the Pacific and in the Atlantic. (There are some additional specific features in the Indian Ocean.) Upwelling, with corresponding differences in dissolved gases in the water and in the sediment, is unfavourable for many benthic animals. Communities in upwelling regions may be rich in biomass but are always poor in species (Karpinski, 1985, 1996). Regional changes in location of upwelling along the east side of the Pacific and the Atlantic over long intervals of geological time may be a factor that has eliminated many benthic taxa (Zezina, 1985a).

In the western Indian Ocean there is a large upwelling zone near the continental slope, and as a consequence the centre of the warm-water fauna is not situated

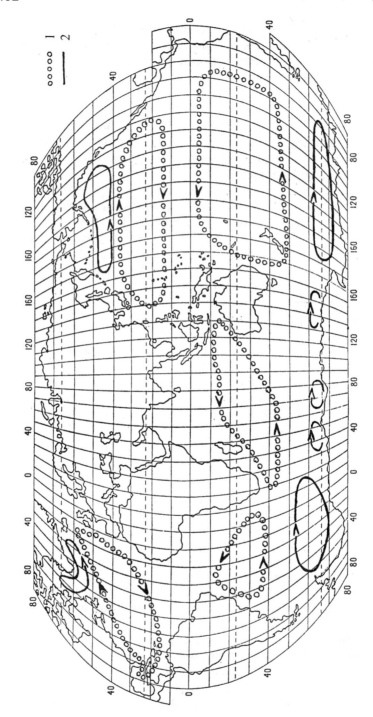

Figure 6   The main gyres of the World Ocean surface waters (after Zezina, 1976a, simplified). 1, Subtropical gyres; 2, subpolar gyres. The gyres in the Pacific Ocean are shown after Burkov (1968); in the Atlantic, after Bulatov (1971); in the Indian Ocean, after Neyman (1970) and Tscherbinin (1971).

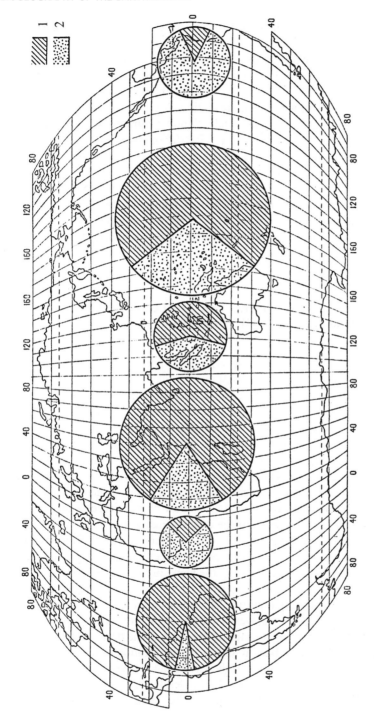

Figure 7 Species richness of warm-water species of Articulate (1) and Inarticulate (2) brachiopods on the shelves and the slopes in the west and the east of the oceans (after Zezina, 1983).

close to the continent, but rather on the slopes of islands and underwater ridges (Zezina, 1985a, 1987).

The global circulation pattern also presents obstacles to distribution of faunistic elements from the western oceanic centres of speciation (Figure 8) which thus tend to be isolated to a considerable extent from the shelves and slopes on the opposite side of the ocean. This separation is more obvious in the Pacific than in the Atlantic and the Indian Oceans. Groups with amphiatlantic geographical ranges show that cross-ocean dispersal has occurred. In the Pacific the occurrence in Hawaii and Easter Is of species from the West Pacific also points to cross-ocean transport (Parin et al., this volume). Evidently some enrichment of the eastern shelves and slopes from the western centres of speciation has occurred in the past. In addition, the East Pacific tropical fauna has Caribbean elements and the East African fauna includes some taxa derived from the west of the Indian Ocean (Zezina, 1985a, 1981c,d, 1987). Such a spread indicates past changes in patterns of continents and ocean circulation.

5. BATHYAL REGIONS WHERE THE SHALLOW-WATER FAUNA MIGHT PENETRATE TO THE DEEP SEA

The asymmetry in shelf and slope biogeography provides clues to regions of the bathyal zone where elements of the shelf and bathyal fauna might penetrate to the deep sea. The usual idea is that the easiest route for colonization of the deep sea by shallow-water animals is in high latitudes, where animals adapted to the cold-water regime might be able to invade the deep sea if they can acquire the necessary biochemical adaptation for life at high pressures (Hochachka and Somero, 1973; Kuznetsov, 1977, 1989). In actual fact, in high-latitude regions shallow-water species are confluent with deep-water species over a rather broad vertical range, including the continental slope (see Figures 3 and 4). On the slope the two faunistic elements can co-exist, so that high latitudes can indeed be presumed to be suitable gateways to the depths (Figure 9).

However, a corresponding situation can be found at lower latitudes in the western parts of the oceans (see Figure 3), where shallow-water species are contiguous with deep-sea species along the slope in the bathyal zone. There are indeed many families of benthic animals that live only in low latitudes and in the deep sea, but are absent in high-latitude, shallow-water regions. This suggests that bathyal faunistic centres in the western Pacific, the western Atlantic and the western Indian Ocean may be contributing directly to the deep-sea fauna. Not all suitable pathways may have been realized, but penetration to the deep sea from the Antarctic, the western Pacific and the Caribbean region is evident in Isopoda (Kussakin, 1973; Hessler and Thistle, 1975; Hessler et al., 1979), Pycnogonida (Turpaeva, 1975), Echinoidea (Mironov, 1980, 1982), Brachiopoda (Zezina,

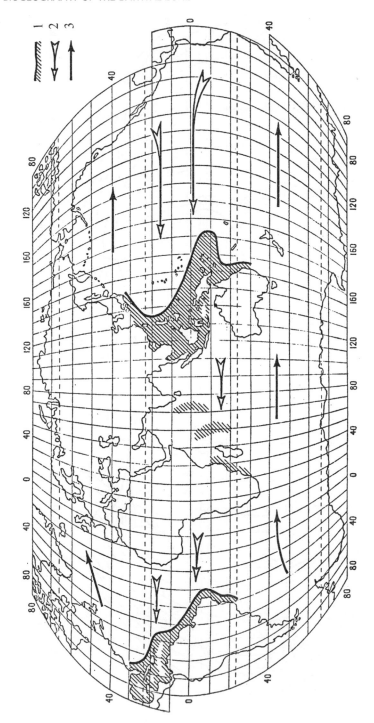

Figure 8 Sublittoral and bathyal regions, enriched by warm-water species (1) and the mean directions of transoceanic surface-water drifts in the low (2) latitudes and the high (3) latitudes (after Zezina, 1983).

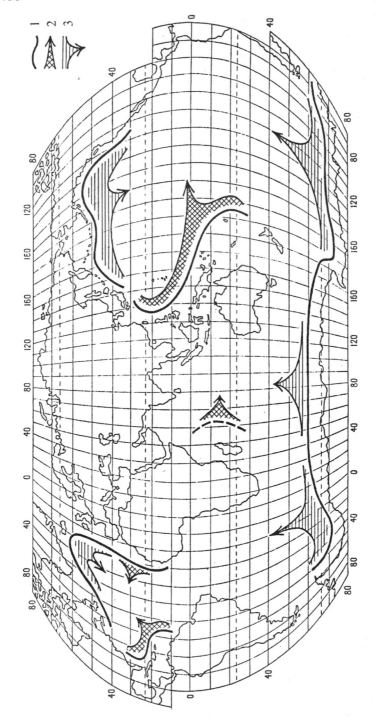

Figure 9 Bathyal regions where bathyal species may be able to colonize the abyssal zone (after Zezina, 1983). 1, Bathyal regions where shallow-water and deep-sea species are mixed; 2, low-latitude colonization potential; 3, high-latitude colonization potential.

1985a) and Holothurioidea (Gebruk, 1990, 1994). Sometimes these regions (especially the northern Atlantic and the northern Pacific) are considered to be places of emergence, where deep-sea species are able to extend into the bathyal and even sublittoral zones. Recent examples of such emergence are known among the Tanaidacea (Kudinova-Pasternak, 1990, 1993) and Holothurioidea (Gebruk, 1994).

6. THE BATHYAL ZONE AS A REFUGE FOR "RELICTS" AND "LIVING FOSSILS"

The concept of relict species is of great interest to biogeographers. The term "relict" was introduced into the zoological literature of the last century by Lovén (1862) to explain findings of marine-type mysid crustaceans in freshwater lakes. Strakhov (1932) defined a fauna as relict (or residual) when it is different from the predominating present-day fauna but has similarities to a fauna that existed in a past epoch, later extinguished over the major part of its range.

The ecological and geographical criteria of recent relicts were analysed in detail by Birshtein (1947). According to this author, those taxa which have remained almost unchanged in time (or which are characterized by a delayed rate of evolution) are considered as relict taxa. If the taxa of recent fauna were initially described from palaeontological objects the extant specimens are then called "living fossils". Interrelations between relicts and living fossils have been discussed by Eldredge (1975) and by Eldredge and Stanley (1984) using many examples from various groups of land, freshwater and marine animals. We will note here that living fossils are the relicts whose history is not reflected in nature or not shown by stratigraphy. This implies that if a taxon of recent fauna had been described first from fossils then the taxon probably had wider distribution or greater abundance in the palaeocommunity than in recent communities (Zezina, 1994).

Relict species have been known for a long time from the sublittoral zone at mid to low latitudes where environmental conditions are probably less variable than in high latitudes. The most famous shallow-water relicts are: the brachiopods of the family Lingulidae; the xiphosurids or "horse shoe crabs" of the family Limulidae; the reef-forming coral *Heliopora coerulea*; and the gastropod mollusks *Neritopsis radula*, *Campanile symbolicum*, *Diastoma melanoides* and *Gourmya gourmyi*.

Other taxa can now be added to these examples, including species from both the shelf and the bathyal zones; for example, bivalve molluscs of the family Trigonidae (five species of the genus *Neotrigonia* distributed from the low-water level to 400 m); cephalopod molluscs of the family Nautiloidae (four species of the genus *Nautilus* known from the depths 5–500 m); the bryozoan *Nellia tenella* found from low water level to 1000 m); primitive crustaceans of the class Cephalocarida (11 species of five genera, mostly sublittoral, but known also to 1600 m); and benthopelagic crustaceans of the family Nebaliidae (15 species of four genera, mostly bathyal, but distributed from sublittoral zone to the abyss). If all recent priapulid worms can be considered as relicts (Malakhov and Adrianov, 1995; Adrianov and Malakhov, 1996), this taxon can also be regarded as eurybathic.

After the shallow-water relicts became known some deep-sea (abyssal) relict groups were discovered, then some of them were found later on the slopes and shelves. Thus, the recent species of the ancient mollusc group Monoplacophora were at first considered to be abyssal relicts, but then further species became known from bathyal depths. For example, *Vema* (*Laevipilina*) *hyalina* was found off California in 129–388 m (McLean, 1979) and some unidentified examples were collected in the North Atlantic and the Antarctic at 160–2000 m (Filatova *et al.*, 1969, 1975).

Many of the species inhabiting localized areas of hydrothermal venting and dependent on benthic bacterial autotrophic production are also relict species not known in present-day shallow-water habitats (Newman, 1979; McLean, 1981; Jones, 1985; Laubier, 1986; Newman and Hessler, 1989; Zevina and Galkin, 1989, 1992; Galkin, 1992). These hydrothermal relicts include several major taxa, and some of the genera and species are the most primitive known for their group (Newman, 1979, 1985, 1986; McLean, 1985; Newman and Hessler, 1989; Yamaguchi and Newman, 1990), as for example *Neolepas zevinae* among lepadomorphs, *Neoverruca brachylepadoformis* among verrucomorphs, *Bathypecten vulcani* among pectinid bivalves and *Neomphalus fretteri* among mesogastropods. There is evidently a similar retention of relict species in both bathyal hydrothermal and bathyal non-hydrothermal habitats (Newman, 1980; Newman and Foster, 1983, 1987; Yamaguchi and Newman, 1990; Zezina, 1994).

In the case of the vestimentiferan tube-worms (subclass Afrenulata or Obturata of the phylum Pogonophora), these cannot now be regarded as purely deep-sea relicts (Belyaev, 1989) and are better characterized as bathyal–abyssal, since they can occur as shallow as 80–200 m where suitable hydrothermal conditions exist (Hashimoto *et al.*, 1993). Likewise, other Pogonophora (Frenulata or Perviata) are also now known from a wide range of depths (22–9735 m), and are thus to be regarded as eurybathyal (Belyaev, 1989).

The distribution of many shallow-water and deep-sea relicts can be related to the bathyal zone. The role of the bathyal zone as a favoured habitat for survival of relicts and living fossils became evident from the time when the first living crossopterygian fish was found on the slope off South Africa (Smith, 1931) and later these fishes were studied near the Comoro Is. As studies of the bathyal zone progressed, more apparent relict species began to appear (Table 3). Two new species of living fossils among articulate brachiopods (*Sphenarina ezogremena*

Table 3   Invertebrate marine relicts confined to the bathyal zone.

Taxa	Depth range, m
Crustacea, Cirripedia (Scalpellida)	
Scillaelepas, 11 species (fide Newman, 1980)	340–3000
Calantica (?) moskalevi (Zevina and Galkin, 1989)	1410–1540
Crustacea, Cirrpedia (Balanomorpha, Chionelasmatida)	
Chionelasmus darwini (Pilsbry, 1907)	420
Eochionelasmus ohtai Yamaguchi 1990	1990–2500
Crustacea, Isopoda (Cirolanidae)	
Bathynomus giganteus Milne-Edwards, 1902	310–2140
Crustacea, Decapoda (Glypheidae)	
Neoglyphea inopinata Forest and Saint Laurent, 1975	181–210
Mollusca, Gastropoda (Padiculariidae)	
Cypraeopsis superstes Dolin, 1991	435–1005
Mollusca, Pleurotomariacea (Pleurotomariidae)	
Pleurotomaria diluculum (Okuani, 1979)	Deep water
Entemnotrochus rumphii (Schepman, 1879)	Deep water
Entemnotrochus adansonianus (Crosse and Fischer, 1861)	Deep water
Perotrochus africanus (Tomlin, 1949)	366
Perotrochus hirasei Pilsbry, 1903	91 and more
Perotrochus quoyanus (Fischer and Bernardi, 1856)	Deep water
Perotrochus teramachii Kuroda, 1955	
Perotrochus amabillis Bayer, 1963	Deep water
Perotrochus lucaya Bayer, 1967	Deep water
Perotrochus midas Bayer, 1965	Deep water
Perotrochus atlanticus Rios and Matthews, 1968	200
Mikadotrochus schmalzi Shikama, 1961	
Mikadotrochus salmianus (Rolle, 1899)	Deep water
Mikadotrochus notialis Lema and Penna, 1969	
Brachiopoda, Ecardines (Valdiviathyridae)	
Valdiviathyris quenstedti Helmcke, 1940	672
Neoancistrocrania norfolki Laurin, 1992	233–250
Brachiopoda, Testicardines (Hispanirhynchiidae)	
Sphenaria ezogremena Zezina, 1981	240
Brachiopoda, Testicardines (Aulacothyropsidae)	
Septicollarina hemiechinata Zezina, 1981	240
Septicollarina oceanica Zerina, 1990	270–485
Echinodermata, Crinoidea (Millericrinidae)	
Proisocrinus ruberrimus Clark, 1910	1254
Echinodermata, Crinoidea (Incertae)	
Guillecrinus reunionensis Roux, 1985	Bathyal
Guillecrinus neocaledonicus Bourseau, Ameziane-Cominardi,	700–1300
Avocat, Roux, 1991	
Echinodermata, Crinoidea (Hemicrinidae)	
Gymnocrinus richeri Bourseau, Ameziane-Cominardi,	470
Roux, 1987	
Echinodermata, Crinoidea (Holopodidae)	
Holopus rangi Orbigny, 1837	Bathyal
Holopus alidis Bourseau, Ameziane-Cominardi, Avocat, Roux, 1991	355–470
Cyathidium foresti Cherbonnier and Guille, 1972	580–1140

and *Septicollarina hemiechinata*) were described (Zezina, 1981a) from the collection of Thomas Mortensen, taken at 240 m off Bali (Figure 10). Both these species have been included in the first Invertebrate Red Data Book (Wells *et al.*, 1983). Recently, another new species, *Septicollarina oceanica*, was described from the 470–485 m on the Sala-y-Gómez Ridge (Zezina, 1990). Inarticulate (hingeless) brachiopods with carbonate shells are also found as living fossils: *Valdiviathyris quenstedti* off St Paul Is in the Indian Ocean at the depth of 672 m (Helmcke, 1940) and *Neoancistrocrania norfolki* (Figure 10) from the underwater ridge between the Tasman Sea and the Fiji Sea at the depths of 233–250 m (Laurin, 1992). Relatives of these forms were known before only from the Late Cretaceous of Europe and Middle Asia (Zezina, 1985b, 1994). At the border between sublittoral and bathyal zones (at 181–210 m) some relict decapod crustaceans of the family Glypheidae are found. This group was regarded as becoming extinct after the Mesosoic until the recent new species *Neoglyphea inopinata* were found in 1908 and described in 1975. The giant cirolanid isopod *Bathynomus giganteus* is another living fossil, known from the bathyal zone in the tropical Atlantic and in the Indian Ocean at depths from 310 to 2140 m, more recently taken also in the South China Sea near the Philippines at 900 m (Kussakin, 1984). Recent intensive studies of bottom invertebrates in the bathyal zone especially by means of submersibles and unmanned vehicles have presented opportunities to discover other relicts and living fossils. Use of manned submersibles opened new opportunities for visual observations (photo, ciné and video recording) in the bathyal zone, and the list of bathyal relicts is now rapidly growing. Among sea lilies, according to Roux *et al.* (1991), recent *Proisocrinus* from the depth 1276 m off the New Caledonia is similar to the Jurassic forms, and recent *Guillecrinus* from the same region is similar to Paleozoic species. Another species of *Guillecrinus* was found in the Indian Ocean off the Reunion Is also at the bathyal depths (Roux, 1985); *Gymnocrinus richeri* (Figure 10) described from the depth 470 m off the New Caledonia (Bourseau *et al.*, 1987) belongs to a Jurassic genus which is united together with the genera *Cyrtocrinus* and *Hemicrinus* in the Jurassic–Early Cretaceous family Hemicrinidae. There are also two recent relict sea lilies, *Holopus rangi* and *Cyathidium foresti* (Figure 10), known from the Caribbean Sea, off Bermuda and Azores, and also from seamounts of the central and eastern Atlantic at the depths 580–1140 m (Moskalev *et al.*, 1983; Zezina, 1994). These aberrant forms possessing stone-like bodies are related to the order Cyrtocrinida, known from the Early Jurassic, that flourished in the Late Jurassic and the Early Cretaceous (Arendt, 1984).

Bathyal relicts are also known among gastropods. The family Pleurotomariidae (Archaeogastropoda, Pleurotomariacea) is represented by 17 living species of three genera (*Entemnotrochus*, *Perotrochus* and *Mikadotrochus*) and is related to a primitive predominantly Paleozoic group which flourished in the Late Paleozoic. Now these molluscs occur below 200 m on underwater banks and on the slopes of reef islands in tropical and subtropical waters from eastern Japan to Taiwan, as

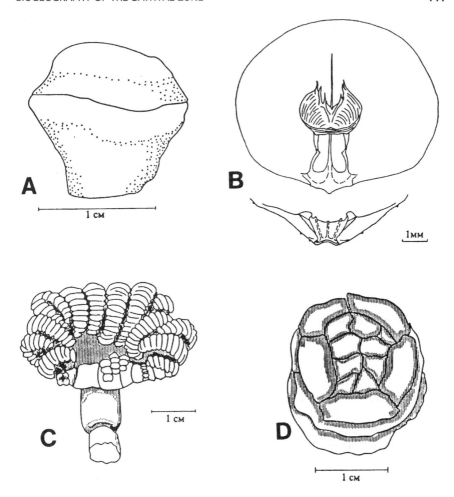

Figure 10 Some "exotic" relict species. A. The inarticulate brachiopod *Neoan-cistrocrania norfolki* from bathyal depths of the Tasman Sea. Drawing from photograph of the holotype described by Laurin (1992) (after Zezina, 1994). B. The articulate brachiopod *Septicollarina hemiechinata* from bathyal depths to the north of Bali Is, drawn from the holotype (after Zezina, 1981a): upper, interior of the brachial valve with brachidium; lower: dorsal view of the beak where foramen, pedicel collar and small irregular nodules on the outer surface of the pedicle valve are seen. C. The crinoid *Gymnocrinus richeri* from the bathyal depths off New Caledonia, drawn from photograph of Bourseau *et al.* (1987) (after Zezina, 1994). D. The crinoid *Cyathidium foresti* from bathyal depths on the Great Meteor Seamount in the Atlantic, drawn from photograph of the specimen found by Dr A. Mironov in a dredge haul (Moskalev *et al.*, 1983). After Zezina, 1994).

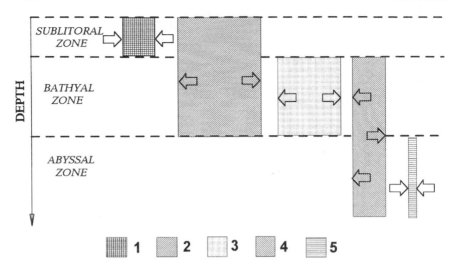

Figure 11   Relict species grouped according to depth, and their increased recognition during the recent period of deep-sea investigations (original figure). 1, sublittoral/shelf; 2, shelf-bathyal; 3, bathyal; 4, bathy-abyssal; 5, abyssal.

well as off South Africa and in the Caribbean (Oliver, 1983). Gastropods of the genus *Cypraeopsis* (*Padiculariina*) were known based on three fossil species from Miocene–Oligocene of Europe and from Miocene of southeastern Asia. The recent species *Cypraeopsis superstes* is found off the New Caledonia and off the Reunion Is at 24 stations in the range of depths 435–1005 m (Dolin, 1991). Extensive lists of bryozoans (Hayward, 1988) and crabs (Guinot, 1984) have been described from the bathyal zone off Reunion and Mauritius, including some new relict taxa.

It appears from recent knowledge of relict species that the bathyal zone is comparable to, and even surpasses, tropical shallow waters and hydrothermal regions of the abyss in its ability to provide a refuge for ancient faunistic elements (Figure 11). The relative richness of the bathyal zone in relicts and living fossils may be related to a number of causes. A primary cause may be the vertical downward displacement of ancient primitive organisms from the shelves to the depths by newer, faster-evolving taxa with which they cannot compete. The bathyal zone is not so poor in nutritional resources as the abyssal zone, so it may be more suitable for the refugees. The second cause may be the fact that the intermediate water masses which are in contact with the slopes have rather lower ranges of temperature than present-day surface water masses (Kuksa, 1983), which may well encourage survival of the displaced, less-adaptable species.

A third driving force towards preservation of eurybathic species on the continental slopes may be changes of sea level on the shelves and variations in temperature and dissolved gases in the abyssal zone, both resulting from global tectonic and hydrological cataclysms. In such periods the shallow-water fauna and the abyssal fauna may be able to find refuge in the bathyal zone and survive there. In periods between global cataclysms we may suppose the bathyal zone can become a source for replenishment of both abyssal and shallow-water biocoenoses with relict species. The capacity of the bathyal zone to be either a refuge for species or a source for recolonization gives it an important role in the history of the marine fauna (Zezina, 1985a).

7. MANAGEMENT AND CONSERVATION OF THE BATHYAL ZONE

The division of the shelf seas of the World Ocean between coastal states in the form of Exclusive Economic Zones (EEZ) has accentuated the bathyal problem. The parts of slopes near continents and islands outside the limits of the EEZ, and especially seamounts and ridges, including the mid-oceanic ridges, became an objective for uncontrolled exploitation of marine resources. Unfortunately, the proposed or actual exploitation of these areas is ahead of scientific studies of life in the bathyal zone (Sasaki, 1978, 1986; Grigg, 1984, 1986). The shelf-free part of the bathyal zone in the open ocean was called "thalassobathyal" by Andriyashev (1979), "thalassial modification of the bathyal zone" by Parin (1982, 1984) and "oceanic faunistic bathyal zone" by Mironov (1983a,b, 1985). A recent review has focused attention on the bathyal zone of seamounts and the dangers of commercial exploitation, including fisheries (Rogers, 1994). We should note that the mid-oceanic parts of the bathyal zone possess very important biological characteristics. They contain many endemic species, whereas most species living on the main continental slopes are absent near oceanic islands and in the thalassobathyal.

These mid-oceanic regions are usually poorer in species diversity when compared with the near-continental parts of the bathyal zone. For instance, among recent brachiopods (Zezina, 1981b) the mid-oceanic bathyal zone of the western Pacific possesses only 18% of the species number known from the near-continental regions. There is the same trend in other systematic groups of the benthos (Pasternak, 1981, 1991, 1994; Mironov, 1985, 1994; Wilson et al., 1985; Wilson and Kaufmann, 1987; Grassle and Maciolek, 1992). Some species niches of the mid-oceanic bathyal zone are represented by very small and sometimes neotenic forms (Zezina, 1981b). It can be suggested that the mid-oceanic regions form a marginal area of distributional ranges or possibly even an expatriation area for many bathyal species usual on continental slopes. An important feature of mid-oceanic benthos is the prevalence of oligo-mixed communities with numerous specimens of a few species. This last feature is very attractive for fishery exploitation but dangerous for the natural community. The species occurring on summits and slopes of isolated underwater ridges are especially vulnerable because of the many hindrances to recruitment. Oligo-mixed associations are sometimes characterized by endemic species which can be rare, relicts, and even living fossils. Their isolated populations cannot be replenished if their abundance falls beyond a critical level and uncontrolled exploitation could destroy what nature has preserved for millions of years.

ACKNOWLEDGEMENTS

I thank Professor Alan Southward and Dr Andrey Gebruk for the invitation to write this work as part of the series of Russian contributions on themes of global bio-oceanology. My work was supported by Grant 96–05–65776 from the Russian Fundamental Research Foundation. I thank my colleagues M.N. Sokolova and A.N. Mironov for commenting on the manuscript. I gratefully acknowledge the help of my uncle A.P. Pirogov, architect and painter, in drawing the charts and schemes, and also my colleague N. Bagirov for modernizing the schemes by computer for the English version.

REFERENCES

Adrianov, A.V. and Malakhov, V.V. (1996). The phylogeny, classification and zoogeography of the class Priapulida. II. Revision of the family Priapulidae and zoogeography of priapulids. *Zoosystematica Rossica* **5**(1), 1–6.

Andriyashev, P.P. (1979). On some problems of vertical zonality of the marine bottom fauna. *In* "Biological Resources of the World Ocean" (S.A. Studenetsky, ed.), pp. 117–138, Nauka, Moscow. (In Russian).

Arendt, Yu.A. (1984). Sea lilies Cyrtocrinids. *Trudy Paleontologicheskogo Instituta Akademii Nauk SSSR* **144**, 1–251. (In Russian with English abstract).

Atlas of the Oceans. Terms, concepts, informational tables. (1980). Navy, Leningrad. 156 pp. (In Russian).

Babin, Cl. and Roux, M., eds (1982). Biogéographie et tectonique des plaques. (Séance specialisée de la Société géologique de France et de l'Association paleontologique francaise, tenue a Brest, le 1–er mars, 1982). *Bulletin de la Société géologique de France*, Série 7, **24**(5–6), 871–1168.

Beklemishev, C.W. (1967). Biogeographical division of the Pacific Ocean within surface and intermediate waters. *In* "The Pacific Ocean. Biology of the Pacific Ocean. Book 1. Plankton" (V.G. Kort, ed.), pp. 98–169. Nauka, Moscow. (In Russian).

Beklemishev, C.W. (1969). "Ecology and Biogeography of the Open Ocean". Nauka, Moscow. (In Russian with English abstract). English translation (1976). Naval Oceanographic Office translation NNOT-25, Washington, DC.

Beklemishev, C.W. (1982). On the nature of the biogeographical evidences. *In* "Marine Biogeography" (O.G. Kussakin, ed.), pp. 5–11. Nauka, Moscow. (In Russian).

Beklemishev, C.W. and Zezina, O.N. (1972). Actualistic and historical views of oceanic biogeography. *Isvestia Akademii Nauk SSSR,* seria geologicheskaya **11**, 147–151. (In Russian).

Beklemishev, C.W., Neyman, A.A., Parin, N.V. and Semina, H.J. (1972). Le biotope dans le milieu marin. *Marine Biology* **15**(1), 57–73.

Beklemishev, C.W., Parin, N.V. and Semina, G.I. (1977). Pelagic biotope (Pelagial). *In* "Oceanology, Biology of the Ocean 1" (A.S. Monin and M.E. Vinogradov, eds), pp. 219–261. Nauka, Moscow. (In Russian).

Belyaev, G.M. (1989). "Deep-sea Oceanic Trenches and their Fauna." Nauka, Moscow. (In Russian).

Belyaev, G.M., Vinogradova, N.G., Levenshtein, R.Ya., Pasternak, F.A., Sokolova, M.N. and Filatova, Z.A. (1973). Distribution patterns of deep-water bottom fauna related to the idea of the biological structure of the ocean. *Oceanology* **13**(1), 114–120.

Birshtein, Ya.A. (1947). The term "relic" in biology. *Zoologicheski Zhurnal* **26**(4), 313–330. (In Russian with English summary).

Bogorov, V.G. (1959). Biological structure of the Ocean. *Doklady Akademii Nauk SSSR,* Seria biologicheskaya **128**(4), 819–826. (In Russian, there is an English translation).

Bogorov, V.G. (1967). Biomass of zooplankton and productive areas in the Pacific Ocean. Geographical zonation of the Ocean. *In* "The Pacific Ocean. Biology of the Pacific Ocean. Part 1. Plankton" (V.G. Kort and V.G. Bogorov, eds), pp. 221–233. Nauka, Moscow. (In Russian).

Bogorov, V.G. and Zenkevich, L.A. (1966). Biological structure of the Ocean. *In* "Ecologia vodnikh organizmov" (L.A. Zenkevich, ed.), pp. 3–14. Nauka, Moscow. (In Russian).

Bourseau, J.-P., Ameziane-Cominardi, N. and Roux, M. (1987). Un Crinoide pedonculé nouveau (Echinodermes), representant actuel de la famille jurassique des Hemicrinidae: *Gymnocrinus richeri* nov. sp. des fonds bathyaux de Nouvelle-Caledonie (SW Pacifique). *Comptes rendu hebdomadaire des sèances de l'Academie des sciences, Paris,* Serie III **305**, 595–599.

Brey, T., Peck, L.S., Gut, J., Hain, S. and Arntz, W.E. (1995). Population dynamics of *Magellania fragilis,* dominating a mixed-bottom macrobenthic assemblage on the Antarctic shelf and slope. *Journal of the Marine Biological Association of the United Kingdom* **75**, 857–869.

Briggs, J.C. (1974). "Marine Zoogeography". McGraw-Hill, New York.

Bulatov, R.P. (1971). Circulation of waters of the Atlantic Ocean in different space-time scales. *Oceanological Researches, Soviet Geophysical Committee, Results of Researches on the International Geophysical Projects* **22**, 7–93. (In Russian with English abstract).

Burkov, V.A. (1968). Water circulation. *In* "The Pacific Ocean, Hydrology of the Pacific Ocean" (V.G. Kort and A.D. Dobrovolsky, eds), pp. 206–281. Nauka, Moscow. (In Russian).

Cairns, S.D. and Keller, N.B. (1993). New taxa and distributional records of Azooxanthellata Scleractinia (Cnidaria, Anthozoa) from the tropical South-West Indian Ocean, with comments on their zoogeography and ecology. *Annals of the South African Museum* **103**, 213–292.

Carney, R.S., Haedrich, R.L. and Rowe, G.T. (1983). Zonation of fauna in the deep sea. *In* "The Sea. Vol. 8. Deepsea Biology" (G.T. Rowe, ed.), pp. 371–398. Wiley, London.

Colin, P.L., Devaney, D.M., Hillis-Colinvaux, L., Suchanek, T.H. and Harrison, J.T. (1986). Geology and biological zonation of the reef-slope 50–360 m depth at Enewetak atoll, Marshall Islands. *Bulletin of Marine Science* **38**, 11–128.

Conan, G., Roux, M. and Sibuet, M. (1981). A photographic survey of a population of the stalked crinoid *Diplocrinus* (*Annacrinus*) *wyvillethomsoni* (Echinodermata) from the bathyal slope of the Bay of Biscay. *Deep-Sea Research* **28A**, 441–453.

Cutler, E.B. (1975). Zoogeographical barrier on the continental slope off Cape Lookout, North Carolina. *Deep-Sea Research* **22**, 893–901.

Day, D.S. and Pearcy, W.G. (1968). Species associations of benthic fishes on the continental shelf and slope of Oregon. *Journal of the Fisheries Research Board of Canada* **25**, 2665–2675.

Djakonov, A.M. (1950). Echinoderm fauna of the Malayan Archipelago and its relations to the past and the present times of the World Ocean. *Trudy Leningradskogo obchestva estestvoispytateley*, **70**(4), 109–126. (In Russian).

Dolin, L. (1991). Mollusca Gastropoda: *Cypraeopsis superstes* sp. nov., Padiculariina relique du bathyal de Nouvelle-Caledonie et de la Réunion. *Memoirs du Muéeum national d'Histoire naturelle* Ser. A **150**, 179–186.

Ekman, S. (1935). "Tiergeographie des Meers". Academisches Verlagt, Leipzig.

Ekman, S. (1953). "Zoogeography of the Sea". Sidgwick & Jackson, London.

Eldredge, N. (1975). Survivors from the good old, old, old days. *Natural History, New York* **84**(2), 60–69.

Eldredge, N. and Stanley, S.M., eds (1984). "Living fossils". Springer, New York.

Filatova, Z.A., Sokolova, M.N. and Levinstein, R.Ya. (1969). On a finding of the mollusc from the Cambrian-Devonian Class Monoplacophora in the central part of the North Pacific. *Dokladi Akademii Nauk SSSR* **185**(1), 192–194. (In Russian).

Filatova, Z.A., Vinogradova, N.G. and Moskalev, L.I. (1975). The molluscs *Neopilina* in the Antarctic. *Okeanologiya* **15**(1), 143–145. (In Russian).

Gage, J.D. (1977). Structure of the abyssal macrobenthos in the Rockall Trough. *In* "Biology of Benthic Organisms" (E.F. Keegan, P. O'Ceidigh and P.J.S. Boaden, eds), pp. 247–260. Pergamon Press, Oxford.

Galkin, S.V. (1992). *Eochionelasmus ohtai* (Cirripedia, Balanomorpha) from hydrothermal vents in the Manus back-arc basin. *Zoologicheski zhurnal* **71**(11), 139–143. (In Russian).

Gardiner, F.P. and Haedrich, R.L. (1978). Zonation in the deep benthic megafauna. *Oecologia* **31**, 311–317.

Gebruk, A.V. (1990). "The Deep-sea Holothurians of the Family Elpididiidae". Nauka, Moscow. (In Russian with English abstract).

Gebruk, A.V. (1994). Two main stages in evolution of the deep-sea fauna of elasipodid holothurians. *In* "Echinoderms Through Time" (B. David, A. Guille, J.-P. Feral and M. Roux, eds), pp. 507–514. Balkema, Rotterdam.

Golikov, A.A. (1990). Amphipods of the Laptev Sea and adjacent waters. *Issledovaniya Fauny Morey* **37**(45), 235–257. (In Russian with English abstract).

Grassle, J.F. and Maciolek, N.J. (1992). Deep-sea species richness: regional and local diversity estimates from quantitative bottom samples. *The American Naturalist* **139** (2), 313–341.

Grassle, J.F., Sanders, H.L., Hessler, R.R., Rowe, G.T. and McLellan, T. (1975). Pattern and zonation: a study of the bathyal megafauna using the research submersible "Alvin". *Deep-Sea Research* **22**, 457–481.

Grigg, R.V. (1984). Resource management of precious corals: a review and application to shallow water reef building corals. *Marine Ecology* **5**(1), 57–74.

Grigg, R.V. (1986). Precious corals: an important seamount fisheries resource. *In* "The Environment and Resources of Seamounts in the North Pacific. Proceedings of the Workshop on the Environment and Resources of Seamounts in the North Pacific" (R.N. Uchida, S. Hayasi and G.W. Boehlert, eds), pp. 43–44. US Department of Commerce,

NOAA Technical Report, NMFS **43**, 43–44.

Griggs, G.B., Carey, A.G. and Kulm, L.D. (1969). Deep-sea sedimentation and sediment – fauna interaction in Cascadia Channel and on Cascadia Abyssal Plain. *Deep-Sea Research* **16**, 157–170.

Guinot, D. (1984). Crabes bathyaux de l'ile de la Réunion. *Publications du Commission national française Recherches antarctique* **5**, 7–31.

Guryanova, E.F. (1964). Zoogeographical division of the continental shelf of the World Ocean by the bottom fauna. *In* "Physiko-geographycheskij atlas mira". Karta 68 D, pp. 291–292. Isdatelstvo Akademii nauk SSSR, Leningrad. (In Russian).

Guryanova, E.F. (1972). Zoogeographical division of the sea (Zoogeographicheskoye rajonirovanie morya). *Issledovaniya Fauny Morei, Nauka, Leningrad* **10**(18), 8–21. (In Russian).

Gutt, J. (1988). On the distribution and ecology of sea cucumbers (Holothurioidea, Echinodermata) in the Weddell Sea (Antarctica). *Berichte zur Polarforschung* **41**, 1–87.

Gutt, J., Gorny, M. and Arntz, W.E. (1991). Spatial distribution of Antarctic shrimps (Crustacea: Decapoda) by underwater photography. *Antarctic Science* **3**, 363–369.

Haedrich, R.L., Rowe, G.T. and Pollini, P.T. (1975). Zonation and faunal composition of epibenthic populations on the continental slope south of New England. *Journal of Marine Research* **33**, 191–212.

Haedrich, R.L., Rowe, G.T. and Polloni, P.T. (1980). The megabenthic fauna in the deep sea south of New England, USA. *Marine Biology* **57**, 165–179.

Hashimoto, J., Miura, T., Fujikera, K. and Ossaka, J. (1993). Discovery of vestimentiferan tube-worms in the euphotic zone. *Zoological Science* **10**, 1063–1067.

Hayward, P.J. (1988). Mauritian cheilostome Bryozoa. *Journal of Zoology* **215**(2), 269–356.

Hecker, B. (1990). Variation in megafaunal assemblages on the continental margin south of New England. *Deep-Sea Research* **37A**, 37–57.

Helmcke, J.G. (1940). Die Brachiopoden der Deutschen Tiefsee-Expedition. *Wissenschaftliche Ergebnisse der Deutsche Tiefsee-Expedition auf dem Dampfer "Valdivia" (1898–1899)* **29**, 215–316.

Hessler, R.R. and Thistle, D. (1975). On the place of origin of deep-sea isopods. *Marine Biology* **32**(2), 155–165.

Hessler, R.R., Wilson, G.D. and Thistle, D. (1979). The deep-sea isopods: a biogeographic and phylogenetic overview. *Sarsia* **64**(1/2), 67–75.

Hinz, C. and Schmidt, M.C. (1995). Studies of benthic assemblages in the Laptev Sea. "Modern State and Perspectives of the Barentz, Kara and Laptev Seas Ecosystems Researches. Abstracts of the Papers Submitted to the International Conference 10–15 October, 1995, Murmansk, Russia" (G.G. Matishov, ed.), pp. 92–93. Russian Academy of Sciences, Murmansk. (In Russian with English title).

Hochachka, P.W. and Somero, G.N. (1973). "Strategies of Biological Adaptations". Saunders, Philadelphia.

Jones, M.L., ed. (1985). "The Hydrothermal Vents of the Eastern Pacific: an Overview". *Bulletin of the Biological Society of Washington* **6**, 1–537.

Jumars, P.A. (1975). Environmental grain and polychaete species diversity in a bathyal benthic community. *Marine Biology* **30**, 253–266.

Karpinsky, M.G. (1985). "Peculiarities in benthos distribution on Peruvian shelf" VNIRO, Moscow. In Russian.

Karpinsky, M.G. (1996). Opportunistic benthic populations on the Peruvian shelf and El-Nino. *In* "Hydrobiological Studies in Marine and Oceanic Fishing Grounds" (A.A. Neyman and M.I. Tarverdieva, eds), pp. 29–34. VNIRO, Moscow. (In Russian).

Koltun, V.M. (1964). Investigation of bottom fauna in The Greenland Sea and the central part of Arctic Basin. *Trudy Arcticheskogo i Antarcticheskogo nauchno-issledovatelskogo Instituta Glavnogo Upravlenia Hydrometeorologicheskoy Sluzby pri Sovete Ministrov SSSr* **259**, 13–78. Isdatelstvo "Transport", Moscow–Leningrad. (In Russian).

Kucheruk, N.V. (1995). Upwelling and benthos: gild structure of impoverished communities. *Doklady Akademii Nauk, Moskva* **345**(2), 276–279. (In Russian with English abstract).

Kucheruk, N.V. and Savilova, T.A. (1985). Quantitative and ecological characteristics of bottom fauna in the shelf and upper slope of the North-Peruvian upwelling region. *Bulletin Moskovskogo Obtchestva Ispytateley Prirody, Moskva, Seria biologicheskaya* **90**(6), 70–79. (In Russian with abstract in English).

Kudinova-Pasternak, R.K. (1990). Tanaidacea (Crustacea, Malacostraca) of the Southeastern part of Atlantic Ocean and the region to the North off Mordvinov (Elephant) Island. *Trudy Instituta Okeanologii Akademii nauk SSSR* **126**, 90–107. (In Russian with English abstract).

Kudinova-Pasternak, R.K. (1993). Tanaidacea (Crustacea, Malacostraca) collected in the 43d cruise of the R/V "Dmitri Mendeleev" in the South-Western Atlantic and the Weddell Sea. *Trudy Instituta Okeanologii Rossijskoi Akademii Nauk* **127**, 134–146. (In Russian with English abstract).

Kuksa, V.I. (1983). "The World Ocean Intermediate Waters". Moscow, Gydrometeoisdat. (In Russian with English abstract).

Kussakin, O.G. (1973). Peculiarities of the geographical and vertical distribution of marine isopods and the problem of deep-sea fauna origin. *Marine Biology* **23**(1), 19–34.

Kussakin, O.G. (1984). A giant isopod crustacean (Gigantski ravnonogi rak). *Priroda Moskva* **10**, 41–42. (In Russian).

Kuznetsov, A.P. (1977). Effect of hydrostatic pressure on marine animals. *In* "Okeanologia, Biologia Okeana, tom 1" (M.E. Vinogradov, ed.) pp. 35–40. Nauka, Moscow. (In Russian).

Kuznetsov, A.P., ed. (1985). "Bottom Fauna of the Open-Oceanic Elevations (Northern Atlantic)". Nauka, Moscow. (*Trudy Instituta Okeanologii Academii nauk SSSR* **120**). (In Russian with English abstracts).

Kuznetsov, A.P. (1989). Deep-sea fauna, its adaptive principles, the history of formation. *Trudy Instituta Okeanologii Akademii Nauk SSSR* **123**, 7–22. (In Russian with English abstract).

Kuznetsov, A.P. and Mironov, A.N., eds (1981). "Benthos of the Submarine Mountains Marcus–Necker and Adjacent Pacific Regions". P.P. Shirshov Institute of Oceanology, Academy of Sciences of the USSR, Moscow. (In Russian with English abstract).

Kuznetsov, A.P. and Mironov, A.N., eds (1994). "Bottom Fauna of Seamounts". Nauka, Moscow. (*Trudy Instituta Okeanologii Rossiyskoy Akademii Nauk* **129**). (In Russian with English abstracts).

Lambert, B. and Roux, M., eds (1991). "L'Environnement Carbonate Bathyal en Nouvelle-Caledonie (Programme Envimarges)". Documents et Travaux, Institut Geologique Albert-de-Lapparent (IGAL), Paris.

Lampitt, R.S., Billett, D.S.M. and Rice, A.L. (1986). Biomass of the invertebrate megabenthos from 500 to 4100 m in the northeast Atlantic Ocean. *Marine Biology* **93**, 69–81.

Lampitt, R.S., Raine, R.C.T., Billett, D.S.M. and Rice, A.L. (1995). Material supply to the European continental slope: a budget based on benthic oxygen demand and organic supply. *Deep-Sea Research* **42A**(11–12), 1865–1880.

Laubier, L. (1986). "Des Oasis au Fond des Mers". Le Rocher, Monaco.

Laubier, L. and Monniot C., ed. (1985). "Peuplements Profonds du Golf de Gascogne". Institut Français de Recherch pour l'Exploitation de la Mer, Brest.

Laubier, L. and Sibuet, M. (1977a). Ecology of the benthic communities of the deep North East Atlantic. *Ambio, Special Report* **6**, 37–42.

Laubier, L. and Sibuet, M. (1977b). Résultats de campagnes BIOGAS 3 aout 1972–2 novembre 1974. Publications du Cenre National pour l'Exploitation des Oceans (C.N.E.X.O). *Résultats des Campagnes à la Mer* **11**, 1–77.

Laurin, B. (1992). Découvert d'un squelette de soutien du lophophore chez un brachiopode inarticule. *Comptes rendu hebdomadaire de Seances de l'Academie des sciences, Paris*, Ser. 3 **314**, 343–350.

Levi, C. and Levi, P. (1982). Spongiaires Hexactinellid du Pacifique Sud-Ouest (Nouvelle-Caledonie). *Bulletin du Muséum National d'Histoire Naturelle, Paris*, Série 4, **4A**(3–4), 283–317.

Levi, C. and Levi, P. (1983a). Eponges Tetractinellides et Lithistides bathyales de Nouvelle-Caledonie. *Bulletin du Muséum National d'Histoire Naturelle, Paris*, Série 4 **5A**(1), 101–168.

Levi, C. and Levi, P. (1983b). Demosponges bathyales recoltées par le N/O "Vauban" au sud de la Nouvelle-Caledonie. *Bulletin du Muséum National d'Histoire Naturelle, Paris*, Série 4 **5A**(4), 931–997.

Levi, C. and Levi, P. (1988). Nouveaux spongiaires Lithistides bathyaux a affinites cretacées de la Nouvelle-Caledonie. *Bulletin du Muséum National d'Histoire Naturelle, Paris, s* Série 4 **10A**(2), 241–263.

Lovén, S. (1862). Om Crustaceer i Venern och Vettern. *Ofversigt Kongelige vettenskaps – akademiens fohrhandlingar* **18**, 285–314.

Lukjanova, T.S. (1974). Valuation of bottom fauna biomass in different vertical zones of the ocean. *Vestnik of Moscow State University. Seria geographycheskaya* **4**, 84–87. (In Russian with English abstract).

Malakhov, V.V. and Adrianov A.V. (1995). "Cephalorhynchia – a New Phylum of the Animal Kingdom". KMK Scientific Press, Moscow.

McLean, J.H. (1979). A new monoplacophoran limpet from the continental shelf off southern California. *Scientific Contributions Natural History Museum, Los Angeles* **307**, 1–19.

McLean, J.H. (1981). The Galapagos rift limpet *Neomphalus*: relevance to understanding the evolution of a major Paleozoic–Mesosoic radiation. *Malacologia* **21**(1–2), 291–336.

McLean, J.H. (1985). Preliminary report on the limpets at hydrothermal vents. *Bulletin of the Biological Society of Washington* **6**, 159–166.

Messing, C.G. (1985). Submersible observation of deep-water crinoid assemblages in the tropical western Atlantic Ocean. *In* "Proceedings of the Fifth International Echinoderm Conference", pp. 185–193. Balkema, Rotterdam.

Messing, C.G., Neumann, A.C. and Lang, J.C. (1990). Biozonation of deep-water lithoherms and associated hard grounds in the northeastern straits of Florida. *Palaios* **5**, 15–33.

Mironov, A.N. (1980). Two ways of formation of deep-sea echinoid fauna. *Okeanologia* **20** (4), 703–708. (In Russian with English abstract).

Mironov, A.N. (1982). Role of the Antarctic in formation of deep-sea fauna of the World Ocean. *Okeanologia* **22**(3), 486–493. (In Russian with English abstract).

Mironov, A.N. (1983a). Specific features of sea urchins distribution in midoceanic regions. Abstracts of papers at the 5th All-Union Symposium on echinoderms (Lvov, 1983), pp. 47–48 Gosudarstvenny prirodovedchesky musey Akademii nauk Ukrainy, L'vov. (In Russian).

Mironov, A.N. (1983b). Accumulation effect in distribution of sea urchins. *Zoologicheskii zhurnal* **52**(8), 1202–1208. (In Russian with English abstract).

Mironov, A.N. (1985). Circumcontinental zonality in distribution of sea urchins in the Atlantic. *Trudy Instituta Okeanologii Akademii Nauk SSSR* **120**, 70–95. (In Russian with English abstract).

Mironov, A.N. (1989). Meridional asymmetry and marginal effect in distribution of sea urchins in open ocean regions. *Okeanologia* **29**(5), 845–854. (In Russian with English abstract).

Mironov, A.N. (1994). Bottom faunistic comlexes from oceanic islands and seamounts. *Trudy Instituta okeanologii Rossiyskoy Akademii Nauk* **129**, 7–16. (In Russian with English abstract).

Mironov, A.N. and Rudjakov, J.A., eds. (1990). "Plankton and Benthos from the Nazca and Sala-y-Gómez Submarine Ridges". Nauka, Moscow. (*Trudy Instituta Okeanologii Akademii nauk SSSR* **124**). (In Russian with English abstracts).

Mortensen, T. (1951). "A Monograph of the Echinoidea" **5**(2), 1–593. Copenhagen.

Moskalev, L.I., Mironov, A.N., Levenstein, R.Ya. and Keller, N.B. (1983). A study of the bottom fauna during the 2nd voyage of RV "Vityaz" in the Central Atlantic. *Zoologicheski Zhurnal* **62**(10), 1598–1599.

Newman, W.A. (1979). A new scalpellid (Cirripedia); a Mesozoic relic living near an abyssal hydrothermal spring. *Transactions of the San Diego Society of Natural History* **19**(11), 153–167.

Newman, W.A. (1980). A review of extant *Scillaelepas* (Cirripedia: Scalpellidae) including recognition of new species from the North Atlantic, Western Indian Ocean and New Zealand. *Tethys* **9**(4), 379–398.

Newman, W.A. (1985). The abyssal hydrothermal vent invertebrate fauna, a glimpse of antiquity? *Bulletin of the Biological Society of Washington* **6**, 231–242.

Newman, W.A. (1986). Origin of the Hawaiian marine fauna: dispersal and vicariance as indicated by barnacles and other organisms. *In* "Crustacean biogeography" (R.H. Gore and K.L. Heck, eds), pp. 21–49. Balkema, Rotterdam.

Newman, W.A. and Foster, B.A. (1983). The rapanuian faunal district (Easter and Sala y Gómez): in search of ancient archipelagos. *Bulletin of Marine Science* **33**(3), 633–644.

Newman, W.A. and Foster, B.A. (1987). Southern hemisphere endemism among the barnacles: explained in part by extinction of northern members of amphitropical taxa. *Bulletin of Marine Sciences* **41**(2), 361–377.

Newman, W.A. and Hessler, R.R. (1989). New abyssal hydrothermal verrucomorphan (Cirripedia: Sessilia): the most primitive living sessile barnacle. *Transactions of the San Diego Society of Natural History* **21**(16), 259–273.

Neyman, A.A., Zezina, O.N. and Semenov, V.N. (1977). Bottom fauna of the shelf and the continental slope. *In* "Okeanologia. Biologia Okeana" Vol. 1 (A.S. Monin and M.E. Vinogradov, eds), pp. 269–281. Nauka, Moscow. (In Russian).

Neyman, V.G. (1970). New charts of water streams in the Indian Ocean. *Doklady Akademii Nauk SSSR* **195**(4), 948–951.

Ohta, S. (1983). Photographic census of large-sized benthic organisms in the bathyal zone of Suruga Bay, Central Japan. *Bulletin of the Ocean Research Institute, University of Tokyo* **15**, 1–244.

Oliver, A.P.H. (1983). "Shells of the World". Hamlyn, Barcelona.

Parin, N.V. (1968). "Ichthyofauna of the Open Ocean". Nauka, Moscow. (In Russian).

Parin, N.V. (1979). Some features of pelagic fish distribution in the ocean. *In* "Biological Resources of the World Ocean" (S.A Studenetskiy, ed.), pp. 102–112. Nauka, Moscow. (In Russian).

Parin, N.V. (1982). Biotopic groups of oceanic fish and some tasks of their study. *In* "Abstracts of Papers at the Second All-Union Congress of Oceanologists (Yalta, 1982). No. 5. *Marine Biology. 3–4*". Marine Hydrophysical Institute, Sevastopol.

Parin, N.V. (1984). Oceanic ichthyogeography: an attempt to review the distribution and origin of pelagic and bottom fishes outside continental shelves and neritic zones. *Archiv für Fischereiwissenshaft, Berlin* **35**(1), 5–41.

Parin, N.V. and Beklemishev, C.V. (1966). Importance of long standing alternations in Pacific waters circulation for the geographical distribution and disposition of pelagic animals. *Hydrobiologicheski Zhurnal, Kiev* **2**(1), 3–10. (In Russian).

Pasternak, F.A. (1981). Alcyonacea and Gorgonacea. *In* "Benthos of the Submarine Mountains Marcus–Necker and Adjacent Pacific Regions" (A.P. Kusnetsov and A.N. Mironov, eds), pp. 40–55. P.P. Shirshov Institute of Oceanology, Academy of Sciences of the USSR, Moscow. (In Russian with abstract in English).

Pasternak, F.A. (1991). Species composition, distribution patterns and recruitment of the bottom fauna of isolated underwater rises. 6th Deep-Sea Biology Symposium, Copenhagen, 1991. Abstract, p. 49. Zoological Museum, Copenhagen.

Pasternak, F.A. (1994). Octocorals and their role in the formation of underwater seascapes of the rift zone of the Reykjanes Ridge (Northern Atlantic). Seventh Deep Sea Biology Symposium, Crete, 1994, Abstracts. Institute of Marine Biology of Crete.

Pfannküche, O. (1985). The deep-sea meiofauna of the Porcupine Seabight and abyssal plain (N.E. Atlantic). I. Population structure, distribution pattern and standing stock. *Oceanologica Acta* **8**, 343–353.

Pogrebov, W.B., Kijko, O.A. and Goncharova, E.G. (1995). Composition, structure and distribution of macrobenthos in Santa Anna Trench. *In* "Modern State and Perspectives of the Barentz, Kara and Laptev Seas Ecosystems Research". Abstracts of the Papers Submitted to the International Conference 10–15 October, 1995, Murmansk, Russia (G.G. Matishov, ed.), pp. 75–76. Russian Academy of Sciences, Murmansk. (In Russian with English title of the book).

Preston, F.M. (1962). The canonical distribution of commonness and rarity. *Ecology* **43**(2), 185–213; **43**(3), 410–432.

Rex, M.A. (1983). Geographical patterns of species diversity in the deep-sea benthos. *In* "The Sea", Vol. 8 (G.T. Rowe, ed.), pp. 453–473. Wiley, New York.

Rice, A.L., Aldred, R.G., Billett, D.S.M. and Thurston, M.H. (1979). The combined use of an epibenthic sledge and deep-sea camera to give quantitative relevance to macrobenthos samples. *Ambio Special Report* **6**, 59–72.

Rice, A.L., Aldred, R.G., Darlington, E. and Wild, R.A. (1982). The quantitative estimation of the deep-sea megabenthos: a new approach to an old problem. *Oceanologica Acta* **5**, 63–72.

Richer de Forges, B. (1992). Les campagnes d'exploration de la faune bathyale dans la zone economique de la Nouvelle-Caledonie. *Memoirs Muséum Nationale d'Histoire Naturelle, Paris* Ser. A **145**, 9–54.

Rogers, A.D. (1994). The biology of seamounts. *Advances in Marine Biology* **30**, 305–350.

Romanova, N.N. (1972). Distribucion de benthos en la platforma y en el talud continental de la costa Peruana. *Instituto del Mar del Peru Informes, Callao.* Ser. Espec. No. **IM-128**, 127–132.

Romanova, N.N. and Karpinksy, M.G. (1979). Peculiarities in benthos distribution on the Peruvian shelf. *14th Pacific Science Congress (Khabarovsk, August 1979)* Section FII (Marine Biology), pp. 191–194. Academy of Sciences of the USSR, Moscow.

Romanova, N.N. and Karpinsky, M.G. (1983). Characteristic of peculiarities in benthos distribution on Peruvian shelf. *In* "Bioproductivity of Upwelling Ecosystems" (M.E.

Vinogradov, ed.), pp. 156–162. Institute of Oceanology, Academy of Sciences of the USSR, Moscow. (In Russian).

Roux, M. (1982). De la biogeographie historique des oceans aux reconstitutions paleobiogèographique: tendances et problèmes illustrés par des exemples pris chez des Echinodermes bathyaux et abyssaux. *Bulletin de la Société geologique de France, 7-e série* **24**(5–6), 907–916.

Roux, M. (1985). Découverte d'un representant actuel des crinoides pedonculés paleozoique dans l'étage bathyal de l'ile de la Réunion. *Comptes rendu hebdomadaire de séances de l'Academie des Sciences, Paris*, Sér. III **301**(10), 503–506.

Roux, M., Bouchet, Ph., Bourseau, J.-P., Gaillard, Ch., Grandperrin, R., Guille, A., Laurin, B., Monniot, C., Richer de Forges, B., Rio, M., Segonzac, M., Vacelet, J. and Zibrowius, H. (1991). L'étagement du benthos bathyal observé a l'aide de la soucoupe "Cyana". *Documents et Travaux Institut geologique Albert-de-Lapparent, Paris* **15**, 151–165.

Rowe, G.T. and Haedrich, R.L. (1979). The biota and biological processes on the continental slope. *The Society of Economic Paleontologists and Mineralogists (SEPM) Special Publication* **27**, 49–59.

Rowe, G.T., Polloni, P.T. and Haedrich, R.L. (1982). The deep-sea macrobenthos on the continental margin of the northwest Atlantic Ocean. *Deep-Sea Research* **29A**, 257–278.

Sanders, H.L., Hessler, R.R. and Hampson, G.R. (1965). An introduction to the study of deep-sea benthic faunal assemblages along the Gay Head–Bermuda transect. *Deep- Sea Research* **12**, 845–867.

Sasaki, T. (1978). The progress and current status on exploration of seamounts fishing grounds. *Bulletin of the Japanese Society of Fisheries and Oceanography* **33**, 51–53.

Sasaki, T. (1986). Development and present status of Japanese trawl fisheries in the vicinity of seamounts. *In* "The Environment and Resources of Seamounts in the North Pacific" (R.N. Uchida, S. Hayasi and G.W. Boehlert, eds), pp. 21–30. *Proceedings of the Workshop on the Environment and Resources of Seamounts in the North Pacific* **43**.

Sirenko, B.I. and Petryashov, V.V. (1995). Bottom ecosystems of the Laptev Sea and adjacent regions of the Central Antarctic Basin. *In* "Modern State and Perspectives of the Barentz, Kara and Laptev Seas Ecosystems Researches." Abstracts of the Papers Submitted to the International Conference October 10–15, 1995, Murmansk, Russia (G.G. Matishov, ed.), pp. 86–87. Russian Academy of Sciences, Murmansk. (In Russian).

Sirenko, B.I. and Piepenburg, D. (1994). Current knowledge on biodiversity and benthic zonation patterns of Eurasian Arctic shelf seas, with special reference to the Laptev Sea. *Berichte zur Polarforschung* **144**, 69–77.

Sirenko, B.I., Petryashov, V.V., Rachov, E. and Hinz, K. (1995). Bottom biocoenoses of the Laptev Sea and adjacent areas. *Berichte zur Polarforschung* **176**, 211–221.

Smith, J.L.B. (1931). A living coelacanthid fish from South Africa. *Transactions of the Royal Society of South Africa* **28**(1), 1–106.

Sokolova, M.N. (1976a). Large scale division of the World Ocean by trophic structure of deep-sea macrobenthos. *Trudy Instituta Okeanologii Akademii nauk SSSR* **99**, 20–30. (In Russian with English abstract).

Sokolova, M.N. (1976b). Trophic zonality of the deep-sea macrobenthos as an element of the biological structure of the ocean. *Okeanologia* **16**(2), 336–342. (In Russian).

Sokolova, M.N. (1978). Trophical classification of distributional types of deep-sea macrobenthos. *Doklady Akademii Nauk SSSR* **241**(2), 471–474. (In Russian).

Sokolova, M.N. (1979). The global distribution of trophic zones on the ocean floor. *Doklady Akademii Nauk SSSR* **246**(1), 250–252. (In Russian. There is an English translation).

Sokolova, M.N. (1986). "Pitanie y Trophicheskaya Struktura Glubokovodnogo Macrobentosa". Nauka, Moscow. (In Russian). See English translations of the book: Sokolova, 1994, 1997.

Sokolova, M.N. (1994). "Feeding and Trophic Structure in the Deep-sea Macrobenthos". Smithsonian Institution Library, Washington, DC.

Sokolova, M.N. (1997). "Feeding and Trophic Structure in the Deep-sea Macrobenthos". Oxonian Press, New Delhi (in press).

Sokolova, M.N., Vinogradova, N.G. and Turpaeva, E.P. (1996). Trophic structure of megabenthos in the bathyal zone of the Norwegian Sea. In "Benthos of the Northern Eurasian Seas" (A.P. Kuznetsov and O.N. Zezina, eds), pp. 12–27. Institute of Oceanology, Russian Academy of Sciences, Moscow. (In Russian with English abstract) (in press).

Stepanov, V.N. (1965). Principal types of structure in the World Ocean waters. Okeanologiya 5(5), 798–802. (In Russian with English abstract).

Stepanov, V.N. (1974). "The World Ocean". Znanie, Moscow. (In Russian).

Stepanov, V.N. (1983). "Okeanosphera". Mysl, Moscow. (In Russian).

Strakhov, N.M. (1932). "Tasks and Methods of Historical Geology". Uchpodgiz, Moscow–Leningrad. (In Russian).

Sverdrup, H.U., Johnson, M.W. and Fleming, R.H. (1942). "The Oceans, their Physics, Chemistry and General Biology". Prentice-Hall, New York.

Tabachnick, K.R. (1994). Distribution of recent Hexactinellida. In "Sponges in Time and Space" (R.W.M. Van Soest, T.M.G. Van Kempen and J.-Cl. Braekman, eds), pp. 225–232. Balkema, Rotterdam.

Thiel, H. (1978). Benthos in upwelling regions. In "Upwelling Ecosystems" (R. Boje and M. Tomczak, eds), pp. 124–138. Springer, Berlin.

Thiel, H. (1981). Benthic investigations in the Northwest African upwelling area. Report of the cruises 26, 36, 44 and 53 of R.V. "Meteor". "Meteor" Forschungen Ergebnisse 33, 1–15.

Thiel, H. (1982). Zoobenthos of the CINECA area and other upwelling regions. Rapport et Procès-verbaux des Réunions Conseil permanent international pour l'Exploration de la Mer 180, 323–334.

Tscherbinin, A.D. (1971). On relations between circulation and water structure in the Indian Ocean. Doklady Akademii Nauk SSSR 199(6), 1413–1416.

Turpaeva, E.P. (1975). On some deep-water pantopods (Pycnogonida) collected in northwestern and south-eastern Pacific. Trudy Instituta Okeanologii Akademii Nauk SSSR 103, 230–246. (In Russian with English abstract).

Valentine, J.W. and Jablonski, D. (1982). Major determinants of the biogeographic pattern of the shallow-sea fauna. Bulletin de la Société géologique de France, Série 7 24(5–6), 893–899.

Vinogradov, M.E., Sagalevich, A.M. and Khetagurov, S.V. (eds), (1996). "Oceanographic Research and Underwater Technical Operations on the Site of Nuclear Submarine 'Komsomolets' Wreck". Nauka, Moscow. (In Russian with English titles).

Vinogradova, N.G. (1956). Zoogeographical subdivision of the abyssal of the World Ocean. Doklady Akademii Nauk SSSR 111, 195–198. (In Russian).

Vinogradova, N.G. (1969). The geographical distribution of the deep-sea bottom fauna. In "The Pacific Ocean, Biology of the Ocean, Book 2. The Deep-sea Bottom Fauna and the Pleuston" (V.G. Kort, ed.), pp. 154–181. Nauka, Moscow. (In Russian).

Vinogradova, N.G. (1976). Large face of the Ocean. In memory of L.A. Zenkevich. Priroda, Moskva 11, 94–105. (In Russian).

Vinogradova, N.G. (1977). Bottom fauna of abyss and ultraabyss. In "Okeanologiya. Biologiya Okeana 1" (A.S. Monin and M.E. Vinogradov, eds), pp. 281–298. Nauka,

Moscow. (In Russian).

Vinogradova, N.G. and Zezina, O.N. (1996). Zoogeographical division of the bottom in the World Ocean. *In* "World Atlas of Resources and Environment". Map 77. Geographical Institute of Russian Academy of Sciences and Zarubezhgeologia, Moscow. (In Russian with titles and abstracts in English).

Vinogradova, N.G., Galkin, S.W., Kamenskaya, O.E., Romanov, V.N. and Savilova, T.A. (1990). The characteristic of the bottom fauna from the continental coast of Namibia in the Benguela upwelling region according the data of the 43-d cruise of R/V "Akademik Kurchatov". *Trudy Instituta okeanologii Akademii nauk SSSR* **126**, 45–61.

Wells, S.M., Pyle, R.M. and Collins, N.M. (1983). "Invertebrate Red Data Book", International Union for Conservation of Nature and Natural Resources, Conservation Monitoring Centre, Cambridge. Unwin, Gland.

Weston, J.F. (1985). Comparison between recent benthic foraminiferal faunas of the Porcupine Seabight and Western Approaches continental slope. *Journal of Micropaleontology* **4**, 156–183.

Wilson, R.R. and Kaufmann, R.S. (1987). Seamount biota and biogeography. *In* "Seamounts, Islands and Atolls" (B.H. Keating, P. Fryer, R. Batiza and G.W. Boehlert, eds), pp. 319–334. *Geophysical Monograph* **43**. American Geophysical Union, Washington, DC.

Wilson, R.R., Smith, K.L. and Rosenblatt, R.H. (1985). Megafauna associated with bathyal seamounts in the central North Pacific Ocean. *Deep-Sea Research* **23**(10), 1243–1254.

Yamaguchi, T. and Newman, W. (1990). A new primitive barnacle (Cirripedia: Balanomorpha) from the North Fiji Basin abyssal hydrothermal field, and its evolutionary implications. *Pacific Science* **44**(2), 135–155.

Zenkevich, L.A. (1948). Biological structure of the Ocean. *Zoologicheski Zhurnal* **27**(2), 113–124. (In Russian with English abstract).

Zenkevich, L.A. (1968). Problema bathyali (The problem of the bathyal zone). *Rybnoe Khozyaistvo* **10**, 4–6. (In Russian).

Zenkevich, L.A. (1973). Problem of investigation of the bathyal zone in the oceans and seas. *Trudy Instituta okeanologii Akademii nauk SSSR* **91**, 10–13. (In Russian with English abstract).

Zevina, G.B. and Galkin, S.V. (1989). New species of cirripeds (Cirripedia, Thoracica) from thermal waters. *Zoologicheski Zhurnal* **68**(3), 134–136.

Zevina, G.B. and Galkin, S.V. (1992). *Altiverruca beringiana* sp. n. (Crustacea, Cirripedia) – a first find of Verrucomorpha in boreal Pacific. *Zoologicheski Zhurnal* **71**(8), 140–144.

Zezina, O.N. (1970). Brachiopod distribution in the Recent Ocean with references to problems of zoogeographical zoning. *Paleontologicheskii Zhurnal Moskva* **2**, 3–17. (In Russian. There is an English translation pp. 147–160).

Zezina, O.N. (1971). "Distribution and some biological features of recent brachiopods". Candidate thesis. Dissertacia kandidata biologicheskikh nauk. Institut Okeanologii Akademii nauk SSSR. Moscow.

Zezina, O.N. (1973). Biogeographical division of the benthic area of the ocean by brachiopods. *Proceedings of All-Union Research Institute of Marine Fisheries and Oceanography (Trudy VNIRO)* **84**(4), 166–180. (In Russian with abstract in English).

Zezina, O.N. (1975). Recent Caribbean deep-sea brachiopod fauna, the sources and the conditions of its formation. *Trudy Instituta Okeanologia Akademii Nauk SSSR* **100**, 188–195.

Zezina, O.N. (1976a). "Ecology and Distribution of the Recent Brachiopods". Nauka, Moscow. (In Russian).

Zezina, O.N. (1976b). Biogeography of the oceanic bathyal zone based on the distribution of the recent brachiopods. "International Geography '76, Publications of XXIII-d International Geographical Congress, USSR, Moscow, 1976, 3. Geography of the Ocean". Moscow State University, Moscow. (In Russian, pp. 44–46 and in English, pp. 46–48).

Zezina, O.N. (1978). On symmetry and its distortions in the latitudinal faunistic zoning at the World Ocean shelves as shown by brachiopods. *In* "Regularities of Distribution and Ecology of Coastal Marine Biocenosis" (O.A. Scarlato and A.N. Golikov, eds), pp. 174–176. USSR–USA Symposium on the Program "Biological Productivity and Biochemistry of the World Ocean", Leningrad (November 30–December 4, 1976). Nauka, Leningrad.

Zezina, O.N. (1979). On the formation of the recent fauna of brachiopods on the shelves and slopes of the World Ocean. *Bulletin Mokovskogo Obchestva Ispytateley Prirody Moskva. Ondel biologicheski* **84**(5), 52–59. (In Russian with English abstract).

Zezina, O.N. (1981a). Recent deep-sea Brachiopoda from the Western Pacific. *Galathea Report* **15**, 7–20.

Zezina, O.N. (1981b). The composition and the ways of formation of the thalassobathyal brachiopod fauna. *In* "Benthos of the Submarine Mountains Marcus–Necker and Adjacent Pacific Regions" (A.P. Kuznetsov and N.A. Mironov, eds), pp. 141–149. P.P. Shirshov Institute of Oceanology, Academy of Sciences of the USSR, Moscow. (In Russian with English abstracts).

Zezina, O.N. (1981c). On amphioceanic differences in recent warm water faunas as it is seen in brachiopods. *In* "Paleontologya, Paleobiogeographia, Mobilism" (G.Ya. Krymholz and C.V. Simakov, eds), pp. 161–166. Trudy 21–oj sessii Vsesojuznogo Paleontologicheskogo Obstchestva Akademii Nauk SSSR, Leningrad, January, 1975. Magadanskoye Kniznoe Isdatelstvo, Magadan. (In Russian).

Zezina, O.N. (1981d). On the formation of recent brachiopod fauna at the shelves and the slopes of the Pacific Ocean. *In* "Deep-Sea Biology of the Pacific Ocean" (N.G. Vinogradova, ed.), pp. 20–23. Papers of 14th International Pacific Science Congress, Khabarovsk, 1979, Section "Marine Biology", 1. Institut Biologii Morya, Dalnevostochni Nauchni Centr Akademii Nauk SSSR, Vladivostok. (In Russian with English abstracts).

Zezina, O.N. (1982). A species, a genus and a family in the biogeographical analysis of the oceanic bottom fauna. *In* "Morskaya biogeographiya" (O.G. Kussakin, ed.), pp. 18–26. Nauka, Moscow. (In Russian).

Zezina, O.N. (1983). "Brachiopods of the World Ocean and the formation of the bathyal fauna." Doctoral thesis. Dissertacia doktora biologicheskikh nauk. Institut Okeanologii Akademii Nauk SSSR, Moscow. (In Russian).

Zezina, O.N. (1985a). "Recent Brachiopods and Problems of the Bathyal Zone of the Ocean". Nauka, Moscow. (In Russian).

Zezina, O.N. (1985b). Rare species of recent brachiopods and the problem of "living fossils". *In* "Ecology of the Ocean Coastal Zone Benthic Fauna and Flora" (A.P. Kuznetsov, ed.), pp. 9–14. P.P. Shirshov Institute of Oceanology, Academy of Sciences of the USSR, Moscow. (In Russian with abstract in English).

Zezina, O.N. (1987). Brachiopods collected by "BENTHEDI"-Cruise in the Mozambique Channel. *Bulletin du Muséum National d'Histoire Naturelle, Paris* Série 4 **9**, Section A, no. 3, 551–562.

Zezina, O.N. (1990). Composition and distribution of articulate brachiopods from the underwater rises of the Eastern Pacific. *Trudy Instituta okeanologii Akademii Nauk SSSR* **124**, 264–269. (In Russian with English abstract).

Zezina, O.N. (1993). New findings of recent deep-sea brachiopods from the Weddell Sea and nearby the Orkney Trench. *Trudy Instituta okeanologii Rossijskoy Akademii nauk* **127**, 198–200. (In Russian with English abstract).

Zezina, O.N. (1994). Bathyal zone of the ocean as a suitable store for zoological relics. *Okeanologia* **34**(3), 404–440. Nauka, Moscow. (In Russian with abstract in English).

Zezina, O.N., Sokolova, M.N., and Pasternak, F.A. (1986). On the biotopic base of geographical ranges of bottom animals in different vertical zones. *Trudy Vsesojuznogo Hydrobiologicheskogo Obtchestva Akademii Nauk SSSR, Nauka, Moskva* **27**, 64–75. (In Russian.)

Zezina, O.N., Neyman, A.A. and Sokolova, M.N. (1993). Trophic zones and landscapes at the ocean bottom. *In* "Feeding of Marine Invertebrates in Different Vertical Zones and Latitudes" (A.P. Kuznetsov and M.N. Sokolova, eds), pp. 65–77. P.P. Shirshov Institute of Oceanology, Russian Academy of Sciences, Moscow. (In Russian with English abstract).

Trophic Structure of Abyssal Macrobenthos

M.N. Sokolova

P.P. Shirshov Institute of Oceanology, Russian Academy of Sciences, Nakhimovsky Prospekt 36, Moscow 117851, Russia

ABSTRACT

The influence of feeding conditions on the large-scale distribution patterns of bottom-living invertebrates larger than 5 mm has been investigated in 228 bottom trawl samples collected by Russian research vessels from depths of 3000 to 6000 m in the Pacific and Indian Oceans. Comparisons were made with data from

ADVANCES IN MARINE BIOLOGY VOL. 32
ISBN 0-12-026132-4

the Atlantic Ocean and additional data from published sources: in all 836 samples were assessed, from 750 stations. The animals were grouped into deposit-feeders, suspension-feeders and carnivores. The ratio of the weights of the three groups indicates the predominant trophic group in a sample. Such data, combined with information on geographic distribution, were then compared with feeding conditions including: sediment type and sedimentation rate; organic carbon content; degree of transformation of organic matter on and in the sediment; redox potential, biochemical oxygen demand; and state of the sediment microflora.

In the Pacific and Indian Oceans it has been possible to define eutrophic and oligotrophic areas. The eutrophic areas correspond to regions of high production in surface waters and extend over the peripheral and equatorial parts of the oceans. Sediments in eutrophic areas contain enough digestible organic matter to allow deposit-feeders to predominate. Oligotrophic areas are restricted to the open parts of the oceans, outside the equatorial belt, where sedimentation rates are low, organic matter is scarce and suspension-feeders predominate, although their population density is very low.

In contrast to the Pacific, conditions on the Atlantic Ocean floor are more generally eutrophic. This seems to be related to the stronger currents near the bottom in Atlantic abyssal depths, particularly near the continents. In the Near-Continental areas suspension-feeders (such as sponges, stalked barnacles, bivalves and ascidians) predominate in the macrobenthos. The South Atlantic Near-Continental Area is characterized by the occurrence of many carnivorous invertebrates, particularly ophiuroids, that feed on a "rain of dead bodies" from the sea above (euphausiids, pteropods and small fishes). This "rain" exerts a strong influence on the trophic structure of the macrobenthos in this part of the Atlantic.

The recognized deep-sea trophic areas differ in the frequency of occurrence of deposit-feeding invertebrates and their degree of restriction to the area, in the size, number and species composition of the macrobenthic invertebrates and in sedimentation conditions. Differences in trophic pattern between the Pacific, Indian and Atlantic oceans are closely related to the size and shape of each ocean. There is evident latitudinal and circumcontinental zonation in the distribution of trophic areas of all three oceans.

The impact of trophic factors on the distribution patterns of deep-sea macrobenthos is shown in the three main types of species distribution: (i) Near-Continental; (ii) Oceanic (for species penetrating from the Near-Continental Area into the Equatorial Area as well as into the extraequatorial oceanic areas); and (iii) Panthalassic (for species which penetrate from eutrophic into oligotrophic areas). Species abundance decreases from the near-continental eutrophic areas to the Oceanic and the Equatorial areas and is least in the Northern and Southern oligotrophic areas.

1. INTRODUCTION

Deep-sea ecological research in Russia began in the 1950s with the collection of deep-sea organisms undertaken by the pioneering expeditions of the RV "Vityaz". One of the major problems studied has been how the benthic invertebrates are nourished. Observations on gut contents have provided knowledge of the food sources and methods of feeding employed by hundreds of organisms from various systematic groups. Most of the results from these studies have been published in the Russian literature (Sokolova, 1956, 1957a,b, 1958, 1977, 1982, 1986, 1987, 1988, 1990a, 1994, 1994a,b, 1995; Litvinova and Sokolova, 1971; Bordovsky *et al.*, 1974; Kozjar *et al.*, 1974). Data on the life style of major mega- and macrobenthos representatives allowed analysis of their distribution in relation to feeding conditions in the deep sea.

The Russian expeditions of the 1950s–1970s covered most of the World Ocean. In addition to collection of benthic organisms, attention was given to ocean floor topography, sediment layers and the water column. In combination, these studies allowed analysis of the distribution of the benthic fauna together with the factors likely to control feeding conditions in different parts of the ocean (Bezrukov, 1962, 1964a,b; Bordovsky, 1964a,b; Bogdanov, 1965; Skornyakova and Murdmaa, 1968; Romankevitch, 1970, 1971, 1977; Lisitsyn, 1971; Koblentz-Mishke, 1977). Among the results of these studies was the discovery of "deserts" on the ocean floor beneath anticyclonic gyres, termed "oligotrophic" areas. The concepts of eutrophic and oligotrophic areas in the ocean were originally proposed for the Pacific where they are strongly expressed. Later studies were made in the Indian Ocean and finally in the Atlantic, where true oligotrophic conditions were not found (Sokolova, 1960b, 1964, 1965, 1968, 1969, 1972, 1976a,b, 1978, 1979, 1981a,b, 1984, 1985, 1986, 1990b, 1993a,b).

The mega- and macrobenthos considered here include both the epifauna and the infauna. Many of the animals live within the sediment (Sipuncula, Echiura, Bivalvia, most Polychaeta, Echinoidea, porcellanasterid Asteroidea, some Holothurioidea and Ophiuroidea). The larger organisms appear to be very sensitive to food supply, and hence they show the most obvious reactions to changing conditions. For this reason they present a convenient objective for large-scale ecological research in the deep sea.

2. SUBJECT OF RESEARCH AND MATERIALS

2.1. Research Objectives

The macrobenthos inhabiting the ocean floor, the subject of the present study, comprises large invertebrates (over 5 mm length), grouped into two size categories corresponding to the sampling gear and the sieving technique. Invertebrates about

5 mm long constitute the "grab macrobenthos". Larger invertebrates, as a rule
> 10 mm, constitute the "trawl macrobenthos". The latter has been the main
objective of studies of nutrition and types of distribution. Odum (1975) regarded
large species as better indicators than small ones. The trawl macrobenthos to a
great extent (but not entirely) corresponds to the "megafauna" in the terminology
of researchers in other laboratories. The size range of megafauna is not always
defined in publications, but is often taken to comprise animals caught in trawl-nets
(Hessler and Jumars, 1974), the quantitative sampling of which is impossible
because of their rarity (Hessler, 1974; Thiel, 1975); they have also been defined
as metazoans easily distinguished on photographs (Grassle *et al.*, 1975). Thus, it
appears that the typical megafauna corresponds to trawl macrobenthos in our
material. The majority of authors class fish among metazoan megafauna as well
as large invertebrates. In this review, only invertebrates are discussed and the size
range of trawl macrobenthos is extended by inclusion of smaller animals, by
sieving trawl samples through the 0.45-mm mesh used for grab samples. Small
invertebrates of about 5 mm do not contribute much to trawl samples in terms of
weight and hence their presence does not influence essentially the calculations of
weight indices. Such invertebrates do not match all the characteristics of
megafauna, partly because they are not easily seen on photographs, and also
because they include infaunal species living in the sediment. Consequently,
macrobenthos considered here includes, in addition to all invertebrates belonging
to megafauna, an intermediate size class of benthic animals closer to macrofauna,
according to definitions used in other countries. Macrofauna is usually defined as
invertebrates caught by quantitative sampling gear, such as the box-corer, retained
on a 0.42-mm mesh (Sanders *et al.*, 1965; Rowe and Menzel, 1971; Greenslate
et al., 1974), or sometimes 0.297-mm mesh (Hessler and Jumars, 1974). The grab
benthos considered in the present work corresponds mainly to this size class; but
is slightly larger because of the use of a mesh size of 0.45 mm. The upper size limit
of macrofauna, separating it from megafauna, has not been defined precisely in the
literature. Based on the practice of deep-sea benthos sampling by grabs and trawls,
Sokolova (1976a,b) earlier took an arbitrary 5 mm length as the upper limit for the
grab benthos. Larger animals are generally rare in deep-sea grab samples. In the
present work, the "megafauna" has not been distinguished as a separate entity;
instead, the concept of grab and trawl macrobenthos is used. The latter is the main
object of the present review, whereas the former is used to evaluate the abundance
of life in terms of biomass per unit area. The usage of such size groups of benthos
for feeding and trophic structure research has been recommended by the Scientific
Committee on Ocean Research (SCOR) working group (Rice *et al.*, 1994).

2.2. Research Areas

As noted above, the present review deals with macrobenthos of the part of the
ocean floor termed the abyssal zone. The abyssal zone lies within the depth range

from 3000 to 6000 m according to the scheme of vertical biological zonation proposed by Belyaev *et al.* (1959, 1973), Zenkevich (1963) and Vinogradova (1969). The abyssal zone is thus one of the principal bathymetric regions of the world ocean, covering 77% of the bottom area (Atlas of Oceans, 1980). The usual scheme of vertical zonation includes a transitional horizon at the upper boundary of the abyssal zone, at 2500–3000 m, corresponding to what some workers call the continental rise. Samples from this depth range have also been considered in this review, making it possible to extend the boundaries of the area studied so as to include the Gulf of Alaska and the Bay of Bengal.

2.3. Environmental Factors

Organic matter in the sediment and in suspension close to the bottom is the main food source for most benthic invertebrates. Hence, as a primary characteristic of feeding conditions, the organic matter content of the sediment and the absolute amount of buried organic carbon has been assessed. Other parameters of sediment formation, such as the overall rate of accumulation, redox potential and the thickness of the upper oxidized layer were also analysed. In addition to sediment factors, bottom topography and the abundance of phytoplankton were also considered. These data were usually obtained on the same expeditions and in the same areas as those where benthic sampling was carried out. This has allowed direct comparison between fauna distribution and the characteristics of the sediment and the water column.

2.4. Methods of Assessment

To examine trophic factors, groups of species with a similar type of feeding were assessed. The organisms from each sample were classed into feeding types, at first without regard to taxonomic affinity and then with it. The advantage of groups of species being used as indicators, compared to individual species, is discussed by Odum (1975). The following parameters were used to analyse the influence of trophic factors at each station: (i) weight proportions of trophic groups in trawl samples; (ii) distribution of supra-species taxonomic groups with similar type of feeding; and (iii) conditions of feeding, evaluated on the basis of (a) organic carbon content of sediment and its distribution (Romankevich, 1977) and (b) the lithologic–genetic (not diagenetic) type of the sediment including the values of organic carbon accumulation rates mg/square cm/1000 years and the rate of sediment accumulation. Records of the abundance of shark teeth (Belyaev and Glikman, 1970) on the ocean floor were used as an additional indicator of low rates of sediment accumulation. The data from (i) resulted in a scheme of distribution of trophic groups of macrobenthos on the ocean floor. The data from (ii) showed which of the taxa cause a change in the proportion of trophic groups. The data from (iii) compared with that from the first two, revealed factors restricting the distribution of abyssal invertebrates.

To calculate the proportions of trophic groups without systematic affinity, only original data were used. For analysis of the distribution of certain systematic groups corresponding to different trophic groups, data from other deep-sea expeditions were included (see Table 1).

Estimation of the abundance of benthic fauna based on trawl macrobenthos is difficult. To compare the abundance of fauna in different deep-sea trenches, Belyaev (1966) used the total number of specimens per catch without correlation to the area sampled. Haedrich *et al.* (1975) and Haedrich and Rowe (1977) have calculated the biomass of megafauna collected by trawls by extrapolating the average wet weight of a single specimen (obtained by dividing the total weight of the catch by the number of specimens) to the absolute number of specimens per unit area. The absolute number of animals was estimated from bottom photographs taken at the site of trawling. It was concluded that the range of the megafauna biomass is similar to that of macroinfauna. These investigations were conducted on the continental slope in the West Atlantic. In the same area, numerical data were obtained on the average size of individuals of dominant megafaunal taxa (echinoderms and crustaceans) from trawl samples and on the average size of macrobenthic invertebrates from box-corer samples (Polloni *et al.*, 1979; Haedrich *et al.*, 1980). The former correspond to the trawl macrobenthos and the latter to the grab macrobenthos sieved on 0.42-mm mesh. The average weight of individuals in the samples obtained by both methods was determined by dividing the wet weight by the total number of individuals. For large animals from trawl samples, the taxa were considered in calculations, whereas for smaller animals from the box-corer samples entire samples were used. Wet weight can be a reliable indicator of abundance, as demonstrated by Rowe and Menzel (1971), who compared it to dry weight and organic carbon content of the infauna.

In the present work, the abundance of trawl macrobenthos was evaluated using an approach similar to that of Haedrich *et al.* (1980), who calculated the average weight for vertical zones identified by species composition. Lack of data on the biomass of megafauna forced us to use the average weight of individuals and their number per trawl sample. Following Belyaev (1966), it was found possible to compare various biotopes by using the number of individuals per trawl sample as indicators of relative abundance. The average weight of trawl macrobenthic specimens was calculated for every region based on the proportion of trophic groups.

The average weight of deep-sea invertebrates in different areas of the ocean floor is important not only for evaluation and comparison of the abundance of macrobenthos in these areas but also for discussion of the size structure of deep-sea macrobenthos. Thiel (1975, 1979), using original data on meiofauna and data on macrofauna from the Atlantic and the Pacific (Rowe, 1971a,b; Rowe and Menzel,

1971), proposed the hypothesis of a decrease in animal size with increasing depth leading to a small average size in deep-sea benthic communities. Thiel considered that this reduction was due to trophic factors and that communities experiencing constant food shortage consist, on average, of smaller individuals (Thiel, 1975). The megafauna was not considered in this hypothesis, although the observed pattern of a size decrease with depth was suggested also for the part of megafauna that feeds on the sea floor (Thiel, 1979). Haedrich *et al.* (1980) analysed the change of average weight of megafauna individuals in the same area at depths from 40 to 5000 m and concluded that at the maximum depths the largest animals have disappeared. It is important to note that Haedrich *et al.* included fish with invertebrates in the megafauna. Lampitt *et al.* (1986) came to a similar conclusion having studied the megafauna of the slope in the northeast Atlantic, from 500 to 4100 m. These authors also explained the disappearance of the largest animals in the deepest zone as a result of diminished food supply compared to that in the overlying zones. However, almost the whole region they studied corresponds to the continental margin, with the deepest part at the continental rise. Their observations, therefore, may relate only to deficit in food supply in this zone by comparison with the shelf and possibly the upper part of the slope. It is this part of the deep sea that is richest in food. The largest amount of food for deep-sea benthic animals (expressed as organic carbon in the sediment) is found at the base of the continental slope and on adjacent parts of the ocean floor (Romankevich, 1977). Of course, within this narrow belt bounding all continents, differences in feeding conditions on the bottom are undisputed, but they are most significant between the shelf and the ocean floor at the base of the slope, whereas the difference between the underlying zones is less. A decrease in the quantity of large megafaunal organisms corresponds to the relatively small change in trophic conditions at the transition from the lower part of the slope to the abyssal plain (at the base of the continental rise). These results, when related not only to fish but to megafaunal invertebrates, suggest a high sensitivity of the size of megabenthos to feeding conditions (Lampitt *et al.*, 1986).

In the present work the average weight of invertebrates of trawl macrobenthos (including the same "megafauna" taxa) is compared with the principal changes in feeding conditions, not with increasing depths along the slope, but with the transition from the edge of the ocean floor to the central parts and in relation to contrasting changes in surface production and conditions of sediment accumulation.

2.5. Volume of Material

Tables 1–3 present the number of trawl and grab samples used for dividing up the ocean floor into regions with different feeding conditions. A total of 836 samples from Russian and other expeditions obtained at 750 oceanographic stations were

Table 1   Material used to determine the trophic structure of the deep-sea benthos.

	Eutrophic areas						Oligotrophic areas		Total	
	Near-continental		Oceanic		Equatorial					
Vessel	Trawls and dredges	Bottom samplers (=grabs)	Trawls and dredges	Bottom samplers (=grabs)	Trawls and dredges	Bottom samplers (=grabs)	Trawls and dredges	Bottom samplers (=grabs)	Trawls and dredges	Bottom samplers (=grabs)
1	2	3	4	5	6	7	8	9	10	11
Pacific Ocean										
"Vityaz", "Akademik Kurchatov", "Dmitrii Mendeleev"	44	42	35	69	14 (16)*	66 (81)	21 (42)	63 (73)	137	240 (265)
"Ob"	1	–	2	6	–	–	–	7	3	18
"Challenger"	7	–	12	–	–	–	–	–	28	–
"Galathea"	11	–	–	–	–	–	–	–	11	–
"Albatross"	17	–	–	–	8	–	3	–	28	–
"Eltanin"	7	–	–	–	–	–	–	–	7	–
Total for the ocean	87	42	49	75	22 (27)	66 (81)	24 (51)	70 (80)	214 (214)	258 (283)
Indian Ocean										
"Vityaz", "Akademik Kurchatov", "Dmitrii Mendeleev"	48 (59)	83	26 (36)	53 (70)	–	–	5	8	79 (100)	144 (161)
"Ob"	3	14	3	18			–	1	6	33
"Challenger"	2	–	5	–			–	–	7	–
"Galathea"	17	–	–	–			–	–	17	–

Table 1 Continued

Vessel 1	Eutrophic areas						Oligotrophic areas		Total	
	Near-continental		Oceanic		Equatorial					
	Trawls and dredges 2	Bottom samplers (=grabs) 3	Trawls and dredges 4	Bottom samplers (=grabs) 5	Trawls and dredges 6	Bottom samplers (=grabs) 7	Trawls and dredges 8	Bottom samplers (=grabs) 9	Trawls and dredges 10	Bottom samplers (=grabs) 11
				Indian Ocean						
"Valdivia"	3	–	1	–	–	–	–	–	4	–
"Valdivia II"	–	–	1	–	–		–	–	1	–
"Investigator"	6	–	–	–	–		–	–	6	–
"Albatross II"	1	–	–	–			–	–		
"Gauss"	3	–	–	–						
Total for the ocean	83 (94)	97	36 (46)	71 (88)			5	9	120 (143)	177 (194)
Total for two oceans	170 (181)	139	85	146 (163)			34 (56)	79 (89)	334 (357)	435 (477)

*Number of samples, when they are more numerous than the number of stations, is shown in parentheses.

used. The quantitative data for the Atlantic Ocean are based mostly on material from Russian expeditions, but some published data from local studies on deep-sea benthos have also been included, comprising 256 stations from the "Challenger", "Valdivia", "Ingolf", "Albatross", "Albatross II", "Talisman", "Princess Alice", "Galathea", "Vema" and some other expeditions.

These stations are shown on the figures. For comparisons of identified regions in terms of the average weight and number of specimens of trawl macrobenthos, 180 samples from the collection of the Institute of Oceanology were used. Analysis of species composition is based on 132 species for which the number of records exceeded 1000.

Groups with a particular type of feeding were distinguished by gut content, following Turpaeva (1953). For each individual, the food particles were examined with a light microscope at 32 and 65× magnification. The percentage and frequency of occurrence of every component of the food mass, a general index of fullness of gut and the proportion of individuals with and without food were determined for each species. The significance of these indices is discussed in Section 2. Table 4 presents the information on the number of species of each taxa examined. The number of specimens is noted in describing the results, but for several species this number was small, owing to the small number of deep-sea samples available for analysis and the need to retain examples for systematic studies.

Most of the samples used in the gut content studies were obtained from the continental slope and slopes of deep-sea trenches where the abundance of animals is relatively high. In total, 1949 specimens of invertebrates belonging to 118 species from 162 stations were dissected. Additional published data on 54 deep-sea species were used.

Table 2   Material used to determine the trophic structure of the deep-sea benthos in the Atlantic Ocean.

		Samples	
Research vessel	Cruise	Trawl	Bottom-sampler
"Akademik Kurchatov"	11	1	2
	14	12	11
	40	10	13
	43	17	15
"Vityaz"	65	4	2
"Mikhail Lomonosov"	5	—	8
"Akademik Mstislav Keldysh"	1	—	4
"Dmitrii Mendeleev"	1;43	7	2
Total		51	57

Table 3 Material used to determine trophic divisions of the Atlantic Ocean floor (non-Russian sources).

Region of investigation	Material investigated	Fishing gear	Data considered in the present work	Source
Gulf of Mexico	Small infauna and large epifauna	Anchor dredge photographs, trawl	At 5 stations on the abyssal plain. C_{org} in the sediment 1.2–0.30%; biomass from 0.155 to 0.009 g m^{-2} (calculated from dredge hauls and photographs)	Rowe and Menzel, 1971
North American Basin (Bermuda section)	Macrofauna	Anchor dredge and epibenthic sled	At 7 stations from a depth of 2873–5001 m. C_{org} in the sediment 0.56–0.24%. Numerous and varied invertebrate infauna and epifauna. Samples with hundreds of deposit feeders: bivalve molluscs (*Malletia, Tindaria*)	Sanders *et al.*, 1965; Hessler and Sanders, 1967; Sanders and Hessler, 1969
Iberian Basin (42°N, 13–15°W)	Meiobenthos, macrobenthos	Agassiz trawl, photographs	Six trawls and 6 bottom sampler stations from depths of 5272–5347 m. Predominance of deposit-feeders in the macrofauna. Abundance of actinarians, 36–72 m². Total biomass of macrofauna in trawl 0.1 g m^{-2}	Thiel, 1972
Bay of Biscay (44–47°N, 4–10°W)	Macrobenthos	Trawls, photographs	Abundance and diversity of benthos, 100% occurrence of holothurians at 27 stations, depths 3800–4825 m. Large holothurians *Psychropotes*,	Laubier and Sibuet, 1977
	Megabenthos, holothurians	Trawls, photographs	21 cm long, in trawls at 3 stations, calculation of their density from trawls and photographs	Sibuet and Lawrence, 1981
Canary Basin (20°N, 21°W)	Megabenthos and macrobenthos	Epibenthic trawl-sledge and photographs	Hundreds of Demospongiae sponges of size 1–2 cm in each catch; on photographs and in catches large deposit feeding holothurians *Psychropotes*, asteroids *Styracaster* and carnivores *Munidopsis*	Rice *et al.*, 1979
Angola Basin (18–22°S, 7–11°E)	Meiobenthos	Bottom sampler	Population of meibenthos 1.5–2 times as high as in the Iberian Basin	Dinet, 1973

Table 4 Material used to study gut contents of marine benthic invertebrates.

Systematic group (order, class)	Number of species	Number of specimens investigated	Number of stations
Actiniaria	3	236	3
Bivalvia	1	1	1
Polychaeta	8	284	14
Echiuroidea	3	3	3
Crustacea			
Cirripedia	1	24	1
Isopoda	6	30	5
Decapoda	12	488	54
Sipuncula	2	6	4
Ascidiacea	4	18	4
Holothurioidea			
Elasipodida	10	37	9
Aspirochirotida	6	16	7
Dendrochirotida	2	15	2
Molpadonida	5	13	6
Apodida	2	12	2
Asteroidea	11	87	14
Ophiuroidea	39	702	30
Echinoidea	3	13	3
Total	118	1985	162

3. FOOD SOURCES FOR DEEP-SEA BENTHOS AND FEEDING GROUPS

3.1. Food Sources

Trophic factors undoubtedly limit deep-sea benthos (Birstein and Vinogradov, 1971; Belyaev *et al.*, 1973; Thiel, 1979; Vinogradova, 1980). Throughout the ocean the salinity and temperature of benthic waters vary only slightly. There are exceptions to this rule in hydrothermal regions, but the relative extent of such places is small compared with the ocean bed. With physico-chemical factors being rather uniform, feeding conditions of deep-sea animals are relatively more variable as a result of uneven availability of organic matter on the ocean floor.

On the ocean floor most of the organisms depend on organic matter derived from the surface photosynthetic zone. In other words, feeding conditions for deep-sea benthos depend on surface primary production (i.e. the quantity of inital organic matter) and the way it is transported to the seabed together with conditions of deposition and transformation in the sediment (Vinogradov and Lisitsyn, 1981).

3.1.1. Different Forms of Organic Matter on the Bottom

The major part of the organic matter synthesized in the surface layer is broken down and mineralized in the same layer (Bishop *et al.*, 1977, cited from Lisitsyn and Vinogradov, 1982). Below the euphotic zone dead organic matter (detritus) commonly predominates in suspension (Hamilton *et al.*, 1968; Romankevich, 1977). The transport to the bottom of organic matter remaining after mineralization at the surface occurs either passively, as pellets and a "rain of corpses", or actively, by vertical migration of plankton. Transport of organic matter from the surface to deeper layers of the ocean in the form of zooplankton faecal pellets was noted by Beklemishev (1964). It is now known that faecal pellets comprise up to 80–99% of the flux of detritus (Lisitsyn and Vinogradov, 1982). These conclusions are based on a higher sedimentation rate of faecal pellets and their slower decomposition (mineralization) compared to dead organisms of the same size, and are also supported by analysis of material collected in sediment traps. For example, Honjo (1978) reported a predominance of faecal pellets in trap samples at a distance of only 114 m from the bottom at 5363 m in the Sargasso Sea. Robison and Bailey (1981) studied the rate of sinking and dissolution of faecal matter of seven species of bathypelagic fish off southern California and noted the dominant role of faecal pellets in the transport of organic matter from pelagic communities to the bottom. Similar conclusions were obtained in the study of production and sedimentation of salp faecal pellets (Madin, 1982): they sink faster (up to $2238 \, \text{m day}^{-1}$) than the pellets of all other members of zooplankton. The composition of the pellets is diverse and includes primarily phytoplankton cells and microflora (Gowing and Silver, 1983).

Lebedeva *et al.* (1982) showed that a decrease of primary production by an order of magnitude caused an equivalent decrease in the amount of detritus per area unit of the surface zone. The proportion of the organic production utilized within the euphotic zone in eutrophic and mesotrophic waters is less (82–89%) than in oligotrophic waters in central regions of the ocean (95% and more). As a result, the detrital flux to depth is more intense in richer regions of the ocean (Lisitsyn and Vinogradov, 1982). Active transport of organic matter by migrating animals slightly increases the proportion of detritus reaching the bottom from the surface zone. This transport is greater in productive and less in oligotrophic waters, strengthening the difference between parts of the ocean with different primary productivity (Lisitsyn and Vinogradov, 1982).

Also of interest and major importance is the recently demonstrated rapid seasonal input of phytodetritus to the ocean floor in highly productive areas such as the North Atlantic (Billett *et al.*, 1983, 1988; Lampitt, 1985; Rice *et al.*, 1986; Gooday, 1988; Lochte and Turley, 1988; Thiel *et al.*, 1988/1989; Turley *et al.*, 1988; Riemann, 1989). Shortly after arrival on the seabed this phytodetritus is redistributed by near-bottom currents, forming patches (Lampitt, 1985). Phytodetritus is enriched with plankton organisms from the euphotic zone such as cyanobacteria, small chlorophytes, diatoms, coccolithophorids, silicoflagellates, dinoflagellates, tintinnids, radiolarians and foraminiferans. It also contains crustacean exuviae and a great number of small faecal pellets. The labile organic matter content is high and it is supposed that many benthic invertebrates are able to use it. Different benthic organisms such as nematodes and foraminiferans are often associated with phytodetritus.

Another important form of organic flux from the euphotic zone to the ocean floor is in the form of "marine snow". This is usually rich in highly nutritive phytodetritus (Trent, 1985), and includes remains of gelatinous pelagic animals (e.g. salps, including their larval homes) and diatoms (Honjo *et al.*, 1984; Smetacek, 1985; Alldredge and Silver, 1988; Alldredge and Gotschalk, 1989). The "snow" is populated by bacteria and microzooplankton and plays an important role in transportation and distribution of these organisms (Fowler and Knauer, 1986). Aggregates of "marine snow" sinking from the euphotic zone adsorb phytoplankton cells and small faecal pellets thereby increasing the sinking speed (Krank and Milligan, 1988).

The significance of dead bodies in the transport of organic matter to deeper layers is secondary compared to the principal pathways described above. Tseitlin (1981) found that the formation of faecal matter in surface layers is three to six times less than that of detritus, whereas the downward velocity of faecal matter is three to five times higher than the average downward velocity of organisms of the same size. Summing up experimental and computational data, Vinogradov and Lisitsyn (1981) concluded that dead bodies of small plankton animals (1–2 mm and less) do not sink deeper than 1000–2000 m and do not reach the bottom sediment, whilst dead bodies of mesoplankton animals, for example large copepods, possibly reach depths of thousands of metres. This conclusion is supported by Stockton and Delaca (1982), who reviewed the role of "falling" food accumulations in feeding of deep-sea benthic animals and emphasized the non-uniformity of such accumulations in time and space.

Dead euphausiiids are consumed by various benthic invertebrates in the South Atlantic (Sokolova, 1993a,b, 1994b). These organisms are very abundant in this part of the ocean and it is possible that their sinking corpses form a continuous flux. Body fragments of nekton animals are also important in the nutrition of deep-sea scavengers (Smith, 1985, 1986), and such fragments can be expected along fish migratory pathways (Tyler, 1988). Additionally, macrophytes (including *Sargassum*), marine grass and terrestrial plants can serve as additional food sources for deep-sea invertebrates (Wolff, 1979, 1980; Suchanek *et al.*, 1986; Grassle and Morse-Porteous, 1987; Reichardt, 1987; Alongi, 1990).

3.1.2. *Organic Matter on the Bottom*

As noted above, only a small part of the initial organic matter produced at the surface layer reaches the bottom, though quantitative estimates vary. According to Menzel (1974), less than 3% of primary production reaches abyssal depths (> 3000 m) (cited after Lisitsyn and Vinogradov, 1982). Honjo (1978) reported that in the Sargasso Sea, at 5367 m, the organic carbon on the floor constitutes about 2% of the entire primary production in this region. According to Bogdanov (1980), in the Pacific Ocean the quantity of suspended organic matter reaching the seabed (in organic carbon accumulation rates mg/square cm/1000 years) constitutes from 5 to 15% of primary production. It was shown earlier (Bogdanov, 1965; Bogdanov *et al.*, 1971), that in the Pacific more than 10% of organic matter reaches the floor at depths of over 3000 m in temperate and high latitudes of the northern hemisphere and close to Antarctica, and less than 5% in tropical and temperate latitudes of the southern hemisphere.

At the seabed the organic matter undergoes complex transformations. It is consumed by benthic organisms, dissolved and mineralized by bacteria, so that only a small part of primary production becomes buried in the sediment. This may constitute only 0.4% of surface production (Lisitsyn and Vinogradov, 1982). In the Pacific, from 97.9 to 99.9% of organic matter reaching the floor disappears before burial (Bogdanov *et al.*, 1979; Bogdanov, 1980). With increasing primary production, there is an increase in the mass of organic matter reaching the floor and the mass of organic matter buried in the sediment (Bogdanov *et al.*, 1979; Lisitsyn and Vinogradov, 1982). Dissolved organic matter in the sediment is absorbed by some invertebrates, but is not a major source of nourishment (Southward and Southward, 1982).

The principal patterns of distribution of organic matter on the ocean floor corrrespond to circumcontinental and latitudinal (climatic) zonation of sediment accumulation (Bezrukov, 1962). The distribution of organic carbon in the sediment is inherited from the distribution of organic carbon in suspension (Bogdanov *et al.*, 1971; Lisitsyn, 1974; Romankevich, 1977) which, in turn, depends on the distribution of its producers in ocean waters. The replacement rate of suspended organic matter is controlled by primary production (Bogdanov *et al.*, 1971, 1979; Romankevich, 1977).

3.1.3. *Circumcontinental and Latitudinal Zonation in Distribution of Organic Matter*

Areas of near-continental and pelagic sediment formation differ significantly in conditions of organic matter production and transformation. The border between the two areas is characterized by changes in the rate of sediment accumulation, the ratio and accumulation rate of organic carbon (Murdmaa *et al.*, 1976; Murdmaa,

1979a,b; Bogdanov, 1980). Circumcontinental zonation results in increased concentration of organic matter in the near-continental region determined by increased productivity of neritic and coastal waters and intensive sediment accumulation resulting from autochthonous (biogenic) and allochthonous (terrigenic) sedimentary material (Romankevich, 1968, 1970, 1977; Bogdanov *et al.*, 1971; Lisitsyn, 1974). For the Pacific, the maximum organic carbon accumulation rate (more than $10 \text{ mg cm}^{-2} \ 1000 \text{ year}^{-1}$) are confined to the periphery of the ocean (Bogdanov *et al.*, 1979; Bogdanov, 1980). In the zone of near-continental sediment formation with high values of primary production and terrigenous sediment accumulation, there is a high rate of organic carbon accumulation containing labile components (Bordovsky, 1974; Romankevich, 1974, 1977). In the zone of near-continental sediment formation the coefficient of organic carbon burial (the per cent of buried organic carbon out of the total reaching the floor) may exceed significantly the maximum values typical for regions of oceanic sediment formation (Bogdanov *et al.*, 1979).

As a result of latitudinal zonation of sediment accumulation in the ocean, the region of pelagic sediment formation is divided into belts, the boundaries between which are marked (as at the boundary with the region of near-continental sediment formation) by changes in the rate of sediment action, percentage content and organic carbon accumulation rate (Murdmaa, 1987). Latitudinal zonation is shown by the increased content of organic matter in fine sediments of the equatorial and high-latitude humid zones and the decreased content in tropical–subtropical arid zones. This results from latitudinal differences in the productivity of surface waters and the rate of biogenic sediment accumulation, which are sharply reduced in central parts of subtropical circulation systems (in arid zones) and are more or less increased beyond them. According to the data for the pelagic region of the Pacific (Bogdanov *et al.*, 1979; Bogdanov, 1980), all three latitudinal belts (northern, equatorial and southern) confined to highly productive zones are characterized by increased values of organic carbon accumulation rates mg/square cm/1000 years supply (over $0.5 \text{ mg cm}^{-2} \ 1000 \text{ year}^{-1}$). Between these areas, within the desert zones at the floor of the basins, the organic carbon supply is low, less than $0.5 \text{ mg cm}^{-2} \ 1000 \text{ year}^{-1}$, an order of magnitude below organic carbon accumulation rates in the near-continental region.

Bogdanov has estimated the coefficient of burial for the region of pelagic sediment formation in the Pacific, where the organic carbon accumulation rates mg/square cm/1000 years reaching the floor is many times higher than that buried in the sediment. The lowest values of the coefficient (less than 0.1%) were observed in deep-sea basins of the desert zones. Closer to the periphery of the ocean, within the same zones, the coefficient often exceeded 0.25%. In humid zones, the coefficient of burial is always more than 0.1%. For regions of pelagic sediment formation, the coefficient can reach 0.25–0.5% in the northern humid and equatorial zones, and 0.1–0.25% in the southern humid zone. Closer to the periphery of the ocean, within the humid zone, the coefficient of burial may reach 0.5–1%. There appears to be greater loss of organic carbon at the sediment-water interface than in the water column during descent, so that processes of burial in the seabed have a major role in controlling organic matter accumulation. In pelagic desert zones with low primary production (less than $100 \, \mathrm{mg \, cm^{-2} \, day^{-1}}$) and minimum rates of sediment accumulation, any organic matter with labile components that reaches the seabed becomes completely metamorphosed in the uppermost sediment layer. Only the refractory ("inert") organic matter, resistant to biochemical oxidation, is buried in the sediment (Bordovsky, 1964a,b, 1966a,b, 1974; Romankevich, 1970, 1974; Bogdanov et al., 1979).

Thus, trophic conditions for deep-sea benthic organisms are significantly different in different parts of the ocean, owing to high variations in organic matter accumulation in the sediment. The maximum differences in the quantity and the quality of organic matter in the sediment are observed between the regions of near-continental and pelagic sediment formation, and, within the latter, between the humid and arid zones in central parts of ocean. In arid zones, under subtropical anticyclonic gyres, the concentration of organic matter in the sediment is minimum and has a low food value. General patterns of organic matter distribution in deep-sea sediments are also related to the associated microflora and microfauna. Deficiencies in analytical technique mean that the microbial population of the sediment has been included in analysis of organic matter content. Thus, the general index of organic matter content in the sediment also includes microflora and microfauna which may be a food source for macrobenthos.

3.2. Feeding Groups: Concept and Methods

In analysing the influence of trophic factors on the composition of benthos, we use the concept of trophic groups of invertebrates that rely on similar food sources, an aspect of the more generalized concept of a "life form", as discussed by Gebruk (1992). Classification into trophic groups depends on the objectives of investigation. Trophic groups used in the present work are based on the concepts developed for aquatic animals in general by Petersen and Boysen-Jensen (1911), Hunt (1925), Thamdrup (1935), Zernov (1949) and Turpaeva (1953). This was further applied to deep-sea macrobenthos by Sokolova (1956, 1960a, 1962, 1965).

The following trophic groups have been used for deep-sea macrobenthic invertebrates: (i) suspension-feeders using detritus (triptonophages) or animal food (zoophages); and (ii) bottom-feeders (benthophages) using the sediment detritus (deposit-feeders) or animal food (carnivores). Deposit-feeders depend solely on feeding conditions of the sediment surface and upper layer, whereas suspension-feeders are not so dependent. In this respect, suspension-feeders, taken as a whole, are opposed to deposit-feeders in synecological analysis. Feeding in deep-sea species has been compared with the corresponding data for shallow-water invertebrates from the same trophic groups.

The specificity of deep-sea material determined the methods and numerical indices used to characterize types of feeding. Since experimental work with deep-sea organisms is very limited, their nutritional status has to be derived from gut contents and quite often on only a small number of specimens. Numerical indices used include the frequency of occurrence of food components and their ratio in the food mass, determined usually by volume rather than weight because of disintegration of the food consumed. The intensity of feeding was estimated using the number of stomachs containing food, related to the total number of stomachs, and the index of stomach fullness. The proportion of the average number of full and empty stomachs is highly variable between groups, reflecting variations in the feeding intensity of invertebrates from different trophic groups which result from principal differences in the nutritive value of the food material they use (in general, the nutritive value is high in carnivores and low in deposit-feeders which take lots of inorganic material (ballast)).

The index of fullness was estimated as the weight of the gut content in relation to the body weight of the animal expressed in decimille (1:10 000). This index was introduced by Zenkevich *et al.* (1931) in studies of fish feeding and intensively developed later by Shorygin (1939, 1952) who noted in particular that the index of fullness relates the food consumed to the size of the organism and characterizes the physiological state of the organism and its satiation. The index of fullness was successfully used to evaluate daily, seasonal and age-related changes in the nutrition of various species of carnivorous decapods from the Caspian Sea (Cherkashina, 1972), the Sea of Okhotsk and the Bering Sea (Tarverdieva, 1974, 1978, 1979, 1981). Using the index of fullness, Tarverdieva (1974) also identified the biotopes in which the feeding of carnivorous crabs is most intensive. Sokolova (1982) used a similar approach to examine the feeding of the deep-sea polychaete *Harmothoe derjugini*.

Turpaeva (1953) used the mean index of fullness to characterize different trophic groups of benthic invertebrates and showed that it increased in deposit-feeders, depending on the mineral component in the food: the higher the mineral content the more food is required to assimilate an equal quantity of nutrients. Similar factors determine the differences between the mean indices of fullness in carnivores and deposit-feeders. The intensity of feeding depends on the quality of a food material: the more concentrated the food, the smaller is the amount needed to satisfy the organism. Higher assimilation of animal food compared to detritus has been demonstrated in many groups of aquatic invertebrates (Tsikhon-Lukanina and Soldatova, 1973; Tsikhon-Lukanina, 1982). As a result, the index of fullness in carnivores consuming energy-rich organisms never reaches such high values as in deposit-feeders which extract food from the mass of inert matter passing through the intestine. In general, indices of fullness in different species from one trophic group vary less than between different trophic groups. At the same time, the index may vary significantly between species from the same group, as well as in specimens of one species of different ages.

3.2.1. *Carnivores*

Carnivores are here defined as invertebrates feeding on any animal prey or dead bodies or both. Thus, carnivores include both predators and scavengers, as well as omnivores that use both food sources according to availability.

3.2.1.1. *Systematic groups*   Carnivorous species are known among most, though not all, taxa of deep-sea macrobenthos. Some groups, for example Actiniaria and Pantopoda (Pycnogonida), are entirely adapted for such food, whereas in others only some forms are carnivores, for example Asteroidea in which species of seven families (see below) are carnivorous whilst all species of the family Porcellanas-teridae feed on sediment detritus. On the other hand, deep-sea holothurians represent a group that lacks carnivorous forms. Holothurians have been recorded on deep-sea photographs near baited traps (Dayton and Hessler, 1972) but scavenging remains unproven.

Lack of data on food preferences in deep-sea benthic invertebrates has resulted in several hypotheses about an increased food plasticity developed convergently in different taxa as a result of oligotrophic conditions at great depths. Thus, Dayton and Hessler (1972) considered it was difficult to divide deep-sea benthic invertebrates into carnivores and deposit-feeders, suggesting that all deep-sea communities comprise foraging animals, except for suspension-feeders and bacteria. Animals such as holothurians, ophiuroids, asteroids, some polychaetes and decapod crustaceans, were assumed to consume anything smaller than themselves, some of them using a broad food spectrum ranging from sporadic highly nutritive (i.e. animal food) to relatively constant low nutritive (sediment detritus). Later, Hessler and Jumars (1974) accepted that deep-sea carnivores had developed omnivory to minimize the time and energy spent searching for food, in accordance with the views of MacArthur (1972). These assumptions were examined by comparing the feeding in deep-sea carnivores and shallow-water counterparts (Tables 5–7). (The number of carnivores exceeds those listed in Table 5.) Data on carnivorous asteroids of the families Astropectinidae, Benthopec-tinidae, Goniasteridae, Pterasteridae, Solasteridae, Brisingidae and Zorasteridae are based on the composition of the food of 21 species (Eichelbaum, 1910; Sokolova, 1957a; Carey, 1972; Korovchinskii, 1974). Carnivores are also known among other deep-sea groups: polychaetes of the families Onuphidae and Phyllodocidae, some Gastropoda, some Amphipoda, and all Pantopoda and Actiniaria. Kucheruk (1979, 1981) identified polychaetes of the families Onuphidae and Phyllodocidae as active predators and scavengers, and *Hyalinoecia* was taken in a baited trap in the Pacific at 2000 m (Dayton and Hessler, 1972).

Deep-sea gastropods that feed on polychaetes and bivalve molluscs were

Table 5   List of deep-sea and shallow-water carnivores in which the gut contents were investigated.

Sl. No. 1	Species 2	No. of specimens examined 3	No. of stations used 4
	Actiniaria		
1	gen. sp. No. 1	62	2
2	gen. sp. No. 2	156	1
3	gen. sp. No. 3	23	1
	Polychaeta		
4	*Laetmonice wyvillei*	83	3
5	*Eunoe nodosa**	38	1
6	*Harmothoe rarispina**	22	1
7	*H. derjugini*	86	5
	Crustacea		
	Isopoda		
8	*Storthyngura bicornis*	10	1
9	*Storthyngura* sp.	2	1
10	Eurycope magna	7	1
	Decapoda		
11	*Chionoecetes angulatus*	45	5
12	*Munidopsis beringana*	14	3
13	*Hyas coarctatus alutaceus*	12	1
14	*Oregonia bifurca*	10	1
15	*Parapagurus pilosimanus*	50	1
16	*Sclerocrangon zenkevitschi*	48	4
17	*Sclerocrangon abyssorum*	26	7
18	*Sclerocrangon derjugini*	28	3
19	*Sclerocrangon communis**	69	7
20	*Sclerocrangon salebrosa**	34	7
21	*Nectocrangon dentata*	75	12
22	*Nectocrangon lar**	77	11
	Echinodermata		
	Asteroidea		
23	*Psilaster pectinatus*	57	2
24,25	fam. Astropectinidae (2 sp.sp)	14	2
	Ophiuroidea		
26	*Ophiura bathybia*	11**	4
27	*Ophiura loveni*	7	1
28	*Ophiomusium* sp.	2	1

Table 5 Continued

Sl. No. 1	Species 2	No. of specimens examined 3	No. of stations used 4
29	*Amphiophiura convexa*	192	8
30	*Amphiophiura pachyplax*	2	1
31	*Amphiophiura concava*	26	1
32	*Amphiophiura bullata pacifica*	215	4
33	*Amphiophiura vitjiazi*	66	1
34	*Bathypectinura heros*	2	1
35–63	fam. Ophiuridae (29 sp.p.)	141	7

*Shallow-water species.
**In addition, 612 specimens were investigated, in which the digestive tract was empty.

described by Rex (1976) from the northwest Atlantic. It can be inferred from morphological peculiarities, similar to those known in shallow-water forms, that prey on hydroids and actiniarians (Turpaeva, 1971a,b, 1973, 1974, 1975) that the deep-sea species of Pantopoda can also be classed as carnivores, presumably attacking coelenterates. This relationship seems to be confirmed by the 100% coincidence of occurrence of pantopods and coelenterates in deep-sea trawl samples.

As a group, the Amphipoda are rare in trawl catches, possibly as a result of

Table 6 Proportion of individuals with food and the mean index of fullness of their digestive tracts.

Systematic group and number of species 1	Total number of specimens 2	Specimens with food, % of the total		Mean index of fullness, $^0/_{000}$ 5
		Mean 3	Limits for species 4	
Decapoda 12 species	488	53	91-33	124
Polychaeta 4 species	176	25	52-13	287
Asteroidea 1 species	41	43	–	112
2 species*	68	39	34-45	–
Ophiuroidea 4 species	498	14	30-13	560
Actiniaria 3 species	241	5	9-0.5	246

*From Carey (1972); the remaining data in the table are from personal material.

Table 7 Index of fullness of digestive tracts of deep-sea and shallow-water carnivores.

Species	Index of fullness, %ₒₒₒ	Number of specimens with food	Number of samples	Depth, m	Source
1	2	3	4	5	6
Actinaria sp. No. 1	280	3	2	1440–3330	Sokolova, 1954
No. 2	166	1	1	3820	Sokolova, 1954
No. 3	293	2	1	1440	Sokolova, 1954
Laetmonice wyivillei	211	5	1	4130	Sokolova, 1956
Eunoe nodosa	387	7	1	87	Sokolova, 1956
Harmothoe rarispina	269	3	1	87	Sokolova, 1956
H. derjugini	281	45	5	2700–3100	Sokolova, 1981, 1986
Sclerocrangon zenkevitschi	50	17	4	2995–4130	Sokolova, 1957b
S. abyssorum	80	14	7	1070–3330	Sokolova, 1957b
S. derjugini	60	20	3	278–900	Sokolova, 1957b
S. communis	131	18	7	48–165	Sokolova, 1957b
S. salebrosa	26	14	7	58–196	Sokolova, 1957b
Nectocrangon dentata	70	36	12	45–1000	Sokolova, 1957b
N. lar	59	47	11	29–278	Sokolova, 1957b
Pandalus borealis	28			20–200	Turpaeva, 1953
Sabinea septemcarinata	58			20–500	Turpaeva, 1953
Pagurus pubescens	96			20–200	Turpaeva, 1953
Parapagurus pilosimanus	112	22	1	2720	Sokolova, 1957b
Munidopsis beringana	89	12	3	3800	Sokolova, 1957b
Oregonia bifurca	316	3	1	1080	Sokolova, 1957b
Hyas coarctatus alutaceus	365	6	1	1080	Sokolova, 1957b
Chionoecetes angulatus var. angulatus	139	41	5	1440–2720	Sokolova, 1957b

Table 7 Continued

Species	Index of fullness, %$_{ooo}$	Number of specimens with food	Number of samples	Depth, m	Source
1	2	3	4	5	6
Paralithodes camtschatica	4.7	114	—	—	Tarverdieva, 1976, 1978
P. platipus	7.7	—	—	40–50	Tarverdieva, 1979
	4.5	91	—	27–200	Tarverdieva, 1979
Chionoecetes bairdi	5.5–7.6	114	—	40–140	Tarverdieva, 1979
	7.5–28.0	114	—	Shelf	Tarverdieva, 1976, 1981
C. opilio	8.7–49.4	—	—	60–130	Tarverdieva, 1976, 1981
		214	—	Shelf	Tarverdieva, 1981
C. opilio	7.4–16.8	18	2	2200	Sokolova, 1957b
Psilaster pectinatus	112	26	7	3600–5000	Litvinova and Sokolova, 1971
Amphiophiura convexa	660				
A. bullata var. *pacifica*	618	28	4	5240–6240	Litvinova and Sokolova, 1971
A. vitjiazi	508	10	1	6810	Litvinova and Sokolova, 1971
A. concava	455	8	1	802	Litvinova and Sokolova, 1971

escape, and catch data are inadequate to determine their food preferences (Kamenskaya, 1977, 1984). However, deep-sea scavenging amphipods have been taken with baited traps and recorded photographically (Dayton and Hessler, 1972; Hessler *et al.*, 1972; Schulenberger and Hessler, 1974; Isaacs and Schwartzlose, 1975; Schulenberger and Barnard, 1976; Rice *et al.*, 1979; Thurston, 1979; Smith and Baldwin, 1982; Hargrave, 1985).

3.2.1.2. *Food objects*   The food contents in individual species from the abyssal, slope and shelf zones, including polychaetes (five species), shrimps (seven species), crabs (13 species), asteroids (30 species) and ophiuroids (22 species) have been compared using original and published data (Table 8). All of the deep-sea carnivores examined have a reduced variety of food objects, owing to the absence of minor prey and casual items. (Certain food objects, which are not found or are less important in shallow-water carnivores, appear or acquire a significant importance in deep-sea forms.)

With regard to prey species, polychaetes and echinoderms are generally consumed equally from the shelf to the abyssal zone. The taxonomic composition of consumed crustaceans changes significantly. In the deep sea, crustaceans become less important, except for isopods. Benthic molluscs (gastropods and bivalves) in most cases lose their primary significance as food objects, except for deep-sea gastropods from the northwest Atlantic, which actively prey on bivalves (Rex, 1976). Both in deep-sea and shallow-water carnivores, any sediment found in the food mass appears to enter as an admixture with the animal components.

3.2.1.3. *"Rain of corpses"*   Sinking dead bodies as a food source for deep-sea carnivores deserve special attention. Feeding on such items, most often on the remains of fish, has been reported for many carnivores from the shelf (Cherkashina, 1972; Tarverdieva, 1976; Litvinova, 1979). There are fewer reports of the consumption of identifiable remains of plankton and fish at great depths by benthic species; Table 8 sums up records outside the shelf zone. Of the 53 species in which such remains were found, 37 belong to ophiuroids, two to polychaetes, four to asteroids and one to decapods. Distinct remains of dead plankton organisms in the food of benthic invertebrates were found at 31 deep-sea stations (Figure 1): 11 of these in the Pacific, seven in the adjacent Barents Sea and the Sea of Japan, four in the Indian Ocean and nine in the Atlantic (one of them in the Caribbean Sea).

Thirteen stations were located on the slope (continental or trench), at 425–7000 m; the remaining 18 stations were situated on the ocean floor at 2760–5790 m and in most cases at some distance from the slope. A common feature of nearly all of the regions where these stations were situated is high plankton productivity. Seven of the slope stations were close to frontal or upwelling zones.

The importance of different plankton groups and fish in the food of deep-sea carnivores was estimated from the gut content of 23 species (Table 9). In terms of frequency of occurrence, dead Pteropoda dominate, followed by dead Copepoda

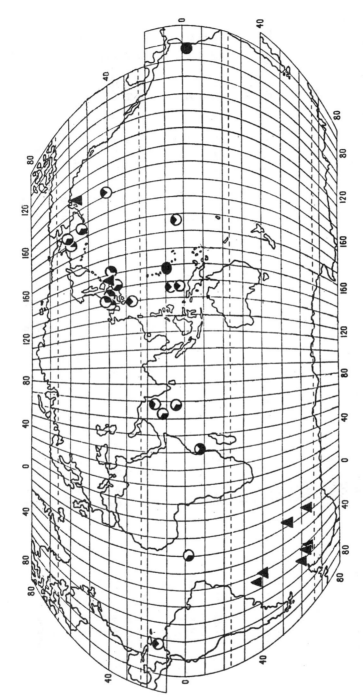

Figure 1 Records of the remains of planktonic invertebrates and nekton found in the digestive tracts of different carnivorous invertebrates, beyond the shelf zone. The circles show the percentage occurrence, as black segments, in the gut contents; the filled triangles show places where euphausid remains are dominant components of gut contents (original map).

Table 8 Consumption of dead planktonic invertebrates and fish by different species of benthic carnivores beyond the shelf zone.

No.	Consumer species	Station (depth (m))	Food object	Source
1	*Laetmonice wyvillei*	2209 (3960)	Crustacea Mysidacea (*Pseudomma* sp.)	Sokolova, 1956
2	*Harmothoe derjugini*	6663 (2760)	Crustacea Amphipoda	Sokolova, 1982
		6652 (3350)	Copepoda	
		6658 (3360)	Copepoda	
3	*Chionoecetes angulatus*	531 (2670)	Pisces	Sokolova, 1957a
		1030 (2133)	Crustacea, Decapoda (*Hymenodora* sp.)	
		542 (1400)		
4	*Nearchaster aciculosus*	near Oregon	Pisces	Carey, 1972
5	*Amphiophiura convexa*	28–29 (3600–3660)	Pisces	Litvinova and Sokolova, 1971
		4851 (4610)		
		4712 (3560)	Pteropoda including *Cavolina uncinata*	
		6276 (5240)		
6	*Amphiophiura pachyplax*	3678 (1660)	Pisces	Litvinova and Sokolova, 1971
7	*Amphiophiura convexa*	4680 (802)	Pisces, Crustacea Mysididacea, Cyclopoida	Litvinova and Sokolova, 1971
8	*Amphiophiura bullata pacifica*	3198 (5790)	Pteropoda including *Cavolina uncinata*	Litvinova and Sokolova, 1971
9	*Amphiophiura vitjiazi*	3528 (6810)	Pteropoda, *Cavolina longirostris, Diacria trispinosa major*	Litvinova and Sokolova, 1971
10	*Ophiura bathybia*	4104 (5435)	Pteropoda	Author's unpublished data
11	*Ophiura loveni*	1013 (6430)	Pisces	Author's unpublished data
12	*Ophiomusium* sp.	3996 (4600)	Heteropoda, *Atlanta* sp.	Author's unpublished data (Identification of the ophiuran by Litvinova)
13	*Ophiura leptoctenia*	6092 (1050)	Heteropoda, *Atlanta* sp. Pteropoda, *Limacina* sp. Crustacea, Euphausiacea, Amphipoda, Copepoda	Litvinova, 1979
14	*Ophiura flagellata*	6669 (425)	Crustacea: Euphausiacea, Amphipoda	Litvinova, 1979
15	*Ophiura sarsi vadicola*	6669 (425)	Crustacea: Amphipoda family Hyperiidae, Euphausiacea	Litvinova, 1979

Table 8 Continued

No.	Consumer species	Station (depth (m))	Food object	Source
16	*Ophiura* sp.	308 (2150)	Pteropoda: *Cavolina uncinata, Clio pyramidata*	Litvinova, 1979
17	*Ophiura irrorata*	308 (2150)	Pteropoda: *Cavolina uncinata, Clio pyramidata, Limacina helicina* + Crustacea Amphipoda family Hyperriidae	Litvinova, 1979
18	*Ophiomusium multispinum*	308 (2150)	Same, Pteropoda +*Cavolina longirostris* + Crustacea Copepoda (*Calanus*)	Litvinova, 1979
19	*Ophiomusium planum*	1224 (2600)	Pteropoda: *Cavolina uncinata, Stiliola subula*	Litvinova, 1979
20	*Ophiomusium* sp.	308 (2150)	Pisces	Litvinova, 1979
21	*Homalophiura madseni*	7291 (7000)	Pteropoda: *Cavolina uncinata, Diacria trispinosa, Limacina helicina*	Litvinova, 1979
22	*Oiphiozonoida* sp.	308 (2150)	Crustacea Copepoda	Litvinova, 1979
23	*Dytaster spinosus*	Indian Ocean, 4800 m	Pisces	Bruun and Wolff, 1961
24 — 52	Ophiuroidea family Ophiuridae 29 sp. sp.	4086 4090 4097 4100 4104 4107 4109 (3100–6290	Crustacea: Euphausiacea, Copepoda, Amphipoda Hyperiidae, Pteropoda, Pisces	Sokolova, 1993a,b; 1994b
53 — 54	Asteroidea family Astropectinidae 2 sp. sp.	4097 4109 (3100–5255)	Crustacea Euphausiacea, Varia, Pisces	Sokolova, 1995

Table 9 Frequency of occurrence of various food objects regarded as "dead bodies rain" in the stomach contents of 23 species of benthic carnivorous invertebrates beyond the limits of the shelf.

Food objects in the "dead bodies rain"	Frequency of occurrence, %
Mollusca	
Pteropoda	39
Heteropoda	9
Crustacea	
Copepoda	18
Mysidacea	9
Amphipoda	21
Euphausiacea	13
Decapoda	4.3
Pisces	34

and Amphipoda (predominantly Hyperiidae). Other plankton groups are rarer, these include crustaceans, Mysidacea, Euphausiacea and Decapoda (the latter being rarest) and heteropod molluscs. Quantitative estimates of the role of plankton and fish remains in the food of deep-sea benthic invertebrates were made by Litvinova and Sokolova (1971) for some ophiuroids from the continental slope and the ocean floor, and for one species of polychaete from the deep-sea basin of the Sea of Japan (Sokolova, 1982).

The phenomenon of intensive euphausiid "dead body rain" representing a main food source for abyssal benthos was initially discovered in the South Atlantic, where large populations of commercially exploitable euphausiids are known (Sokolova, 1994b). The highest level of consumption of euphausiid fragments was observed in the Weddell and the Scotia Seas. In these regions, euphausiid remains served as practically the only food source for ophiuroids. Ophiuroids from the Scotia Sea, especially small forms, generally ate euphausiid eyes (macerated tissue of eyes is well marked and discernible from the ommatidia). In five species (with discs 6 mm and less), ommatidia and tissue of crustacean eyes constituted from 19 to 100% of the food.

In the southern part of the Argentine Basin (up to 41°S), the role of euphausiids (measured as a volume ratio) declined to half that in the Weddell and Scotia Seas. At the northern Argentine Basin station, the proportion of euphausiid remains was low, whereas fish and pteropod shells increased. In the Orkney Trench, sediment enriched with euphausiid remains was the main food source for ophiuroids at the depths mostly exceeding 6000 m. Chitinous euphausiid remains were found in 52% of the stomachs dissected, although their volume ratio was low. Additionally, trench ophiuroids used the remains of holothurians and molluscs, as well as unidentified agglutinating foraminiferans (Sokolova *et al.*, 1995).

As with the carnivorous ophiuroids, one-third of the food found in two abyssal species of *Astropecten* from the Scotia Sea and the Argentine Basin consisted of the remains of fish and plankton organisms: euphausiids in the Scotia Sea and other crustaceans in the Argentine Basin (Sokolova, 1995).

3.2.1.4. *Food spectrum* The comparison reviewed above of feeding in deep-sea and shallow-water carnivores from different systematic groups (Sokolova, 1986) shows that there is a group of carnivores among deep-sea invertebrates, and that most deep-sea carnivores do not show the increased euryphagy required by the hypothesis of Dayton and Hessler (1972) and Hessler and Jumars (1974).

There are variations in food spectrum in deep-sea carnivores, relative to shallow-water forms. In some groups the food spectrum increases with depth, in others decreases, in the third it remains stable. Widening of the food spectrum from the shelf to the slope and further to the ocean floor can be observed in four families of ophiuroids. Brisingiid asteroids that do not occur on the shelf have a similar spectrum in slope and ocean-floor species. A narrowing of food spectrum was observed in abyssal polychaetes of the families Aphroditidae and Polynoidae, shrimps of the family Crangonidae, crabs of the families Majidae and Galatheidae, hermit crabs of the family Paguridae and asteroids of the families Astropectinidae, Benthopectinidae, Goniasteridae, Pterasteridae, Solasteridae and Zoroasteridae.

Changes in food spectrum in deep-sea species apparently reflect changes in availability of food. This was suggested for shrimps (Sokolova, 1957b) and ophiuroids from different biotopes (Litvinova, 1979). A narrowing of the food spectrum depth is related to impoverishment of the food supply. Indirect indices of feeding intensity (proportion of stomachs with food and the index of fullness) in various deep-sea and shallow-water carnivores demonstrate that the feeding intensity does not change in a majority of deep-sea carnivores. An increased ability to withstand starvation has been reported for carnivorous deep-sea asteroids (Carey, 1972) and increased food portions are known in deep-sea crabs. These features are probably adaptations of deep-sea carnivores to scarce food supply.

3.2.1.5. *Distribution* The frequency of occurrence and distribution of several groups of macrobenthic carnivores on the ocean floor in the Pacific have been studied (Sokolova, 1986). The rarest were decapods, crabs and hermit crabs, which are confined to the peripheral parts of the ocean close to the continental slope base. Next in frequency were pantopods, which do not occur in distant central regions, except for the near-equatorial area. Polychaetes of two families, aphroditids and polynoids, were 1.5 times more frequent than pantopods; since some were present in the central regions. Asteroids of seven families were three times more frequent than pantopods and almost two times more common than aphroditids and polynoids (including central regions). Finally, sea-anemones (actiniarians) had the highest frequency of occurrence, with nearly 20% of all records from remote central oceanic regions.

The distribution of carnivorous asteroids and actiniarians in the Indian Ocean is similar to that in the Pacific (Sokolova, 1986). Some data were published on the distribution of carnivorous polychaetes and amphipods. Thus, Kucheruk (1978, 1981) reported a limited distribution of onuphid and phyllodocid polychaetes, restricted to the near-continental regions of the ocean floor. In contrast, carnivorous amphipods, like actiniarians, are distributed widely. Schulenberger and Hessler (1974) have confirmed the earlier assumption (Bowman and Manning, 1972; Hessler and Jumars, 1974) about the wide-spread distribution and abundance of large mobile necrophagous amphipods in the abyssal benthos. According to Kamenskaya (1977, 1980, 1984), the benthopelagic species of carnivorous amphipods are able to inhabit poor central oceanic regions owing to their rapid detection of food with the help of highly developed receptors and their storage of reserves (Dahl, 1979; Smith and Baldwin, 1982).

According to original and published data, the distribution of large crawling carnivores is usually limited and associated with more or less favourable feeding conditions on the ocean floor. The distribution of sessile (actiniarians) and fast-swimming carnivores (benthopelagic amphipods) is wider. Usually they are the only large benthic carnivores in the central oligotrophic parts of the ocean. One feeding strategy in this habitat is based on fast detection of scattered food, consumption of large food portions, and storage of food reserves. Alternatively, actiniarians can survive prolonged starvation awaiting a rare live prey. Necrophagy is known in both crawling and fast-swimming carnivores from the slope and peripheral parts of the ocean, but only among fast-swimming forms in the depauperate central parts. The ability of actinians to use dead remains has not been proven, but seems possible.

3.2.2. Deposit-Feeders (Detritivores)

3.2.2.1. *Food objects* Deposit-feeders are defined as invertebrates feeding on the sediment organic matter. This organic matter includes a non-living part (detritus) and live components (microflora and microfauna of the sediment), which are inseparable by the analytical techniques used for determining organic carbon (C_{org}) concentration. The relative role of the two components in the food of deposit-feeders (except for some particular cases), has not been evaluated precisely and remains controversial. Apparently, their roles are different in different groups.

Detritus represents matter of organic origin, unable to reproduce (Strickland, 1960) and subject to decomposition (Odum, 1975), the latter implying the presence of bacteria.

It has long been considered that shallow-water deposit-feeders utilize detritus together with associated micro-organisms (Hunt, 1925); that is, bacteria and protozoans (Turpaeva, 1953; Rodina, 1966). It has also been pointed out (Yonge, 1956) that the direct significance of detritus as a food is not clear. It could be indirect; that is, only as a nutrient medium for the development of bacteria and protozoa. However, it has been shown (Moriarty, 1982) that in some shallow-water deposit-feeders, pure detritus without satellite organisms could constitute 60–80% of the consumed organic matter. On the other hand, selective feeding on bacteria from the sediment has been demonstrated in a number of shallow-water deposit-feeders (Gerlach, 1977; Moriarty, 1982). It seems likely that bacteria are among important organic components of sediment for deep-sea deposit-feeders. Commensal bacteria colonizing intestines of deep-sea deposit-feeders (Allen and Sanders, 1966) are not considered in the present work.

Bacteria in the intestines of deep-sea organisms were first detected by ZoBell (1954), who proposed that bacterial proteins and fats could be used as a food source. Later, Morita (1979) proposed that bacteria provide some amino acids (methionine) for deep-sea benthos and also serve as a principal food source since they are rich in proteins, carbohydrates and lipids. Bacteria are better digested and nutritionally richer than other components of the sediment. Together with diatoms, protozoans and nematodes they could satisfy the protein requirements of deposit-feeders (Phillips, 1984).

However, the role of free-living sediment bacteria in nutrition of deep-sea benthic organisms is still not well known. Jannasch and Wirsen (1973) suggested that only commensal bacteria colonizing digestive tracts may play a significant role in the nutrition of deep-sea benthic animals, since free-living deep-sea bacteria were thought to be generally absent. This view was based on exceptionally low metabolic activity of benthic bacteria observed during prolonged (1 year) exposure of solid organic matter (starch, agar-agar and gelatin) at the seabed at the depth of 1830–5300 m in the North Atlantic (Jannasch et al., 1970; Jannasch and Wirsen, 1973). Despite this conclusion, the idea of utilization of deep-sea bacteria by benthic deposit-feeders continues to be discussed. For instance, Khripounoff and Sibuet (1980) suggested the utilization of bacteria by deep-sea deposit-feeding holothurians.

Thiel (1972, 1973, 1975, 1979) analysed the deep-sea benthic ecosystem in general and pointed out that substances synthesized by bacteria serve as a food for other organisms. Free-living bacteria may be eaten by deep-sea deposit-feeders together with amorphous detritus present in the sediment, and particularly with faecal pellets (Sokolova, 1986), as has been described for shallow-water deposit-feeders. Digestion of sediment bacteria by the holothurian Stichopus japonicus (Levin and Voronova, 1979) suggests that there may be a similar phenomenon in deep-sea holothurians, especially as bacteria have been found in their digestive tracts (Rodina, 1966; Branchi et al., 1979). More recently, direct evidence has been obtained suggesting utilization by deep-sea holothurians of sediment organic matter transformed by benthic bacteria and of bacteria themselves (Akhmet'eva et al., 1982). These conclusions are based on chromatographic analysis of carbohydrate fractions of the holothurian intestine content and the surrounding sediment. The authors suggest that holothurians select material of bacterial origin with a chemical composition differing sharply from the average composition of the sediment organic matter. Selective feeding by means of chemoreception by tentacle apical buds has been demonstrated in shallow-water holothurians (Bouland et al., 1982). Detailed studies of feeding in three abyssal holothurian species have demonstrated that the microflora concentration in the holothurian gut content is four to five times higher than that in the sediment (Moore et al., 1995).

Selective use of bacteria associated with faecal pellets is known for *Hydrobia* (Newell, 1965) which indicates there may be such a type of feeding in deep-sea deposit-feeders. Enrichment of faecal pellets with microflora was shown by Lopez and Levinton (1987) and Jumars et al. (1990). Tendal (1979) has also suggested that deep-sea deposit-feeders are able to use the microflora developing on the excretory products of deep-sea protozoan sarcodines. Finally, Gage and Tyler (1991) came to the conclusion that deposit-feeders use the associated bacteria rather than the sediment organic matter.

Yeast-like cells present in deep-sea sediments could also play an important role as a food source, as Burnett (1981) concluded after investigating the microflora of the surface sediment layer at 1200 m in the San Diego Trough. He showed that these cells (yeast or fungi) dominated the microflora, with diverse protozoans, mostly flagellates and amoebae, also present. The yeast-like cells colonized a thicker sediment layer (up to 100 mm) than protozoans (up to 60 mm). Earlier, Burnett (1977) reported an exceptional abundance of sarcodines in the uppermost 1-mm sediment layer from the central Pacific. Possibly, protozoans represent another food source for deep-sea deposit-feeders. Selective feeding of deposit-feeders on protozoans associated with phytodetritus was shown by Billett et al. (1988) and Patterson (1990).

The use of metazoans ingested together with the sediment by deposit-feeders is also a disputed question. Many authors discussing nutrition in large benthic deposit-feeders point to utilization of small metazoan invertebrates and remains of large dead animals in addition to sediment organic matter. Various small metazoans have been recorded together with detritus in the gut content of shallow-water benthic deposit-feeders (Cohnheim, 1901; Eichelbaum, 1910; Blegvad, 1914; Hunt, 1925), and sometimes a decrease in their number has been reported in the hind gut (Walter, 1973). At the same time, a limited significance of meiobenthos in feeding of shallow-water deposit-feeders has also been suggested (Gerlach, 1977; Renaud-Mornant and Helleouet, 1977). Regarding deep-sea benthic deposit-feeders, this question remains largely hypothetical. There is a suggestion that where food is scarce on the ocean floor large deposit-feeders are forced to become omnivorous ("generalists"), capable of eating and assimilating any material with a nutritive value (Madsen, 1961; Carey, 1972; Dayton and Hessler, 1972; Thiel, 1979). Photographs of deep-sea holothurians on the ocean floor not far from a baited trap were used to support this point of view (Dayton and Hessler, 1972).

Utilization of metazoan meiobenthos and remains of macrofauna by large deep-sea deposit-feeders was studied in detail in four Atlantic holothurian species by Khripounoff and Sibuet (1980), who have convincingly demonstrated negative selectivity for live metazoans, as well as the rare occurrence in the food of large remains of macrofauna (polychaetes, holothurians and ophiuroids) in the last stage of decomposition. The authors suggested that organo-mineral aggregates, possibly formed with the help of bacteria and common in the guts, are the principal food for these deposit-feeders.

The results of these special studies of food composition in typical deep-sea deposit-feeders throw doubt on the hypothesis of necrophagy in deep-sea holothurians and their feeding on live metazoan meiobenthos. Presumably, in this case it would be erroneous to extrapolate to deep-sea deposit-feeders the results on food composition in shallow-water members of this group, which are known to have small invertebrates mixed with the sediment in their guts. The presence of metazoans in the gut content of shallow-water deposit-feeders may only be a reflection of their abundance in the surrounding sediment since their actual digestion remains unconfirmed.

3.2.2.2. *Systematic groups* Deposit-feeders are known in the majority of groups of deep-sea benthos, for example various bivalves, a number of polychaete families, echiurids, sipunculids, some amphipods and isopods, a majority of holothurians and irregular echinoids, asteroids (mainly of one family), and individual species of ophiuroids. Feeding in 48 species of deep-sea deposit-feeders representing different systematic groups has been examined (Table 10), mostly based on the food composition. Food collection methods, selectivity and position of deposit-feeders in the sediment, discussed earlier (Sokolova, 1958, 1977), are only considered in part in the present study.

Deposit-feeders in general are adapted to utilize a low-calorie food diluted by mineral material. Their adaptations include, first of all, taking in large masses of the sediment, exceeding half the body weight of the animal in some species. The ratio of intestine length to body length averages 2.7 in holothurians (30 measurements, eight species), 4.7 in echinoids (five measurements, one species) and 6.0–7.0 in sipunculids (two measurements, two species). Lengthening of the hind gut has been reported in deep-sea deposit-feeding bivalves of the genus *Abra* (Allen and Sanders, 1966) and in Protobranchia (Allen, 1971). This, according to Allen (1979), is the main distinguishing feature of deep-sea deposit-feeding bivalves. Similar dependence of the intestine length on the type of food is known in fish. Adult fish which feed on detritus (plus bacteria) and vegetation have an intestine more than three times the length of the body, whereas in carnivorous fish this ratio is always about one or less (Borutskii and Verigina, 1961).

The deep-sea deposit-feeding macrobenthos includes some other groups in which deposit-feeding can be assumed by analogy with related shallow-water animals.

Gastropods: deposit-feeding in deep-sea gastropods was described by Rex (1976), who identified deposit-feeders among gastropods of the northeastern

Table 10   List of deep-sea deposit-feeders in which gut contents were investigated.

Sl. No.	Species	Stations No. of RV "Vitya"	Number of specimens	Index of fullness, $\%_{000}$	
				Average	Limits (range)
1	2	3	4	5	6
	Polychaeta				
1	*Travisia profundi*	617	10	2507	1510–3677
2	*T. forbesii*	542	2	4647	
3	*Brada sp.*	946	3	453	441–533
4	*Amphicteis mederi*	162	5	1422	366–2245
	Echiuroidea				
5	*Jakobia birsteini*	162	1		
6	*Ikedella achaeta*	140	1		
7	*Prometor grandis*	23	1		
	Crustacea order Isopoda				
8	*Antharcturus hirsutus*	2078	5		
9	*Antharcturus sp.*	2208	5		
10	*gen. sp. fam. Ischnomesidae*	973	5		
	Sipunculoidea				
11	*Phascolion lutense*	618	2	1332	869–3000
12	*Golfingia margaritacea*	149, 177, 357	4	1878	
	Holothurioidea order Eliasipodida				
13	Fam. Elpidiidae sp. "a"	956	2	1332	1785–2500
14	sp. "b"	956	3	1954	
15	*Elpidia birsteini*	2216, 5612	8	1509	824–1909

Table 10 Continued

Sl. No.	Species	Stations No. of RV "Vitya"	Number of specimens	Index of fullness, $\%_{000}$	
				Average	Limits (range)
1	2	3	4	5	6
16	*E. kurilensis*	162	5	1855	1224–2426
17	*E. hanseni*	2217, 2218	10	1635	900–2674
18	*Scotoplanes kurilensis*	5637	1		
19	*Peniagone* sp. "a"	5633	2		
20	*Scotoplanes hanseni*	5633	2		
21	*Peniagone* sp. "b"	5637	1		
22	**Fam. Psychropotidae** *Psychropotes raripes*	140 5624	1 1	3850	Single count
23	**Order Aspidochirotida** *Paelopatides solea*	591	1		
24	**Order Dendrochirotida** *Sphaerothuria bitentaculata*	541, 618, 2209	9	1291	166–2800
25	*Cucumaria abyssorum?*	2209	8	1391	437–2077
26	**Order Molpadonida** **Fam. Molpadiidae** aff. *Trochostoma* sp.	618	3	4892	6292–6250
27	sp. "a"	603	2	4997	
28	sp. "b"	603	3	5572	4313–7255
29	sp. "c"	539, 541	3	5315	4020–6890
30	sp. "d"	23	2		

Table 10 Continued

Sl. No.	Species	Stations No. of RV "Vitya"	Number of specimens	Index of fullness, %/000	
				Average	Limits (range)
1	2	3	4	5	6
	Fam. Gephyrothuriidae				
31	sp. "a"	5617	2		
32	sp. "b"	5633	2		
33	Pseudostichopus sp.	2120,2208	5	4834	4505–4873
34	Molpadiodemas sp.	2209	5	4280	2322–5387
	Order Apodida				
35	Fam. Synaptidae sp. "a"	603	4	6120	5529–6470
36	sp. "b"	591	8	6108	3809–6844
	Asteroidea				
37	Eremicaster tenebrarius	539,956,618,956	11	4147	2704–5312
38	E. vicinus	5633	11		
39	Eremicaster sp.	5617	1		
40	Thoracaster sp.	2074,2116	2	5979	Single count
41	Vitjazaster djakonovi	2119	1	5247	
42	Ctenodiscus crispatus	542	4	2500	
43	Thorocaster magnus	5624	1		
44	Styracaster sp.		1		
	Ophiuroidea				
45	Amphiophiura sculptilis	4582,5249,4630	38		
	Echinoidea				
46	Aeropsis fulva	534	3	5741	1211–1968
47	Urechinus loveni	2220	5	1500	
48	Brisaster latifrons	132	5	1588	

Atlantic, by the intestine filled with ooze, seen through the shell. These included 20 species of Archaeogastropoda, 15 species of Mesogastropoda and five species of Opisthobranchia from the continental slope and abyssal plain between 478 and 4860 m depth.

Bivalve molluscs: deep-sea bivalves of the superfamily Nuculacea (Protobranchia) and species of the family Scrobiculariidae in the superfamily Tellinacea are deposit-feeders (Knudsen, 1970, 1979; Allen, 1979). Deposit-feeders dominate among deep-sea bivalves in terms of the number of species (Knudsen, 1970). Parallel adaptations have been observed in different families, which distinguish them from the related shallow-water forms. Most important of these is the elongation of the hind section of the intestine, where a commensal microflora is associated with the faecal matter and bacterial decomposition of faeces is presumed to supply additional food for the molluscs (Allen and Sanders, 1966; Allen, 1979).

Polychaetes: deep-sea polychaetes of the families Fauveliopsidae, Maldanidae and Terebellidae may be considered to be deposit-feeders because they are similar to shallow-water species of these families that undoubtedly use this type of feeding (Fauchald and Jumars, 1979).

Isopods: deep-sea macrobenthic isopod species of the genus *Antharcturus* and species of the family Ischnomesidae can be regarded as deposit-feeders (Sokolova, 1958). The gut contents in deposit-feeding species of *Antharcturus* demonstrate sorting of detritus particles and so do the differences in the composition of intestinal content compared with the composition of benthic sediments. In members of the family Ischnomesidae there is a higher degree of detritus sorting than in arcturids, which brings these isopods close to polychaetes of the family Ampharetidae in the composition of their gut content. Unlike arcturids and ischnomesid isopods, the genera *Storthyngura* and *Eurycope*, regarded earlier as deposit-feeders, are more correctly considered as omnivores, varying from deposit-feeding to predation.

Amphipods: Deposit-feeding amphipods in the deep-sea macrobenthos were examined by Kamenskaya (1977, 1978, 1984), who noted a high index of intestinal fullness. At abyssal depths, Kamenskaya noted among benthic deposit-feeders two species of the genus *Neochela* of the family Corophiidae from the Pacific Ocean, and among benthopelagic species, *Bathyceradocus stephenseni* of the family Gammaridae from the Pacific and Indian oceans, and species of the genus *Epimeria* of the family Paramphithoidae from the Pacific Ocean. Treatment of amphipod collections from the ocean floor is still incomplete, so the number of deposit-feeding species known to inhabit the ocean floor may grow in the future.

Ophiuroids: only a few species of deep-sea ophiuroids can be regarded as deposit-feeders. The majority, owing to high trophic plasticity, are included among omnivorous or carnivorous invertebrates in which the animal component of the food is either considerable or dominant. Results available so far indicate that *Perlophiura profundissima*, whose food composition was investigated by

Litvinova (1979), and *Amophiura sculptilis* are deposit-feeders (Litvinova and Sokolova, 1971).

3.2.2.3. *Food composition*   The results of my analysis of gut content of deep-sea deposit-feeders from various systematic groups, with published data on food composition in deep-sea holothurians (Khripounoff and Sibuet, 1980), show that in the abyssal macrobenthos there is a distinct group of invertebrates feeding on the detrital organic matter contained in the sediment. Possibly the bacteria colonizing the detritus, and the excretory products of metazoans and some protozoans, play a significant role as a source of food. Live or dead invertebrates are scarcely used for food by deposit-feeders. The hypothesis of conversion of deep-sea deposit-feeders to omnivorous invertebrates ("generalists"), for whom the animal component of food acquires great significance (Dayton and Hessler, 1972; Thiel, 1979), has not been confirmed.

The overall analysis of intestinal contents in deep-sea deposit-feeding polychaetes (four species), echiurids (three species), sipunculids (two species), holothurians (nine species), asteroids (three species) and echinoids (two species) showed in all except one, the rare occurrence of solitary fragments of large metazoans and an absence of intact small metazoans. The exception was one species of asteroid, *Eremicaster vicinus* (Porcellanasteridae), in which small invertebrates of several groups were found. Published data on the food composition in four species of deep-sea holothurians (Khripounoff and Sibuet, 1980) show that they do not select metazoan meiobenthos. These authors also suggested the significance of integumental remains of large invertebrates in the food of holothurians. In my opinion, the state of these remains in the last stage of decomposition makes this assumption unlikely. The rarity of occurrence and smaller numbers of intact small metazoan invertebrates as well as fragments of large dead invertebrates in the gut content should be considered as a general feature of food composition in deep-sea deposit-feeders compared with their shallow-water counterparts from different systematic groups.

Detritus and tests of planktonic diatoms are two principal components of the gut content of all the investigated species of deep-sea deposit-feeders and in almost all species the third principal component is mineral particles. Organic matter in the detritus is apparently the principal source of food, according with the opinion of Khripounoff and Sibuet on the feeding of deep-sea holothurians mainly on organo-mineral aggregates. Other components of detritus, found only in some deposit-feeders and constituting generally a small fraction of the volume of the gut content, have some importance in the feeding of deposit-feeders. Faecal pellets of agglutinated sediment, being a substrate for bacteria, are presumably important in the feeding of deep-sea deposit-feeders. They have been found in nearly half of the species investigated, some of whom showed an increased concentration of pellets indicating selectivity. It is likely that the food is associated bacteria, not the pellets themselves, since their concentration does not change from foregut to hindgut in deposit-feeders.

Protozoan sarcodines of the subclass Xenophyophoria, found in a whole range of species, may be used in the same way as faecal pellets, as carriers of bacteria (Tendal, 1979). Sarcodines of the subclass Granuloreticulosia, family Komokiidae, found in significant quantity in some species, constitute, in individual cases, up to 40% of the gut contents. Available data do not permit a definite answer as to whether the abundance of komokiids in the gut reflects their abundance on the sea floor or is the result of selection. Secretory foraminiferans are found in almost all investigated species of deposit-feeders and constitute up to 3–5% of the food volume; agglutinated foraminiferans are found in half of the species but in much lower quantity than the secretory ones. A comparison of the number of foraminiferans with plasma (Rose Bengal stain) in the foregut and hindgut in certain species of holothurians did not reveal a decrease in number of stained foraminiferans as they passed through the intestines. This indicates incomplete digestion and only partial utilization of foraminiferan plasma in the intestines of deposit-feeding holothurians. The role of foraminiferans in nutrition of deep-sea invertebrates was analysed in detail by Davies (1987) and Gooday *et al.* (1992).

The spore and pollen analysis of the gut content in two species of asteroids and one species of holothurian in parallel with the analysis of benthic sediments from their biotopes showed increased concentration of spores and pollen of higher plants in these deposit-feeders. A comparison of the number of intact and crushed pollen grains in the foregut and hindgut of holothurians revealed an increase in number of crushed pollen grains after their passage through the intestines. Together with the highly nutritive properties of spores and pollen these data allow us to consider spores and pollen as one of the possible additional sources of food for deep-sea deposit-feeders (Kozjar *et al.*, 1974).

In deep-sea deposit-feeders, as in their shallow-water counterparts, we observe adaptations to the use of low-calorie food diluted with mineral particles of the benthic sediment. These include the elongation of intestines, which is typical of the majority of deposit-feeders, and an increase in the rate of feeding resulting in non-stop feeding (even during reproduction) and also in high indices of intestinal fullness. The average weight of the gut contents constitutes one third of the body weight of a deposit-feeder. The indices of fullness in deep-sea deposit-feeders from different systematic groups differ because of different selectivity. Indices are highest in the animals with lowest selectivity, as was shown earlier for shallow-water deposit-feeders (Turpaeva, 1953). Among deep-sea and shallow-water sipunculids, asteroids and holothurians, judging from the indices of fullness of intestines, the intensity of feeding does not differ substantially. In the deep-sea deposit-feeding bivalve molluscs, according to Allen (1979), elongation of the intestine (and consequently the increase of ratio of its content to the body weight) is a specific feature distinguishing the deep-sea deposit-feeders from shallow-water ones, and is presumably linked to the development of commensal bacteria in the hindgut.

3.2.2.4. *Distribution*   A comparison of the distribution of different systematic groups of large benthic deposit-feeders on the floor of the Pacific and Indian oceans, based on the data from trawling (Sokolova, 1986), showed one common feature in all groups: all of them, to some extent, are absent from open tropical parts of ocean remote from the coast, and are confined to the ocean periphery and to the equator. This tendency is most evident in deposit-feeding asteroids (family Porcellanasteridae); irregular echinoids (families Pourtalesiidae, Urechinidae, Hemiasteridae, Salenidae, Brissidae and Aeropsidae); holothurians of the family Molpadiidae (Molpadonida) and family Synaptidae (Apodida); it is somewhat less distinct in bivalve molluscs of the families Malletiidae, Nuculanidae and holothurians from the families Gephyrothuriidae (Molpadonida), Synallactidae (Aspidochirotida), Elpidiidae, Deimatidae and Laetmogonidae (Elasipodida). Less restricted distribution is observed in the groups feeding on the surface layer of sediments; that is, bivalve molluscs and holothurians (Aspidochirotida and Elasipodida). Similar trends of distribution have been reported for other systematic groups of deep-sea deposit-feeders.

In polychaetes an affinity to the periphery of the oceans was demonstrated by Levenstein (1970) for the families Fauveliopsidae and Opheliidae and by Kucheruk (1981) for the families Fauveliopsidae and Ampharetidae.

Among sipunculids, only some small forms of the family Golfingiidae occasionally penetrate the central remote regions of the oceans (Murina, 1979). Kamenskaya (1977) examined the distribution pattern of the deposit-feeding amphipod family Paramphithoidae (genus *Epimeria*) in the Pacific Ocean and found them mainly along the periphery of the ocean, apparently correlated with feeding conditions.

Mironov (1975) demonstrated the wide distribution of the smallest echinoids of the family Pourtalesiidae, in the genera *Echinosigra* and *Pourtalesia*, best adapted to burrowing, along the periphery of the ocean and suggested they had lower demands for the quantity and quality of organic matter in the sediment than others in the family. Thus, among large benthic deposit-feeders there is no widely distributed species similar to fast-moving or sessile carnivores inhabiting remote central regions of the ocean.

3.2.3. *Suspension-feeders (Sestonophages)*

It is known that the concept of "seston" includes plankton, tripton and mineral suspensions (Zernov, 1949). Thus, suspension-feeders include invertebrates feeding on organic detritus suspended in the water and organisms present in the water. Some suspension-feeders are triptonophagous, feeding mostly on detritus, other suspension-feeders are zoophagous, feeding on planktonic as well as small benthic invertebrates floating above the sea floor. Thus, the concept of "suspension-feeder", developed for aquatic animals in general by Zernov (1949) and Turpaeva (1953), has been subdivided according to the composition and source of the food. The group of suspension-feeders includes invertebrates with different mechanisms of feeding. According to Jørgensen (1966) the majority of suspension-feeders filter water through special structures, retaining suspended particles. They include Spongia, Bivalvia, Tunicata, Cirripedia and others.

A minority of suspension-feeders are non-filtering but trap suspended particles moving along their body surface. A key role in this trapping is played by cilia and mucus. Such "sedimentators" include Brachiopoda, Bryozoa and also members of Cnidaria, Polychaeta and Echinodermata.

Recently, some Spongia of the group Cladorhizidae have been found to have the ability to seize small victims from water (Vacelet and Boury-Esnault, 1995). These Spongia live in conditions similar to the deep sea. They have no morphological structures for filtration of water, and they capture passively floating victims. This feeding mechanism is similar to those of Cnidaria.

The group of deep-sea suspension-feeders includes both filterers and sedimentators for which every unit of food does not require individual capture.

3.2.3.1. *Source of food* The source of food of deep-sea benthic suspension-feeders cannot be studied experimentally, and hence the discussion is based on the specific conditions of their habitat. The composition of deep-sea seston has been studied by Mel'nikov (1975a) in the eastern Pacific, where at a depth of about 4000 m organic detritus (10–50 μm particles) constituted 98–99% of seston by weight. In the macroplankton size group only olive-green cells (products of excretion of salps according to Silver and Alldredge, 1981) and a small amount of bacterial aggregates were observed. The concentration of the suspended organic matter decreased with depth. Similar observations on the size composition of deep-water suspended matter have been reported by other authors. According to Honjo (1978), in the Sargasso Sea in a sediment trap at a depth of 5367 m, 114 m above the sea bottom, about 90% of the particles were finer than 62 μm, with large particles being rare. According to Rowe and Gardner (1979), on the northwestern slope of the Atlantic Ocean (depth 2200–3650 m), sediment traps at 518 m above the ocean bottom and close to the bottom, collected roughly 80% of organic particles not exceeding 63 μm. Detritus, apparently, has a greater significance in the food of deep-water suspension-feeders than in those from the photic zone, since phytoplankton is virtually excluded from their food spectrum. Deep-sea suspension-feeders, feeding on large particles (several millimetres), utilize "marine snow", studied in detail by Silver and Alldredge (1981). These authors point out that in Tunicata (according to Fournier (1973)) large quantities of the components of "marine snow" are found in the intestine. "Marine snow" contains traces of proteins, carbohydrates and lipids (Fowler and Knauer, 1986) testifying to its nutritional value.

The content of microflora in suspended particles is comparable to that of deposited particles (Baird et al., 1985; Karl et al., 1988). Bacteria, small metazoans and their larvae (Aller, 1989) are transferring with suspended mud above the bottom sediment. Transport of phytodetritus by the same mode (Lampitt, 1985) has been mentioned above.

3.2.3.2. *Systematic groups*   Among the macrobenthic invertebrates inhabiting the ocean floor, suspension-feeders include members of the following systematic groups: Spongia, Cnidaria (Pennatularia, Antipatharia, Madreporaria), Bivalvia, Polychaeta, Crustacea (Cirripedia, Amphipoda), Bryozoa, Brachiopoda, Ascidiacea, Crinoidea and Holothurioidea (Dendrochirotida). The gut contents have been studied in a small number of deep-sea suspension-feeders, including Bivalvia (fragments), Cirripedia, Ascidiacea and Dendrochirotida (Table 11). Knudsen (1970) reported on some Bivalvia, and Kamenskaya (1977, 1984) studied Amphipoda of the Pacific Ocean. The occurrence of suspension-feeding in deep-sea members of remaining groups has been decided by analogy with their shallow-water relatives.

3.2.3.3. *Food composition*   Analysis of available data on the food composition in deep-sea suspension-feeders of different systematic groups has revealed that they can be differentiated into triptonovores, catching detritus and its satellite components, and zoophages, catching mainly near-bottom plankton and minute benthos floating over the bottom. Triptonovores include filter-feeders such as bivalve molluscs of the family Arcidae, stalked ascidians and non-filtering sediment-collectors such as amphipods of the family Ampeliscidae and Dexaminidae, brachiopods and holothurians of the order Dendrochirotida. In addition, by analogy with shallow-water suspension-feeders of corresponding groups, filtering sponges and bivalve molluscs (Limopsidae and Mytilidae) and sedimentators such as serpulomorph polychaetes, Bryozoa and most likely crinoids can be included among triptonovores. Zoophages include plankton-filtering cirripede crustaceans of the family Scalpellidae and catchers of plankton and floating minute benthos such as bivalves of the families Pectinidae, Verticordiidae, Cuspidariidae and Poromyidae. These molluscs are intermediate between sedimentators and waiting predators. Pennatularia, Antipatharia, and most likely Madreporaria, feed mostly on animal food and can be included in this group by analogy with shallow-water forms.

In deep-sea triptonovores the ingested suspension differs from that eaten by shallow-water forms in containing fewer intact organisms and their remains, reflecting the composition of deep-sea suspended matter. Among deep-sea zoophagous suspension-feeders scalpellids are characterized by a smaller size range of food objects compared with shallow-water Lepadomorpha, because the feeding apparatus is weaker in deep-sea cirripedes (Zevina, 1981). In molluscs of the families Cuspidariidae and Poromyidae the food of deep-sea members differs from that of shallow-water forms in the species of crustaceans collected, as a result of a fauna change with depth (Knudsen, 1967, 1970).

3.2.3.4. *Distribution*   Suspension-feeders of the various systematic groups are not uniformly distributed on the ocean floor. Studies of their distribution (Sokolova, 1986), mainly in the Pacific, show that the most common and widely

Table 11 Deep-sea suspension-feeders examined for gut contents.

Sl. No.	Species	Station No.	Number of dissected specimens	Number of specimens with food	Index of fullness of intestine, %ₒₒₒ	
					Average	Limits
1	Crustacea: Cirripedia					
	Neoscalpellum eltanini	1636	24	20	41	17–83
	Ascidiacea: Stolidobranchiata					
2	*Culeolus murrayi*	5621	6	5	157	57–297
3	*Culeolus tenuis*	3363	5	5	373	191–548
4	*Culeolus robustus*	5608	2	2	291	—
5	*Culeolus* sp.	908	5	5	187	158–234
6	Holothurioidea: Dendrochirotida					
	Cucumaria abyssorum	524	5	3	70	35–106
7	*Psolus* sp.	542	10	4	—	—

distributed groups in the Pacific and the Indian oceans are sponges and serpulid polychaetes. Madreporaria, Bryozoa, Cirripedia, bivalve molluscs of the order Septibranchia and the families Arcidae and Pectinidae, some brachiopods and Antipatharia are almost as widely distributed on the Pacific Ocean floor but are less common. Pennatularia, crinoids and molluscs of the family Limopsidae have considerably narrower distributions while stalked ascidians and holothurians of the order Dendrochirotida have the narrowest distributions. Thus, the most widely distributed groups are characterized by a considerable selectivity of feeding whilst those with the narrowest distribution show the least selectivity. Some groups with a limited distribution may be confined to the most productive regions of the ocean by a dependence on high productivity, as known for their shallow-water relatives. For example, crinoids are abundant and diverse even in shallow-waters and are known (Meyer, 1973) to be limited by food supply; that is, the level of primary productivity. The feeding activity of crinoids requires stimulation by an adequate number of food particles in the water (Jørgensen, 1966), otherwise the feeding apparatus remains inactive. This feature may distinguish crinoids from other suspension-feeders, for example sponges, some cnidarians, serpulomorph polychaetes and ascidians, in which transport and collection of particles do not depend strictly on the concentration of particles in the water and are, as a rule, continuous (Jørgensen, 1966).

The limited distribution of other groups of suspension-feeders, such as large stalked ascidians and holothurians of the order Dendrochirotida, may be a result of their low feeding selectivity, since they use resuspended material (ascidians) or because of their ability to feed above the bottom (holothurians). As a result, these suspension-feeders depend on the quality of the sedimented food material, its accumulation and transformation. To meet their food requirements, high sedimentation rates of organic material are essential in order to obtain the required minimum of assimilable matter from the mass of inorganic material. Selective suspension-feeders (selecting either particle size or quality) are more widely distributed. The limited data available on deep-water seston composition are insufficient to distinguish regions with different productivity. Measurement of the quantity and the size of suspended detritus particles and bacterial aggregates at a depth of 4000 m in a moderately productive region (Mel'nikov, 1975a,b, 1976) suggests that even under relatively favourable feeding conditions, a benthic suspension-feeder must feed actively, independently of suspension concentration, and it must be adapted to feed on very small tripton particles.

No data are available on the quantity of near-bottom deep-sea zooplankton available as food for scalpellids, but their relatively wide distribution on the ocean floor is indirect proof of an adequate quantity of small zooplankton in the near-bottom waters. The distribution of bivalve molluscs, Madreporaria and Antipatharia that feed on plankton and suspended small benthos supports this conclusion.

4. TROPHIC AREAS

4.1. Notes on Methodology

The following analysis of spatial distribution of macrobenthos on the ocean floor omits continental slopes, deep-sea trenches and hydrothermal vents because their wide range of depths and/or physico-chemical characteristics could mask the influence of feeding conditions. By excluding the influence of vertical zonation, it is possible to observe the large-scale influence of trophic factors within the framework of latitudinal and circumcontinental zonation, in different climatic zones and at different distances from the continents. The only indication of vertical zonation in the area under consideration is carbonate accumulation above the calcium carbonate compensation depth.

For calculation of the proportion by weight of trophic groups in the trawl catches, all invertebrates with a recognized type of feeding were included, without exception.

Earlier works have revealed (Sokolova, 1960a, 1969, 1976a,b) the important role of deposit-feeders among the deep-sea macrobenthos. Therefore, deposit-feeders have been considered as the principal indicator in the current investigation. An index has been selected of predominance of deposit-feeders when they contribute > 50% of the total weight of the invertebrate catch (parallel calculations are made for suspension-feeders and carnivores). Use of this index reveals regions where a predominance of deposit-feeders is found at a majority of stations and regions where such predominance is rare. Intermediate regions are characterized by such predominance at about half the stations. The biological essence of the index indicates the advantage of feeding from the sediment over feeding from the water and also shows the relative ratings of trophic groups (deposit-feeders, suspension-feeders and carnivores).

The ocean floor has been divided into trophic areas according to the proportions of the various trophic groups at each station, the records of indicator deposit-feeders at individual stations and a subsequent comparison with geological information for the same stations.

Near the continents, deposit-feeders constitute about 70% of the deep-sea macrobenthos. A sharp increase in the thickness of the upper oxidized layer of the sediment marks a transition from the near-continental to the pelagic type of sedimentation where the predominance of deposit-feeders decreases to 47%, with a decrease of sediment organic carbon and in rate of sedimentation. This is the boundary of large-scale distribution of Fe–Mn nodules. Where the C_{org} content is minimal, deposit-feeders constitute only 5% of the macrobenthos and the very low sedimentation rates are indicated by the massive accumulation of shark teeth on the ocean floor. This is now known to apply to both the Pacific and Indian Oceans. Earlier, Sokolova (1970) showed that the biomass of bottom sampler benthos could serve as an indicator of boundaries between regions of the Pacific Ocean with different feeding conditions. When other data were absent, biomass could be used to determine boundaries between the regions.

Subdivision of the major eutrophic and oligotrophic regions has been based on the frequency of occurrence of large deposit-feeders on different types of sediment, using the number of records of these animals on a particular type of sediment, as a percentage of the total number of records of the animals. Mapping sediments with the greatest frequency of occurrence of large deposit-feeders revealed zones of trophic optimum at the periphery of the ocean (Sokolova, 1976a,b).

For additional definition of each of the trophic groups in all regions, the frequency of occurrence of large taxa comprising these groups has been estimated.

The next task was to estimate the abundance of trawl macrobenthos in regions revealed according to structural features. For each of the trawl catches, mean indices of weight of a single specimen and the number of specimens in the trawl catch were calculated for each region, for the two oceans combined and for each ocean separately. When calculating the mean fresh weight of a single specimen 3 mg was taken as the lower limit for invertebrate macrobenthos, following Strel'tsov *et al.* (1974). In the present study, all entire animals > 3 mg were weighed to obtain the total wet weight for calculation of the weight of a single specimen.

Statistical comparison of the abundance of macrobenthos in each of the trophic regions was based on the mean weight values for single specimens, the number of specimens per trawl sample and the biomass per square metre from the bottom sampler. The reliability of differences between the means for the regions was checked with the *t*-test, which showed 95 and 99% confidence limits (Sokolova, 1986).

4.2. General Descriptions of Trophic Areas

4.2.1. *Eutrophic and Oligotrophic Conditions*

Non-uniform distribution of organic matter and differences in its transformation in sediment deposits create substantial differences in feeding conditions. Eutrophic conditions are defined as those where organic matter in the deposits is sufficient for the survival and predominance of large bottom-feeding deposit-feeders in the macrobenthos. Oligotrophic conditions occur where nutritive material is scarce in the deposit, restricting the distribution and predominance of large deposit-feeders in the macrobenthos. Oligotrophic feeding conditions arise where there is minimum influx of organic matter, which decomposes mostly on the surface of the deposit and is buried in a highly transformed inactive state, comprising stable, insoluble, high molecular weight residual compounds

(Bordovsky, 1964b, 1966a,b, 1974; Romankevich, 1970, 1974; Bogdanov et al., 1971, 1979).

Oligotrophic conditions are characterized by the following indices:

• Low content of organic matter in the sediment – less than 0.25% of dry weight (Romankevich, 1970, 1977; Bogdanov et al., 1971).
• Extremely low rates of sediment accumulation – 1–3 mm 1000 year^{-1}, or less (Lisitsyn, 1971, 1974).
• Lowest organic carbon accumulation rates – 0.1–0.5 mg cm^{-2} 1000 year^{-1} and less (Romankevich, 1968, 1977; Bogdanov et al., 1971; Bordovsky, 1974).
• Lowest values of the coefficient of burial of C_{org} mostly less than 0.1–0.15% (Bogdanov et al., 1979; Bogdanov, 1980).

Oligotrophic conditions are observed in areas of pelagic sediments, mainly eupelagic clays in regions where Fe–Mn nodules accumulate (Skornyakova and Murdmaa, 1968; Murdmaa et al., 1976; Murdmaa, 1979a,b, 1987). Such conditions are confined to deep-sea basins of the central tropical parts of the ocean, in regions of minimum surface productivity (Lisitsyn, 1974, 1981). The presence of nodules themselves is not proof of oligotrophy. Different morphogenetic types of nodules reflect changes in formation processes in different conditions (Skornyakova, 1983, 1986; Skornyakova and Murdmaa, 1986; Murdmaa, 1987). In oligotrophic conditions on eupelagic clays the nodules are all of one type (Murdmaa, 1987), whereas on radiolarian oozes at the periphery of the equatorial zones of the Pacific and Indian oceans the nodules are variable in morphogenetic types and contents. There they do not indicate oligotrophic conditions.

Eutrophic conditions are found in the remaining peripheral and equatorial parts of the ocean floor under highly productive surface waters. They arise where organic matter is not completely oxidized on the surface of the deposit and is buried in an active state (Bordovsky, 1964b, 1966a,b). Eutrophic conditions differ from oligotrophic in the following indices:

• High content of organic matter in the sediments – from 0.25 to 1–2% (Romankevich, 1970, 1974, 1977).
• Usually higher rates of sedimentation, from 3–4 to 30 mm 1000 year^{-1} or more (Lisitsyn, 1971, 1974).
• Higher organic carbon accumulation rates in the sediment – from 0.5 to 10 mg cm^{-2} 1000 year^{-1} or more (Romankevich, 1968, 1977; Bogdanov et al., 1971, 1979; Bordovsky, 1974).
• Higher values of the coefficient of burial of C_{org}, > 0.25% to 1% and more (for the periphery of the pelagic region) (Bogdanov et al., 1979; Bogdanov, 1980).

Eutrophic conditions are observed on hemipelagic sediments, also on myopelagic clays (Murdmaa et al., 1976; Murdmaa, 1979a,b, 1982) and pelagic biogenic siliceous carbonate sediments (Bezrukov, 1964a,b; Bezrukov *et al.*, 1970). Transition from unfavourable oligotrophic conditions to favourable eutrophic conditions is gradual.

4.2.2. Differences in the Role of Trophic Groups in Macrobenthos under Different Feeding Conditions

In eutrophic conditions deposit-feeding invertebrates frequently predominate in the macrobenthos but they rarely predominate in oligotrophic conditions. Examination of 100 trawl samples from the Pacific Ocean (Figure 2), showed that in eutrophic conditions large deposit-feeders predominated by weight in 55% of trawl catches, while in oligotrophic conditions they predominated in 3–14% of catches. In fact, the value of 14% is probably too high for oligotrophic conditions, since half the trawl samples were deliberately taken in depressions, where large deposit-feeders are probably more common because of increased accumulation of food material.

Individual systematic groups with different trophic strategies may show peculiarities of distribution which reveal changes in the composition of populations indicating cardinal differences in the proportion of trophic groups in eutrophic and oligotrophic conditions (Sokolova, 1986).

Deposit-feeders of various systematic groups tend to avoid oligotrophic parts of the ocean and are more common near the ocean periphery and in the equatorial region, whereas suspension-feeders of many systematic groups are distributed extensively in many parts of the oceans. Passive, attached carnivores, such as actiniarians of the family Actinostolidae; and the most mobile necrophages (amphipods of the family Lysiannasidae) are widespread in the open parts of oceans (Kamenskaya, 1977, 1980, 1984). The presence of swimming scavengers over the entire expanse of the deep ocean floor has been revealed in photographs taken by a "monster-camera" on the sea floor in the peripheral and open parts of the Pacific Ocean (Dayton and Hessler, 1972; Schulenberger and Hessler, 1974; Schulenberger and Barnard, 1976). Decreases in their number and diversity were observed along a transect from the periphery to the central part of the Pacific. At the farthest point from the periphery of the northeastern basin, not one animal was recorded near baited traps within half an hour of the arrival of the trap, whereas at the periphery of the basin, in the same latitudes, a very large number of diverse scavengers was recorded in the same time and circumstances (Dayton and Hessler, 1972).

4.2.3. Trophic Areas and their Spatial Disposition in Oceans

Areas on the ocean floor with eutrophic conditions and eutrophic structure of macrobenthos can be separated from areas with oligotrophic conditions and

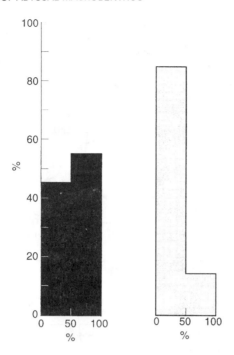

Figure 2 Proportion of deposit-feeders in trawl-caught macrobenthos in eutrophic (black) and in oligotrophic (grey) conditions in the Pacific Ocean. Ordinates % of the total samples; abscissae % weight of deposit-feeders (after Sokolova, 1984).

oligotrophic structure. Near-Continental, Oceanic and Equatorial eutrophic areas, and Northern and Southern oligotrophic areas can be identified (Figures 3–5). In nature the borders between the areas with different trophic conditions are not linear. They have a scalloped form because of local variations in conditions, such as near-bottom currents and local relief. For example, eutrophic spots may be wedged into the border of an oligotrophic area and vice versa. This is a particular feature of the Equatorial Area of the Pacific Ocean.

Among eutrophic areas the Near-Continental Area is characterized by more favourable feeding conditions than the Oceanic and Equatorial ones, while among oligotrophic areas the Southern Area has the least favourable feeding conditions. The Near-Continental Area occupies roughly 30%, the Oceanic and Equatorial 53% and Oligotrophic 17% of the floor area of all oceans taken together (excluding the North Polar Basin). These areas are differentiated by the nature of their macrobenthos: the predominance of deposit-feeders; the degree of restriction of deposit-feeders and carnivores of different systematic groups; the mean weight and number of invertebrates in trawl catches; and the biomass of the bottom

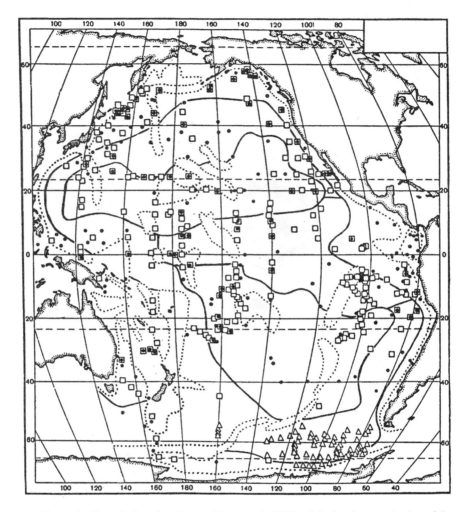

Figure 3 Boundaries of major trophic areas (solid lines) in the deep-sea basins of the Pacific Ocean, and positions of sampling stations. Filled circles, trawls; squares, bottom samplers (grabs); triangles, photo-stations. The dotted lines show the 3000-m isobath (after Sokolova, 1986, with additions).

sampler benthos. In each of these areas eutrophic or oligotrophic conditions arise in a unique manner depending on the position of the area in the ocean.

Differences (primarily associated with their individual shape and size) are observed in the spatial disposition of the Trophic Areas in each ocean.

Pacific Ocean. The full complement of trophic areas is found in the Pacific Ocean: three eutrophic (Near-Continental, Oceanic and Equatorial) and two

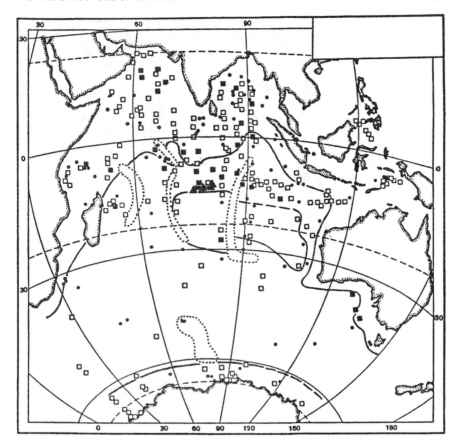

Figure 4   Boundaries of trophic areas (solid lines) in the deep-sea basins of the Indian Ocean, and positions of sampling stations. Filled circles, trawls; open squares, bottom samplers (grabs); filled squares, trawls together with grabs; filled triangles, "SAFARI II" stations. The dotted lines show underwater rises with depths less than 3000 m (after Sokolova, 1986, with additions).

oligotrophic (Northern and Southern) to the north and south of the equator (Figure 3). The Near-Continental Eutrophic Area encircles the oceans, interrupted only in the southwest. The Oceanic Eutrophic Area forms a ring within the Near-Continental Area. It is interrupted near the equator, separating the Oceanic Area in the north and south from the "base" of the Equatorial Area "girdling the ocean". The oligotrophic areas are situated under the central parts of the anticyclonic subtropical gyres north and south of the Equatorial Eutrophic Area.

Indian Ocean. Only two eutrophic areas are observed in the Indian Ocean, Near-Continental and Oceanic (Figure 4). The Equatorial region merges with the

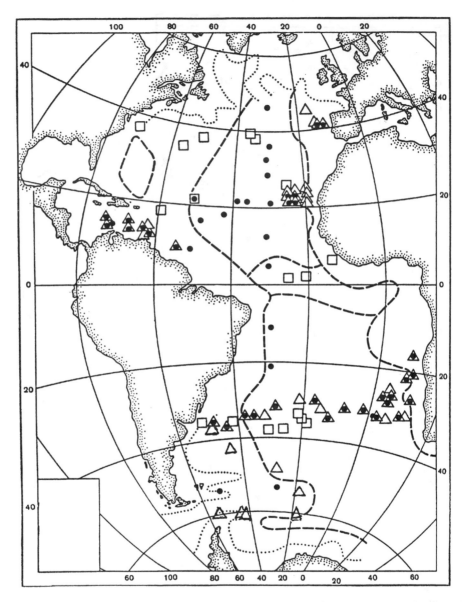

Figure 5 Boundaries of trophic areas (broken lines) in the deep-sea basins of the Atlantic Ocean and positions of sampling stations. Open triangles, trawling stations of Russian expeditions; filled circles, bottom sampler (= grab) stations of Russian expeditions; open squares, stations of the "Challenger" expedition with description of catches. The 3000-m isobaths to the north and the south are shown by dotted lines (after Sokolova, 1990, modified).

Near-Continental one in the north of the ocean. The boundary separating this combined Near-Continental Area from the Oceanic Area passes roughly along 5–7°N. There is only one oligotrophic area in the south. The Northern Oligotrophic Area is absent. The Near-Continental Area forms a semi-circle, separated from its near-Antarctic sector by the Oceanic Area in the east roughly between 50 and 60°S and in the west between 40 and 60°S. The Oceanic Eutrophic Area is circular, enclosing the only oligotrophic area.

Atlantic Ocean. In the Atlantic Ocean, apparently, all three eutrophic areas can be observed: Near-Continental, Oceanic and Equatorial (Figure 5). Oligotrophic areas have not been discovered.

Twice as much organic matter has been found in the sediments of the Atlantic Ocean as in Pacific Ocean sediments. This is explained by the relatively high productivity of the phytoplankton, the higher rate of organic matter accumulation in the sediments, and also by the larger area of the shelf, slope and continental rise in the Atlantic Ocean in comparison with the Pacific Ocean (Romankevich, 1977). On a global scale, trophic areas become part of giant trophic zones. To illustrate global zonation, schematically (Figure 6), the trophic areas identified in the present work have been shown as zones.

A single circumcontinental zone is formed by the Near-Continental Eutrophic Area, which to a large extent corresponds with the region of near-continental sediment formation. Of the five latitudinal zones, three represent eutrophic areas corresponding with the three humid zones of pelagic sediment formation; the other two latitudinal zones represent oligotrophic areas in the two arid zones of pelagic sediment formation (Lisitsyn, 1974, 1981) under the central part of the anticyclonic gyres. Vertical trophic zonation depending on the distribution of food material in relation to the bottom relief is on a small scale. The vertical trophic zonation described originally for deep-sea macrobenthos (Sokolova, 1960a) was followed and developed later by the investigators of various recent and ancient marine basins (Neyman, 1961, 1963, 1971, 1975, 1988; Kuznetsov, 1963, 1964, 1974, 1980; Tarasov, 1978, 1982; Kuznetsov et al., 1979; Beklemishev et al., 1982).

4.3. Description of Trophic Areas in each Ocean

4.3.1. *Trophic Areas in the Atlantic Ocean*

The Near-Continental Eutrophic Area in the Atlantic Ocean is separated from the Oceanic Eutrophic Area roughly along the boundary of hemipelagic sediment formation (Murdmaa, 1979a,b, 1987). Terrigenous sediments of the bordering part of the floor of the North American Basin have the same organic content as the

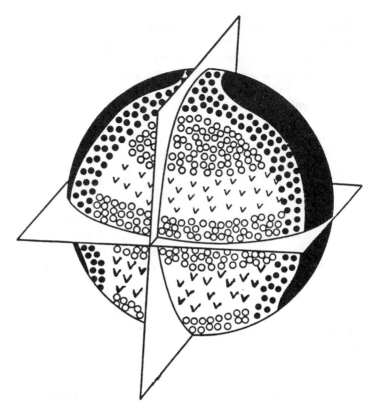

Figure 6 Diagrammatic representation of global trophic zones of deep-sea macrobenthos. Open circles, latitudinal eutrophic zone; V, latitudinal oligotrophic zone; black circles,
near-continental zone. The land masses are shown in solid black (after Sokolova,
1976b).

hemipelagic sediments of the Pacific Ocean. The organic carbon accumulation
rates mg/square cm/1000 years of terrigenous and terrigenous–calcareous
sediments in the Atlantic Ocean is one and a half times as high as in the Pacific
Ocean (Romankevich, 1977). Such data confirm favourable feeding conditions in
the Near-Continental Eutrophic Area which occupies almost half the floor of the
Atlantic Ocean. It forms a narrow semi-circle, interrupted in the southeast by the
Oceanic Eutrophic Area, and accounts for a large part of the ocean floor of the
central southeastern Atlantic. The pelagic clays of the central region of the floor
of the ocean, where oligotrophic conditions might be expected for the benthos,
contain as much organic matter as the sediments of the Oceanic Eutrophic Area
of the Pacific Ocean. In organic carbon accumulation rates, pelagic clays of the
Atlantic are 14 times richer than the myopelagic clays of the Pacific Ocean

(Romankevich, 1977). The rate of sediment accumulation correspondingly differs by 3–10 times. It follows that the central parts of the floor of the Atlantic are close to the Oceanic Eutrophic Area of the Pacific and Indian Oceans in conditions. The Oceanic Area in the Atlantic is divided by the Equatorial Eutrophic Area into two halves. It is charcterized by high rates of sediment accumulation (Lisitsyn, 1974) and higher concentration of C_{org} in the sediments (Romankevich, 1977; Emel'yanov and Romankevich, 1979) as well as by higher number and biomass of phytoplankton in the surface waters above the region (Volkovinskii et al., 1972; Zernova, 1974) than in the adjoining regions.

More than half the area of the Atlantic Ocean floor is covered by biogenous calcareous sediments (30–50% $CaCO_3$), which is not the case in any other ocean. Terrigenous sediments occupy second place in terms of area and are usually weakly calcareous or weakly siliceous (Emel'yanov et al., 1975). Pelagic clays have a limited distribution. A large number of the trawl and grab samples used in this work were obtained on calcareous sediments.

In addition to the predominance of calcareous sediments, the Atlantic Ocean differs in that it is open to ingress of cold high-latitude near-bottom water, which initiates a more active general near-bottom circulation (Hollister et al., 1984). Hence in the Near-Continental Area of the Atlantic Ocean, there are strong "contour" currents (Murdmaa, 1987), particularly in the western Atlantic. Near-bottom currents are important in the distribution of sediments in the western Atlantic, and also along the margins of Africa and Antarctica. Turbulent mixing of waters predominates in the bottom mixed layer from about 1 cm to about 100 m above the bottom (Hollister et al., 1984). During localized underwater storms in the northwest Atlantic, the velocity of the near-bottom current can increase from 5 to >50 cm s^{-1} and the sediment load can rise to 100 times background. At the end of the storm the velocity drops sharply (to 5 cm s^{-1}) and the surface of the layer of suspended particles sinks. Very fine suspended particles may then be transported long distances by weaker currents. Following such a storm, a 0.5 cm-thick layer of sediment can be deposited, corresponding to 500-year normal accumulation. In these conditions the benthic animals are likely to be fairly sedentary and better adapted for feeding on suspended material than for burrowing in search of food (Hollister et al., 1984). Therefore, a predominance of suspension-feeders may be predicted in the Near-Continental Eutrophic Area of the Atlantic.

Large deposit-feeders of various systematic groups (primarily holothurians) are universally distributed in the North Atlantic, indicating the predominance of eutrophic conditions for the macrobenthos (Sokolova, 1990a). On the Atlantic Ocean floor as a whole, large deposit-feeders dominate by mass in 37% of trawl catches and suspension-feeders dominate in 56%. Most of the catches dominated by suspension-feeders are from the ocean fringes, in the Near-Continental Area, where the water movement is strongest; suspension-feeders predominate in 68% of trawl catches and deposit-feeders predominate in 32%. In the Pacific Ocean the situation is reversed, deposit-feeders form more than half the mass in 70% of catches in the Near-Continental Area.

In the Oceanic Eutrophic Area of the Atlantic Ocean deposit-feeders predominate in 44% of catches, which is similar to the situation in the Pacific Ocean (Sokolova, 1986).

In different parts of the Atlantic Ocean deposit-feeders, suspension-feeders and carnivores from different systematic groups have restricted distributions. Trawl catches from various parts of the ocean differ in the number and weight of the invertebrates and the grab catches differ in the biomass per unit area. As in the Pacific Ocean, the Near-Continental Eutrophic Area of the Atlantic shows a greater restriction to biotope of most of the invertebrates under consideration than does the Oceanic Area (Table 12). The number of records of a taxon in a given biotope calculated as a percentage of the total number of records is here termed the restriction index, or index of confinement (Beklemishev, 1961). The mean restriction index for the Near-Continental Area for deposit-feeders is 70% in the Atlantic and 61% in the Pacific.

In the Atlantic, 88% of the trawl catches from the Near-Continental Area contained over 100 invertebrate individuals, whereas in the Oceanic Area such numbers were found in only 39% of the catches. The mean weight per specimen in the catch was also less in the Oceanic Area. Comparison with similar data from the Pacific (Sokolova, 1986) indicates that the Atlantic macrobenthos is generally smaller-sized. Secondly, the difference between the two eutrophic areas in terms of biomass of invertebrates is much less in the Atlantic Ocean (Sokolova, 1990a). Biomass of bottom-sampler benthos in the Atlantic Near-Continental and Oceanic areas differs on average by one order of magnitude (Table 13). In the Pacific Ocean, the Near-Continental and Oceanic areas have 2.64 and 0.176 g m^{-2}, respectively (Sokolova, 1986). In the Near-Continental Area of the Atlantic the biomass on biogenic calcareous sediments is three times higher than in terrigenous sediments (Table 13), whereas in the Oceanic Area, the biomass in calcareous sediments is almost 10 times that in myopelagic clays. Thus, in both areas calcareous sediments contain the higher biomass.

In the southern Atlantic Ocean (Scotia Sea, Weddell Sea and Argentine Basin) the macrobenthos has a eutrophic structure corresponding to that in the Near-Continental areas of the Pacific and Indian Oceans. It is characterized by a predominance of large invertebrate deposit-feeders in trawl catches. In this respect the southern Near-Continental Area differs from other Near-Continental parts of the Atlantic Ocean where suspension-feeders prevail. It is also characterized by diverse carnivorous invertebrates, particularly ophiuroids. The gut contents of these ophiuroids show that they feed on "dead body rain" from the water column above, notably planktonic crustaceans (mainly euphausiids), pteropods and small fishes. Thus, in the South Atlantic Ocean the "dead body rain" has a considerable influence on the trophic structure of the deep-sea macrobenthos.

Table 12 Restriction of invertebrates in terms of their frequency of occurrence in the trophic areas on the Atlantic Ocean floor.

Trophic group 1	Systematic group 2	Near-continental area Number of reports 3	Near-continental area Affinity % 4	Oceanic area Number of reports 5	Oceanic area Affinity % 6	Total number of reports 7
Deposit-feeders	Holothurioidea	93	71	38	29	131
	Echinoidea	17	65	8	35	25
	Asteroidea, Porcellanasteridae	29	60	20	40	49
	Bivalvia, Protobranchia	49	67	24	33	73
	Sipunculoidea	51	87	7	13	58
	Mean index		70		30	
Suspension-feeders	Spongia	47	67	22	33	69
	Pennatularia	23	56	18	44	41
	Actiniaria	23	57	17	43	40
	Madreporaria	17	70	7	30	24
	Polychaeta	16	47	18	53	34
	Cirripedia	52	72	20	28	72
	Bivalvia	49	57	37	43	86
	Bryozoa	25	50	25	50	50
	Brachiopoda	48	74	17	26	65
	Crinoidea	14	65	7	44	21
	Ascidiacea	75	82	16	18	91
	Mean index		69		35	
Carnivores	Pantopoda	17	68	8	32	25
	Asteroidea	24	54	21	46	45
	Mean index		61		39	

Table 13 Biomass of bottom-sampler (or grab) benthos in different trophic areas and different types of sediments in the Atlantic Ocean.

| Trophic area | Type of sediment | Biomass, g cm^{-2} | | Number of samples | Investigated area (ocean basin) |
		Mean	Range		
Near-continental	Calcareous clays	2.708	15.120–0.232	15	Canary, Angola, Cape, South Antilles, Guyana, Colombian, Gulf of Mexico
	Terrigenous clays	0.990	2.956–0.110	11	Iberian, Argentine, Venezuelian, Grenadian
	Overall	1.981		26	
Oceanic	Calcereous clays	0.271	0.810–0.005	20	Canary, Cape Verde, Angola, Cape, North American, Newfoundland
	Myopelagic clays	0.038	0.085–0.001	10	Canaries, Brazil
	Overall	0.193		30	

4.3.2. Trophic Areas in the Pacific and Indian Oceans

4.3.2.1. Near-Continental Eutrophic Area   The Near-Continental areas with the best feeding conditions stretch from the bases of the continental slopes over sloping and undulating accumulative plains on the floor of the ocean basins. Their benthic populations show great quantitative and qualitative richness. The Eutrophic Area is distinguished by the highest rates of accumulation of benthic sediments and organic matter on the ocean floor (Romankevich, 1977; Murdmaa, 1979a,b, 1987). In the Near-Continental Area hemipelagic sediments accumulate, mainly comprising terrigenous clays as well as terrigenous–biogenous siliceous (diatomaceous) and calcareous (coccolith–foraminiferous) oozes. In the Indian Ocean, where the Equatorial Area merges with the Near-Continental Area, hemipelagic sediments under the equatorial zone are rich in biogenous material. The organic carbon accumulation rates (Bordovsky, 1974) in the equatorial zone of the Indian Ocean has the same value ($4 \, \mathrm{mg \, cm^{-2} \, 1000 \, year^{-1}}$) as in the calcareous oozes of the western part of the Near-Continental Area of this ocean.

The boundary of the Near-Continental Eutrophic Area coincides roughly with the boundary of the region of near-continental sediment formation, where there is a sharp increase in the thickness of the upper oxidized layer of the sediment (Murdmaa, 1979a,b, 1982, 1987). The sediment cover of the abyssal plains of this region is continuous and uniform over large distances. Fe–Mn nodules are rare and confined to large submarine rises (Skornyakova and Zenkevich, 1976).

In the Near-Continental Area the frequency of occurrence and restriction indices for the majority of groups under consideration are typical: in the Pacific Ocean (Table 14) the average indices are one and a half times higher than for other eutrophic regions for all deposit-feeders.

Deposit feeders frequently predominate, feeding on settled and buried organic matter. In the Pacific Ocean deposit-feeders dominate in 70% of samples from the Near-Continental Area. The mean weight of one specimen of macrobenthos from this area is > 1 g in 65% of samples, greater than in samples from other areas. The number of invertebrate specimens in trawl catches is also rather high (Table 15), > 50 specimens in 92% of samples.

The mean bottom-sample biomass in the Near-Continental Area is 8.5 times higher than in the Oceanic Eutrophic Area and 53 times higher than in the Oligotrophic Area.

The Pacific Near-Continental Area is richer than that of the Indian Ocean: the mean weight of one invertebrate specimen in trawl catches is 1.5 times higher (Table 16) and the mean number of macrobenthos specimens per trawl catch is twice as high. The mean biomass of bottom sampler benthos is three times as high in the Pacific Ocean as in the Indian Ocean (Tables 15 and 16).

In the Pacific, 62% of grab samples contain a biomass greater than $1 \, \mathrm{g \, m^{-2}}$, while in the Indian Ocean 84% of samples have a biomass less than $1 \, \mathrm{g \, m^{-2}}$.

Table 14 Restriction of trophic types and systematic groups to different eutrophic areas in the Pacific Ocean based on the occurrence of invertebrates.

Trophic group	Systematic group	Near-continental eutrophic area (80 stations)		Oceanic and Equatorial eutrophic area (72 stations)		Total number of records of groups
		Number of records in the area	Affinity, %	Number of records in the area	Affinity, %	
Deposit-feeders	Holothurioidea	54	60	35	40	89
	Echinoidea	41	63	24	37	65
	Asteroidea	47	64	26	36	73
	Bivalvia	35	57	26	43	61
Carnivores	Actiniaria	32	51	31	49	63
	Asteroidea	47	77	14	23	61
Suspension-feeders	Spongia	67	53	59	47	126
	Polychaeta	9	24	29	76	38
	Cirripedia	21	64	12	36	33

Table 15 Mean weight of one specimen and number of specimens in trawl catches from different trophic areas of the Pacific and Indian Oceans.

Trophic area	Number of samples	Weight of one specimen in trawl catch, g		Number of samples	Number of specimens in trawl catch	
		Mean	Limits		Mean	Limits
Pacific Ocean						
Near-continental eutrophic	35	3.73	25.25–0.22	25	260	1482–10
Oceanic eutrophic	32	0.92	5.86–0.05	32	50	324–9
Northern and Southern oligotrophic	25	0.25	2.28–0.005	24	9	28–2
Indian Ocean						
Near-continental eutrophic	46	2.23	11.49–0.08	34	116	924–6
Oceanic eutrophic	26	0.84	3.53–0.007	26	19	83–4

Table 16 Biomass of bottom-sampler benthos in different eutrophic areas of the Pacific and Indian oceans.

	Pacific			Indian		
Trophic areas	Number of samples	Mean, $g\,m^{-2}$	Limits	Number of samples	Mean, $g\,m^{-2}$	Limits
Near-continental	48	2.64	15.92–0.14	97	0.755	15.80–0.027
Oceanic	73	0.176	0.885–0.024	88	0.145	0.730–0.004

However, in the Pacific Ocean almost all samples with a biomass > 1 g (49% samples of 62) were obtained from the temperate latitudes of the northern hemisphere, which are absent from the Indian Ocean.

4.3.2.2. *Oceanic Eutrophic Area*   The Oceanic Eutrophic areas are far from the continents, mostly on undulating abyssal plains in the ocean basins and their populations are considerably poorer in biomass and diversity than the population of the Near-Continental Area. They lie in regions of pelagic sediment accumulation, predominantly within zones of humid sediment formation with high rates of accumulation (Lisitsyn, 1974). Myopelagic clays, biogenous siliceous, siliceous–calcareous and calcareous sediments (at supercritical depths) are present (Murdmaa *et al.*, 1976; Murdmaa, 1979a,b, 1987). The cover of myopelagic clays may be interrupted by exposures of bedrock on abyssal hills. Iron–manganese nodules are unevenly distributed, relatively uncommon, and confined to higher elevations (Skornyakova and Zenkevich, 1976; Murdmaa, 1979a,b, 1987; Murdmaa *et al.*, 1989).

The macrobenthos of the Oceanic Area is depauperate in comparison with the Near-Continental Area (Table 14). This depauperation is most noticeable among deposit-feeders, which seldom predominate among the macrobenthos and exceed half the weight in only 45% of trawl catches. The mean weight of one specimen in trawl samples from the Oceanic Area is less than one-third of that in the Near-Continental Area (Table 15). It is > 1 g in only 21% of trawl samples from the Oceanic Area, that is one-third as often as in the Near-Continental Area. The average number of specimens in trawl catches from the Oceanic Area is one-fifth of that found in the Near-Continental Area (Table 15). The number of specimens is > 50 in one-third of samples from the Oceanic Area, three times less than in the Near-Continental Area. The average bottom-sample biomass in the Oceanic Area is in 8.5 times less than in the Near-Continental Area (Table 16). The total amplitude of fluctuation of biomass in the Oceanic Area is not more than $1\ g\,m^{-2}$.

Comparisons of data from the Oceanic areas of the Pacific and Indian Oceans show less difference than is found between the Near-Continental areas in these two oceans. The average number of individuals per catch in the Pacific Ocean is at least twice that in catches from the Indian Ocean, but the average biomass of bottom-sampler benthos shows practically similar values in both Oceans (Table 16).

4.3.2.3. *Equatorial Eutrophic Area*   The Equatorial Area covers the low-hilly and undulating plains to the north and south of the Equator, at the bottom of the ocean basins. This region is present in both the Pacific and Atlantic oceans, but its deep-sea macrobenthos has been investigated only in the Pacific Ocean. The equatorial belt, as a whole, is distinguished by high rates of pelagic sediment accumulation ($3-30$ mm 1000 year^{-1}) and by high accumulation rates of total C_{org} (Romankevich, 1977). Biogenous sediments (siliceous, diatomaceous–radiolarian oozes) are widely distributed, and at depths less than 4–4.9 km there are carbonate, coccolith–foraminiferan siliceous-carbonate oozes with high rates of sedimentation and higher organic carbon accumulation rates mg/square cm/1000 years, which decreases along the equator from east to west and on both sides of the axis of the belt. The equatorial belt, especially its periphery, is characterized by sharp fluctuations in sedimentation rates: ranging from $1-3$ to 10 mm > 1000 year^{-1}, with numerous areas of zero accumulation (especially in the north), associated with erosion and redeposition of sediments by the near-bottom currents (Murdmaa *et al.*, 1976; Murdmaa, 1979a,b, 1987). Feeding conditions are, consequently, heterogeneous and the benthic population is not uniform in distribution. The mean mass and number of invertebrate macrobenthos in trawl catches (Table 17) and the biomass of bottom-sampler benthos for the entire region (Table 18), are close to the means for the Oceanic Area. They have been combined with the latter when comparing different trophic areas.

Large invertebrates, more than 1 g in weight, are found in the Equatorial Area in trawl catches in the eastern half of the axial enriched band of carbonate and siliceous–carbonate oozes. They constitute on average 16% of the total number of species of macrobenthos in the trawl samples. Suspension-feeders (sponges) are the most frequent and abundant macrobenthos. Carnivores (Zoantharia and decapod crustaceans) and deposit-feeders (echinoids) are less important than suspension-feeders. Calculations of the numerical characteristics of the macrobenthos of the Equatorial Area include samples from a "rich spot" situated south of the Equatorial belt within the limits of the Southern Oligotrophic region.

At the periphery of the Equatorial Eutrophic Area a combination of suspension-feeding epifauna on nodules and deposit-feeders on the surface of soft sediment between nodules is usual in the Pacific (Greenslate *et al.*, 1974; Thiel and Schneider, 1988; Thiel and Schriever, 1989, 1990; Thiel, 1991; Thiel *et al.*, 1993). Benthos abundance at the northern periphery of the Equatorial Eutrophic Area in the Pacific was described by Tilot (1992), and the benthos at the southern periphery of the Equatorial Eutrophic Area of the Indian Ocean was described by Monniot (1984).

The trophic structure of megafauna at the northeast periphery of the Equatorial

Table 17 Mean weight of one specimen and number of specimens in trawl catches from the Equatorial Area of the Pacific Ocean.

Number of samples	Weight of one specimen, g		Number of samples	Number of specimen	
	Mean	Limits		Mean	Limits
14	0.325	1.269–0.041	15	22	176–4

Table 18 Biomass of bottom-sampler benthos in the Equatorial Eutrophic Area of the Pacific Ocean.

Region	Number of samples	Biomass, g m^{-2}	
		Mean	Limits
Northern periphery	21	0.124	0.769–0.004
Southern periphery	17	0.037	0.088–0.005
Axis-like part	42	0.312	0.800–0.040
Region as a whole	80	0.205	—

Eutrophic Area of the Pacific was analysed by Tilot (1992) from photographs. Some similar material was obtained from almost the same site by Foell and Pawson (1986). Tilot (1992) described in detail the region (130° 00'W–130 10'W, 13° 56N'–14° 08 N') lying at a mean depth of 4950 m. She found there a rich fauna (240 taxa), a quantitative abundance of population and a mosaic pattern in distribution of trophic groups. Suspension-feeders were more abundant than other trophic groups. There are siliceous oozes (3–8 cm 1000 year^{-1}) and near-bottom currents (from 3–5 cm s^{-1} to 10 cm s^{-1}) in this region. Polymetallic nodules of varied sizes and forms cover 30, 40 and 50% of the surface in different facies. There are areas without nodules and recent sediments. Therefore the conditions in the region studied by Tilot correspond to the conditions described in our data (Sokolova, 1986) for the Equatorial Eutrophic Area.

Data obtained by the SAFARI II Expedition to the Central Basin of the Indian Ocean (Monniot, 1984) supplement the picture of trophic division (Sokolova, 1979, 1986). These data confirm that the composition and abundance of the benthic population near of the northwestern border of the Oligotrophic Area correspond to the same characteristics of the adjacent Equatorial Area. Russian expeditions have no data from the region surveyed by SAFARI II. The Madagascar Basin is related to the Eutrophic Area and SAFARI II data on benthos supplement knowledge of the area.

4.3.2.4. *Oligotrophic areas*   Oligotrophic areas lie under the central parts of the subtropical anticyclonic gyres on the wide hummocky plains of the ocean floor. They are in the region of pelagic sediment accumulation, in zones with the arid type of sediment formation (Lisitsyn, 1974, 1981). In the majority of invertebrate groups under consideration a decrease is observed in the frequency of occurrence in the oligotrophic areas in comparison with the neighbouring eutrophic areas (Table 19). For all echinoderms among the deposit-feeders, and cirripedes among the suspension-feeders, this decrease in frequency may be considered as definite and steep.

Most data come from a special investigation of the oligotrophic areas of the Pacific Ocean. Extra data on the benthos of the oligotrophic areas of the Indian Ocean have been added to determine the characteristics common to oligotrophic areas in general. These are the discontinuity and the thinness of the cover of soft sediments and the massive accumulation of iron–manganese nodules (Skornyakova and Zenkevich, 1976; Bezrukov *et al.*, 1981; Bussau *et al.*, 1995). In the Northern Oligotrophic Area of the Pacific Ocean the sediments are predominantly eupelagic clays; in the peripheral parts there are regions with myopelagic clays, and in individual instances at depths of 3500–4600 m there are carbonate oozes. The rates of accumulation of eupelagic clays in the Northern Oligotrophic Area are < 1 mm 1000 year^{-1}; the mean C_{org} content is 0.22% (Murdmaa *et al.*, 1976; Murdmaa, 1979a,b, 1987), and the organic carbon accumulation rate is 0.15 mg cm^{-2} (Romankevich, 1977).

In the Southern Oligotrophic Area the sediments are eupelagic and zeolitic clays, and at depths of 3200–4500 m, carbonate oozes, in the Southern, Peruvian and Central basins. In the Peruvian Basin metalliferous ore sediments have been found. In the Southern Oligotrophic Area the rates of accumulation of pelagic sediments and their organic matter content are lower than in the Northern Oligotrophic Area. The rate of accumulation of eupelagic clays in the Southern Basin is, on the average, 0.4 mm 1000 year^{-1} (Murdmaa *et al.*, 1976). The benthos rarely contains a predominance of large deposit-feeders in the oligotrophic areas (in 3–14% of catches only).

The mean weight per specimen in trawl catches from oligotrophic areas is lower than from all other trophic areas (Table 15). It is > 1 g per specimen in only 7% of cases; that is, almost one-tenth the frequency found in the Near-Continental eutrophic areas and one-third that in Oceanic eutrophic areas. The number of specimens per trawl catch from the oligotrophic areas is, on average, one-twentieth of that in the Near-Continental and one-fourth that in the Oceanic eutrophic areas of both Oceans (Table 15). In the oligotrophic areas, the majority of samples (92%) contained < 25 individual invertebrates while 8% contained from 25 to 50 and no sample contained more than 50 specimens.

The average biomass of bottom-sampler benthos in the oligotrophic areas is one-six of that in the adjacent Oceanic eutrophic areas (Table 20). The total amplitude of biomass fluctuation in samples from the oligotrophic areas is about

Table 19 Frequency of occurrence of different systematic groups of benthos in trawl catches from oligotrophic and adjacent eutrophic areas in the Pacific Ocean.

Trophic groups	Systematic group	Oceanic and Equatorial Eutrophic Areas			Northern and Southern Oligotrophic Areas		
		1	2	3	1	2	3
Deposit-feeders	Holothurioidea	73	39	53	36	6	17
	Asteroidea Fam. Porcellanasteridae		26	36		2	6
	Echinoidea		25	34		2	6
	Bivalvia Fam. Nuculanidae and Malletiidae		26	36		6	17
Carnivores	Asteroidea		14	19		8	22
	Actiniaria		32	44		14	39
Suspension-feeders	Spongia		60	82		21	58
	Cirripedia		12	40		6	17
	Polychaeta Fam. Serpulidae		29	16		14	39

1, Number of samples; 2, number of records; 3, frequency of occurrence, %.

Table 20 Biomass of bottom-sampler benthos in different oligotrophic areas of the Pacific Ocean.

Trophic area	Number of samples	Biomass, $g\,m^{-2}$ Mean	Limit
Northern Oligotrophic	35	0.031	0.107–0
Southern Oligotrophic	44	0.019	0.070–0

$100\,mg\,m^{-2}$. The spatial non-uniformity observed in other trophic areas in biomass of the bottom-sampler benthos is absent in the oligotrophic areas. A difference is seen only between the Northern and Southern oligotrophic areas of the Pacific Ocean. The mean biomass in the Southern oligotrophic areas is almost half that in the Northern Area (Table 20). A comparison of the mean biomass on different types of sediment within the Northern and Southern oligotrophic areas, does not reveal differences between them, such as are observed in a similar comparison of biomass on different sediments in the Oceanic Area. Evidently, pelagic sediments at the bottom of the basins in the zone of arid sediment formation are poor in organic matter. Moreover, in the Southern Oligotrophic Area such depauperation is greater than in the Northern Area, which is confirmed by geological records.

Larger invertebrates (> 1 g) are rare in trawl catches from the oligotrophic areas: only four out of 24 samples from the oligotrophic areas of the Pacific Ocean contained one large invertebrate per sample. These included suspension-feeders: a glass sponge (over 5 g); a bivalve mollusc of the genus *Arca* (*c.* 3 g); a carnivorous actiniarian (*c.* 2 g) and a deposit-feeding holothurian (*c.* 6 g). The latter was found in the Southern Oligotrophic Area during trawling on the polygon (Station 6298–62 of R/V "Vityaz") in the depression at the foot of Petelin Mt. Its discovery in unfavourable feeding conditions with negligible rates of accumulation of sediment and organic matter is, apparently, explained by natural redistribution of sedimentary material. There can be a relative increase in accumulation rate of pelagic clays in depressions (Murdmaa *et al.*, 1976; Murdmaa, 1979, 1987) and, in view of this, there may be some food for large deposit-feeders in the sediment. From this example it is seen that, in oligotrophic conditions, the number and composition of large invertebrates (with weights > 1 g) enable us to visualize individual cases of small-scale non-uniformity.

4.3.2.4.1. *Epifauna* In oligotrophic areas unfavourable feeding conditions for benthic populations may be accompanied by optimum conditions for attachment of epifauna because of the abundance of iron–manganese nodules. Beyond the limits of the oligotrophic areas, in the Oceanic and Equatorial eutrophic areas, such nodules are scattered on siliceous oozes and myopelagic clays, and are rarely in high concentrations; they are found on elevations of the bottom but are often absent on level substrata (Skornyakova and Zenkevich, 1976). In the Near-Continental areas nodules and other solid substrata are rare or absent. This fact, in the opinion of some researchers (Millar, 1958; Koltun, 1970), can restrict the distribution in oceanic depths of species not capable of surviving on soft sediments.

In oligotrophic areas nodules are also present on the slopes and tops of abyssal hills and in depressions. Nodules of varied shapes often lie on the bottom in a single layer, their sizes generally varying from 0.5 to 10 cm in diameter. To the north and south of the Equator, Skornyakova and Zenkevich (1976) have distinguished latitudinal zones with high concentrations of nodules. In the northern zone nodules cover, on average, 32% of the surface area and in the southern zone, about 50%. High concentrations of nodules are formed where sedimentation rates are minimal (< 1 mm 1000 year^{-1}). Between these zones, high concentrations of nodules are rare and they are equally rare at the periphery of the oligotrophic areas in the adjacent Oceanic Eutrophic Area. Shark teeth can also serve as hard substrata for attachment in oligotrophic areas. According to Belyaev and Glikman (1970), major regions of large-scale accumulation of shark teeth are confined to oligotrophic areas. The number of teeth > 0.5 cm long may be several hundred to > 1000 in a trawl catch. In this way, invertebrates may have as much solid substratum as in shallow waters.

The epifauna includes a majority of deep-sea suspension-feeders and some carnivorous invertebrates. Some lie freely on the surface of soft sediment, or are partially embedded and rise above its surface. They constitute the soft-bottom epifauna.

In the regions under study the invertebrates attached to hard substrata are not particularly large, their body length, if it is not a branching colony, generally does not exceed 1–2 cm.

A calculation of the number of invertebrates on each type of substratum in samples from the oligotrophic and adjacent eutrophic areas of the Pacific Ocean shows that iron–manganese nodules are frequently used as substratum, half as frequently in the eutrophic area as in the oligotrophic area. In the Northern Oligotrophic Area, in addition to nodules, many invertebrates were found on shark teeth; in the Oceanic Eutrophic Area, debris of volcanic rocks, pumice and gravelly–pebbly material of glacial origin are widely used as substratum. Iron–manganese nodules should not be considered the only substratum for attachment of invertebrates. Evidently, their choice depends on their abundance.

In comparing the frequency of occurrence of different systematic groups in the epifauna from oligotrophic and eutrophic areas, the records in both the oligotrophic areas fall into three groups (Antipatharia, Actiniaria and Brachiopoda) in comparison with eutrophic areas. Frequency of occurrence tends to decrease in the Southern Oligotrophic Area compared with the Northern Area. This is observed not only in Antipatharia, Actiniaria and Brachiopoda but also in three more groups: serpulids, cirripedes, and particularly in bivalve molluscs of the family Arcidae. In the Southern Oligotrophic Area, where coverage with nodules reaches the highest values for the ocean, several characteristic groups of epifauna decrease.

To compare quantitative characteristics of epifauna of different trophic regions average values were taken for the total weight of all sedentary invertebrates in trawl samples (Sokolova, 1986). The average weight of epifauna in samples from both oligotrophic areas was very low, less than 200 mg. In the Equatorial Eutrophic Area it is undoubtedly more than 300 mg. In the Oceanic Eutrophic Area it is more than 1 g, that is, much higher than in the other areas. However, in both eutrophic areas epifauna constituted less than 10% of the total weight of the catches, while in the oligotrophic areas it was 30–50% of the total. In the oligotrophic areas in nearly 70% of samples invertebrates attached to hard substrata were dominant among suspension-feeders, while in eutrophic areas they were dominant in only 30% of samples. This allows us to consider that, with shortage of food and abundance of a hard substratum, the proportion of sedentary invertebrates could increase in the macrobenthos but not their absolute number.

Sedentary invertebrates comprise nearly half the total number of specimens in trawl samples from the Southern Oligotrophic Areas and nearly one-third in the Northern Oligotrophic and both eutrophic areas (Sokolova, 1986). A comparison of the mean weight and number of specimens of sedentary invertebrates in catches from different areas shows that the weight of individual specimens of epifauna in eutrophic conditions is higher than in oligotrophic conditions. A decrease is observed in the mean weight of individuals of the epifauna in unfavourable conditions, which could arise in each of the groups either with a decrease in the size of the individual of a species or with elimination of species of large-bodied invertebrates from the population in oligotrophic areas. For example, in the brachiopod *Pelagodiscus atlanticus*, Zezina (1976b) reported a size decrease with transition from favourable to unfavourable conditions. An example of elimination of large species shows miniaturization in ascidians at oligotrophic areas among which, large-stalked forms disappear and only small solitary and colonial forms survive.

4.3.2.4.2 *Remarks*   In reviewing the literature on deep-sea bottom fauna, Gage and Tyler (1991) noted that Spongia in oligotrophic conditions are usually small in size and in the impoverished North Oligotrophic Area of the Pacific are represented by such forms as *Cladoriza* and *Asbestopluma*. The occurrence of Actiniaria in central parts of the ocean is pointed out. Gage and Tyler (1991) also believe that large-bodied deposit-feeders are absent in the most oligotrophic regions of the ocean floor. Large-bodied fauna is represented by very motile scavenging opportunists or predators. The existence of suspension-feeders such as Spongia (Hexactinellidae and Demospongia) in oligotrophic areas can be explained by the fact that they are "caloric dwarfs". Lipps and Hickman (1982) in discussing Antarctic deep-sea bottom fauna (agglutinating rhizopod protozoa) assume that animals with complicated skeletal structures are able to live in food deficit conditions because they have relatively small amounts of protoplasmatic tissue, slow growth rates and low metabolic requirements.

In Russian works on the distribution of deep-sea invertebrates in eutrophic and oligotrophic conditions, the differences between large- and small-sized animals were noted early on. It has been shown that small-sized deposit-feeders such as nematodes, polychaetes, isopods, tanaids and others successfully inhabit the eutrophic and the oligotrophic areas of the ocean (Sokolova, 1988). This is explained by their ability to exist at the expense of the thinnest surface layer of a sediment. In this layer, organic matter may have some food value even in oligotrophic areas (Sokolova, 1972).

At the transition from eutrophic area to oligotrophic area the trophic structure of the meiobenthos remains the same. It consists mainly of deposit-feeders (Sokolova, 1972, 1977). A study of deep-water bottom fauna in the Northern Oligotrophic Area of the Pacific was based on material collected by box-corer (Hessler, 1974; Hessler and Jumars, 1974). The objects of the research were small-sized invertebrate animals which have not been especially considered in Russian works. Hessler (1974) reported that a large part of the sample is meiofauna (passes through a 1.0-mm mesh). Common meiofaunal metazoan taxa – Nematoda, Ostracoda, Copepoda – together are more numerous (2.26 times) than all macrofaunal taxa. Among the latter 85% of the specimens are Polychaeta, Tanaida, Bivalvia and Isopoda. The taxa of this size-class show high species diversity: 174 specimens in 54 species.

Hessler and Jumars (1974) confirmed the numerical domination of meiofaunal animals (in particular Nematoda) as 1.5–3.9 times more than macrofaunal animals, and the high species diversity of small-sized invertebrates. Most species were met only once. The density of small-sized invertebrates was 84–160 specimens m^{-2}. Among the polychaetes motile deposit-feeders totalled 51.7%, predators were in second place and sestonophages comprised less than 3% of specimens.

In oligotrophic conditions motile forms prevail among small-sized deposit-feeders because, where food is scarce, only a few invertebrate animals are able to exist within the limits of their tubes and burrows (Hessler and Jumars, 1974; Jumars and Fauchald, 1977). The absence of a decline in species diversity of most small macrobenthos in the oligotrophic area was confirmed by Wilson and Hessler (1987). In the same area mega- and macrobenthos are scarce, according to data from photos, anchor dredge and epibenthic sled (Hessler and Jumars, 1974). Slow-moving megafaunal deposit-feeders ("croppers") are usually absent. The time-lapse "monster camera" has shown that in oligotrophic conditions the adaptative feeding strategy for large invertebrates is scavenging and possibly generalism, with high mobility. Scavenging amphipods crowd near traps in the oligotrophic area (Dayton and Hessler, 1972; Hessler, 1974; Hessler and Jumars, 1974). The diet of generalists reflects the scarcity of food in extreme oligotrophic conditions (Jumars and Gallager, 1982). These studies confirm and supplement one another, showing that the distributions of different size groups are not the same, because of the different needs and abilities if animals of different weights.

4.4. Influence of Deposit-feeders on Benthic Sediments in Different Trophic Areas

Relations between deposit-feeders and deep-water sediments have been discussed by Gage and Tyler (1991). Transformation of sediment organic matter in the digestive tracts of deep-sea deposit-feeders (echinoderms) has received some attention (Bordovskii et al., 1974; Khripounoff and Sibuet, 1980; Akhmet'eva et al., 1982; Akhmet'eva, 1987). Another equally important consequence of the activity of deposit-feeders is the redistribution of ingested sediment particles, concentrating particles either by size or by quality with varying degrees of selectivity.

The deep-sea macrobenthic deposit-feeders ingest mainly the fine aleurite fraction of sediments. The proportion of such particles in the gut content of seven forms of Echinodermata from the Bering Sea was found to be two to three times greater than in the sediments, owing to their feeding behaviour. It has also been shown that the food reworking process in the guts of invertebrates is responsible for diminution in size of sediment mineral particles (Sokolova, 1987).

The faecal products of deposit-feeders tend to contain a higher concentration of organic matter than in the natural sediment, with different composition, as assessed by Bordovsky et al. (1974) and Akhmet'eva et al. (1982). The concentration of organic matter in the intestines of several deep-sea holothurians is seven to eight times as high as in the surrounding sediment. As food passes through the intestine some organic matter is assimilated, and the concentration decreases but it remains higher than in the natural sediment so that the faeces seem enriched with organic matter. According to these authors, the intestinal contents of holothurians are one or two orders of magnitude richer in lipoid substances (bituminoid) compared with the surrounding sediment.

In two species of deep-sea holothurians a lower concentration of organic carbon is found in the hindgut in comparison with that in foregut (Moore et al., 1995). Deep-sea holothurians utilize only 15% of organic carbon and 22% of organic nitrogen (Khripounoff and Sibuet, 1980). The low efficiency of utilization of carbon and nitrogen by these holothurians is a result of the lower nutritive value of the organic matter of deep-sea sediments, when it is subject to enzymatic hydrolysis.

As already mentioned, there are no similar estimates for deep-sea deposit-feeders. Only indirect indices have been obtained regarding the extent of their activity. Tentative results for the alteration of deep-sea sediments by deposit-feeders per unit of time are available for the northern periphery of the Equatorial

Eutrophic Area of the Pacific Ocean (Paul *et al.*, 1978) where time-lapse photography of the ocean floor was conducted at a depth of 4873 m for 202 days. Here, echinoderms (holothurians) are most abundant (35%) among benthic populations. The population density of macrofauna was about three individuals/100 m^{-2}. Here changes in microtopography caused by the activity of organisms continued for several months while alterations to deposit-feeders' faeces took weeks. Faecal cords of holothurians (Elasipoda) on the ocean floor decreased considerably during 137 days because of their utilization by bacteria or meiobenthos. In deep-sea trenches (where feeding conditions are optimum, as in the Near-Continental Area) photographs showed that echinoderms and particularly holothurians were important among deposit-feeders influencing the benthic sediments (Lemche *et al.*, 1976). The population density of 10 species of holothurians, determined from the photographs, was 1–3 individuals 100 m^{-2} for six species, which agrees with the results cited above for the ocean floor in the Equatorial Area. Three species of holothurians showed 10 times this density (i.e. 1 individual 10 m^{-2}).

Evidently, in the extent of their influence on benthic sediment holothurians should be regarded as one of the most important groups of deposit-feeders. From their example one can attempt a tentative estimate of the scale of influence of deep-sea feeders on the ocean floor sediments (Moore *et al.*, 1995).

From indirect evidence (index of fullness of intestines) the intensity of feeding in deep-sea deposit-feeders is equal to the intensity of feeding in shallow-water forms. Some data even suggest that this intensity could be higher in the deep-sea forms, in view of the feeding conditions at depths where loss of diurnal signals means that round-the-clock activity is possible. Thus, in estimating the role of deep-sea benthic deposit-feeders in the transformation of benthic sediments (Bordovsky *et al.*, 1974) it was suggested that they ought to put the sediment through their digestive tract continuously, to ensure the necessary energy. This suggestion arose from the analysis of the chemical composition of the intestinal contents showing utilization of only a small part of the sediment organic matter as it passes through. The indices of lower assimilation of C_{org} and N_{org} obtained for deep-sea holothurians by Khripounoff and Sibuet (1980) agree with this suggestion. Thus, the intensity of feeding may be considered invariantly low and in later estimates it may not be taken in consideration.

Sibuet and Lawrence (1981) have estimated the ratio of population density of shallow-water and deep-sea holothurians in the Bay of Biscay. The population density of holothurians on the abyssal plain is lower than in the shallow water, in the sublittoral zone, by 10^4 according to the results of trawl catches and by 10^3 according to photographs. There are correspondingly from 3 to 7 individuals per 1000 or 10 000 m^{-2}; that is, the density of deep-sea holothurians determined in this region from photographs seems considerably lower than the density determined by the same means on the ocean floor in the Equatorial Area (1 individual per 100 m^{-2}) and in the deep-sea trenches (1–10 individuals per

$100 \, \text{m}^{-2}$). For extrapolation, here we have taken the densities of deep-sea holothurians, determined from the photographs taking into account data scatter, and consider them to vary in the range from 1 individual per $100 \, \text{m}^{-2}$ to 1 individual per $1000 \, \text{m}^{-2}$ on the ocean floor in eutrophic conditions. Published results (Crozier, 1918; Hyman, 1955; Thorson, 1966; Hauksson, 1979; Mosher, 1980; Hammond, 1982) for seven different species of littoral and sublittoral holothurians, indicate that the quantity of sediment worked by their population amounts to several tens of kilograms dry weight $\text{m}^{-2} \, \text{year}^{-1}$. Reducing this value by 10^3 or 10^4 according to the reduction of the population density of deep-sea holothurians in comparison with the shallow-water ones, 10 g (or less) of dry weight $\text{m}^{-2} \, \text{year}^{-1}$ is worked in the deep sea.

Optimum eutrophic conditions on the deep ocean floor are found in the Near-Continental areas at the base of the continental slopes (Sokolova, 1981b, 1982) and deep-sea deposit-feeders are largely confined to this area (restriction index 61%); in 70% of catches they are predominant in the macrobenthos. There are many large forms and the average weight of one specimen (2.9 g) is comparable with the average size of common shallow-water holothurians (Hauksson, 1979). In view of these figures the maximum working of sediments by holothurians is to be found in the Near-Continental Area, amounting to not less than 10 g dry weight $\text{m}^{-2} \, \text{year}^{-1}$.

In the Oceanic and Equatorial eutrophic areas, with a decrease in the total population of macrobenthos to roughly one-fifth, deposit-feeders become sparse and dominate in only 45% of catches, and large forms are smaller (average weight of one specimen, 0.9 g). Correspondingly, in these areas one may expect the quantity of worked sediment to be proportional to the lower population density of holothurians; possibly 1–3 g dry weight $\text{m}^{-2} \, \text{year}^{-1}$.

In the central oligotrophic areas holothurian populations are extremely sparse. They dominate only in rare cases and large forms are practically absent. The total population of macrobenthos, as well as the average weight of a single specimen, is 10 times lower than that in the Near-Continental Eutrophic Area. Therefore, it can be suggested that even the maximum quantity of the sediment worked by holothurians in the oligotrophic areas is about 1 g and the average values are near 1 mg dry weight $\text{m}^{-2} \, \text{year}^{-1}$. Thus, the quantity of sediment worked by deep-sea holothurians decreases by 2 orders of magnitude (and sometimes even more) from the Near-Continental Area of the ocean to its central oligotrophic areas.

Several estimates have been made of the thickness of the fully worked layer of abyssal sediments but there are no estimates of the time scale. All measurements have been made in regions with eutrophic conditions. In the eastern equatorial Pacific this layer is 2–4 cm (Arrhenius, 1952), in the Atlantic 5 cm (Ericson et al., 1961), in the Near-Continental Area of the eastern Pacific north of California 5 cm thick (Griggs et al., 1969). Thus, the thickness of the layer worked completely, since in the latter region traces of activity of invertebrates have been observed in the sediment to a depth of 50 cm. On the ocean floor, the thickness of the sediment layer suitable for feeding of deposit-feeders (except for millimetre-sized forms or those in the superficial thin layer) and exposed to their effect is associated with global changes in conditions of sediment formation. Major changes occur at the transition from near-continental sediment formation to pelagic sediment formation (i.e. at the border of the Near-Continental Eutrophic Area) and also at the transition from the humid zone of pelagic sediment formation to the arid zone (i.e. at the boundaries of the Oceanic and Equatorial eutrophic areas with the oligotrophic areas). These changes in the thickness of the sediment layer suitable for feeding are associated with different organic carbon accumulation rates mg/square cm/1000 years, its coefficient of burial and the rates of sediment accumulation in general. Lacking data for various parts of the ocean, tentative estimates of this layer are: tens of centimetres in the Near-Continental Eutrophic Area; a few centimetres in the Oceanic and Equatorial eutrophic areas and only a few millimetres in the oligotrophic areas (Sokolova, 1986).

4.5. On the Species Composition in Different Trophic Areas

It is interesting to know something of the species composition of the populations of the various trophic areas, particularly their similarities and differences. Comparisons can be based on a random selection of species from different systematic groups. Species identified from the deep-sea fauna collections gathered during expeditions of the Institute of Oceanography, Academy of Sciences of the USSR allow the examination of the fauna from the central parts of the oceans. The species identification of different systematic groups is not uniform, which has to a great extent determined the number of species examined from different higher taxa and also influenced the choice of the taxa themselves. Moreover, in selecting for analysis species of deposit-feeders known to be confined to the Near-Continental Area were not included (i.e. the species serving as indicators of change in feeding conditions at the boundaries of the trophic areas). One hundred and thirty-two species were chosen, constituting nearly 7% of the total number of species known from the abyssal zone (Vinogradova, 1977): 15 species of sponges (Koltun, 1970), 13 species of Pennatularia and Antipatharia (Pasternak, 1958, 1961a,b, 1962, 1964, 1970, 1973, 1975a,b, 1976, 1977; Grasshoff, 1981), six species of solitary Madreporaria (Keller, 1976, 1982), eight species of scaphopod molluscs (Chistikov, 1982), seven species of bivalve molluscs (Filatova, 1958, 1976; Oliver and Allen, 1980; Filatova and Shileiko, 1984), 37 species and subspecies of sipunculids (Cutler, 1977; Murina, 1979; Cutler and Cutler, 1980), 17 species and subspecies of pantopods (Turpaeva, 1971a,b,c, 1973, 1975), 14 species of cirriped crustaceans (Zevina, 1970, 1973a–c, 1975a,b, 1976), 12 species of brachiopods (Zezina, 1965, 1973, 1975a,b, 1976a, 1981), one species of ascidian (Kott, 1969; Vinogradova, 1970) and two species of ophiuroids (Belyaev and Litvinova, 1972, 1976). The type of feeding of the species and subspecies examined here is varied. Many of them, 63 species, are suspension-feeders, 44 deposit-feeders, 17 carnivores and eight species could not be reliably included in any one of these groups. These are scaphopod molluscs, whose feeding has not been adequately studied as yet. The available data confirm that they make some use of benthic foraminifers and detritus (Davies, 1987).

Analysis of the distribution of these 132 species of macrobenthos has made it possible to distinguish three types of distribution, determined by the characteristics of species present either in the most favourable feeding conditions or in any favourable conditions or finally in any conditions, from the most favourable to the most unfavourable.

4.5.1. *Types of Species Distribution*

- The Near-Continental type of distribution is identified for species and subspecies confined only within the limits of the Near-Continental Eutrophic Area, living in the most favourable conditions in the abyssal depths of the ocean (groups I–III). The distribution of species within the Near-Continental Area can be more or less wide; that is, the species may be found in the entire "ring" around the periphery of the ocean, in half of this "ring" or in only one sector. The species may be confined to one, or spread through two or three oceans.
- The Oceanic type of distribution is distinguished for species and subspecies inhabiting not only the Near-Continental Eutrophic Area but also in the adjoining Oceanic Eutrophic Area such as the Equatorial Eutrophic Area (groups IV–VI). Many such species are more widespread in the Near-Continental Eutrophic Area than in the Oceanic Area. These species may be found almost all round the entire "ring" of the Near-Continental Area in three oceans but in only one Oceanic Area. Some species with the oceanic type of distribution can be found even in the oligotrophic areas on sedimentary accumulations near island shelves and submarine mountain structures, especially if the latter protrude as "tongues" from a region with eutrophic conditions into a region with oligotrophic conditions. Here, eutrophic conditions can be conserved within the boundaries of the accumulations. The oceanic type includes three species and one subspecies (group VI) that are not found in the Near-Continental Eutrophic Area but are known only from the Oceanic and the Equatorial areas.
- The Panthalassic type of distribution is distinguished for species and subspecies inhabiting the eutrophic as well as the oligotrophic areas (groups VII, VIII, VIIIa), though many of them may reach only one of the Oceanic areas. The term "Panthalassic" has been taken from plankton classification where it is used for the most widely distributed group of species above the principal pycnocline. Here, the Panthalassic species are the most widely distributed on the ocean floor in abyssal depths.

All three types of distribution are found among deep-sea and eurybathyal species which are common in abyssal depths. Among the eurybathyal species, which are rare in abyssal depths, only two types of distribution (Near-Continental and Oceanic) have been identified, that is, they are greatly affected by the restrictive influence of the trophic factor. On the whole, among Near-Continental species eurybathyal species predominate (63%), while among Oceanic species there are more deep-sea species, constituting 57%, while among Panthalassic species there are still more deep-sea species (66%).

The dependence on feeding conditions decreases from the Near-Continental to the Panthalassic species. The majority of species and subspecies among those examined (86%) are characteristic of Near-Continental (42%) and Oceanic (44%) types of distribution while the remaining 14% have Panthallasic distribution. Among the species and subspecies found beyond the Near-Continental Area, most of the Oceanic ones (83%) exhibit restriction, in terms of frequency of occurrence, to the Near-Continental Area for which a large number of records are available. Most of the Panthalassic species (71%) do not exhibit such restriction, but are spread, more or less uniformly, over various eutrophic areas.

The population of megabenthic invertebrates of the Near-Continental Eutrophic Area is the richest in terms of species number because there are not only Near-Continental but most of the Oceanic (93%) and Panthalassic (94%) species and subspecies in the area. The population of the oligotrophic areas, in terms of species number, is only a small part (a quarter) of the population of both the Oceanic and Equatorial areas. Of the total number of species occurring in the Near-Continental Area, 56% are distributed in the Oceanic and Equatorial areas and only 13% are able to penetrate the oligotrophic areas (Figure 7).

4.5.2. Species Groups

Data on the distribution of the species in the eight groups identified are given in Table 21. The 56 species constituting groups I and III are restricted to the Near-Continental Eutrophic Area; 54 species (groups IV, V and VI) inhabit the Near-Continental Eutrophic Area and enter the Oceanic or the Equatorial areas; four species (group VIa) are found only in the Oceanic and Equatorial areas, and 18 species penetrate from the eutrophic areas to the oligotrophic areas (groups VII, VIII and VIIIa).

Panthalassic species:

• *Group VII*. Beyond the limits of the eutrophic areas, only five eurybathial species under consideration (see the list) enter the oligotrophic areas. Four of them (1–4) are widely distributed in the Near-Continental Area on the ocean floor areas as well as areas on the upper part of continental slopes and trenches, while No. 3 is also common on the shelf. The fifth species is found in the

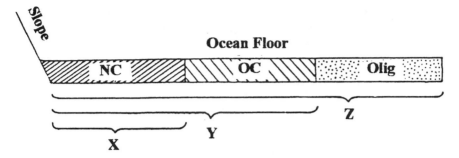

Figure 7 Fundamental pattern of deep-sea trophic areas, shown as meridional transect of the ocean floor. NC, The Near-Continental Eutrophic Area; OC, the Oceanic Eutrophic Area; Olig, the Oligotrophic Area; X, near-continental type of species distribution; Y, oceanic type of species distribution; Z, panthalassic type of species distribution (X, 100% of species; Y, 56% of species; Z, 13% of species) (after Sokolova, 1991, modified).

Near-Continental Area only on the slope (abyssal and bathyal depths). Among the five species of group VII, Nos 1, 3 and 4 have roughly the same number of records in the Oceanic and the Equatorial areas as in the Near-Continental Area.

    1. *Arca (Bentharca) asperula* (Dall)
    2. *Scalpellum vitreum* Hoek
    3. *Golfingia (Nephasoma) improvisa* (Theel)
    4. *Pelagodiscus atlanticus* (King)
    5. *Ophiopyrgus wyvillethomsoni* Lyman

• *Group VIII.* Twelve deep-sea species under consideration have been found in the oligotrophic areas (see the list). All are found from the Near-Continental Eutrophic Area to the oligotrophic areas in one or more oceans and almost all (except No. 11) deeper than 2000 m. Four (Nos 2, 4, 5 and 7) are found in the Northern and in the Southern Oligotrophic areas, five species (Nos 6, 8, 9, 11 and 12) are found only in the Northern Oligotrophic Area, and the remaining 3 (Nos 1, 3 and 10) only in the Southern Oligotrophic Area.

    1. *Cladorhiza rectangularis* Ridley and Dendy
    2. *Cladorhiza longipinna* Ridley and Dendy
    3. *Asbestopluma biseralis* Ridley and Dendy
    4. *Bathypathes lyra* Brock
    5. *Costentalina tuscarorae subcentralis* Chistikov
    6. *Arca orbiculata* Dall
    7. *Spinula calcar* Dall
    8. *Spinula oceanica* Filatova
    9. *Ledella crassa* Knudsen
    10. *Scalpellum regium* W. Thomson

Table 21   Groups of species and subspecies with different distribution in the trophic areas.

Investigated taxon	Number of species	Identified group									
		I	II	III	IV	V	VI	VIa	VII	VIII	VIIIa
Spongia	15	2	1	2	3		4			3	
Pennatularia	11	3	2	3	2		1				
Antipatharia	2				1					1	
Madreporaria	6			2	1		2				1
Scaphopoda	8			1	2		2	2		1	
Bivalvia	7			2					1	4	
Pantopoda	17	6	1	5	2		3				
Cirripedia	14	1	1	5	2		2		1	2	
Sipuncula	37	5	9	1	4	6	9	2	1		
Brachiopoda	12		4		2		5		1		
Ascidiacea	1						1				
Ophiuroidea	2								1	1	
Total	132	17	18	21	19	6	29	4	5	12	1

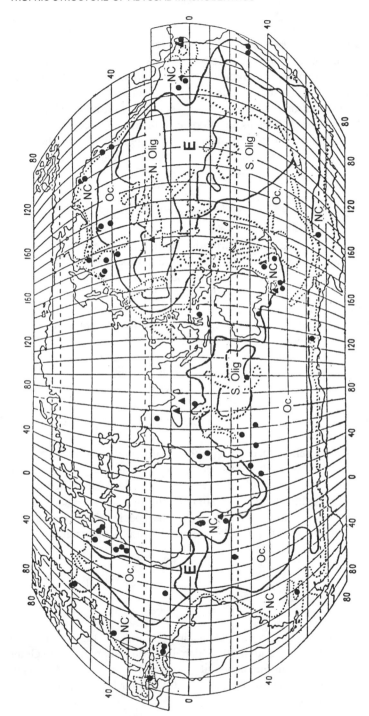

Figure 8 Records of echiuroid worms in different trophic areas. The 3000-m isobath is shown as a dotted line; the solid lines enclose the recognized trophic areas. NC, Near Continental Eutrophic; Oc, Oceanic Eutrophic; E, Equatorial Eutrophic; N. Olig, Northern Oligotrophic; S. Olig, Southern Oligotrophic. The solid circles show those identified to species; the solid triangles are unidentified echiurans (after Sokolova, 1995, modified).

11. *Scalpellum abyssicola* Hoek

12. *Perlophiura profundissima* Beljev and Litvinova

• *Group VIIIa.* Besides the above12 deep-sea species in the oligotrophic areas, one species of solitary madreporarian coral (*Deltocyathus parvulus* Keller) has been found, which unlike the remaining 12 has not been observed in the Near-Continental Eutrophic Area. Most records are from the Oceanic Eutrophic Area where it inhabits the floor except for one occurrence on a submarine mountain (depth 1950 m). Under oligotrophic conditions there are single records of this coral for the Northern and Southern areas of the Pacific Ocean.

4.5.3. *Echiura Distribution*

We can see a typical example in the distribution of echiuran-worms at the ocean floor linked to the feeding conditions. For the first time the distribution of echiuran-worms is related to trophic conditions and the species of these worms are classified by trophic areas (Sokolova and Murina, 1995). Forty species of the families Bonelliidae and Thalassematidae as well as 10 unidentified forms of the family Bonellidae are used for the analysis. All these worms (except *Kurchatovus* species) are obligate deposit-feeders, sorting detritus from the bottom surface. *Kurchatovus* species are specialized forms which live and feed on sunken plant remains. Almost all these species (including species of *Kurchatovus*) are found only in eutrophic conditions. Most of the species have the Near-Continental (72%) or the Oceanic (26%) types of distribution. Only one unidentified bonellid (*Sluiterina* sp.) occurs also in the Northern Oligotrophic Area, thus giving it a Panthalassian distribution (Figure 8).

Echiura are more closely associated with the Near-Continental Trophic Area than the mean linkage (42%) to this area calculated earlier for 12 groups of other invertebrates with different types of feeding (Sokolova, 1984, 1986). The Echiura are very dependent on sediment organic matter and the Near-Continental Trophic Area is the richest on the ocean floor. *Kurchatovus* species are linked to the Near-Continental Trophic Area because they depend on sunken plant remains from the continents.

ACKNOWLEDGEMENTS

I thank Professor Alan Southward and Dr Andrey Gebruk for the invitation to prepare this review, the preparation of which has been supported by Grant 95-05-14191 from the Russian Fundamental Research Foundation. I also thank Olga Zezina, Jury Rogkoff, Natasha Detinova and Alexandr Evgrafov for help in completing the manuscript.

REFERENCES

Akhmet'eva, E.A. (1987). The transformation of organic matter by deposit-feeders. *In* "Feeding of Marine Invertebrates and its Significance in Formation of Communities" (A.P. Kuznetsov and M.N. Sokolova, eds), pp. 46–54. Institut Okeanologii Akademii Nauk SSSR, Moscow. (In Russian with English summary).

Akhmet'eva, E.A., Smirnov, B.A. and Bordovskii, O.K. (1982). Some special features of the composition of organic matter in intestinal contents of benthic deposit feeders. *Okeanologiya* **22**(6), 1021–1024. (In Russian with English abstract).

Alldredge, A.L. and Gotschalk, C.C. (1989). Direct observations of mass flocculations of diatom blooms: characteristics, settling velocities and formation of diatom aggregates. *Deep-Sea Research* **36A**, 159–173.

Alldredge, A.L. and Silver, M.W. (1988). Characteristics, dynamics and significance of "marine snow". *Progress in Oceanography* **20**, 41–82.

Allen, J.A. (1971). Evolution and functional morphology of the deep-water protobranch bivalves of the Atlantic. *In* "Proceedings of the Joint Oceanographic Assembly, 1970", pp. 251–253, Tokyo.

Allen, J.A. (1979). The adaptations and radiation of deep-sea bivalves. *Sarsia* **64**, 19–27.

Allen, J.A. and Sanders, H.L. (1966). Adaptations of abyssal life as shown by the bivalve *Abra profundorum* (Smith). *Deep-Sea Research* **13**(6), 1175–1184.

Aller, J.Y. (1989). Quantifying sediment disturbance by bottom currents and its effect on benthic communities in a deep-sea western boundary zone. *Deep-Sea Research* **36A**, 901–934.

Alongi, D.M. (1990). Bacterial growth rates, production and estimates of detrital carbon utilization in deep-sea sediment of the Solomon and Coral Seas. *Deep-Sea Research* **37A**, 731–746.

Arrhenius, G.O.S. (1952). Sediment cores from the East Pacific. *Report Swedish Deep-sea Expedition* **5**, 1–227.

Atlas of the Oceans. "Terms, Concepts, Informational Tables" (1980). Navy, Leningrad. (In Russian).

Baird, B.H., Nivens, D.E., Parker, J.H. and White, D.C. (1985). The biomass, community structure and spatial distribution of the sedimentary microbiota from a high energy area of the deep-sea. *Deep-Sea Research* **32A**, 1089–1099.

Beklemishev, V.N. (1961). Terminology and concepts necessary for the quantitative study of populations of ectoparasites and nidicoles. *Zoologicheskii Zhurnal* **40**(2), 147–158. (In Russian).

Beklemishev, K.V. (1964). Superfluous feeding in zooplankton and the problem of food sources of benthic animals. *Trudy Vsesoyuznogo Hidrobiologicheskogo Obtchestva Akademii Nauk SSSR, Nauka, Moskva* **8**, 354–358. (In Russian).

Beklemishev, K.V., Semenova, N.L. and Malyutin, O.I. (1982). The structure of biological boundary of the pseudobathyal zone in the White Sea. *Zhurnal Obshchey Biologii* **3**, 366–373. (In Russian).

Belyaev, G.M. (1966). "Hadal Bottom Fauna of the World Ocean". Institute of Oceanology, USSR Academy of Sciences, Moscow. (In Russian. English translation published 1972 by Israel Program for Scientific Translations, Jerusalem).

Belyaev, G.M. and Glikman, L.S. (1970). Shark teeth on the floor of the Pacific Ocean. *Trudy Instituta Okeanologii Akademii Nauk SSSR* **88**, 252–276. (In Russian with English title).

Belyaev, G.M. and Litvinova, N.M. (1972). New genera and species of deep-sea ophiuroids. *Bulletin Moskovskogo Obtchestva Ispytateley Prirody, Moskva, Seria biologicheskaya* **77**(3), 5–20. (In Russian with English abstract).

Belyaev, G.M. and Litvinova, N.M. (1976). Rare new species of deep-sea ophiuroids. *Trudy Instituta Okeanologii Akademii Nauk SSSR* **99**, 126–139. (In Russian with English abstract).

Belyaev, G.M., Birstein, Ya.A., Bogorov, V.G., Vinogradova, N.G., Vinogradov, M.E. and Zenkevich, L.A. (1959). On the scheme of vertical biological zonation of the ocean. *Doklady Akademii Nauk SSSR* **129**(3), 658–661. (In Russian).

Belyaev, G.M., Vinogradova, N.G., Levenstein, N.G., Pasternak, F.A., Sokolova, M.N. and Filatova, Z.A. (1973). Distribution patterns of deep-water bottom fauna related to the idea of the biological structure of the ocean. *Oceanology* **13**, 114–120.

Bezrukov, P.L. (1962). Some problems of the zonation of sediment formation in the World Ocean. *Trudy Okeanographycheskoy Komissii* **10**(3), 3–9. (In Russian).

Bezrukov, P.L. (1964a). Sediments of the northern and central parts of the Indian Ocean. *Trudy Instituta Okeanologii Akademii Nauk SSSR* **64**, 182–202. (In Russian with English abstract).

Bezrukov, P.L. (1964b). Sediment formation in the northern and central parts of the Indian ocean. *In* "Trudy Mezhdunarodnogo Geologicheskogo Kongressa. XXII Sessia. Doklady Sovetskyh Geologov: Geologia Dna Okeanov i Morei", pp. 41–42, Nauka, Moscow. (In Russian).

Bezrukov, P.L., Petelin, V.P. and Aleksina, I.A. (1970). Types of sediments, their distribution and composition. *In* "The Pacific Ocean. Sedimentation in the Pacific Ocean", Part 1 (Kort, V.G., ed.), pp. 170–237. Nauka, Moscow. (In Russian with English title).

Bezrukov, P.L., Skornyakova, N.S. and Murdmaa, I.O. (1981). Main features of distribution and composition of oceanic iron–manganese concretions fields. *Litologiya i poleznye Iskopaemye* **5**, 51–63. (In Russian with English title).

Billett, D.S.M., Lampitt, R.S., Rice, A.L. and Mantoura, R.F.C. (1983). Seasonal sedimentation of phytoplankton to the deep-sea benthos. *Nature* **302**, 520–522.

Billett, D.S.M., Llewellyn, C. and Watson, J. (1988). Are deep-sea holothurians selective feeders? *In* "Echinoderm Biology" (R.D. Burke, M.P. Bacon and W.W. Silker, eds), pp. 421–429. Balkema, Rotterdam.

Birstein, Ya.A. and Vinogradov, M.E. (1971). The role of trophic factors in taxonomic identification of marine benthic fauna. *Bulletin Moskovskogo Obschestva ispytatelei Prirody, Moskva, Seria biologicheskaya* **76**(3), 59–92. (In Russian with English abstract).

Bishop, J.K., Edmond, J.M., Ketten, D.R. Mladendorf, P.V., Lambert, P. and Parsley, R.L. (1977). The chemistry, biology and vertical flux of particulate matter from the upper 400 m of the equatorial Atlantic Ocean. *Deep-Sea Research* **24**(6), 511–548.

Blegvad, H. (1914). Food and conditions of nourishment among the communities of invertebrate animals found on or in the sea bottom in Danish waters. *Report of Danish Biological Station* **22**, 3–56.

Bogdanov, Yu.A. (1965). Suspended organic matter in the waters of the Pacific Ocean. *Okeanologiya* **5**(2), 286–297. (In Russian with English title and abstract).

Bogdanov, Yu.A. (1980). "Pelagic sedimentation process in the Pacific Ocean". Author's Abstract of Dissertation for the Award of the Degree of Doctor of Geological and Mineralogical Sciences, Institut Okeanologii Akademii Nauk SSSR, Moscow. (In Russian).

Bogdanov, Yu.A., Lisitsyn, A.P. and Romankevich, E.A. (1971). Organic matter of suspensions and benthic sediments of seas and oceans. *In* "Organicheskoe Veshchestvo Sovremennykh i Iskopaemykh Osadkov", pp. 35–40, Nauka, Moscow. (In Russian).

Bogdanov, Yu.A., Gurvich, E.G. and Lisitsyn, A.P. (1979). A model for the accumulation of organic carbon in the benthic sediments of the Pacific Ocean. *Geokhimya* **1**, 918–927. (In Russian with English title).

Bordovsky, O.K. (1964a). "Accumulation and Transformation of Organic Matter in Marine Sediments". Nedra, Moscow. (In Russian).

Bordovsky, O.K. (1964b). On the characteristics of organic matter in the benthic sediments of the Pacific and Indian Oceans. *In* "Environment and Processes of Petroleum Formation", pp. 100–128, Nauka, Moscow. (In Russian).

Bordovsky, O.K. (1966a). Processes of accumulation and ways of transformation of organic matter in the benthic sediments of the ocean. *In* "Khimicheskie Protsessy v Moryakh i okeanakh", pp. 42–47, Nauka, Moscow. (In Russian).

Bordovsky, O.K. (1966b). Organic matter in the benthic sediments of marine and ocean basins. *In* "Biologicheski Resursy vodoemov, Puti ikh Rekonstruktsii i Isopol'zovanya". pp. 42–53, Nauka, Moscow. (In Russian).

Bordovsky, O.K. (1974). "Organic Matter in Marine and Oceanic Sediments in the Stage of Early Diagenesis". Nauka, Moscow. (In Russian).

Bordovsky, O.K., Sokolova, M.N., Smirnov, B.A., Akhmet'eva, E.A. and Zezina, O.N. (1974). Evaluation of the role of the benthic population in the transformation of organic matter of sediments (with reference to deep-sea benthic deposit feeders of Kuril–Kamchatka Trench). *Okeanologiya* **14**(1), 161–166. (In Russian with English abstract). English translation: *Oceanology* **14**(1), 128–132.

Borutskii, E.V. and Verigina, I.A. (1961). Anatomical characteristics of fish determining the type of feeding. *In* "Metodicheskoe Posobie po Izucheniyu Pitaniya i Pishchevykh Otnoshenii Ryb v Estestvennykh Usloviyakh", pp. 10–23. Nauka, Moscow. (In Russian).

Bouland, C., Massin, C. and Jangoux, M. (1982). The fine structure of the buccal tentacles of *Holothuria forskali* (Echinodermata, Holothurioidea). *Zoomorphology* **101**(2), 133–149.

Bowman, T.E. and Manning, R.B. (1972). Two arctic bathyal crustaceans: the shrimp *Bythocaris cryonesus* new species and the amphipod *Eurythenes gryllus*, with *in situ* photographs from Ice Island t-3. *Crustaceana* **23**(2), 187–201.

Branchi, A.J., Branchi, M., Scoditti, P.M. and Bensoussan, M.G. (1979). Distribution des populations bacteriènnes heterotrophes dans les sediments et les tractus digestifs d'animaux benthiques recueillis dans la faille Vema et les plaines abyssales de Demerara et de la Gambie. *Vie marine. Marseille* **1**, 7–12.

Burnett, B.R. (1977). Quantitative sampling of microbiota of the deep-sea benthos. 1. Sampling techniques and some data from the abyssal central North Pacific. *Deep-Sea Research* **24**(8), 781–789.

Burnett, B.R. (1981). Quantitative sampling of nanobiota (microbiota) of the deep-sea benthos. III. The bathyal San Diego Trough. *Deep-Sea Research* **28A**(7), 649–663.

Bussau, C., Schriever, G. and Thiel, H. (1995). Evaluation of abyssal metazoan meiofauna from a manganese nodule area of the eastern South Pacific. *Vie Mileu* **45**(1), 39–48.

Carey, A.G. (1972). Food sources of sublittoral, bathyal and abyssal asteroids in the northeast Pacific Ocean. *Ophelia* **10**, 35–47.

Cherkashina, N.Ya. (1972). Feeding of the crustaceans *Astacus leptodactylus eichwaldi* Batt and *Astacus pachypus* (Rathke (fam. Astacidae) in the Turkmenian waters of the Caspian Sea. *Proceedings of All-Union Research Institute of Marine Fisheries and Oceanography, Moscow* **90**, 55–71. (In Russian.)

Chistikov, S.D. (1982). The present-day scaphopod mollusks of the Family Entalinidae (Schaphopoda: Gadilida). 2. Subfamily Heteroschimoidinae. *Zoologicheskii Zhurnal* **61**(9), 1309–1323. (In Russian with English summary).

Cohnheim, O. (1901). Versuche über Resorption, Verdauung und Stoffwechsel von Echinodermen. *Zeitschrift für Physiologiche Chemie* **33**, 9–54.

Crozier, W.J. (1918). The amount of bottom material ingested by holothurian (*Stichopus*). *Journal of Experimental Zoology* **26**, 379–389.

Cutler, E.B. (1977). The bathyal and abyssal Sipuncula. *Galathea Report* **14**, 135–156.

Cutler, E.B. and Cutler, N.J. (1980). Deep-water Sipuncula from the Gulf of Gascogne. *Journal of the Marine Biological Association of the United Kingdom* **60**, 449–459.

Dahl, E. (1979). Deep-sea carrion feeding amphipods: evolutionary patterns in niche adaptation. *Oikos* **33**(2), 167–175.

Davies, G.D. (1987). "Aspects of the biology and ecology of deep-sea Scaphopoda". PhD Thesis, Heriot-Watt University, Edinburgh.

Dayton, P. and Hessler, R. (1972). Role of biological disturbance in maintaining diversity in the deep-sea. *Deep-Sea Research* **19**(3), 199–208.

Dinet, A. (1973). Distribution quantitative du meiobenthos profound dans la région de la dorsale de Walvis (Sud-Ouest Africain). *Marine Biology* **20**(1), 20–26.

Eichelbaum, E. (1910). Uber Narung und Ernährungsorgane von Echinodermata. *Wissenschaftliche Meeresuntersuchungen* N.F.**11**, *Abt. Kiel*, 187–275.

Emel'yanov, E.M. and Romankevich, E.A. (1979). Organic matter. *In* "Geochemistry of the Atlantic Ocean. Organic Matter and Phosphorus" (A.P. Lisitsyn, ed.), pp. 32–102. Nauka, Moscow. (In Russian).

Emel'yanov, E.M., Lisitsyn, A.P. and Il'in, A.V. (eds) (1975). "Types of Bottom Sediments of the Atlantic Ocean". Results of Researches on the International Geophysical Projects. Academii Nauk SSSR, Kaliningrad. (In Russian).

Ericson, D.B., Ewing, M., Wollin, G., Heezen, B.C. (1961). Atlantic deep-sea sediment cores. *Bulletin of the Geological Society of America* **72**, 193–285.

Fauchald, K. and Jumars, P. (1979). The diet of worms: a study of the polychaete feeding guilds. *Oceanography and Marine Biology: An Annual Review* **17**, 193–284.

Filatova, Z.A. (1958). Some new species of bivalve molluscs from the northwestern Pacific. *Trudy Instituta Okeanologii Akademia Nauk SSSR* **27**, 280–218. (In Russian).

Filatova, Z.A. (1976). Species composition of the deep-sea bivalve mollusk genus *Spinula* (Dall, 1908) (Malletiidae) and their distribution in the World Ocean. *Trudy Instituta Okeanologii Akademii Nauk SSSR* **99**, 219–240. (In Russian with English summary).

Filatova, Z.A. and Shileiko, A.A. (1984). The size, structure and distribution of the deep-sea family Ledellidae (Bivalvia, Protobranchia) in the World Ocean. *Trudy Instituta Okeanologii Akademii Nauk SSSR* **119**, 106–144. (In Russian with English summary).

Foell, E.J. and Pawson, D.L. (1986). Photographs of invertebrate megafauna from abyssal depths of the north-eastern equatorial Pacific Ocean. *Ohio Journal of Science* **86**, 61–68.

Fournier, R.O. (1973). Studies on pigmented microorganisms from aphotic marine environments. Pt. 3. Evidence of apparent utilization by benthic and pelagic Tunicata. *Limnology and Oceanography* **18**(1), 38–43.

Fowler, S.W. and Knauer, G.A. (1986). Role of large particles in the transport of elements and organic compounds through the oceanic water column. *Progress in Oceanography* **16**, 147–194.

Gage, J.D. and Tyler, P.A. (1991). "Deep-sea Biology: A Natural History of Organisms at the Deep-sea Floor". Cambridge University Press, Cambridge.

Gebruk, A.V. (1992). The problem of the life form conception with reference to the Echinodermata. *Uspekhy sovremennoy biologii, Akademia Nauk Rossii, Moskva* **53**(2), 176–185. (In Russian with English title).

Gerlach, S.A. (1977). On food-chain and productivity relationships of macrofauna, meiofauna and bacteria. *Third International Meiofauna Conference, Hamburg*, p. 1.

Gooday, A.J. (1988). A response by benthic Foraminifera to the deposition of phytodetritus in the deep-sea. *Nature* **332**, 70–73.

Gooday, A.J., Levin, L.A., Linke, P. and Heeger, T. (1992). The role of benthic Foraminifera in deep-sea food webs and carbon cycling. *In* "Deep-Sea Food Chains and the Gobal Carbon Cycle" (G.T. Rowe and V. Pariente, eds), pp. 63–91. Kluwer Academic, The Netherlands.

Gowing, M.M. and Silver, M.W. (1983). Origins and microenvironments of bacteria mediating fecal pellet decomposition in the sea. *Marine Biology* **73**, 7–16.

Grasshoff, M. (1981). Die Gorgonaria, Pennatularia und Antipatharia des Tiefwassers der Biskaya (Cnidaria, Anthozoa). Ergebnisse der französichen Expeditionen BIOGAS, POLYGAS, GEOMANCHE, INCAL, NORATLANTE und fahrten der "Thalassa". II. Taxonomischen Teil. *Bulletin du Muséum Nationale d'Histoire Naturelle, Paris, Zoologie*, Series 4 **3A**, 941–978.

Grassle, J.F., Morse-Porteous, L.S. (1987). Macrofaunal colonization of disturbed deep-sea environments and the structure of deep-sea benthic communities. *Deep-Sea Research* **34A**, 1911–1950.

Grassle, J.F., Sanders, H.L., Hessler, R.R., Rowe, G.T. and McLellan, T. (1975). Pattern and zonation: a study of the bathyal megafauna using the research submersible "Alvin". *Deep-Sea Research* **22**(7), 457–481.

Greenslate, J., Hessler, R.R. and Thiel, H. (1974). Manganese nodules are alive and well in the sea floor. *Proceedings of Marine Technology Society*, Tenth Annual Conference, 171–181.

Griggs, G.B., Carey, A.G. and Kulm, L.D. (1969). Deep-sea sedimentation and sediment-fauna interaction in Cascadia Channel and on Cascadia Abyssal Plain. *Deep-Sea Research* **16**(2), 157–170.

Haedrich, R.L., and Rowe, G.T. (1977). Megafaunal biomass in the deep-sea. *Nature* **269**(5624), 141–142.

Haedrich, R.L., Rowe, G.T. and Polloni, P.T. (1975). Zonation and faunal composition of epibenthic populations on the continental slope south of New England. *Journal of Marine Research* **33**(2), 191–212.

Haedrich, R.L., Rowe, G.T., and Polloni, P.T. (1980). The megabenthic fauna in the deep-sea south of New England, USA. *Marine Biology* **57**(3), 165–179.

Hamilton, R.D., Holm-Hansen, O. and Strickland, J.D.H. (1968). Notes on the occurrence of living microscopic organisms in deep water. *Deep-Sea Research* **15**(6), 77.

Hammond, L.S. (1982). Patterns of feeding and activity in deposit feeding holothurians and echinoids (Echinodermata) from a shallow back-reef lagoon, Discovery Bay, Jamaica. *Bulletin of Marine Science* **32**(2), 549–571.

Hargrave, B.T. (1985). Feeding rates of abyssal scavenging amphipods (*Eurythenes gryllus*) determined by *in situ* time-lapse photography. *Deep-Sea Research* **32A**, 443–450.

Hauksson, E. (1979). Feeding biology of *Stichopus tremulus*, a deposit feeding holothurian. *Sarsia* **64**(3), 155–160.

Hessler, R.R. (1974). The structure of deep benthic communities from central oceanic waters. *In* "The Biology of the Oceanic Pacific" (C.B. Miller, ed.), pp. 79–93. Oregon State University Press.

Hessler, R.R. and Jumars, P.A. (1974). Abyssal community analysis from replicate box cores in the central North Pacific. *Deep-Sea Research* **21**, 185–209.

Hessler, R.R. and Sanders, H.L. (1967). Faunal diversity in the deep-sea. *Deep-Sea Research* **14**, 65–78.

Hessler, R.R., Isaacs, J.D. and Mills, E.W. (1972). Giant amphipod from the abyssal Pacific

Ocean. *Science* **175**(4002), 636–637.

Hollister, C.D., Nowell, A.R.M. and Jumars, P.A. (1984). The dynamic abyss. *Scientific American* **250**(3), 32–43.

Honjo, S. (1978). Sedimentation of materials in Sargasso Sea at a 5367 m deep station. *Journal of Marine Research* **36**(3), 469–492.

Honjo, S., Doherty, K.W., Agrawal, Y.C. and Asper, V.L. (1984). Direct optical assessment of large amorphous aggregates (marine snow) in the deep-ocean. *Deep-Sea Research* **31A**, 67–76.

Hunt, O.D. (1925). The food of the bottom fauna of the Plymouth fishing grounds. *Journal of the Marine Biological Association of the United Kingdom*, **13**(3), 560–599.

Hyman, L. (1955). "The Invertebrates. Vol. 4, Echinodermata". McGraw-Hill, New York.

Isaacs, J.D. and Schwartzlose, R.A. (1975). Active animals of the deep-sea floor. *Scientific American* **233**(4), 89–91.

Jannasch, H.W. and Wirsen, C.O. (1973). Deep-sea microorganisms *in situ* response to nutrient enrichment. *Science* **180**, 641–643.

Jannasch, H.W., Eihnjellen, K., Wirsen, C.O. and Farmanfarmanian, A. (1970). Microbial degradation of organic matter in the deep-sea. *Science* **171**, 672–675.

Jørgensen, C.B. (1966): "Biology of Suspension Feeding". Pergamon Press, Oxford.

Jumars, P.A. and Fauchald, K. (1977). Between-community contrasts in successful polychaete feeding strategies. *In* "Ecology of Marine Benthos" (B.C. Coull, ed.), pp. 1–20. University of South Carolina Press, Columbia, SC.

Jumars, P.A. and Gallagher, E.D. (1982). Deep-sea community structure three plays on the benthic proscenium. *In* "The Environment of the Deep-Sea" (W.G. Ernst, J.C. Morin, eds), pp. 217–255. Prentice-Hall, Englewood Cliffs, NJ.

Jumars, P.A., Mayer, L.M., Deming, J.W., Baross, J.A. and Wheatcroft, R.A. (1990). Deep-sea deposit-feeding strategies suggested by environmental and feeding constraints. *Philosophical Transactions of the Royal Society of London* **A331**, 85–101.

Kamenskaya, O.E. (1977). "Deep-sea Amphipods (Amphipoda, Gammaridea) of the Pacific Ocean". Thesis for the award of the Degree of Candidate of Biological Sciences, Institut Okeanologii Akademii Nauk SSSR, Moscow. (In Russian).

Kamenskaya, O.E. (1978). Quantitative distribution of deep-sea amphipods. *Trudy Instituta Okeanologii Akademii Nauk SSSR* **113**, 22–27. (In Russian with English summary).

Kamenskaya, O.E. (1980). Deep-sea amphipods (Amphipoda, Gammaridea) from the collections of the expedition of the drifting station "Severnyi Polyus-22". *In* "Biologia Tzentral'nogo Arkticheskogo Basseina", pp. 241–251, Nauka, Moscow. (In Russian).

Kamenskaya, O.E. (1984). Ecological classification of deep-sea amphipods. *Trudy Instituta Okeanologii Akademii Nauk SSSR* **119**, 154–160. (In Russian with English summary).

Karl, D.M., Knauer, G.A. and Martin, J.H. (1988). Downward flux of particle organic matter in the ocean: a particle decomposition paradox. *Nature* **332**, 438-441.

Keller, N.B. (1976). Deep-sea madreporarian corals of the genus *Fungiacyathus* from the Kuril-Kamchatka, Aleutian trenches and other regions of the World Ocean. *Trudy Instituta Okeanologii Akademii Nauk SSSR* **99**, 31–44. (In Russian with English summary).

Keller, N.B. (1982). New data on the madreporarian corals of the genus *Deltocyathus* (family Caryophylllidae Gray, 1847). *Trudy Instituta Okeanologii Akademii Nauk SSSR* **117**, 147–150. (In Russian with English summary).

Khripounoff, A. and Sibuet, M. (1980). La nutrition d'echinodermes abyssaux. I. Alimentation des holothuries. *Marine Biology* **60**(1), 17–26.

Knudsen, J. (1967). The deep-sea Bivalvia. *"John Murray" Expedition, 1933–1934,*

Scientific Report **11**(3), 237–343.

Knudsen, J. (1970). The systematics and biology of abyssal and hadal Bivalvia. *Galathea Report* **11**, 236–241.

Knudsen, J. (1979). Deep-sea bivalves. *In* "Pathways in Malacology" (S. van der Spoel, A.C. van Bruggen and J. Lever, eds), pp. 195–224, Bohn, Scheltema, Holkema, Utrecht.

Koblentz-Mishke, O.I. (1977). Primary production. *In* "Oceanologia, Biologia Okeana", Vol. 1 (A.S. Monin and M.E. Vinogradov, eds), pp. 62–64. Nauka, Moscow. (In Russian).

Koltun, V.M. (1970). Deep-water Spongia fauna of the northwestern part of the Pacific Ocean. *Trudy Instituta Okeanologii Akademii Nauk SSSR* **86**, 165–221. (In Russian with English summary).

Korovchinskii, N.M. (1974). "Systematics, distribution and some features of the biology of deep-water asteroids of the family Brisingidae". Moscow State University, Moscow. (In Russian).

Kott, P. (1969). Antarctic Ascidiacea. *Antarctic Research Series* **13**, 1–239.

Kozjar, L.A., Sokolova, M.N. and Zezina, O.N. (1974). Spores and pollen of higher plants in the feeding of marine invertebrates. *Okeanologiya* **4**(3), 522–525. (In Russian with English summary).

Krank, K. and Milligan, T.G. (1988). Macroflocs from diatoms: *in situ* photography of particles in Bedford Basin, Nova Scotia. *Marine Ecology Progress Series* **44**, 183–189.

Kucheruk, N.V. (1978). Deep-sea Onuphidae (Polychaeta) from the collections of the 16th voyage of the research ship the "Dmitry Mendeleev". *Trudy Instituta Okeanologii Akademii Nauk SSSR* **113**, 88–106. (In Russian with English summary).

Kucheruk, N.V. (1979). On zoogeographic zonation of the abyssobenthic zone. *Bulletin Moskovskogo Obshchestva Ispytateley Prirody, Seriya biologicheskaya, Moskva* **84**(5), 60–66. (In Russian with English abstract).

Kucheruk, N.V. (1981) On the regularities of the distribution patterns of deep-sea Polychaeta of the eastern Pacific. *Trudy Instituta Okeanologii Akademii Nauk SSSR* **115**, 37–52. (In Russian with English summary).

Kuznetsov, A.P. (1963). "Benthic Invertebrate Fauna of the Kamchatka Waters of the Pacific Ocean and Northern Kuril Islands". Izdatelstvo Akedemii Nauk SSSR, Moscow. (In Russian).

Kuznetsov, A.P. (1964). Distribution of the benthic fauna of the western part of Bering Sea according to trophic zones and some problems of trophic zonation. *Trudy Instituta Okeanologii Akademii Nauk SSSR* **69**, 98–177. (In Russian with English summary).

Kuznetsov, A.P. (1974). Trophic structure of marine benthic fauna as an indicator of physico-chemical regime in the sea. *In* "Gidrobiologiya i Biogeografiya Shel'fov Umerennykh i Kholodnykh Vod Mirovogo Okeana", pp. 33–34, Zoologicheskyi Institut Akademii Nauk SSSR, Leningrad. (In Russian).

Kuznetsov, A.P. (1980). "Ecology of the Benthic Communities of Shelf Zones of the World Ocean". Nauka, Moscow. (In Russian).

Kuznetsov, A.P., Osipova, A.N. and Gekker, R.F. (1979). Trophic status of benthic population of the Fergana Bay of the Paleogene Tethys Sea and its changes in relation to the variation in paleoecological conditions (paleo-ecological reconstruction from the results of trophological analysis). *In* "E'kologia Donnogo Naseleniya Shel'fovoi Zony" (A.P. Kuznetsov, ed.), pp. 78–87. Institut Okeanologii Akademii Nauk SSSR, Moscow. (In Russian).

Lampitt, R.S. (1985). Evidence for the seasonal deposition of detritus to the deep-sea floor and its subsequent resuspension. *Deep-Sea Research* **32**, 885–897.

Lampitt, R.S., Billett, D.S.M. and Rice, A.L. (1986). Biomass of the invertebrate megabenthos from 500 to 4100 m in the northeast Atlantic Ocean. *Marine Biology* **93**, 69–81.

Laubier, L. and Sibuet, M. (1977). Résultats des campagnes BIOGAS 3 aout 1972–2 novembre 1974. *Résultats des Campagnes à la Mer Centre National pour l'Exploration des Oceans* **11**, 1–11.

Lebedeva, L.P., Vinogradov, M.E., Shushkina, E.A. and Sazin, A.F. (1982). Estimating the rates of detritus formation in marine planktonic communities. *Okeanologiya* **22**, 652–659.

Lemche, H., Hansen, B., Madsen, F.J., Tendal, O.S. and Wolff, T. (1976). Hadal life as analyzed from photographs. *Videnskabelige meddelelser Dansk Naturhistorisk Forening* **139**, 263–336.

Levenstein, R.Ya. (1970). Ecology and zoogeography of some representatives of the family Opheliidae (Polychaeta, Annelida) of the Pacific Ocean. *Trudy Instituta Okeanologii Akademii Nauk SSSR* **88**, 213–226. (In Russian with English summary).

Levin, V.S. and Voronova, E.I. (1979). Assimilation of bacterial food by the Far Eastern trepang. *In* "Materialy 4-go Vsesoyuznogo Kollokviuma po Iglozhim (G.S. Gongadze, ed.), pp. 121–123. Tbilisi. (In Russian).

Lipps, J.H. and Hickman, C.S. (1982). Origin, age and evolution of Antarctic and deep-sea fauna. *In* "The Environment of the Deep Sea" (W.G. Ernst and J.G. Morin, eds), pp. 324–356. Prentice-Hall, Englewood Cliffs, NJ.

Lisitsyn, A.P. (1971). The rate of contemporary sediment accumulation in oceans. *Okeanologiya* **11**(6), 957–968. (In Russian with English summary).

Lisitsyn, A.P. (1974). "Sediment Formation in the Oceans". Nauka, Moscow. (In Russian).

Lisitsyn, A.P. (1981). Zonation of the natural environment and sediment formations in oceans. *In* "Klimaticheskaya Zonal'nost' i Osadkonakoplenie" (A.P. Lisitsyn, ed.), pp. 5–45. Nauka, Moscow. (In Russian).

Lisitsyn, A.P. and Vinogradov, M.E. (1982). Global patterns of distribution of life in the ocean and their reflection in the composition of benthic sediments. *Izvestiya Akademii Nauk SSSR, Seriya Geologicheskaya* **4**, 5–24. (In Russian with English title).

Litvinova, N.M. (1979). On the feeding of ophiuroids. *Zoologicheskii Zhurnal* **58**(10), 1501–1410. (In Russian with English summary).

Litvinova, N.M. and Sokolova, M.N. (1971). On the feeding of deep-sea ophiuroids of the genus *Amphiophiura*. *Okeanologiya* **11**(2), 293–301. (In Russian with English summary).

Lochte, G.R. and Turley, C.M. (1988). Bacteria and cyanobacteria associated with phytodetritus in the deep-sea. *333*, 67-69.

Lopez, G.R. and Levinton, J.S. (1987). Ecology of deposit-feeding animals in marine sediments. *The Quarterly Review of Biology* **62**, 235–260.

MacArthur, R.H. (1972). "Geographical Ecology. Patterns in the Distribution of Species". Harper & Row, New York.

Madin, L.P. (1982). Production, composition and sedimentation of salp fecal pellets in oceanic waters. *Marine Biology* **67**(1), 39–45.

Madsen, F.T. (1961). The Porcellanasteridae: A monographic revision of an abyssal group of sea-stars. *Galathea Report* **4**, 33–176.

Mel'nikov, I.A. (1975a). Microplankton and organic detritus in the southeastern Pacific. *Okeanologiya* **15**(1), 146–156. (In Russian with English summary).

Mel'nikov, I.A. (1975b). The finely-dispersed fraction of the suspended organic matter in the waters of the eastern part of the Pacific Ocean. *Okeanologiya* **15**(2), 266–271. (In Russian with English summary).

Mel'nikov, I.A. (1976). Morphological characteristics of the organic detritus particles. *Okeanologiya* 16(4), 679–702. (In Russian with English summary).

Menzel, D.W. (1974). Primary productivity, dissolved and particulate organic matter and the sites of oxidation of organic matter. "The Sea", Vol. 5 (E. Goldberg, ed.), pp. 659–678. Wiley, New York.

Meyer, D.L. (1973). Feeding behavior and ecology of shallow-water unstalked crinoids (Echinodermata) in the Caribbean Sea. *Marine Biology* 22(2), 105–129.

Millar, R.H. (1958). Asciidacea. *Galathea Report* 10, 7–90.

Mironov, A.N. (1975). Mode of life of pourtalesiid sea urchins (Echinoidea, Pourtalesiidae). *Trudy Instituta Okeanologii Akademii Nauk SSSR* 103, 281–288. (In Russian with English summary).

Monniot, C. (1984). Composition des peuplements benthiques abyssaux; resultats des campagnes SAFARI dans l'Ocean Indien. *Comité National Français des Recherches Antarctiques* 55, 49–68.

Moore, H., Manship, B. and Roberts, D. (1995). Gut structure and digestive strategies in three species of abyssal holothurians. *In* "Echinoderm Research" (R.H. Emson, A.B. Smith and A.C. Campbell, eds), pp. 111–119. Balkema, Rotterdam.

Moriarty, D.J.W. (1982). Feeding of *Holothuria atra* and *Stichopus chloronotus* on bacteria, organic carbon and organic nitrogen in sediments of the Great Barrier Reef. *Australian Journal of Marine and Freshwater Research* 33(2), 255–263.

Morita, R.Y. (1979). Deep-sea microbial energetics. *Sarsia* 64(1), 9–12.

Mosher, C. (1980). Distribution of *Holothuria arenicola* Semper in the Bahamas with observation on habitat, behavior and feeding activity (Echinodermata, Holothurioidea). *Bulletin of Marine Science* 30(1), 1–12.

Murdmaa, I.O. (1979a). Oceanic facies. *In* "Okeanologia. Osadkoobrazovanie i Magmatizm Okeana" (A.S. Monin and P.L. Bezrukov, eds), pp. 269–306. Nauka, Moscow. (In Russian).

Murdmaa, I.O. (1979b). Conditions of accumulation of sedimentary formations. *In* "Geologicheskie Formasii Severo-zapadnoi chasti Atlanticheskogo Okeana", pp. 167–185. Nauka, Moscow. (In Russian).

Murdmaa, I.O. (1982). Near-continental marine and oceanic facies and their paleooceanographic interpretation. *In* "Lavinnaya Sedimentatsiya v Okeane" (A.P. Lisitsyn, ed.), pp. 71–82. Isdatel'stvo Rostovskogo Universiteta, Rostov. (In Russian).

Murdmaa, I.O. (1987). "Ocean Facies". Nauka, Moscow. (In Russian).

Murdmaa, I.O., Skornyakova, N.S. and Agapova, G.V. (1976). The facies status of distribution of iron-manganese concretions in the Pacific Ocean. *Trudy Instituta Okeanologii Akademii Nauk SSSR* 109, 7–25. (In Russian with English summary).

Murdmaa, I.O., Svalnov, V.N. and Skornyakova, N.S. (1989). Facies division of the Indian Ocean. *In* "Zelezo-margancevyi konkrecii Central'noy kotloviny Indijskogo Okeana", Nauka, Moscow. (In Russian).

Murina, V.V. (1979). "Marine sipunculid worms of the World Ocean". Dissertation for the Award of the Degree of Doctor of Biological Sciences, Zoologicheskii Institut Akademii Nauk SSSR, Leningrad. (In Russian).

Newell, R. (1965). The role of detritus in the nutrition of two marine deposit-feeders, the prosobranch *Hydrobia ulvae* and the bivalve *Macoma baltica*. *Proceedings of the Zoological Society of London* 144, 25–45.

Neyman, A.A. (1961). Some patterns of quantitative distribution of benthos in the Bering Sea. *Okeanologiya* 1(2), 294–304. (In Russian with English title).

Neyman, A.A. (1963). Quantitative distribution of benthos on the shelf and upper horizons of the slope of the eastern part of the Bering Sea. *Proceedings of All-Union Research Institute of Marine Fisheries and Oceanography* 48, 145–205. Moscow. (In Russian).

Neyman, A.A. (1971). Benthic population of shelves of the northern part of the Indian Ocean. *Proceedings of All-Union Research Institute of Marine Fisheries and Oceanography* **72**, 56–64. Moscow. (In Russian).

Neyman, A.A. (1975). Distribution of some species of invertebrates on the shelf of the northeastern part of the Sea of Okhotsk. *Bulletin Moskovskogo Obshchestva ispytatelei Prirody, Moskva, Seria biologicheskaya* **80**(3), 42–50. (In Russian with English summary).

Neyman, A.A. (1988). "Quantitative distribution and trophical structure of benthos on the shelves of the World Ocean". Science Report. Informcentre VNIRO, Moscow. (In Russian).

Odum, E.P. (1975). "Fundamentals of Ecology". Mir, Moscow. (Russian translation).

Oliver, G. and Allen, J.A. (1980). The functional and adaptative morphology of the deep-sea species of the Arcacea (Mollusca: Bivalvia) from the Atlantic. *Philosophical Transactions of the Royal Society of London* **B291**, 6–76.

Pasternak, F.A. (1958). Deep-sea Antipatharia of the Kuril–Kamchatka Trench. *Trudy Instituta Okeanologii Akademii Nauk SSSR* **27**, 180–191. (In Russian).

Pasternak, F.A. (1961a). New data on the specific composition and distribution of deep-sea Pennatularia of the genus *Kophobelemnon* in the northern Pacific. *Trudy Instituta Okeanologii Akademii Nauk SSSR* **45**, 240–258. (In Russian with English summary).

Pasternak, F.A. (1961b). Pennatularia (Octocorallia) and Antipatharia (Hexacorallia) collected by the Soviet Antarctic Expedition during 1955–1958. *Trudy Instituta Okeanologii Akademii Nauk SSSR* **46**, 217–230. (In Russian).

Pasternak, F.A. (1962). Pennatularia of the genus *Umbellula* Cuvier (Coelenterata, Octocorallia) from the Antarctic and Subantarctic waters. *In* "Issledovaniya fauny morei", **1**(9), pp. 105–128. Izdatel'stvo Akademii Nauk, Leningrad. (In Russian).

Pasternak, F.A. (1964). Deep-sea pennatularians (Octocorallia) and antipatharians (Hexacorallia) obtained by the R/S Vityaz in the Indian Ocean. *Trudy Instituta Okeanologii Akademii Nauk SSSR* **69**, 183–215. (In Russian with English summary).

Pasternak, F.A. (1970). Sea feathers (Octocorallia. Pennatularia) found in the ultra-abyssal zone on the Kuril–Kamchatka Basin. *Trudy Instituta Okeanologii Akademii Nauk SSSR* **86**, 236–248. (In Russian).

Pasternak, F.A. (1973). Deep water sea-feathers (Octocorallia, Pennatularia) of the Aleutian Trench and Gulf of Alaska. *Trudy Instituta Okeanologii Akademii Nauk SSSR* **91**, 108–127. (In Russian with English summary).

Pasternak, F.A. (1975a). The deep-sea pennatularians of the genus *Umbellula* from the Caribbean Sea and Puerto-Rico Trench. *Trudy Instituta Okeanologii Akademii Nauk SSSR* **100**, 160–173. (In Russian with English summary).

Pasternak, F.A. (1975b). New data on the specific composition and distribution of deep-sea pennatularians (Pennatularia, Octocorallia) of the Peru–Chile region of the Pacific Ocean and the south Atlantic Ocean. *Trudy Instituta Okeanologii Akademii Nauk SSSR* **103**, 101–118. (In Russian with English summary).

Pasternak, F.A. (1976). New data on the composition and distribution of the deep-sea antipatharians (Hexacorallia, Antipatharia) in Pacific, Indian and Atlantic oceans). *Trudy Instituta Okeanologii Akademii Nauk SSSR* **99**, 45–58. (In Russian with English summary).

Pasternak, F.A. (1977). Antipatharia. *Galathea Report* **14**, 157–164.

Patterson, D.J. (1990). *Jakoba libera* (Ruinan, 1938), a heterotrophic flagellate from deep-oceanic sediments. *Journal of the Marine Biological Association of the United Kingdom* **70**, 381–393.

Paul, A.Z., Thorndike, E.M., Sullivan, L.G., Heezen, B.C. and Gerard, R.D. (1978). Observations of the deep-sea floor from 202 days of time-lapse photography. *Nature*

272(5656), 812–814.

Petersen, C.G.J. and Boysen-Jensen, P. (1911). Animal life of the sea-bottom, its food and quantity. *Report of Danish Biological Station* **20**, 3–81.

Phillips, N.W. (1984). Role of different microbes and substrates as potential suppliers of specific, essential nutrients to marine detritivores. *Bulletin of Marine Science* **35**, 283–298.

Polloni, P.T., Haedrich, R.L., Rowe, G.T. and Clifford, C.H. (1979). The size-depth relationship in deep ocean animals. *Internationale Revue der Gesamten Hydrobiologie* **65**(1), 39–46.

Reichardt, W.T. (1987). Burial of Antarctic macroalgal debris in bioturbated deep-sea sediments. *Deep-Sea Research* **34A**, 1761–1770.

Renaud-Mornant, J. and Helleouet, M. (1977). Rapport micro-meiobenthos et *Holodeima atra* (Holothurioidea) dans un lagon Polynésien (Tiahura, Moorea, île de la Société). *Bulletin du Muséum d'Histoire Naturelle, Zoologie* Series 3, **331**, 853–865.

Rex, M.A. (1976). Biological accommodation in the deep-sea benthos: comparative evidence on the importance of predation and productivity. *Deep-Sea Research* **23**(11), 975– 987.

Rice, A.L., Aldred, R.G., Billett, D.S.M. and Thurston, M.H. (1979). The combined use of an epibenthic sledge and a deep-sea camera to give relevance to macro-benthos samples. *Ambio Special Report* **6**, 59–72.

Rice, A.L., Billett, D.S.M., Fry, J., John, A.W.G., Lampitt, R.S., Mantoura, R.F.C. and Morris, R.J. (1986). Seasonal deposition of phytodetritus to the deep-sea floor. *Proceedings of the Royal Society of Edinburgh* **88B**, 265–279.

Rice, A.L., Angel, M.V., Grassle, J.F., Hargrave, B.T., Hessler, R.R., Harikoshi, U., Lochte, K., Sibuet, M., Smith, K.L., Thiel, H. and Vinogradova, N. (1994). Suggested criteria for describing deep-sea benthic communities; the final report of SCOR Working Group 76. *Progress in Oceanography* **34**, 81–100.

Riemann, F. (1989). Gelatinous phytoplankton detritus aggregates on the Atlantic deep-sea bed. Structure and mode of formation. *Marine Biology* **100**, 533–539.

Robison, B.H. and Bailey, T.G. (1981). Sinking rates and dissolution of midwater fish fecal matter. *Marine Biology* **65**(2), 135–142.

Rodina, A.G. (1966). The food value and composition of detritus. *In* "Biologicheskie Resursy Vodoemov, Putikh Rekonstruktsii i Ispol'zovaniya" (V.G. Bogorov, ed.), pp. 35–42. Nauka, Moscow. (In Russian).

Romankevich, E.A. (1968). Distribution of organic carbon and nitrogen in contemporary and Quaternary sediments of the Pacific Ocean. *Okeanologiya* **8**(5), 825–837. (In Russian with English summary).

Romankevich, E.A. (1970). Organic matter in the sediments. *In* "The Pacific Ocean. Sedimentation in the Pacific Ocean", Vol. **2**, pp. 107–159. Nauka, Moscow. (In Russian with English titles).

Romankevich, E.A. (1971). The relationship of organic matter of suspension, benthic sediments and benthos with biological productivity. *Doklady Akademii Nauk SSSR* **198**(5), 1199–1203. (In Russian with English summary).

Romankevich, E.A. (1974). Biological composition of sediments of the Pacific Ocean. *Litologiya i Poleznye Iskopaemye* **1**, 27–40. (In Russian with English title).

Romankevich, E.A. (1977). "Geochemistry of Organic Matter in the Ocean". Nauka, Moscow. (In Russian, English translation, 1984, Springer, Berlin).

Rowe, G.T. (1971a). Benthic biomass in the Pisco, Peru upwelling. *Investigacións Pesqueras* **35**(1), 127–135.

Rowe, G.T. (1971b). Benthic biomass and surface productivity. *In* "Fertility of the Sea", Los Angeles, 2 (J.D. Costlow, ed.), pp. 441–454. Gordon & Breach, New York.

Rowe, G.T. and Gardner, W.D. (1979). Sedimentation rates in the slope water of the north-west Atlantic Ocean measured directly with sediment traps. *Journal of Marine Research* **37**(3), 581–600.

Rowe, G.T. and Menzel, D.W. (1971). Quantitative benthic samples from the deep Gulf of Mexico with some comments on the measurement of deep-sea biomass. *Bulletin of Marine Science* **21**(2), 556–566.

Sanders, H.L. and Hessler, R.R. (1969). Ecology of the deep-sea benthos. *Science* **163**(3874), 1419–1424.

Sanders, H.L., Hessler, R.R. and Hampson, G.R. (1965). An introduction to the study of deep-sea benthic faunal assemblages along the Gay Head-Bermuda Transect. *Deep-Sea Research* **12**(6), 845–867.

Shorygin, A.A. (1939). Feeding, selectivity and trophic interrelationships of some Gobiidae of the Caspian Sea. *Zoologicheskii Zhurnal* **18**(1), 27–53. (in Russian with English title and summary).

Shorygin, A.A. (1952). "Feeding and Trophic Interrelation of Fish of the Caspian Sea." Pishepromizdat, Moscow. (In Russian).

Schulenberger, E. and Barnard, J.L. (1976). Amphipods from an abyssal trap set in the North Pacific Gyre. *Crustaceana* **31**, 241–258.

Schulenberger, E. and Hessler, R. (1974). Scavenging abyssal benthic amphipods trapped under oligotrophic central North Pacific Gyre waters. *Marine Biology* **28**(3), 185–187.

Sibuet, M. and Lawrence, J.M. (1981). Organic content and biomass of abyssal holothuroids (Echinodermata) from the bay of Biscay. *Marine Biology* **65**(2), 143–147.

Silver, M.W. and Alldredge, A.L. (1981). Bathypelagic marine snow. Deep-sea algal and detrital community. *Journal of Marine Research* **39**(3), 501–530.

Skornyakova, N.S. (1983). Regional variations in composition of ferromanganese nodules from the Indian Ocean. *Litologia i Poeznye Iskopaemye* **4**, 117–128. (In Russian with English title).

Skornyakova, N.S. (1986). Local variability of fields of ferromanganese nodules. *In* "Ferromanganese nodules of the central Pacific Ocean" (I.O. Murdmaa and N.S. Skornyakova, eds), pp. 109–179. Nauka, Moscow. (In Russian).

Skornyakova, N.S. and Murdmaa, I.O. (1968). Lithologo-facies types of deep-sea pelagic (red) clays of the Pacific Ocean. *Litologia i Poleznye Iskopaemye* **6**, 17–37. (In Russian with English title).

Skornyakova, N.S. and Murdmaa, I.O. (1986). Formation of ferromanganese nodule ore deposits in the radiolarian belt and geological criteria of their prospecting. *In* "Ferromanganese Nodules of the Central Pacific Ocean" (I.O. Murdmaa and N.S. Skornyakova, eds), pp. 297–316. Nauka, Moscow. (In Russian with English title).

Skornyakova, N.S. and Murdmaa, I.O. (1992). Local variations in distribution and composition of ferromanganese nodules in the Clarion-Clipperton Province. *Marine Geologie* **103**(2), 381–405.

Skornyakova, N.S. and Zenkevich, N.L. (1976). Patterns of spatial distribution of ferro-manganese nodules. *In* "Ferro-manganese Nodules of the Pacific Ocean" (P.L. Bezrukov, ed.), pp. 37–76. Nauka, Moscow. (In Russian).

Smetacek, V.S. (1985). The role of sinking in diatom life-history cycles: ecological, evolutionary and geological significance. *Marine Biology* **84**, 239–251.

Smith, C.R. (1985). Food for the deep-sea: utilization, disperal and flux of neckton falls at the Santa Catalina Basin floor. *Deep-Sea Research* **32A**, 417–442.

Smith, C.R. (1986). Neckton falls, low-intensity disturbance and community structure of infaunal benthos in the deep-sea. *Journal of Marine Research* **44**, 567–600.

Smith, K.L. and Baldwin, R.J. (1982). Scavenging deep-sea amphipods: effects of food odor on oxygen consumption and a proposal metabolic strategy. *Marine Biology* **68**(3), 387–398.

Sokolova, M.N. (1954). "The feeding and food groups of deep-sea benthos of the Far Eastern seas". Dissertation. Instituta Okeanologii Akademii Nauk SSSR, Moscow.

Sokolova, M.N. (1956). Feeding in deep-sea benthos. Feeding in *Laetmonice producta* var. *wyvillei* McIntosh. *Doklady Akademii Nauk SSSR* **110**(6), 1111–1114. (In Russian with English summary).

Sokolova, M.N. (1957a). Feeding in some carnivorous invertebrates of the deep-sea benthos of the Far Eastern Seas and northwestern part of the Pacific Ocean. *Trudy Instituta Okeanologii Akademii Nauk SSSR* **20**, 279–301. (In Russian, English translation.) *In* Nikitin, B.N. (ed.) (1959). "Marine Biology" pp. 227–244, American Institute of Biological Sciences, Washington, DC.

Sokolova, M.N. (1957b). The nourishment of some species of the Crangonidae of Far Eastern seas. *Trudy Instituta Okeanologii Akademii Nauk SSSR* **23**, 269–285. (In Russian).

Sokolova, M.N. (1958). Feeding of deep-sea invertebrates. *Trudy Instituta Okeanologii Akademii Nauk SSSR* **27**, 123–153. (In Russian).

Sokolova, M.N. (1960a). On some pecularities of distribution of benthic communities in the northwestern Pacific. *Trudy Instituta Okeanologii Akademii Nauk SSSR* **34**, 336–342. (In Russian).

Sokolova, M.N. (1960b). The distribution of the groupings (biocenoses) of the bottom fauna of the deep-sea trenches in northwestern Pacific. *Trudy Instituta Okeanologii Akademii Nauk SSSR* **34**, 21–59. (In Russian).

Sokolova, M.N. (1962). On trophic zonation in the distribution of benthos of the northern half of Pacific Ocean. *In* "Voprosy E'kologii", pp. 201–203, Publishing House Kievskogo Universiteta, Kiev. (In Russian).

Sokolova, M.N. (1964). On some regularities involved with the distribution of food groupings in the abyssal benthos. *Okeanologiya* **4**(6), 1079–1088. (In Russian).

Sokolova, M.N. (1965). On the irregularity in the distribution of nutrition groupings of deep-sea benthos in connection with the irregularity of sedimentation. *Okeanologiya* **5**(3), 498–506. (In Russian with English summary).

Sokolova, M.N. (1968). On the relations between the trophic groups of deep-sea macrobenthos and the composition of benthic sediments. *Okeanologiya* **8**(3), 179–191. (In Russian with English summary).

Sokolova, M.N. (1969). Patterns of distribution of deep-sea benthic invertebrates depending on the mode and conditions of their feeding. *In* "Pacific Ocean. Biology of the Ocean", Vol. 2. The Deep-sea Bottom Fauna and the Pleuston" (V.G. Kort, ed.), pp. 182–210. Nauka, Moscow. (In Russian).

Sokolova, M.N. (1970). Weight characteristics of meiobenthos of various regions of the deep-sea Trophic Areas of the Pacific Ocean. *Okeanologiya* **10**(2), 348–356. (In Russian with English abstract).

Sokolova, M.N. (1972). Trophic structure of deep-sea macrobenthos. *Marine Biology* **16**(1), 1–12.

Sokolova, M.N. (1976a). Large scale division of the ocean by the trophic structure of the deep-sea macrobenthos. *Trudy Instituta Okeanologii Akademii Nauk SSSR* **99**, 20–30. (In Russian with English summary).

Sokolova, M.N. (1976b). Trophic zonality of deep-water macrobenthos as an element of the biological structure of the ocean. *Okeanologiya* **16**(2), 336–342. (In Russian with English summary).

Sokolova, M.N. (1977). Bottom animals adaptation to the utilization of organic matter from bottom sediments. *In* "Okeanologiya. Biologiya Okeana" Vol. 1 (A.S. Monin and M.E. Vinogradov, eds), pp. 53–57. Nauka, Moscow. (In Russian).

Sokolova, M.N. (1978). Trophic classification of the type of the distribution of deep-sea macrobenthos. *Doklady Akademii Nauk SSSR* **241**(2), 471–474. (In Russian with English title and summary).

Sokolova, M.N. (1979). The global distribution of trophic zones on the ocean floor. *Doklady Akademii Nauk SSSR* **246**(1), 250–252. (In Russian with English summary).

Sokolova, M.N. (1981a). On characteristic features of the deep-sea benthic eutrophic regions of the World Ocean. *Trudy Instituta Okeanologii Akademii Nauk SSSR* **115**, 5–13. (In Russian with English summary).

Sokolova, M.N. (1981b). Trophic zones on the floor of the ocean and the characterization of their population. *In* "Biologiya Bol'shikh Glubin Tikhogo Okeana" (M.E. Vinogradov, ed.), Papers of the 14th International Pacific Science Congress, Khabarovsk, 1979, Section Marine Biology, 1 pp. 8–14. Institut Biologii Morya, Dal'nevostochni Nauchni Centr Akademii Nauka SSSR, Vladivostok. (In Russian with English summary).

Sokolova, M.N. (1982). On the nutrition of the polychaete *Harmothoe derjugini* in the abyssal zone of the Sea of Japan. *Trudy Instituta Okeanologii Akademii Nauk SSSR* **117**, 76–80. (In Russian with English summary).

Sokolova, M.N. (1984). About species composition of the population of the deep-sea trophic regions. *Trudy Instituta Okeanologii Akademii Nauk SSSR* **119**, 33–46. (In Russian with English summary).

Sokolova, M.N. (1985). Trophic structure of the deep-water benthos. *In* "Oceanology. Biology of Ocean. Vol. 2. Biological Productivity of the Ocean" (M.E. Vinogradov, ed.), pp. 218–227. National Marine Fisheries Service, Northeast Fisheries Center, Woods Hole, MA.

Sokolova, M.N. (1986). "Feeding and Trophic Structure in the Deep-sea Macrobenthos". Nauka, Moscow. (In Russian).

Sokolova, M.N. (1987). Toward an evaluation of the mineral component in the food of deep-sea deposit feeders. *In* "Feeding of Marine Invertebrates and its Significance in Formation of Communities" (A.P. Kuznetsov and M.N. Sokolova, eds), pp. 83–96. Institut Okeanologiii Akademii Nauk SSSR, Moscow. (In Russian with English summary).

Sokolova, M.N. (1988). On the feeding of the sea urchin *Stereocidaris nascaensis* Allison, in the southern part of Nazca Ridge. *In* "Structural and Functional Researches of the Marine Benthos" (A.P. Kuznetsov and M.N. Sokolova, eds), pp. 38–44. Institut Okeanologii Akademii Nauk SSSR, Moscow. (In Russian with English summary).

Sokolova, M.N. (1990a). Feeding of various regular echinoid species in the regions of Nazca Ridge and Sala-y-Gomez rise. *In* "Feeding and Bioenergetics of Marine Bottom Invertebrates" (A.P. Kuznetsov, ed.), pp. 28–47. Institut Okeanologii Akademii Nauk SSSR, Moscow. (In Russian with English summary).

Sokolova, M.N. (1990b). Macrobenthic trophic structure on the Atlantic Ocean floor. *Trudy Instituta Okeanologii Akademii Nauk SSR* **126**, 20–39. (In Russian with English summary).

Sokolova, M.N. (1991). Bottom population of deep trophic oceanic areas and possible ecological effects of large-scale extraction of manganese iron concretions. *In* "Biological Resources: The State, Prospects and Problems of their Rational Exploitation. Biotrophic Basis of Distribution of Commercial and Food Marine Animals: Collected Papers", pp. 24–33. VNIRO, Moscow. (In Russian).

Sokolova, M.N. (1993a). New data on the trophic structure of macrobenthos in the abyss of the Atlantic Ocean. *Trudy Instituta Okeanologii Akademii Nauk Rossii* **127**, 50–64.

(In Russian with English summary).

Sokolova, M.N. (1993b). On the role of the euphausiids in feeding of deep-sea macrobenthos. *In* "Feeding of Marine Invertebrates in Different Vertical Zones and Latitudes" (A.P. Kuznetsov and M.N. Sokolova, eds), pp. 23–31. Institut Okeanologii Akademii Nauk Rossii, Moscow. (In Russian with English summary).

Sokolova, M.N. (1994a). Feeding peculiarities of macrobenthos from seamounts as seen in sea urchins. *Trudy Instituta Okeanologii Akademii Nauk Rossii* **129**, 31–42. (In Russian with English summary).

Sokolova, M.N. (1994b). Euphausiid "dead body rain" as a source of food for abyssal benthos. *Deep-Sea Research* **41A**(4), 741–746.

Sokolova, M.N. (1995). On feeding of some starfishes from the family Astropectinidae. *In* "The Composition and Distribution of Bottom Invertebrata" (A.P. Kuznetsov and O.N. Zezina, eds), pp. 19–32. Institut Okeanologiii Akademiii Nauk Rossii, Moscow. (In Russian with English summary).

Sokolova, M.N. and Murina, V.V. (1995). On the distribution of Echiura worms at the ocean floor in connection to the feeding conditions. *In* "The Composition and Distribution of Bottom Invertebrates" (A.P. Kuznetsov and O.N. Zezina, eds), pp. 33–44. Institute of Oceanology, Russian Academy of Sciences, Moscow. (In Russian with English summary).

Sokolova, M.N., Vinogradova, N.G. and Burmistrova, I.I. (1995). "Rain of dead bodies" as a trophic source for the ultraabyssal macrobenthos in the Orkney trench. *Okeanologiya* **35**(4), 587–591. (In Russian with English summary).

Southward, A.J. and Southward, E.C. (1982). The role of dissolved organic matter in the nutrition of deep-sea benthos. *American Zoologist* **22**, 647–659.

Stockton, W.L. and Delaca, T.E. (1982). Food falls in the deep-sea; occurrence, quality and significance. *Deep-Sea Research* **29**(2), 157–169.

Strel'tsov, V.E., Agarova, I.Ya. and Petukhov, V.A. (1974). Zoobenthos and estimate of organic matter in the marine benthic sediments of sandy littoral zone of Dalnyi Beach (Barents Sea). *In* "Donnye otlozheniya i Biogeotsenozy Barentseva i Belogo morei" (V.E. Strel'tsov, ed.), pp. 129–141. (In Russian).

Strickland, J.D. (1960). Measuring the production of marine phytoplankton. *Bulletin of the Fisheries Research Board of Canada, Ottawa* **122**, 1–172.

Suchanek, T.H., Williams, S.L., Ogden, J.C., Hubbard, D.K. and Gill, I.P. (1986). Utilization of shallow-water seagrass detritus by Caribbean deep-sea macrofauna: $\delta^{13}C$ evidence. *Deep-Sea Research* **32A**, 201–214.

Tarasov, V.G. (1978). Distribution and trophic zonation of the soft-bottom communities in the Vostok Bay (Sea of Japan). *Biologiya Morya* **1978**(6), 16–22. (In Russian with English summary).

Tarasov, V.G. (1982). "Trophic structure and metabolism of benthic communities of soft beds of Vostok Bay of the Sea of Japan". Dissertation for the Award of the Degree of Candidate of Biological Sciences, Vladivostok. (In Russian).

Tarverdieva, M.I. (1974). The distribution and feeding habits of the young king crab *Paralithodes camtschatica* off western Kamchatka. *Trudy VNIRO* **99**, 54–62. (In Russian).

Tarverdieva, M.I. (1976). Feeding of *Paralithodes camtschatica*, *Chionoecetes bairdi* and *C. opilio* in the south-east of the Bering Sea. *Biologiya Morya*, **1976**(1), 41–48. (In Russian).

Tarverdieva, M.I. (1978). Diurnal feeding cycle of the king crab, *Paralithodes camtschatica*. *Biologiya Morya* **1978**(3), 91–95. (In Russian with English summary).

Tarverdieva, M.I. (1979). Feeding of the blue crab *Paralithodes platypus* in the Bering Sea. *Biologiya Morya* **1979**(1), 53–57. (In Russian with English summary).

Tarverdieva, M.I. (1981). On feeding in the snow crabs *Chionoecetes opilio* and *C. Bairdi* in the Bering Sea. *Zoologicheskii Zhurnal* **60**, 991–997. (In Russian with English summary.)

Tendal, O.S. (1979). Aspects of the biology of Komokiacea and Xenophyophoria. *Sarsia* **64**(1), 13–17.

Thamdrup, H.M. (1935). Beiträge zur Ökologie der Wattenfauna auf experimentaller Grundlage. *Meddelelser fra Kommisionen for Danmarks Fiskeri-og Havundersogelser, Serie Fisker* **10**(2), 1–125.

Thiel, H. (1972). Meiofauna und Struktur der benthischen Lebensgeimeinschaft des Iberischen Tiefseeboden. *Meteor Forschungen Ergebnissen*, Reihe, D, **12**, 36–51.

Thiel, H. (1973). Der Aufbau der Lebensgemeinschaft am Tiefseebecken. *Natur und Museum* **103**(2), 39–46.

Thiel, H. (1975). The size structure of the deep-sea benthos. *Internationale Revue des Gesamten Hydrobiologie* **60**(5), 575–606.

Thiel, H. (1979). Structural aspects of the deep-sea benthos. *Ambio Special Report* **6**, 25–31.

Thiel, H. (1991). The requirement for additional research in the assessment of environmental disturbances associated with deep-seabed mining. *In* "Marine Biology, its Accomplishment and Future Prospects" (J. Mauchline and T. Nemoto, eds), pp. 133–144. Elsevier, Amsterdam.

Thiel, H. and Schneider, J. (1988). Manganese nodule-organisms interactions. *In* "The Manganese Nodule Belt of the Pacific Ocean" (P. Halbach, G. Friedrich and U.v. Stakelberg, eds), pp. 102–110. Ferdinand Enke Verlag, Stuttgart.

Thiel, H. and Schriever, G. (1989). Cruise report DISCOL 1, Sonnecruise 61. *Berichte aus dem Zentrum Meeres und Klimaforschung der Universitat Hamburg* **3**, 1–75.

Thiel, H. and Schriever, G. (1990). Deep-sea bed mining, environmental impact and the DISCOL Project. *Ambio* **19**, 245–250.

Thiel, H., Pfannkuche, O., Schriever, G., Lochte, K., Gooday, A.J., Hembleben, C., Mantoura, R.F.C., Turley, C.M., Patching, J.W. and Riemann, F. (1988/1989). Phytodetritus on the deep-sea floor in a central Oceanic region on the Northeast Atlantic. *Biological Oceanography* **6**, 203–239.

Thiel, H., Schriever, G., Bussau, C., and Borowski, C. (1993). Manganese nodule crevice fauna. *Deep-Sea Research* **40A**(2), 419–423.

Thorson, G. (1966). Some factors influencing the recruitment and establishment of marine benthic communities. *Netherlands Journal of Sea Research* **3**, 267–293.

Thurston, M.N. (1979). Scavenging abyssal amphipods from the North-East Atlantic Ocean. *Marine Biology* **51**(1), 55–68.

Tilot, V. (1992). "La structure des assemblages megabenthiques d'une province a nodules polymetalliques de l'Océan Pacifique Tropical Est". Thèse de doctorat en sciences de l'Univeristé de Bretagne Occidentale, Département de l'Environnment Profond, IFREMER, Centre de Brest.

Trent, J.D. (1985). "A study of macroaggregates in the marine environment" Ph.D. Thesis, University of California at San Diego, CA (cited by Thiel *et al.*, 1988/1989).

Tseitlin, V.B. (1981). Estimating the vertical flow of detritus from the surface of the tropical ocean. *Okeanologiya* **21**(4), 713–718. (In Russian with English summary).

Tsikhon-Lukanina, E.A. (1982). Food assimilibility in benthic mollusks. *Okeanologiya* **22**(5), 833–838. (In Russian with English summary).

Tsikhon-Lukanina, E.A. and Soldatova, I.N. (1973). Assimilation of food by aquatic invertebrates. *In* "Trophologiya Vodnykh Zhivotnykh: Itogi, Zadachi" (G.V. Nikolsky and P.L. Pyroznikov, eds), pp. 108–121. Nauka, Moscow. (In Russian).

Turley, C.M., Lochte, K. and Patterson, D.J. (1988). A barophilic flagellate isolated from

4500 m the mid-North Atlantic. *Deep-Sea Research* **35A**, 1079–1092.

Turpaeva, E.P. (1953). Feeding and trophic groups of marine benthic invertebrates. *Trudy Instituta Okeanologii Akademii Nauk SSSR* **7**, 259–299. (In Russian).

Turpaeva, E.P. (1971a). Deep-sea Pantopoda collected in the Kuril–Kamchatka trench. *Trudy Instituta Okeanologii Akademii Nauk SSSR* **92**, 274–291. (In Russian with English summary).

Turpaeva, E.P. (1971b). An addition to the pantopod fauna of deep-sea trenches of the northwestern part of Pacific Ocean. *Trudy Instituta Okeanologii Akademii Nauk SSSR* **92**, 292–297. (In Russian with English summary).

Turpaeva, E.P. (1971c). Genus *Ascorhynchus* (Pantopoda) in the deep-sea fauna of Pacific Ocean. *Bulletin Moskovskogo Obschestva ispytatelei Prirody, Moskva, Seriya Biologicheskaya* **76**(3), 104–110. (In Russian with English abstract).

Turpaeva, E.P. (1973). Pycnogonids (Pantopoda) collected by the expeditions of the Institute of Oceanology in the north-west Pacific. *Trudy Instituta Okeanologii Akademii Nauk SSSR* **91**, 178–191.

Turpaeva, E.P. (1974). The Pycnogonida of the Scotia Sea and the surrounding waters. *Trudy Instituta Okeanologii Akademii Nauk SSSR* **98**, 275–305.

Turpaeva, E.P. (1975). Pantopoda from the northeastern part of the Pacific Ocean. *Trudy Instituta Okeanologii Akademii Nauk SSSR* **103**, 230–246. (In Russian with English summary).

Tyler, P.A. (1988). Seasonality in the deep-sea. *Oceanography and Marine Biology: An Annual Review* **26**, 227–258.

Vacelet, J. and Boury-Esnault, N. (1995). Carnivorous sponges. *Nature* **373**, 333–338.

Vinogradov, M.E. and Lisitsyn, A.P. (1981). Global patterns of distribution of life in the ocean and their reflection in the composition of benthic sediments. Patterns of distribution of plankton and benthos in the ocean. *Izvestiya Akademii Nauk SSSR, Seriya Geologicheskaya* **3**, 5–25. (In Russian with English title).

Vinogradova, N.G. (1969). Vertical distribution of deep-sea benthic fauna. *In* "Pacific Ocean. Biology of the Pacific Ocean. Book 2. The Deep-sea Bottom Fauna and Pleuston" (V. Kort, ed.), pp. 129–152. Nauka, Moscow. (In Russian).

Vinogradova, N.G. (1970). Deep–sea ascidians of the genus *Culeolus* of the Kuril–Kamchatka Trench. *Trudy Instituta Okeanologii Akademii Nauk* **86**, 489–512. (In Russian with English summary.)

Vinogradova, N.G. (1977). Fauna of the shelf, continental slope and abyssal zone. *In* "Okeanologiya. Biologiya Okeana", Vol. 1. (M.E. Vinogradov, ed.), pp. 178–187. Nauka, Moscow. (In Russian).

Vinogradova, N.G. (1980). Academician L.A. Zenkevich's teaching on the biological structure of the ocean. *Okeanologia* **20**(5), 766–773.

Volkonskii, V.V., Zernova, V.V., Semina, G.I., Sukhanova, I.N., Movchan, O.A., Sanina, L.V. and Tarkhova, I.A. (1972). Distribution of phytoplankton in the World Ocean. *E'kspress Informatsiya. Promyslovaya Okeanologia i Podvodnaya Tekhnika* **9**(3), 1–14. (In Russian).

Walter, M.D. (1973). Fressverhalten und Darminhaltsuntersuchungen bei Sipunculiden. *Helgolander Wissenschaftlich Meereuntersuchungen* **25**, 486–494.

Wilson, G.D.F. and Hessler, R.R. (1987). Speciation in the deep-sea. *Annual Review of Ecology and Systematics* **18**, 185–207.

Wolff, T. (1979). Macrofaunal utilization of plant remains in the deep-sea. *Sarsia* **64**, 117–136.

Wolff, T. (1980). Animals associated with seagrass in the deep-sea. *In* "Handbook of Seagrass Biology: An Ecosystem Perspective" (R.C. Phillips and C.P. MaRoy, eds) pp. 119–124. Garland Press, New York.

Yonge, C.M. (1956). Marine bottom substrata and their fauna. *In* "Proceedings of the 14th International Zoological Congress, Copenhagen" **1**.

Zenkevich, L.A. (1963). "Biology of the Seas of the USSR". Publishing House Academy of Sciences USSR, Moscow. (In Russian: English translation, 1963, Allen & Unwin, London.

Zenkevich, L.A., Brotskaya, V.A. and Dekhtereva, A.N. (1931). Materials on feeding in Barents Sea fish. II. *In* "*Doklady 1 Sessii Gosudaratvennogo Okeanograficheskogo Instituta*, **4**". (In Russian).

Zernov, S.A. (1949). "General Hydrobiology". Akademii Nauk SSSR, Leningrad. (In Russian).

Zernova, V.V. (1974). Distribution of the phytoplankton biomass in the tropical Atlantic. *Okeanologiya* **14**(6), 1070–1076. (In Russian with English abstract).

Zevina, G.B. (1970). Cirripeds of the genus *Scalpellum* in the north-west part of the Pacific Ocean. *Trudy Instituta Okeanologii Akademii Nauk SSSR* **86**, 252–276. (In Russian with English summary).

Zevina, G.B. (1973a). Scalpellidae (Cirripedia) of the Indian Ocean. 2. Species of the subgenera *Annandaleum, Mesoscalpellum* and *Neoscalpellum* of the genus *Scalpellum*. *Zoologicheskii Zhurnal* **52**, 1000–1007. (In Russian with English summary).

Zevina, G.B. (1973b). Scalpellidae (Cirripedia) of the Indian Ocean. 1. Species of the subgenera *Scalpellum* and *Arcocalpellum* of the genus *Scalpellum*. *Zoologicheskii Zhurnal* **52**, 843–848. (In Russian with English summary).

Zevina, G.B. (1973c). Scalpellids (Scalpellidae, Cirripedia) of the Gulf of Alaska. *Trudy Instituta Okeanologii Akademii Nauk SSSR* **91**, 136–140. (In Russian with English summary).

Zevina, G.B. (1975a). Cirripedia Thoracica of the American Mediterranean region. *Trudy Instituta Okeanologii Akademii Nauk SSSR* **100**, 233–258. (In Russian with English summary).

Zevina, G.B. (1975b). Cirripedia Thoracica collected by the r/v "Akademik Kurchatov" in the Atlantic sector of the Antarctic. *Trudy Instituta Okeanologii Akademii Nauk SSSR* **103**, 183–193. (In Russian with English summary).

Zevina, G.B. (1976). Abyssal species of barnacles (Cirrepedia, Thoracica) of the North Atlantic. *Zoologicheskii Zhurnal* **55**, 1149–1155. (In Russian with English summary).

Zevina, G.B. (1981). Cirriped crustaceans of the suborder Lepadomorpha of the World Ocean. 1. Family Scalpellidae. *Opredeliteli po Faune SSSR* **127**. Nauka, Leningrad. (In Russian).

Zezina, O.N. (1965). On the distribution of the deep-sea brachiopod species *Pelagidiscus atlanticus* (King). *Okeanologiya* **5**(2), 354–358. (In Russian with English summary).

Zezina, O.N. (1973). Composition and distribution of brachiopods in the benthos of the Gulf of Alaska. *Trudy Instituta Okeanologii Akademii Nauk SSSR* **91**, 192–202. (In Russian with English summary).

Zezina, O.N. (1975a). Recent Caribbean Sea deep-sea brachiopod fauna, the sources and the conditions of its formation. *Trudy Instituta Okeanologii Akademii Nauk SSSR* **100**, 188–195. (In Russian with English summary).

Zezina, O.N. (1975b). Deep-sea brachiopods from the southwestern part of Pacific Ocean and Scotia Sea. *Trudy Instituta Okeanologii Akademii Nauk SSSR* **103**, 247–258. (In Russian with English summary).

Zezina, O.N. (1976a). "Ecology and Distribution of the Recent Brachiopoda". Nauka, Moscow. (In Russian).

Zezina, O.N. (1976b). On determination of the growth rate and production of the brachiopod species *Pelagodiscus atlanticus* (King) from the bathyal and abyss. *Trudy*

Instituta Okeanologii Akademii Nauk SSSR **99**, 85–90. (In Russian with English summary).

Zezina, O.N. (1981). New and rare cancellothyrioid brachiopods from the bathyal and abyssal ocean. *Trudy Instituta Okeanologii Akademii Nauk SSSR* **115**, 155–164. (In Russian with English summary).

ZoBell, C.E. (1954). The occurrence of bacteria in the deep-sea and their significance for animal life. *International Union Biological Science* **16**, 20–26.

An Outline of the Geographical Distribution of Oceanic Phytoplankton

H.J. Semina

P.P. Shirshov Institute of Oceanology, Russian Academy of Sciences, Nakhimovsky Prospekt 36, Moscow 117851, Russia

ABSTRACT

The principal phytogeographical divisions of the ocean are described, based on net phytoplankton (principally diatoms and large dinoflagellates), in relation to large-scale gyres in the oceans. For different types of distribution, each is subdivided into a basal area and an area of expatriation (temporary colonization).

ADVANCES IN MARINE BIOLOGY VOL. 32
ISBN 0-12-026132-4

The influence of temperature, salinity and concentration of nutrients is considered and the degree of endemism of phytoplankton taxa for the different regions discussed.

1. INTRODUCTION

Studies of the geographical distribution of marine phytoplankton began at the end of the 19th century, when Cleve (1897) formulated his "plankton types". He was followed by Gran (1902) who proposed as "plankton elements" what other plankton workers recognize as "indicator species". As work progressed, data on phytoplankton distribution accumulated in the form of species lists for various local regions and also as maps of species occurrence. In the 1930s–1940s the distributions of phytoplankton species became known for larger ocean-scale regions (Hart, 1937; Graham and Bronikovsky, 1944). Smayda (1958) compiled distributional data for diatoms in several regions of the world ocean and discussed the factors underlying these distributions. Beginning in the 1960s, new data were collected during research vessel cruises, and it became possible to compile distributional data that would comprehensively reflect the peculiarities of species distribution in different oceans as well as in the World Ocean as a whole (Semina, 1967, 1974; Hasle, 1968, 1969, 1976, 1986; McIntyre et al., 1970; Makarova, 1971; Beklemishev et al., 1977; Semina et al., 1977; Baars, 1979; Makarova and Semina, 1982; Semina and Makarova, 1983; Semina and Ryzhov, 1985; Semina and Levashova, 1993).

The diversity of distribution of phytoplankton required a method of classification, and Margalef (1961) proposed the concept of "range types", based on "indicator species". He listed these range types on a preliminary basis, drawing on what was then known of the ecology of the species (their relation to temperature, salinity and concentration of nutrients). Margalef provided simplified range patterns in map form, and arbitrarily drew the boundaries in places where data were lacking.

Smayda (1978) noted the types of ranges according to Margalef (1961) without discussing them deeply. The principal types of phytoplankton ranges and the first division of the World Ocean based on phytoplankton were described by Russian scientists (Semina, 1967, 1974; Beklemishev et al., 1977; Semina et al., 1977; Beklemishev and Semina, 1986; Semina and Levashova, 1993). In his book on the biogeography of zooplankton (Beklemishev, 1969; English translation: Beklemishev, 1976) the author used data on phytoplankton to help characterize zooplankton distribution.

Great progress in studies of phytoplankton ranges resulted from utilization of data based on identifications of species by electron microscopy. This method substantially increased the accuracy of determinations (Hasle, 1976; Semina,

1979; Rivera, 1981; Makarova and Semina, 1982; Semina and Makarova, 1983; Makarova, 1988).

In recent years only a few papers dealing with the geography of phytoplankton have been published. Beklemishev and Semina (1986) found only 65 diatom species ranges in the World Ocean but this provided enough occurrence data to draw conclusions about range types. Unfortunately, some scientists have not always informed themselves about data published in other countries, particularly in Russia. For instance, the geography chapter of "Biology of the Diatoms" (Guillard and Kilham, 1977), contained lists of diatom species characteristic of various parts of the World Ocean, but did not provide any generalization of the geographical distribution of phytoplankton.

In analysing the geographical distribution of a species we usually compare it with separate abiotic factors (temperature, salinity, etc.). However, Beklemishev et al. (1972) established that the distribution of a species is influenced by a combination of abiotic factors rather than by separate ones. Thus, it is necessary to relate distributional ranges to major hydrographical features, especially the boundaries of ocean gyres and the vertically distributed properties of the water.

2. METHODS AND MATERIALS

The list of phytoplankton ranges discussed in this section was compiled by Semina and co-authors. Figures 1–5 are compiled from unpublished and published records (about 100 sources and 2000 stations, Semina et al., 1977). For the Pacific (Semina and Levashova, 1993) about 1000 stations were used. For the North Polar Basin and northern seas (Semina and Ryzhov, 1985), the ranges were compiled from 400 stations, using species that could be positively identified by light microscopy. However, the ranges of *Thalassiosira gravida* Cleve and *Thalassiosira pacifica* Gran and Angst are based on identification by scanning electron microscopy (SEM). Only the ranges of the species which are represented by a sufficient number of occurrences have been considered. Seasonal variations in the boundaries of ranges have not been taken into account. For most of the species quoted the taxonomy has been corrected according to Tomas (1996); the authority for a species is given on its first mention in the text.

3. TYPES OF RANGES

As noted in the introduction, similar individual species ranges can be combined into a type of range.

This section deals mostly with the ranges of both oceanic and panthalassic

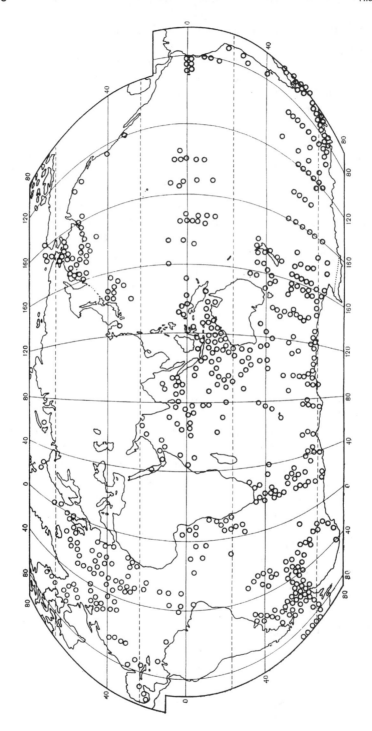

Figure 1   Occurrence of *Rhizosolenia alata* (after Beklemishev *et al.*, 1977).

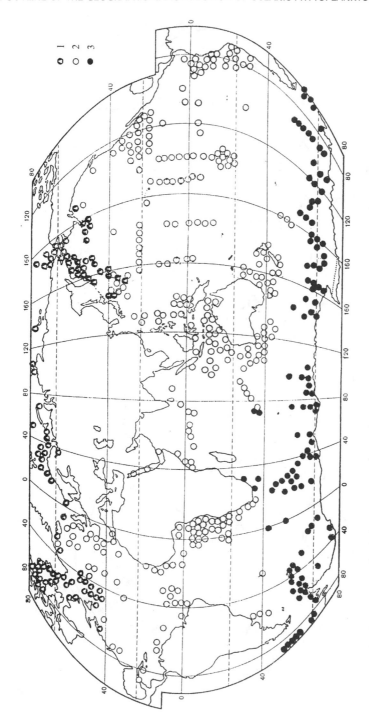

Figure 2 Occurrence of *Thalassiosira nordenskioldii* (1), *Planktoniella sol* (2) and *Chaetoceros neglectus* (3) (after Semina *et al.*, 1977).

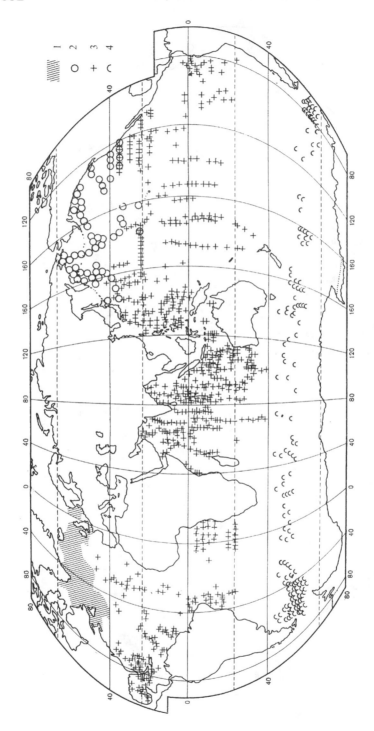

Figure 3 Occurrence of *Ephemera (Navicula) planamembranaceae* (1), *Neodenticula seminae* (2), *Pyrocystis noctiluca* (3) and *Rhizosolenia curvata* (4) (after Semina *et al.*, 1977).

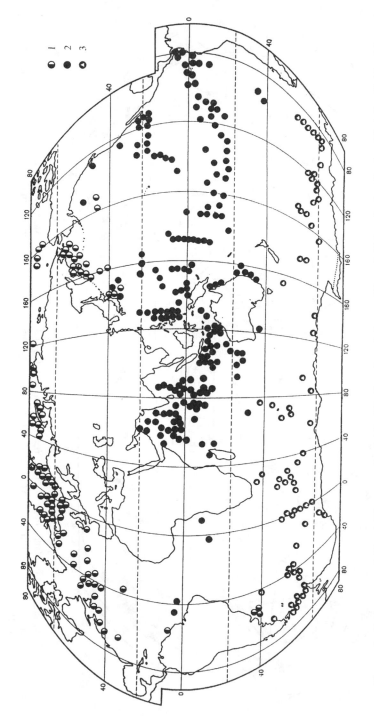

Figure 4 Occurrence of *Coscinidiscus oculus-iridis* (1) *Ceratium deflexum* (2) and *Rhizosolenia simplex* (3) (after Beklemishev *et al.*, 1977).

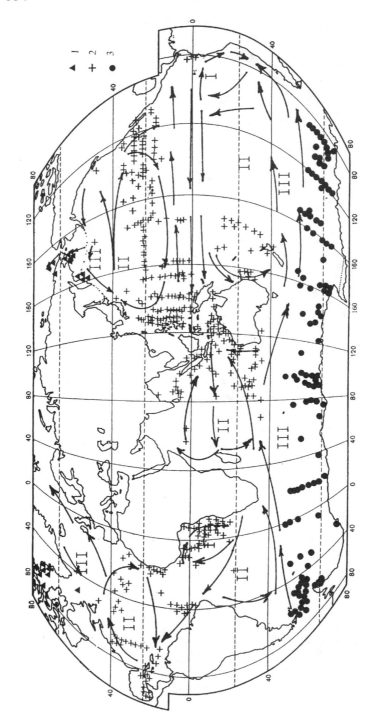

Figure 5 Occurrence of *Navicula granii* (1), *Hemiaulus hauckii* (2) and *Chaetoceros bulbosus* (3), (after Semina *et al.*, 1977). Direction of currents (arrows) and gyres: I, tropical cyclonic; II, subtropical anticyclonic; III, cyclonic subarctic (in the north and Antarctic in the south) (after Stepanov, 1983).

species (i.e. species which occur near the coast as well as in the open sea or ocean). In some types of ranges purely neritic species are also discussed.

3.1. Cosmopolitan Type

Rhizosolenia alata (Brightwell) Sundström may be regarded as an example (Figure 1). This species is common in all the three major oceans, from the Arctic to the high Antarctic.

3.2. Tropical Type

This type of range contains many sub-types. Most of the species have a wide distribution, and are termed widely tropical. Some worldwide examples are *Planktoniella sol* Wallich (Schütt) and *Pyrocystis noctiluca* Murray and Haeckel (Figures 2 and 3). The occurrence of other widely tropical species in the Pacific is shown in Figures 6–8 (*Ceratium carriense* Gourret, *Ceratium belone* Cleve, *Ceratium falcatum* (Kofoid) Jørgensen, *Chaetoceros coarctatus* Lauder, *Rhizosolenia castracanei* H. Pergallo). A narrower range in the tropical Pacific includes *Ceratium incisum* (Karsten) Jørgensen (Figure 9), which lives almost exclusively within the limits of 20°N and 25°S (i.e. an equatorial sub-type). *Ceratium lunula* (Schimper) Jørgensen is more widely distributed in the tropical zone (Figure 10); its range may be defined as "equatorial northern central".

More complex patterns of distribution are those of *Climacodium frauenfeldianum* Grunow (Figure 11) and *Pseudoeunotia doliolus* Grunow (Figure 12). The first is termed "west–equatorial–bicentral" type, while the second is regarded as "distant–neritic–northern–central". *Hemiaulus hauckii* Grunow in Van Heurck belongs to the same range subtype as *Climacodium frauenfeldianum* (Figure 5).

3.3. Bipolar Type

The following species appear to have bipolar ranges: *Thalassiosira antarctica* Comber, *Thalassiosira gravid* (cleve) (Figure 13), *Fragilariopsis cylindrus* (Grunow) Krieger, and *Porosira glacialis* (Grunow) Jørgensen. These species inhabit high and temperate latitudes of the northern hemisphere, but they were found in the Antarctic by Hasle (1976). These species have also been detected in the Bering Sea, the North Pacific, and in the Antarctic by Semina (1981b), Semina and Makarova (1983) and Semina *et al.* (1982).

3.4. Tropical Indo-Pacific Type

As an example, we can select *Ceratium deflexum* (Kofoid) Jørgensen, which is frequently found in the tropical zones of the Pacific and Indian Oceans. Rare findings of this species in the Atlantic (Figure 4) cannot yet be explained.

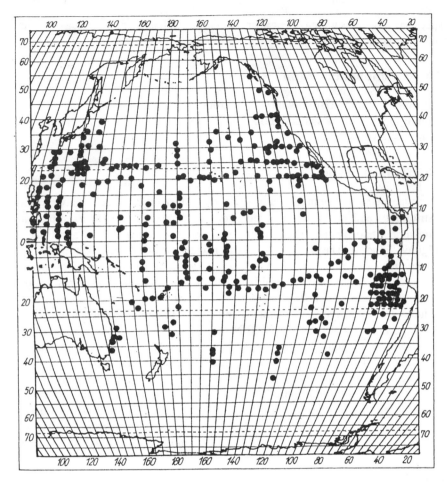

Figure 6   Occurrence of *Ceratium carriense* (after Semina and Levashova, 1993).

3.5. Arcto-boreal Type

Species typical of this range type live in the North Pacific, North Atlantic and in the seas neighbouring the northern parts of these oceans, as well as in the Arctic seas and in the Polar Basin itself. *Thalassiosira nordenskioldii* Cleve is a neritic species that lives in shelf-waters, but it may extend beyond the shelf in the period of the spring bloom (Figure 2). A panthalassic species, *Coscinodiscus oculus-iridis* Ehrenberg, has its greatest frequency of occurrence to the north of 40°N. Single findings of this species in the tropical Atlantic (south to 40°N, Figure 4)

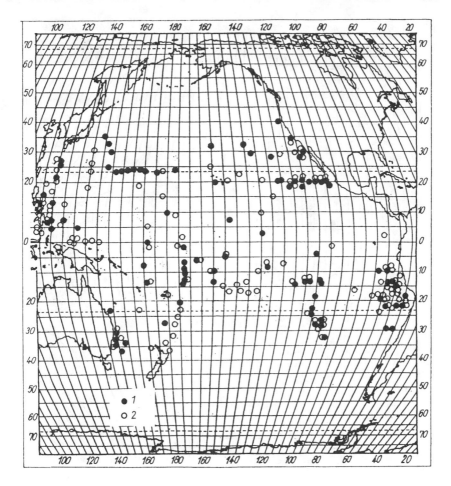

Figure 7 Occurrence of *Ceratium belone* (1) and *Ceratium falcatum* (2), (after Semina and Levashova, 1993).

can be referred to a non-breeding expatriation (colonization) area. Another widely distributed panthalassic species is *Ceratium arcticum* (Ehrenberg) Cleve (*sensu* Graham and Bronikovsky, 1944; Figure 14). Reports of this species not shown in Figure 14 can also be classed as expatriation. Narrow ranges of the arcto-boreal type are those of *Bacterosira bathomphala* (Cleve) Syversten and Hasle (formerly *B. fragilis*), *Amphiprora paludosa* var. *hyperborea* (Grunow) Cleve, *Nitzschia frigida* Gran, and *Navicula granii* Jørgensen. The last two species may be regarded as ice–neritic ones (Figures 5, 15 and 16).

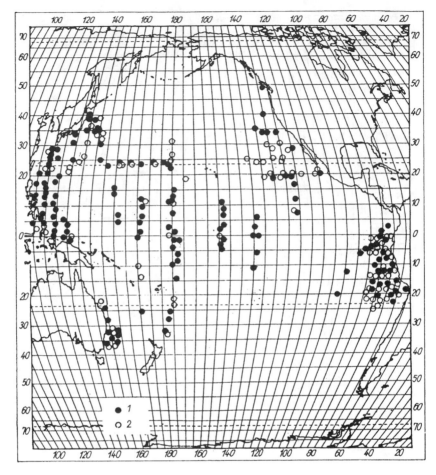

Figure 8   Occurrence of *Chaetoceros coarctatus* (1) and *Rhizosolenia castracanei* (2) (after Semina and Levashova, 1993).

3.6.   Boreal Types

3.6.1.   *Boreal-Pacific Type*

Neodenticula seminae (Simonsen and Kanaya) Akiba and Yanagisawa is an example of this type (Figure 3). Earlier (Semina, 1981a) this species was designated as belonging to the boreal–tropical type. Later it was found that in the tropical zone of the Pacific there was another species, the chains of which are similar to the chains of *N. seminae* so that under the light microscope they might be identified as *N. seminae*. SEM investigation of these colonies showed that their

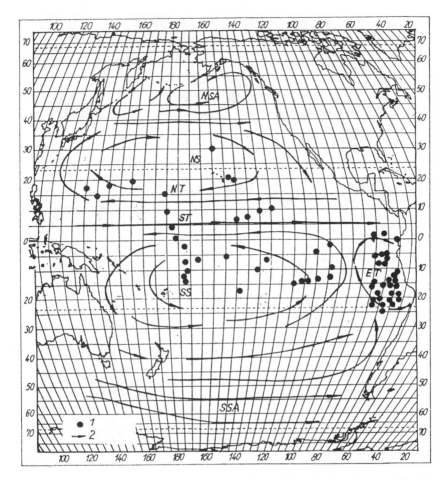

Figure 9 Occurrence of *Ceratium incisum* (1), and boundaries of gyres (2) (after Burkov *et al.*, 1973) Gyres: NSA and SSA, northern and southern cyclonic; NS and SS, northern and northern tropical cyclonic; ST, southern tropical anticyclonic; ET, east-tropical cyclonic (after Semina and Levashova, 1993).

valve structure is essentially different from that of *N. seminae* (*sensu stricto*) which is endemic to the North Pacific and the Bering Sea, the Okhotsk Sea and, in part, the Japan Sea (own observations).

3.6.2. *Boreal-Atlantic Type*

The example for this type of range is *Ephemera* (*Navicula*) *planamembranaceae* Hendey) Paddock (Figure 3).

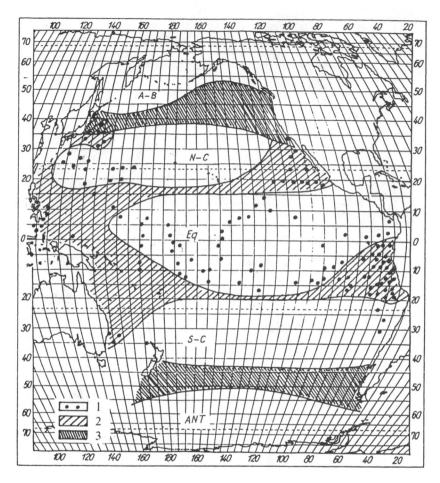

Figure 10 Occurrence of *Ceratium lunula* (1) and phytogeographical regions and provinces in the Pacific (after Semina, 1974): A-B, arcto-boreal; N-C, north central province of tropical region; E, equatorial province of tropical region; SC, central province of tropical region; ANT, notal–antarctic region; 2, transition zone between equatorial and central species; 3, transition zones between tropical and arcto-boreal regions in the north and between tropical and antarctic regions in the south (after Semina and Levashova, 1993).

3.7. Boreal–Tropical Type

Examples for this type of range are *Chaetoceros didymus* Ehrenberg and *Chaetoceras affinis* Lauder (Figure 17). These are also examples of a neritic type of range.

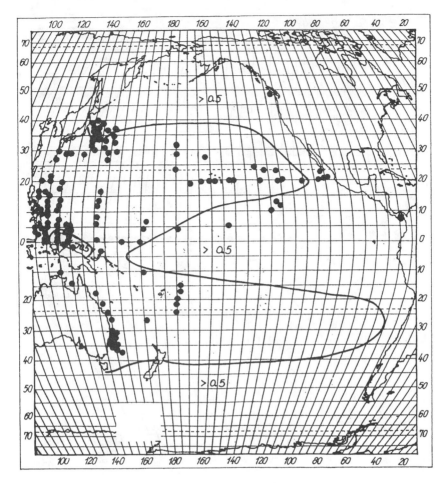

Figure 11   Records of occurrence of *Climacodium frauenfeldianum* (closed circles). The solid line shows dissolved inorganic phosphate equal to 0.5 μg at 1^{-1} (at 100 m, winter of northern hemisphere, according to Ivanenkov, 1979 (after Semina and Levashova, 1993).

3.8. Tropical–Notal–Antarctic Type

An example is *Rhizosolenia cylindrus* (*sensu* Hasle).

3.9. Notal–Antarctic Type

The species of this type of range are widely distributed in the Antarctic and Sub-antarctic. Examples are: *Chaetoceros neglectus* Karsten, *Rhizosolenia*

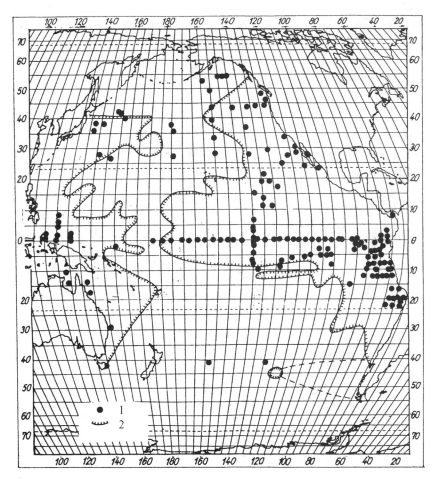

Figure 12   Occurrence of *Pseudoeunotia doliolus* (1, closed circles). The hatched line
(2) shows the limits of neritic species of zooplankton, which are not found in the open ocean
beyond (after Beklemishev, 1969 and Semina and Levashova, 1993).

simplex Karsten, *Chaetoceros bulbosus* (Ehrenberg) Heiden (Figures 2, 4 and 5).
Rhizosolenia curvata Zacharias (Figure 3) has a more limited range in the
subantarctic only. An ice-neritic species, *Actinocyclus* (*Charcotia*) *actinochilus*
(Ehrenberg) Simonsen, is widespread in the high Antarctic (Zernova, 1966; Hasle,
1968).

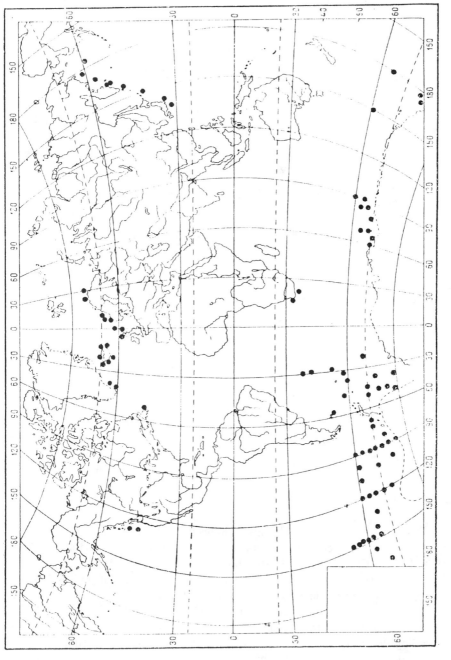

Figure 13 Occurrence of *Thalassiosira gravida* (after Semina and Makarova, 1983).

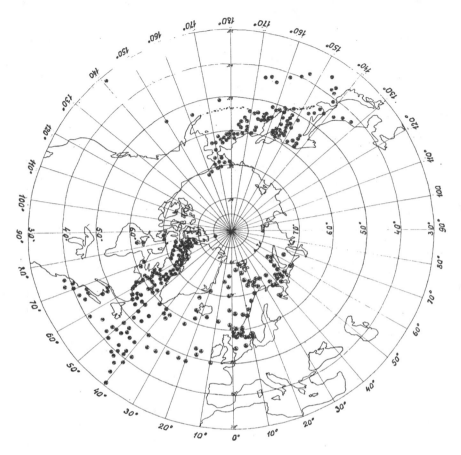

Figure 14   Occurrence of *Ceratium arcticum* (after Semina and Ryzhov, 1985).

3.10.   Boreal–Notal Type

Thalassiosira pacifica Gran and Angst belongs to this type or range. It is a neritic species, found in the temperate waters of the Pacific and near Chile (Hasle, 1976, 1978; Rivera, 1981; Semina, 1981b; Makarova, 1988).

To demonstrate the species possessing different types of ranges, SEM photographs of several diatoms are given in Figures 18–21.

From the foregoing classifications it is possible to revise previous tabulations of phytoplankton regions and ranges. Table 1 shows (left) the types of ranges according to Margalef (1961), rearranged by Sournia, the editor of the "Phytoplankton Manual", in the chapter by Smayda (1978), and (right) the range

Table 1   Types of oceanic phytoplankton ranges.

From Margalef (1961)	From Semina (1974) with alterations
I Cosmopolitan, euryoic	I Cosmopolitan
II Cosmopolitan, warm waters	II Tropical
III Cosmopolitan, cold water	III Bipolar
IV Atlantic, euryoic	Does not exist
V Atlantic, warm waters	Does not exist
VI Indo-Pacific, euryoic	Does not exist
VII Indo-Pacific, warm waters	IV Tropical Indo-Pacific
VIII Boreal, euryoic	V Arcto-boreal
IX Boreal, cold waters	VI Boreal–Pacific
	VII Boreal–Atlantic
X Cosmopolitan, temperate and warm	VIII Boreal–tropical
waters	IX Tropical–notal–antarctic
XI Austral, euryoic	X Notal–antarctic
XII Austral, cold waters	XI Boreal–notal

types according to Semina (1974), with recent updating from the results summarized above. Of the 12 types of ranges proposed by Margalef, only eight can now be recognized in the oceans but an additional three types of ranges (boreal–natal, boreal–Pacific, and boreal–Atlantic) need to be appended to Margalef's list. The existence of the following three types of Margalef has not been confirmed: "Atlantic, euryoic", "Atlantic, warm waters", and "Indo-Pacific, euryoic". Margalef (1961) treated the genera *Histoneis, Triposolenia, Oxytoxum* and *Kofoidinium* as examples of his "Indo-Pacific, warm water type" (in our terminology "tropical Indo-Pacific type"). All these taxa have now been found in the Atlantic (see Semina, 1974). Species such as *Amphisolenia globifera* Stein, *Cladopyxis brachiolata* Stein and *Oxytoxum milneri* Murray and Whitting, which were attributed by Margalef to the "Atlantic, warm waters type", have now been found in the Pacific. The ranges of many other species have also been corrected by more recent researches, and the majority of phytoplankton range types must now be regarded as circum-oceanic.

4.   PECULIARITIES OF THE PHYTOPLANKTON COMMUNITY BIOTOPE

Oceanic phytoplankton show variable biotopes. Those that possess passive buoyancy are able to live in specific parts of the ocean only as a result of environmental factors such as upwelling or turbulence. Autotrophic phytoplankton require sunlight. Thus, the lower boundary of the phytoplankton

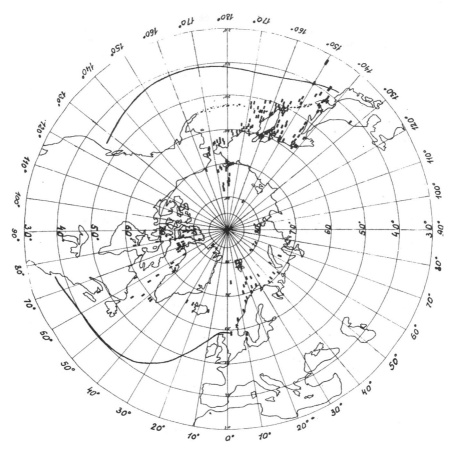

Figure 15   Occurrence of *Bacterosira bathyomphala* (formerly *B. fragilis*). The line shows the south boundary of subarctic gyres according to Burkov *et al.*, 1973 (after Semina and Ryzhov, 1985).

community (outside the shelf seas) in the tropics, and sometimes also in temperate latitudes, is the main vertical discontinuity, the pycnocline. We define the main pycnocline (Semina, 1974) as a permanent (existing during the whole year rather than seasonal) discontinuity layer. The main pycnocline restricts the depth of the upper mixed layer and therefore prevents neutrally buoyant phytoplankton cells from sinking into insufficiently illuminated layers. Thus, the depth of the main pycnocline is important to any consideration of geographical distribution of phytoplankton.

Oceanic phytoplankton is abundant in the layer above the main pycnocline.

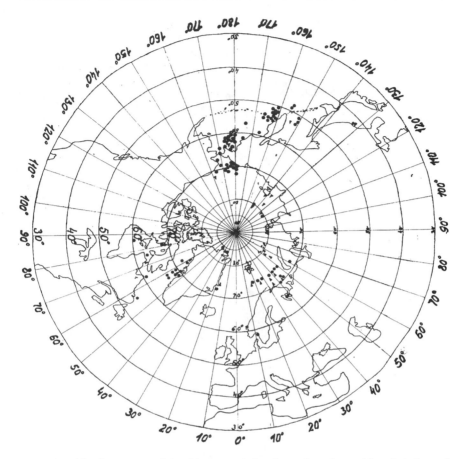

Figure 16   Occurrence of *Amphiprora paludensis* var. *hyperborea* (closed circles and *Nitzschia frigida* (filled triangles) (after Semina and Ryzhov, 1985).

This is the layer (down to 200 m depth) in which large-scale gyres are well developed. These gyres set the limits to biotope areas (Beklemishev *et al.*, 1972). Figure 5 shows these large-scale gyres according to Stepanov (1983), prepared from geostrophic flow calculations (Burkov *et al.*, 1973). The large-scale gyres, including their current systems, are stable objects located in particular water masses. The velocities of the currents are the lowest in the centre of the gyre and highest at the periphery, especially in the western parts of the oceans. At the boundaries of the gyres there are fronts which separate the gyres from one another. Vertical water transport parameters, important for maintenance of phytoplankton populations, are dependent on the character of the gyres. Owing to the individual

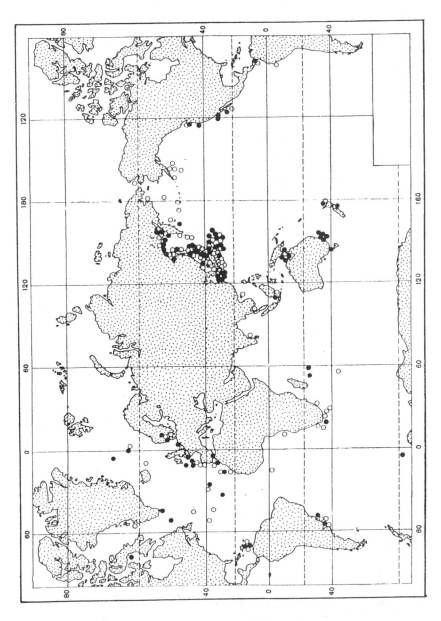

Figure 17 Occurrence of *Chaetoceros affinis* (closed circles) and *Chaetoceros didymus* (open circles) (after Semina, 1967).

Figure 18 Arcto-boreal species of diatoms: *Thalassiosira nordenskioldii* (A), *Thalassiosira hyalina* (Grunow) Gran (C), *Coscinodiscus oculus-iridis* (E). Boreal–Pacific species *Neodenticula seminae* (B, the whole valve; D, part of the same valve). A boreal–notal species, *Thalassiosira pacifica* (F) is also shown. Scale bars: A, B, C, F, 10 μm; D, 2μm; E, 100 μm.

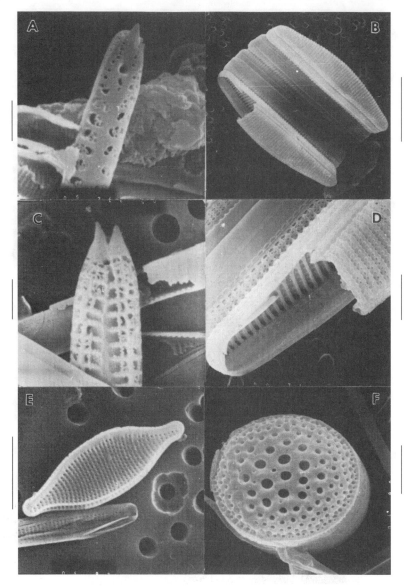

Figure 19 Tropical species of diatoms: *Thalassiothrix karstenii* (A), *Thalassiothrix lanceolata* Hustedt (C), *Nitzschia bicapitata* Cleve (E). *Pseudoeunotia doliolus* (B, the chain; D, cell of the same chain), *Thalassiosira oestrupii* (Ostenfeld) Hasle (F). Scale bars: A, C, 2 μm; D, E, F, 5 μm; B, 20 μm.

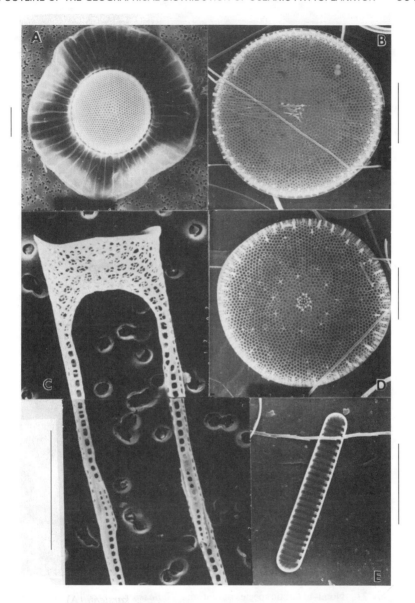

Figure 20   Tropical species of diatoms: *Planktoniella sol* (A), *Hemiaulus hauckii* (C). Bipolar species: *Thalassiosira antarctica* (B), *Thalassiosira gravida* (D), *Fragilariopsis cyclindrus* (E), Scale bars: A, 20 µm; B–E, 10 µm.

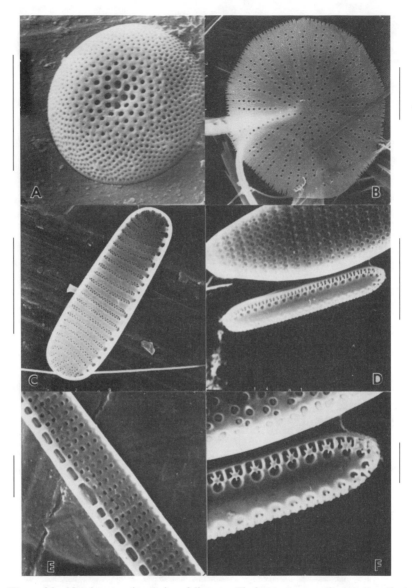

Figure 21 Notal–antarctic species of diatoms: *Thalassiosira gracilis* (A), *Fragilariop-sis curta* (C), *Nitzschia lineola* Cleve (E). High-antarctic species *Actynocyclus* (*Charcotia*) *actinochilus* (B). Cosmopolitan species: *Thalassionema nitzschioides* (Grunow) Grunow ex Hustedt (D, the whole valve; F, part of the same valve). Scale bars: A, B, C, D, 10 μm; E, F, 2 μm.

peculiarities of the gyres, each of them has its own characteristic community of phytoplankton, even though some of the constituent species are circum-oceanic. Thus, the gyres can be regarded as one of the most important factors controlling the geographical distribution of phytoplankton species in the ocean. Environmental conditions for phytoplankton in the subantarctic, subtropical and equatorial regions of the Pacific that relate to large-scale gyres are quite different (Semina, 1974).

McGowan (1986) compared the northern Pacific cyclonic gyre, the central northern Pacific anticyclonic gyre and the eastern tropical Pacific (as a part of circulation system of the tropical Pacific). He found that these regions differ not only in their structure but are also functionally different; this explains why the response of these regions to climate or episodic perturbations is also different. According to Beklemishev (1969), only the large-scale gyres (over 1000 km in diameter), and the abiotic conditions therein, are stable. McGowan (1986) regards these stable large-scale systems, which differ from one another, as having promoted the formation of stable associations of species. "Only the large-scale structures have the degree of difference and temporal persistence to allow evolution to operate in a mobile, moving medium" (McGowan, 1986).

5. THE BASIS OF RANGE AND EXPATRIATION AREA

The influence of ocean currents on the distribution of species has long been recognized, but this topic is often treated erroneously. It is necessary to stress that the currents are constituents of gyres that are important in controlling the ranges of species. Ekman (1953), in discussing the distribution of pelagic life, and quoting Lohmann, refers to the need to distinguish between regions where a species flourishes and those where it just exists. In this contribution the latter is referred to as an expatriation area, following Ekman, while the area where a species is established is referred to as the range basis or basis of range.

The *basis of a range* is more or less a closed gyre where the species can live indefinitely regardless of its presence or absence in the other parts of the ocean (Beklemishev, 1969, 1976); that is, a centre of distribution. The *reproduction area* (Ekman, 1953) is the part of a species range where reproduction can occur with such intensity that the population can maintain itself without the need for inflow of individuals from other parts of the range. Hence, the basis of the range is a part of the reproduction area. The reproduction area is larger where part of the water mass suitable for reproduction constantly flows out of the circulation pattern serving as the basis of the range, and does not return. The part of a reproduction area that is located outside the range basis is designated the non-sterile expatriation area. In such an area environmental conditions make it possible for the species to live and reproduce normally, but the population cannot remain there indefinitely without the inflow of new individuals from the range basis because of continuous transport in one direction.

Outside the range basis lies the sterile expatriation area (Ekman, 1953) which is the area where the individuals cannot pass their entire life cycle, or where reproduction does not compensate for mortality. This situation may be caused by environmental conditions or extreme dispersion of the population. In this area the species gradually dies out, unless replenished from the basic range area.

The first attempt to distinguish basis of range and expatriation area for oceanic phytoplankton was made in 1977 (Semina *et al.*, 1977) and later in other papers (Semina, 1979; Semina and Ryzhov, 1985; Semina and Levashova, 1993). For this section we can accept the large-scale gyres (Figure 5) as the bases of ranges of oceanic and panthalassic species. The occurrence of a species beyond these gyres can be attributed to expatriation. The ranges in Figures 1–5 and 13 are compared below with the gyres shown in Figure 5; the ranges in Figures 6–12 are compared with the gyres in Figure 9. The basis of the ranges for neritic species is shelf sea, and we do not consider this here.

The cosmopolitan species, *Rhizosolenia alata*, is abundant in all the large-scale gyres. Broadly tropical species such as *Planktoniella sol*, *Pyrocistis nocticula*, *Ceratium carriense*, *Ceratium belone*, *Ceratium falcatum*, *Ceratium coarctatus* and *Rhizosolenia castracanei* have the basis of their range in all the tropical gyres (between 40°N and 40°S).

The range basis of *Ceratium incisus* is situated in the northern and in the southern tropical gyres and in the eastern tropical cyclonic gyre. The range basis of *Ceratium lunula* includes the northern and the southern tropical gyres, the eastern cyclonic and the northern subtropical gyres. The last two species do not inhabit the southern subtropical gyre as a whole; this gyre probably plays the role of an expatriation area for them.

The range basis for *Climacodium frauenfeldianum* is located in the northern subtropical gyre and in the seas around Australia (Timor Sea, Arafura Sea, Carpentaria Gulf and Coral Sea). The basis of range of *Pseudoeunotia doliolus* lies in the northern subtropical, eastern cyclonic, and northern and southern tropical gyres, as well as in the seas mentioned for the range basis of *Climacodium frauenfeldianum*. The range of *Hemiaulus hauckii* is similar to that of *C. frauenfeldianum*. In the Pacific the basis of its range is the same as that of *H. hauckii*. In the Atlantic this species occurs in the gyres analogous to those in the Pacific, and in the Indian Ocean in the east of the southern subtropical gyre. For the last three species, it is not possible to explain their occurrence in the southern subtropical gyre.

The range basis of *Ceratium deflexum* is situated in the Pacific in all the tropical gyres, but this species has not been found in the outer central parts of the northern and southern subtropical gyres. In the Indian Ocean this species occurs in the subtropical gyre and in the monsoonic area in the northern part of the ocean, where the direction of currents in the gyres depends upon the season. Two panthalassic species, *Ceratium arcticum* and *Coscinodiscus oculus-iridis*, have their range bases in the gyres of the polar basin and of all the northern seas, as well as in subarctic gyres in the northern Atlantic and Pacific. A bipolar species *Thalassiosira gravida* lives in the sub-arctic gyres of the northern Atlantic and Pacific and in the Antarctic gyre of the Southern Ocean. A boreal-pacific species *Neodenticula seminae* has its range basis in the gyres of the Bering Sea and Okhotsk Sea and in the subarctic gyre in the northern Pacific.

A boreal–Atlantic species *Ephemera (Navicula) planamembranaceae* inhabits the subarctic Atlantic gyre where its range basis is located.

Among the notal–Antarctic species, the most abundant is *Chaetoceros neglectus*. This species demonstrates a good coincidence of the northern boundary of its range and the northern boundary of the antarctic gyre: both boundaries are shifted to the north in the eastern Atlantic and in the Indian Ocean, compared with the Pacific. The range of *Chaetoceras bulbosus* is located somewhat to the south, and the northern boundary of this range is also shifted further to the south in the Pacific than in the two other oceans. The range of a rarer species, *Rhizosolenia simplex*, shows the same regularity in the location of the northern boundary of the range.

Notal–antarctic species have their range bases in the Antarctic gyre.

The findings of species beyond the bases of ranges (outside the gyres) are regarded as examples of an expatriation area. Expatriation areas are discussed here without their division into sterile and non-sterile ones. Transport of species outside their range basis occurs along the currents and across the currents. The last is possible owing to turbulent mixing or formation of meanders with a subsequent formation of rings and their separation from the main current.

Some of the species mentioned above do not show any expatriation areas (for instance, widely-tropical *Ceratium belone*, *Ceratium falcatum* and *Rhizosolenia castracanei*). For some species the expatriation areas are small and are not discussed here. An uncertainty pertaining to certain expatriation areas is mentioned above.

Expatriation areas of tropical species are very pronounced in the northeastern Pacific and Atlantic, where these species are carried from their range bases; by the Gulf Stream/North Atlantic Drift in the Atlantic and by the Alaska current in the Pacific.

In the Atlantic, the tropical species extend to 65°N. This extension is shown by *Planktoniella sol* and also by the following species, as noted by Lucas *et al.* (1973): *Biddulphia sinensis* Greville, *Hemiaulus* spp., *Rhizosolenia acuminata* (H. Pergallo) H. Pergallo, *Rhizosolenia bergonii* H. Pergallo, *Guinarda striata* (= *Rhizosolenia stolterforthii*) Hasle. In the Gulf of Alaska we found a large expatriation area of *Pseudoeunotia doliolus*; this species was also found here by Venrick (1971). In the same place Semina and Tarkhova (1972) discovered tropical species of *Ceratium*.

In the northwestern Atlantic and Pacific, where there are no northwardly directed currents, there have been isolated findings of tropical species (Beklemishev and Semina, 1986; Semina and Levashova, 1993). In the South Atlantic an invasion of tropical species can occur outside the boundaries of the Antarctic gyre (Semina *et al.*, 1982) and also in the Pacific (Zernova, 1966; Hasle, 1976; Sukhanova *et al.*, 1988). The panthalassic arcto-boreal species *Coscinodiscus oculus-iridis* and *Ceratium arcticum* typically have large expatriation areas (especially in the Atlantic), which do not appear to be associated with any particular currents. An expatriation area of *Neodenticula seminae* extends up to the latitude of Honshu Is. In the central Pacific this species is found to 20°S.

A special publication (Semina, 1979) was dedicated to expatriation areas of Antarctic species. The most extended expatriation area of these species is found in the Pacific near the shores of South America where the organisms are carried with the Peru current up to the equator.

Less explicit expatriation areas of Antarctic species are observed near New Zealand, off Madagascar and in the Atlantic, where the organisms are also carried by north-going currents. There is an extensive list of species (at least 20) transported out of the Antarctic, but the number of findings for each species is small and the abundance is also low. Figure 22 illustrates the scheme of range bases and the direction of species expatriation.

The most important point confirming that the occurrence of a species is an example of expatriation is that it does not inhabit the whole gyre but penetrates only a small part.

The expatriation areas of phytoplankton are larger than those of zooplankton. This can be seen comparing the expatriation areas of phytoplankton with the expatriation areas of various groups of plankton animals in the Pacific (Heinrich, 1993). The physiological properties of phytoplankton organisms must fit them better for long-distance transport. For successful expatriation, most animals require the ability to reproduce and the ability to complete their life histories. The life cycle of algae incorporates both sexual as well as vegetative reproduction. It may happen that environmental conditions inhibit sexual reproduction but that vegetative reproduction can take place. In such a situation diatom species will divide until they become so small that their further existence is not possible. It is the vegetative phase that is conveyed for long distances in expatriation.

The bases of ranges have a latitudinal zonation which depends on latitudal zonation of the location of large-scale gyres. This is clearly seen in Figure 22. Circumcontinental zonation is expressed in the North Pacific and Atlantic and most clearly seen for expatriation areas near the coasts at the periphery of the gyres, owing to the direction of the currents.

6.   RANGES AND ENVIRONMENTAL FACTORS

While discussing distribution ranges most authors pay major attention to water temperature. Undoubtedly, the arcto-boreal, notal–antarctic, subantarctic, high-

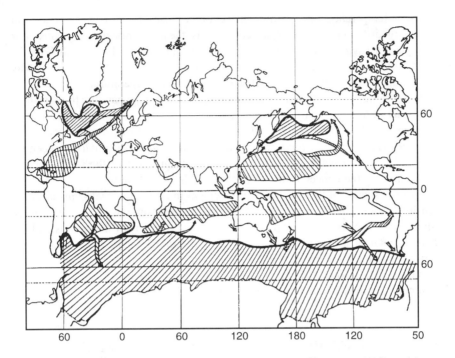

Figure 22   Scheme of range bases and direction of expatriation of species. Range bases of arcto-boreal and notal–antarctic species are shown hatched /////. Range bases of tropical species are shown hatched \\\\\. The arrows show the direction of expatriation of the two groups (after Semina, 1979).

antarctic, and bipolar species live at low temperatures, while tropical species occur at higher temperature. The cosmopolitan species, however, appear able to live at low as well as at high temperature. The range bases of arcto-boreal species at the north of the oceans and those of notal–Antarctic species at the south of the oceans are limited by the isotherms 5–10°C (at 100 m depth).

Species may differ in the temperature limits to which their basic ranges appear adapted. Species of high and temperate latitudes show a range basis related to the temperatures prevailing during the season in which they are most abundant. For example, the spring species such as *Nitzschia frigida*, *Amphiprora paludosa* var. *hyperborea* and *Bacteriosira bathyomphala* (formerly *B. fragilis*) occur in the Bering Sea and in the Pacific near Kamchatka at temperatures from −1.6. to + 4.3°C, while a primarily summer species such as *Ceratium arcticum* occurs at −0.4 to +11°C (Semina and Ryzhov, 1985).

The environmental factors (temperature, salinity and concentrations of the nutrients) in a range basis differ from the same factors in an expatriation area. For 11 tropical species in the Pacific, Semina and Levashova (1993) found that the limits of temperature and salinity in the range bases of these species are not obviously different, but in expatriation areas these factors differ greatly. For example, for *Pseudoeunotia doliolus* in the basis of the range T = 13–28°C (in the layer 0–100 m), while in the expatriation area it was 4–12°C. This explains why one cannot establish the range of species using only temperature.

Most tropical species (Semina and Levashova, 1993) appear able to live where there is a low as well as a high concentration of phosphates and silicates. However, one species (*P. doliolus*) lives only at high concentration of nutrients, while *Hemiaulus hauckii*, and *Climacodium frauenfeldianum* live only at low concentration of these elements. For these species the configuration of the range appears determined by the concentration of nutrients (Figures 11 and 12).

One peculiarity of the biotope that is evident in all the tropical gyres is a high density gradient in the main pycnocline. This promotes the existence of large phytoplankton cells (over 10^5 μm^3) in the tropical regions of the oceans. A high density gradient at the lower boundary of the biotope helps to keep the cells buoyant and inhibits sinking. Such a large density gradient is absent at the lower boundary of the biotope in temperate and high latitudes where large cells are less common (Semina, 1972; Semina and Tarkhova, 1972).

7. PHYTOGEOGRAPHICAL REGIONS OF THE OCEAN AND ENDEMISM OF PHYTOPLANKTON

The geographical regions of the Pacific were specified by Semina (1974). At the north the arcto-boreal region is located, and at the south lies the notal–antarctic region. Between the two named regions is the tropical region. At the boundaries of the species there are stripes designating the transition zones where tropical species are mixed with arcto-boreal at the north and with notal–antarctic at the south. In the tropical region the provinces are marked out: equatorial, northern central and southern central ones; the transition zones also exist between them (Figure 10).

The phytogeographical division of the World Ocean based on Beklemishev *et al.* (1977) is shown in Figure 23. The same three phytogeographical regions that correspond to the Pacific regions are divided into zones (in the arcto-boreal, tropical and notal–antarctic). The last is divided in high-antarctic and low-antarctic zones. The boundaries of the regions in the Pacific slightly differ from those given by Semina (1974). SB on Figure 23 is a mixing zone of arcto-boreal and tropical species, and LA partly includes a mixing zone of notal–antarctic and tropical species.

The endemism of phytoplankton is most pronounced at the species level and endemism is most frequent in the tropics. There are 140 tropical species of diatoms in the Pacific, and over 400 species of peridinians. In the Pacific part of the

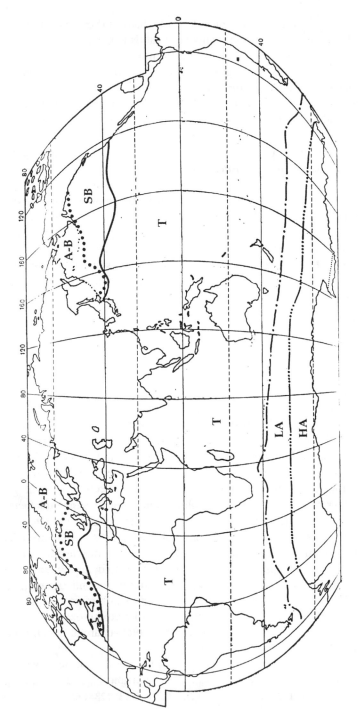

Figure 23 Phytogeographical regions of the World Ocean, AB, arcto-boreal region; SB, mixing zone of arcto-boreal and tropical species; LA and HA, notal–antarctic region; HA, high antarctic; LA, low antarctic; partly including mixing zone of notal–antarctic and tropical species; T, tropical region (after Beklemishev *et al.*, 1977 with alterations).

arcto-boreal region there are 22 endemic species of diatoms (Semina, 1981b); the number of endemic peridinian species is not yet established. In the notal–antarctic region there are about 30 endemic species of diatoms (Hasle, 1968, 1969). Balech (1970) treats 32 species of peridinians as endemic Antarctic ones. A sub-antarctic endemic species is *Rhizosolenia curvata*.

The number of endemic genera of phytoplankton is considerably lower. Among the diatoms, an endemic genus of arcto-boreal region is *Bacterosira*. In the Pacific part of the arcto-boreal region the endemic genus is *Neodenticula*. For the tropical region, it is possible to indicate at least 10 endemic genera of the diatoms: *Asterolampra*, *Bacteriastrum*, *Climacodium*, *Ethmodiscus*, *Gossleriella*, *Hemidiscus*, *Hemiaulus*, *Planktoniella*, *Roperia* and *Streptotheca*. Among peridinians, no endemic arcto-boreal genera are known. In the tropical region, there are the following 12 endemic genera of peridinians: *Amphisolenia*, *Triposolenia*, *Histoneis*, *Chitaristes*, *Ornithocercus*, *Pyrocystis*, *Kofoidinium*, *Noctiluca*, *Podolampas*, *Oxytoxum*, *Cladopyxis* and *Ceratocorys*.

The geography of higher-rank taxons (families and orders) cannot be discussed here since there are insufficient data on geographical distribution of all the species and genera that form the named higher taxons. It is likely that there are no endemics among the phytoplankton taxons of these ranks.

8. CONCLUSIONS

In this review the distribution of the species in the ocean is compared not only with some separate abiotic factors but also with natural areas of the biotope. The typical natural areas for panthalassic and oceanic species are stable large-scale circulations (gyres). The latitudinal zonation of ranges distribution depends on the zonation of the location of the gyres. The circumcontinental zonation is more clearly shown for the expatriation areas of these species. Each large-scale gyre differs from other similar gyres by a whole system of environmental factors, because each of the gyres includes a separate water mass. Owing to these properties of the gyres, no homogeneous phytoplankton exists in the ocean. The evolution of the communities appears to have been in a direction leading to their isolation (though not an absolute separation). This would appear to be the explanation for the ranges of various types that differ in their geographical position.

REFERENCES

Baars, J.W.M. (1979). Autecological investigations on marine diatoms. I. Experimental results in biogeographical studies. *Hydrobiogical Bulletin* **13**, 123–137.

Balech, E. (1970). The distribution and endemism of some Antarctic microplankters. *In* "Antarctic Ecology" Vol I, (M.W. Holdgate ed.), pp. 143–147. Academic Press, London.

Beklemishev, C.W. (1969). Range and biogeographical division (Chapter IV) and Pacific Ocean (Chapter V). *In* "Ecology and Biogeography of the Open Ocean" (W.G. Bogorov, ed.), pp. 96–109 and pp. 110–163, Nauka, Moscow. (In Russian).

Beklemishev, C.W. (1976). "Ecology and Biogeography of the Open Ocean". Naval Oceanographic Office translation NOO T-25, Washington.

Beklemishev, C.W. and Semina H.J. (1986). The geography of planktonic diatoms of high and temperate latitudes of the World Ocean. *Trudy vsesoyuznogo gidrobiolocheskogo obshestva* **27**, 7–23. (In Russian).

Beklemishev, C.W., Neyman, A.A., Parin, N.V. and Semina, H.J. (1972). Le biotope dans le milieu marin. *Marine Biology* **15**, 57–73.

Beklemishev, C.W., Parin, N.V. and Semina H.J. (1977). Biogeography of the open sea. 1. Open sea. *In* "Okeanologia. Biologia Okeana. I. Biologicheskaya Struktura Okeana" (A.S. Monin, ed.), pp. 219–611, Nauka, Moscow. (In Russian).

Burkov, V.A., Bulatov, R.P. and Neyman, V.G. (1973). Largescale features of water circulation in the World Ocean. *Okeanologiya* **13**(3), 395–403. (In Russian).

Cleve, P.T. (1897). "A Treatise on the Phytoplankton of the Atlantic and its Tributaries and on the Periodical Changes of the Plankton of the Skagerak, 1897". Nya Tidnigs Aktiebolags Tryckeri, Uppsala.

Ekman, S. (1953). "Zoogeography of the Sea". Sidgwick & Jackson., London.

Graham, H.W. and Bronikovsky, N. (1944). The genus *Ceratium* in the Pacific and North Atlantic Oceans. *In* "Scientific Results of Cruise VII of the Carnegie during 1928–1929 under the command of Captain J.P. Ault". Biology V. Carnegie Institution, Washington.

Gran, H.H. (1902). Das plankton des Norwegischen Nord-meeres, *Report on Norwegian Fishery and Marine Investigations* **2**(5), 1–222.

Guillard, R.R.L. and Kilham, P. (1977). The ecology of marine diatoms. *In* "The Biology of Diatoms" (D. Werner, ed.), pp. 372–469. California University Press, California.

Hart, T.J. (1937). *Rhizosolenia curvata* Zacharias, an indicator species in the Southern Ocean. *Discovery Reports* **16**, 413–446.

Hasle, G.R. 1968. Distribution of marine diatoms in the Southern oceans. *In* "Primary productivity and benthic marine algae of the Antarctic and Subantarctic" (V.C. Bushell and S.Z. El-Sayed, eds), pp. 6–8. Antarctic Marine Folio Series, American Geographical Society **10**.

Hasle, G.R. (1969). An analysis of phytoplankton of the Pacific Southern Ocean: abundance, composition and distribution during the "Brategg" Expedition, 1947–1948. *Hvalradets Skrifter* **52**, 1–168.

Hasle, G.R. (1976). The biogeography of some marine planktonic diatoms. *Deep-Sea Research* **23**, 319–338.

Hasle, G.R. (1978). Some *Thalassiosira* species with one central process (Bacillariophyceae). *Norwegian Journal of Botany* **25**, 77–110.

Hasle, G.R. (1986). Problems in open ocean phytoplankton biogeography. *UNESCO Technical Papers in Marine Science* **49**, 118–134.

Heinrich, A.K. (1993). Boundaries and structure of the oceanic plankton communities. *In* "Comparative Ecology of Plankton Oceanic Communities" (U.A. Rudyakov, ed.), pp. 6–46, Nauka, Moscow. (In Russian).

Ivanenkov, V.N. (1979). The main nutrient salts. 7.2 The main distribution of nutrient salts in the World Ocean. *In* "Okeanologya. Khimiya Okeana" (A.S. Monin, ed.) pp. 188–229. Nauka, Moscow. (In Russian).

Lucas, C.E. Glover, R.S. and Hardy, A. (eds) (1973). Continuous plankton records: A plankton atlas of the North Atlantic and the North Sea. *Bulletins of Marine Ecology* **7**, 1–174.

McGowan, J. (1986). The biogeography of pelagic ecosystems. UNESCO Technical Papers in Marine Science **49**, 191–200.

McIntyre, A., Be, A.W.H. and Roche, M.B. (1970). Modern Pacific coccolithophorids, a paleontological thermometer. *Transactions of the New York Academy Sciences* Ser. II **32**(6) 720–731.

Makarova, I.V. (1971). About biogeography of the genus *Thalassiosira* Cl. *Botanicheskii zhurnal* **56**(10), 1459–1476 (In Russian).

Makarova, I.V. (1988). Geographical distribution and ecology of living species of genus *Thalassiosira Cl. In* "Diatom Algae of the Seas of USSR: The Genus *Thalassiosira* Cl" (Z.I. Glezer and O.G. Kusakin, eds), pp. 26–35, Nauka, Leningrad. (In Russian).

Makarova, I.V. and Semina, H.J. (1982). New data about morphology of *Thalassiosira hyalina* (Grun.) Gran (Bacillariophyta *Botanicheskii zhurnal* **67**(1), 109–111.

Margalef, R. (1961). Distribucion ecologica y geografica de las especies del Fitoplankton marino. *Investigacion Pesquera* **19**, 81–101.

Rivera, P.R. (1981). Beitrage zur Taxonomie und Verbreitung der Gattung *Thalassiosira* Cleve. *Bibliotheca Phycologica* **56**, 1–220.

Semina, H.J. (1967). Phytoplankton. *In* "Pacific Ocean. Biology of the Pacific Ocean. Vol. I. Plankton" (V.G. Kort, ed.) pp. 27–85. Nauka, Moscow. (In Russian).

Semina, H.J. (1972). The size of phytoplankton cells in the Pacific Ocean. *International Revue der gesampten Hydrobiologie* **57**, 177–205.

Semina, H.J. (1974). "The Phytoplankton of the Pacific Ocean." Nauka, Moscow. (In Russian).

Semina, H.J. (1979). The geography of plankton diatoms of the Southern Ocean. *Nova Hedwigia* **64**, 341–356.

Semina, H.J. (1981a). Morphology and distribution of a tropical *Denticulopsis. In* "Proceeding of 6th Symposium on Living and Fossil Diatoms" (R. Ross, ed.), pp. 179–191 Koeltz, Konigstein.

Semina, H.J. (1981b). Species composition of phytoplankton of the western part of Bering Sea and adjoing part of Pacific. II. Diatoms. *In* "Ecology of Marine Phytoplankton" (in (H.J. Semina, ed.) pp. 6–32. P.P. Shirshov Institute of Oceanology, Moscow. (In Russian).

Semina, H.J. and Levashova, S.S. (1993). The biogeography of tropical phytoplankton species in the Pacific Ocean. *International Revue der gesampten Hydrobiologie* **78** (2), 243–262.

Semina, H.J. and Makarova, I.V. (1983). On morphology and distribution of two bipolar species of the genus *Thalassiosira* (Bacillariophyta). *Botanicheskii zhurnal* **68**(5), 611–618.

Semina, H.J. and Ryzhov, V.M. (1985). The ranges of some phytoplankton species of the Northern Hemisphere. *In* "Izuchenya Okeanicheskogo Fitoplanktona" (H.J. Semina, ed.), pp. 16–27. P.P. Shirshov Institute of Oceanology, Moscow. (In Russian).

Semina, H.J. and Tarkhova, I.A. (1972). Ecology of phytoplankton in the North Pacific Ocean. *In* "Biological Oceanography of the North Pacific Ocean" (S. Motoda, ed.). pp. 117–124. Idemitsu Shoten, Tokyo.

Semina, H.J., Belyaeva, T.V., Zernova, V.V., Movchan, O.A., Sanina, L.V., Sukhanova, I.N. and Tarkhova, I.A. (1977). Distribution of the indicator plankton algae in the World Ocean. *Okeanologiya* **17**, 867–877. (In Russian).

Semina, H.J., Golikova, G.S. and Nagaeva, G.A. (1982). Phytoplankton of the southern Atlantic in November–December of 1971. *Trudy Instituta Okeanologii* **11**, 5–19.

Smayda, T. (1958). Biogeographical studies of marine phytoplankton. *Oikos* **9**(II), 158–191.

Smayda, T. (1978). Quantitative and autecological aspects. Biogeographical meaning; indicators. *In* "Phytoplankton Manual" (A. Sournia, ed.) Section 8. Interpreting the Observations, pp. 225–229. UNESCO, Paris.

Stepanov, V.N. (1983). "Oceanosphaera". Mysl, Moscow. (In Russian).

Sukhanova, I.N., Zhitina, L.S., Mikaelan, A.S. and Sergeeva, O.M. (1988). Phytoplankton: its composition and distribution. 1. The time-space variability of phytoplankton. *In* "Ecosystem of the Sub-Antarctic zone of the Pacific Ocean" (M.E. Vinogradov and A.S. Mikaelan, eds), pp. 124–145. Nauka, Moscow. (In Russian).

Tomas, C.R. (ed.) (1996). "Identifying Marine Diatoms and Dinoflagellates". Academic Press, London.

Venrick, E.L. (1971). Recurrent groups of diatom species in the North Pacific. *Ecology* **52**(4), 614–625.

Zernova, V.V. (1966). Phytogeographical division of the South Ocean and distribution of indicator species of plankton algae. *In* "Atlas of Antarctic", Figure 128. Centralnoe upravlenie geodezii i kartografii SSSR, Ministerstvo geologii, Moscow. (In Russian).

Taxonomic Index

Note: Page references in *italics* refer to Figures; those in **bold** refer to Tables

Subject Index

Note: Page references in *italics* refer to Figures; those in **bold** refer to Tables

Cumulative Index of Titles

Note: **Titles of papers** have been converted into subjects and a specific article may
therefore appear more than once

Cumulative Index of Authors